Water, Cultural Diversity, and Global Environmental Change

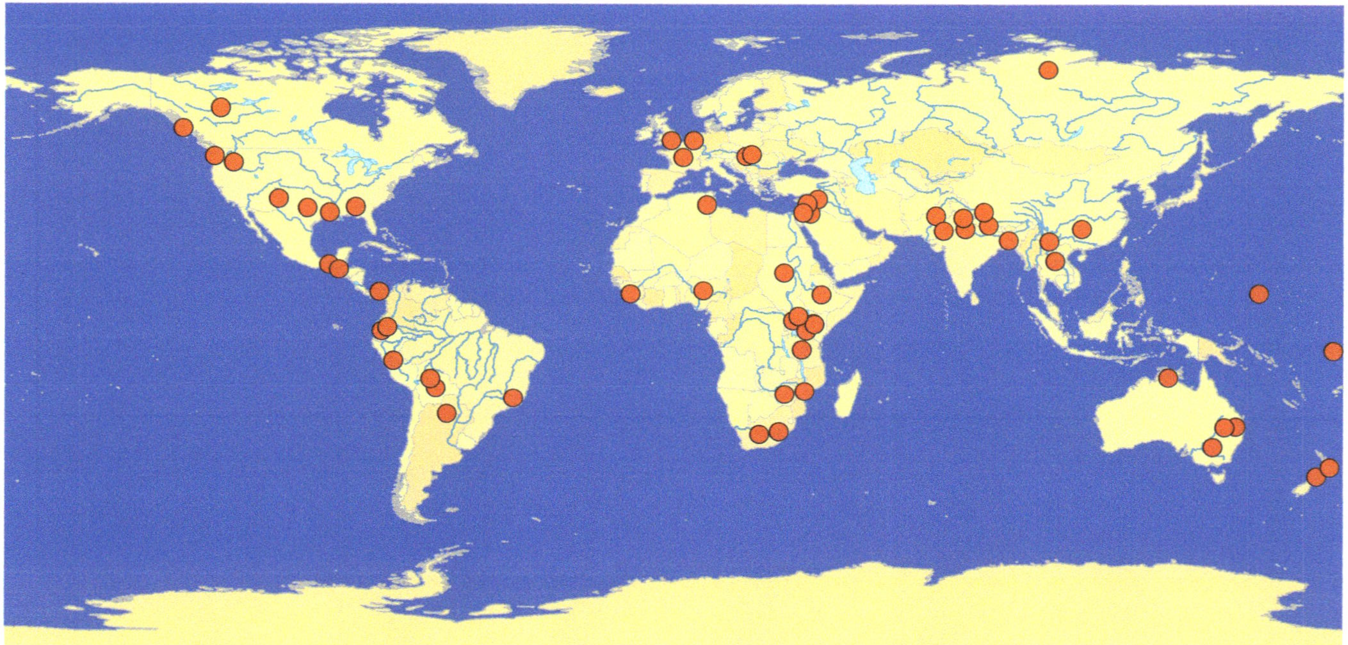
Mayfield, UNT Philosophy of Water Project, 2010

Location of case studies

Barbara Rose Johnston
Editor-in-Chief

Water, Cultural Diversity, and Global Environmental Change

Emerging Trends, Sustainable Futures?

Editorial Board:
Lisa Hiwasaki
Irene J. Klaver
Ameyali Ramos Castillo
Veronica Strang

Springer

United Nations Educational, Scientific and Cultural Organization
International Hydrological Programme

UNESCO-IHP
1, rue Miollis
75732 Paris Cedex 15
France
ihp@unesco.org

CENTER FOR POLITICAL ECOLOGY
HUMAN RIGHTS & ENVIRONMENT INITIATIVE

United Nations
Educational, Scientific and
Cultural Organization

International
Hydrological
Programme

UNITED NATIONS
UNIVERSITY
UNU-IAS
Institute of Advanced Studies
Traditional Knowledge Initiative

UNESCO ISBN 978-602-98372-4-7
Springer paperback ISBN 978-94-007-1866-1
Springer hardback ISBN 978-94-007-1773-2
e-ISBN 978-94-007-1774-9
DOI 10.1007/978-94-007-1774-9
Springer Dordrecht Heidelberg London New York

Library of Congress Control Number: 2011944025

Published jointly by the United Nations Educational, Scientific and Cultural Organization (UNESCO) Office in Jakarta, Jl. Galuh (II) no 5, Kebayoran Baru, Jakarta, 12110, Indonesia, and Springer SBM, Van Godewijckstraat 30, PO Box 17, 3300 AA Dordrecht, The Netherlands.

Cover illustration: Cover art by Olivia Campbell, Canada

Printed on acid-free paper

Springer is part of Springer Science+Business Media (www.springer.com)

Foreword

The Water-culture-environment nexus: Practical lessons from the field

Richard A. Meganck

This book is the product of many experts on all aspects of the water/culture continuum. I am not yet another expert, particularly on the cultural aspects of water. I am a surface water hydrologist who has worked in international water management and water policy development. I spent the last six years of my active United Nations career as an administrator of an institute of higher education to train the next generation of water specialists. UNESCO-IHE has awarded in excess of 15,000 MSc and PhD degrees to professionals from 162 countries in its 51-year history. That experience has given me tremendous insight into the many varied cultures of water that comprise the mosaic of how we use and misuse this most precious of the world's resources. And so it is from this perspective and experience that I address my remarks to you, the reader.

Consider this: A woman drops a bucket into a dry or contaminated well in northern Africa. A family watches its home vanish beneath a new reservoir in Asia. A priest baptizes a newborn in South America. A monk washes an image of Buddha in Thailand. A woman and her daughter set out on the daily search for water in Mongolia. An Indian family disperses the ashes of a loved one into the winds above the Ganges river. A farmer kicks the salt-crusted soil poisoning his fields in Australia and ponders his fragile economic future. Policy makers in Saná, Yemen, debate what response should be forthcoming as the water table of the city drops by more than six metres each year. A man in New Orleans, USA, watches as his home of 50 years is consumed by the rising floodwaters.

In all corners of the globe people confront both opportunity and tragedy related to water. Sometimes too much water, sometimes too little. But increasingly, the presence of or absence of water is chaotic, with less predictable patterns that produce local disasters of all sorts.

Water is the tie that binds every thing, both animate and inanimate regardless of the outcome. It is not an overstatement that a river reflects the life and memory of any country or region. Water is mystical, religious, powerful, revered and feared. That range of descriptors itself defines the very essence of this book – that nexus where water-culture and the broader environment interact. And as I reflect on the importance of forcing these cross-discipline interactions, I can't help but think about how decision-makers, educators and students, will benefit from being exposed to this broad continuum of ideas.

It goes without saying that water and development, and by that I mean water and the advance of the human species, are two sides of the same coin. From the beginning of time, water has played a fulcrum role in where societies migrated and settled and how they interacted. Even to the casual observer it is obvious that the accumulation of knowledge about nature by communities has impacted how we have dealt with the very resources upon which we depend – our water, air and soil, and in turn, how those resources have impacted the evolution of cultures. The historical complexity of this nexus – water, culture and environment – is captured in Pietro Laureano's *The Water Atlas*. Others have documented the results of natural resource management decisions at the societal level. For example, Sandra Postel points out in her 1992 book *Last Oasis* that we are entering a new era of water *scarcity,* but also documents that we have only infrequently and only recently considered the intimate ways in which water affects other aspects of our development such as food security, industrialization, international relations, and the growth of cities. We always assumed water was there for the asking, for the taking by anyone, at almost anytime and in seemingly unlimited quantities. And many people expected water to be provided free of charge. As Postel notes, unless citizens insist on policies, laws and institutions that promote sustainable water use, we will face even more drastic changes than have been forecast. Water stress will turn into water scarcity with more frequency and with greater impact. Water refugees will become an even more common phenomenon. And while water conflicts have been infrequent in the past, they are anticipated to escalate, prompting growth in the numbers of 'hydrological refugees'.

Without question environmental damage, localized climate change, rapid population growth, unpredictable trading partners, and pressures from enemies were all factors in the demise of historically doomed societies, but still our earlier incarnations found solutions to those same problems and persisted. However, while history provides a perspective on the consequences of action (or inaction) in response to local and regional crises, the problem today is that the nature of our crises are now global in scale and require an unprecedented level of cooperation to understand, let alone resolve. The management of a watershed is all the more complex when you consider that there are some 263 major transboundary rivers in the world. Even the oceans were not conceived of as an integral unit by large masses of the population until we saw those incredible photos taken from space for the first time – a short four decades ago.

Ours is indeed the water planet. That scenario is fast changing and has engendered an active debate about the responsibilities of water rich countries to water poor countries, trade in virtual water, water footprints at all geographic scales, a water futures market, among many other new soft and hard technologies to deal with new complexities and urgencies. Related to these debates are the questions of how best to organise ourselves – financially and technically – to attack a problem that is global in scale yet presents so many varied local manifestations? How to organise ourselves to fully identify and comprehend the range of sustainable alternatives? How to structure a process that results in decisions that are implementable?

Answering these questions requires a broad and nuanced understanding of the integrated nature of the major environmental and cultural problems that we humans face.

Let me provide a personal illustration here.

In 1969, after completing my Masters degree, I joined the U.S. Peace Corps and was assigned to work in Colombia, South America on water problems in the Amazon drainage. People lived for millennia with the seasonal changes that resulted in flood and relative drought cycles. But what changed when the forestry industry became a global concern and the demand for precious hardwood insatiable was remarkable. Not only were the flood and drought cycles more pronounced, but traditional patterns of living were abandoned and replaced with a more modern lifestyle. Entire economies were impacted and of course access roads brought in tremendous amounts of investment and environmental damage. And the damage was not only to the resource but to the structure of the affected communities. Poverty became more pronounced and outside influences caused small communities to grow wildly and without forethought or at times disappear. Of course this was going to happen. How could we expect any region to transition at such a fast pace without time for understanding or adjustment?

In 1974, after completing my PhD, I joined the Organization of American States. A few years later I transferred to the United Nations where I was sent to the desert of Northern Mexico to plan an ecological reserve. Yet nobody living there was interested in an ecological reserve. Only the funding agency, the Inter-American Development Bank, was interested in this concept. The only thing the farmers of this region were interested in was water. It took six months for their message of human need to change my way of thinking. Over time we managed to raise the funds for a commercial scale well, and the farmers abandoned their overgrazing and over-tree-harvesting economy to return to what they were really were good at – growing beans. And their bean production tripled in a single year. They got rid of their goats and stopped cutting and selling firewood. Once their livelihood was secure, the farmers had an interest in seeing an ecological reserve established and over the next three years we planted 3 million pine seedlings. Today, there is a seedling survival rate approaching 70% and the watershed is being restored with the active participation of the local community and with the assistance of The Nature Conservancy, a U.S.-based non-governmental organization.

The point of these examples? We humans have some degree of control in the human/environmental equation. We can choose to make decisions that change action and behaviour – and in these changes help restore and sustain the environment. Both of these experiences taught me that you can't assume that making a decision about any individual sector will not impact other parts of the community. Humility and the willingness to observe and listen are two of the most important tools any field scientist can have in his portfolio.

I could mention many other stories such as the impact of women and girls primarily in Africa having to dedicate an average of four hours each day collecting fuel and water – this total resulting in a loss of some 40 million productive working days each year. I could mention that the reason the United Nations Children's Fund has invested so many resources into drilling wells at schools throughout Africa is that the lack of water for personal hygiene is the principal reason why young girls drop out of school once they reach puberty. Imagine the impact to any culture of that trend. I could note the impact that millions of people suffer from having to drink arsenic-laden waters or the fact that more than a billion people must survive drinking contaminated water and more than two billion live without any sanitation services. The statistics can at times be overwhelming.

Rather than continuing along that line, please permit to focus on a few concluding comments and questions that I hope you will keep in mind as you turn your attention to the following chapters in this book.

1. The North-South, South-North and even the South-South continua for sharing experiences still seems to be more of a barrier than a bridge to learning. Yes, there are many bright spots and our ability to network in ways that were inconceivable only a decade ago has helped tremendously. But why is it that we continue to exchange experiences but find it so difficult to adapt our technologies and policies based on the trail and error of others? Donald Worster in his now classic work *Rivers of Empire* referred to this syndrome as the 'hydraulic trap'. Of course he was comparing the promise of engineering to solve all problems and the difficulty of pushing any new technology through a political process. I have heard a similar concept referred to as 'policy arrogance' in UN meetings; the we-know-what-is-best-for-you syndrome. Here I would also add the burden of conditionality demanded by the development banks. And at a time when the North is every bit as vulnerable as the South to the ravages of global change, I maintain that the South may have many of the solutions the North needs.
2. We cannot continue to isolate natural science from social science in debating what is sustainable and expect logical outcomes. We need a common language, and we need the patience and respect to recognise a common language cannot be achieved without broadened our understanding of reality and perspectives of reality. As a retired employee of UNESCO, I might also ask: Are the three keys to sustainability (as promoted by UNESCO) – science, culture and education – achievable when one part of the house has no idea what another part of the house is involved in? This book is meant to encourage a shift in thinking – that sustainable futures require embracing the challenge of cross-discipline and cross-cultural communication. The first step is to listen, with respect.
3. Our ability to analyze massive amounts of trend data provides us with an opportunity to look to the past and indicate alternatives as to the future. We have the very tools at our disposal to begin to comprehend the impacts of decisions taken today on our lives and those of our children. We must begin to use this information in cross disciplinary ways to our benefit, rather than trying to convince decision makers that history can teach us something about the future. That is still a very difficult goal to sell in the marketplace.

4. Hydrological poverty is a reality in many parts of the world and it will drive more change than culture will – given the absolute need for water. Most of the nearly three billion people to be added to the world's population by 2050 will live in countries where water tables are already falling. The international community together with the private sector must address these issues head-on to avoid civil unrest and new 'Rwanda's' catalyzed by a lack of water and protein.
5. Money alone will not solve the problems of water or environmental management, and in many contexts may be the source of old and new evils. Consider the aid and corruption continuum. New technologies such as trading in virtual water, water being considered a vital part of Official Development Assistance, the buying and selling of water futures may, if carefully crafted, offer an opportunity to link development assistance with lending and granting policies while helping to ensure that assistance gets to those who need it.
6. Finally, I continue to believe that solving our environmental problems and improving the understanding of the relationship between water and culture is though education. It must be part of any solution, but in this case we cannot hope to improve water management unless we invest in the future of the sector. If Africa has any hope of complying with the UN Millennium Development Goals regardless of the date, it is estimated that a 300% increase in the number of trained water professionals is required. The situation while a bit better in Asia is still daunting as a 200% increase is projected. Latin America will require a 50% increase in the numbers of trained technicians and professionals. And even the OECD countries are finding that with the wave of retirements planned for the next 10 years they will fall far short of the numbers of trained personnel. How can we have any hope about the future if we do not invest in what is obviously needed today?

I want to end this foreword with a short poem. It comes from the book *Wind, Water and* Stars by Antoine de Saint-Exupery.

Water, you have no taste, no colour, no scent,
one cannot describe you,
but only enjoys you without knowing you.
You are not something that is needed for life; you are life itself.
Your fulfillment of the sense of happiness,
that, cannot be explained by senses only.
With you all the powers from
which we have distinct counsels come back to us.
You are the most precious wealth in the world
and yet you are also the most fragile one.
So pure, emerge from the womb of the earth.

Thank you.

Preface

> Water is the essential lifeblood of our planet, with the power to generate, sustain, receive, and ultimately to unify life (UNESCO-IHP 2009)

> Culture takes diverse forms across time and space. This diversity is embodied in the uniqueness and plurality of the identities of the groups and societies making up humankind. As a source of exchange, innovation and creativity, cultural diversity is as necessary for humankind as biodiversity is for nature. In this sense, it is the common heritage of humanity and should be recognized and affirmed for the benefit of present and future generations (UNESCO Universal Declaration on Cultural Diversity, 2001, Article I).

> Cultural diversity is crucial to environmental sustainability; it generates the multiple human possibilities necessary for generating sustainable adaptations in a changing world (UNESCO-IHP 2009)

The above statements express the core concerns and guiding principles of this book. The essays and cases explored in *Water, Cultural Diversity and Environmental Change: Emerging Trends, Sustainable Futures?* project a view of the world that sees no inherent separation between human cultures and the environment in which they reside. Indeed, water serves as a unifying element: it is a substance (H_20) in gas, liquid and solid form; it is an agent, a transformative force, a realm, a source of life and death; and, in a human sense, water can be an entity, an idea, a source of power and wealth, of misery and woes, or inspiration and joy. Human history and the evolution of culturally diverse understandings and ways of life reflect in many ways our interactions with, need for, and use of water. Because culture is learned, lived, and expressed in broader social relationships, it is a significant dimension in the factors that shape both conflict and consensus in understanding, valuing, using, and managing water. Our collective and individual needs for water form a basis for both collaboration and conflict.

It is with these understandings in mind that we have shaped the core questions of this book: What are the roles that water plays in sustaining diverse forms of human sociocultural life? What roles do diverse human societies and cultures play in valuing, managing, preserving and using water and its associated ecosystems? What are the consequences of these resource relations in sustaining, or undermining the means to sustain, the viability of human communities and their environments? Recognising that the crises we face have been largely generated by human action, particularly human activity guided by capitalist, managerial and technoscientific cultures, at what point and in what contexts are cultural values and behaviour appropriate or inappropriate? The traditional knowledge, stewardship and management systems, and technologies developed by different peoples typically reflect the deep relationship between biological and cultural diversity. Might these time-tested strategies help meet the complex needs of a changing environment?

The core concepts utilized in this book draw upon a larger trend in sustainability science, a recognition of the synergism and analytical potential in utilizing a coupled biological and social systems analysis which is urgently needed, as the functioning viability of nature is both sustained and threatened by humans. Our emphasis on cultural diversity means that this book is deliberately wide in scope and form, sampling an array of case studies, ideas, images, issues and activities from across the globe. We also hope that, by presenting such a collation, we encourage further exploration, both of the specific examples, and of the wider concepts and ideas presented. Alongside its conceptual and empirical contributions, this book is intended as a call to action, but also as a window to a wider series of actions already underway.

The UNESCO-IHP Water and Cultural Diversity Project

As part of the Water and Cultural Diversity project by UNESCO's International Hydrological Programme (UNESCO-IHP), an expert advisory group on Water and Cultural Diversity was formed in January 2008. Over the subsequent 2 years, the group worked to refine UNESCO-IHP's Water and Cultural Diversity project mission, identify goals and implementation initiatives, draft brochures, white papers, and policy statements, participate in meetings and conferences, and assist in the development of an international Community of Practice (CoP) engaged in defining and asserting sociocultural perspectives in water sciences and management. Members of the Water and Cultural Diversity expert advisory group came from both academic and community activist backgrounds and included practitioners with skills in anthropology, community advocacy, development, ecology, forestry, engineering, ethnobotany, geography, hydrology, philosophy and Indigenous studies. Participants came from Australia, China, Canada, India, Japan, Mexico, Nepal, the Netherlands, Paraguay, Sierra Leone, the United Kingdom and the United States, while institutional project partners included the United Nations University-Institute of Advanced Studies Traditional Knowledge Initiative (Darwin, Australia) and the Research Institute for Humanity and Nature (Kyoto, Japan). This book is a major collaborative outcome of the relationships created and the work undertaken during the life of the expert advisory group and of the broader UNESCO-IHP Water and Cultural Diversity project.

Water and Biodiversity: A Global Crisis

In late 2010, a cover story in the journal *Nature* drew renewed attention to the relationship between the use and misuse of rivers and the degenerative consequences for humanity. Among its conclusions are the harsh facts that the cumulative impacts of dams, pollution, agricultural runoff, the conversion of wetlands and the introduction of exotic species have led to a situation in which some 80% of the world's population – nearly five billion people – live in areas where river waters are highly threatened. Much of the industrialized world faces a massive decrease in biodiversity due to loss of habitat, decreasing ecosystemic integrity, and the diminishing quality and availability of water. These impacts are spreading through newly industrializing areas, and water security is now under extreme threat in many regions, destabilizing individual health, cultural integrity and diversity, and the social and political relations within and between nations.

Mapping the threats to water security for both humankind and biodiversity has allowed scientists to graphically demonstrate the intersect between water, biodiversity, and human security, and their conclusions are clear: the world itself is in crisis. This crisis is more than a matter of supply and demand; it is also a crisis in access, with the inequities in access to water further separating the rich from the poor. And it is a complex crisis, as actions to address human needs often come in conflict with broader biodiversity and river health concerns. Remote areas where water and biodiversity is plentiful are attractive for energy development and water security potential, yet such actions not only threaten biodiversity, but also long term global water security and transnational security (Palmer 2010:534; Vörösmarty et al. 2010:555–561).

Other research conducted by the United Nations Environment Programme and the WorldFish Center demonstrate how truly significant freshwater fisheries are in providing food and economic security for tens of millions of the world's peoples, an estimated 100 million people in Africa alone. Inland fisheries are for many of the world's people the primary source of dietary protein and other essential nutrients. Yet, climate change and other forces are generating rapid environmental changes in the world's freshwater systems, changes that are having a marked degenerative effect on fresh-water fisheries (Dugan et al. 2010).

Similar concern for the rapidly escalating rate of species extinction was expressed at the United Nations Convention on Biological Diversity meeting in Nagoya, Japan in October 2010, resulting in the adoption of a consensus agreement on how best to conserve, protect and share the world's biodiversity resources. The Nagoya Protocol includes recognition of the relationship between unsustainable uses of inland water resources and biodiversity, and includes pledges to expand protect areas to 17% of land (including inland waters) and marine protected areas to 10% of the oceans by 2020 (Convention on Biological Diversity 2010).

Achieving the stated goals of this agreement, as is the case for other areas of international concern, requires aggressive action, a commitment to deprioritize short-term benefits (that often meet human needs) over long-term gain (that sustains a functioning ecosystem), and significant political will. Implementing this global environmental agreement will not be easy, given the significant link between biodiversity, water and other critical resources often found in such settings, and the rise of violent conflict. A 2008 study published in the scientific journal *Conservation Biology*, for example, mapped the global location of war over a 50-year period in relation to biodiversity hotspots. The study's conclusion: more than 80% of the world's major armed conflicts during the last half-century have taken place in some of the most biologically diverse and threatened places on Earth (Hanson et al. 2009).

Is it possible to chart a way forward that ensures that human needs and desires and biodiversity concerns are addressed in peaceful and sustainable fashion? Authors in this collection of essays and case-specific vignettes argue that such outcomes are indeed possible, especially when actions taken recognize and respect the role that water plays as the lifeblood of this planet. We argue for a coupled social and biological systems approach to water management, one that acknowledges and sustains the synergistic links between water, biodiversity, and cultural diversity.

Core Concepts and Tools

In keeping with the intent to promote both diversity and constructive action, this book offers an array of ideas, concepts, and tools to understand and manage the sociocultural implications of the growing water crisis, and to suggest alternative pathways to sustainability. Some of these concepts and tools have only been developed recently, whilst others have a considerable history behind them. While there is no one model, precept, or concept that will be universally applicable and effective in every circumstance, there are certain universal conditions and constructs that can be recognized. These are briefly outlined here, and explored in more detail in the case studies themselves.

Cultural Ways of Life

Culture is a universal. All humans have learned behaviour expressed in the patterns we call culture. These learned patterns of behaviour serve many functions, not the least of which is the ability to interact, engage, and adapt to changes in the environment and in society. *Cultural ways of life* are also universal. We all have ways of understanding, engaging, communicating, sharing, and reproducing our knowledge, values, beliefs, and expressions. While all cultures evolve and change, the means to support and sustain a cultural way life may be destroyed, and this destruction can have profound impacts on the ability to utilize and reproduce culturally-distinct knowledge, values and traditions.

All culturally-distinct communities, indigenous and otherwise, are unique, yet also share certain commonalities. Many communities enjoy a way of life that sustains the local environment while nourishing the social relationships and cultural meanings that define their community. The production and reproduction of a way of life involves a way of knowing that is a result of a long-term systemic interaction between people and their surroundings. This knowledge is rooted in cultural practices and spiritual values and enshrined in customary laws. These biocultural experiences and relationships have over the millennia enabled human groups, and the environs on which they depend, to survive and thrive.

Across time and space, human communities – through their varied ways of life – form and shape a vast and complex mosaic of ecosystems. This diversity, and its potential for providing both adaptive capacity and the sustainable use of the resource that is the lifeblood of the planet, may well hold the key to our future. Monocultures for example, may be highly productive in particular ways over short periods but they are inherently vulnerable to unforeseen variability. Such variability, particularly in a climatic sense, is an increasing part of the contemporary world. Yet, while they represent sources of adaptive diversity, local place-based cultures are also vulnerable to large-scale or ill-considered change. While adaptation and change is a normal part of cultural transmission, the sustainability of cultural ways of life depends heavily on the maintenance of some strong continuities. When the means to support and sustain cultural continuity is destroyed, we suggest that destruction is not just immoral and/or illegal; it lessens the diversity and therefore the adaptive capacity of humanity as a whole.

Thus, in exploring the linkages between water, cultural diversity and environmental change, we are especially concerned here with the role of culture in affecting water quality, in shaping patterns of water access and use, and in the complex consequences of these *resource relations*, especially the ways in which culturally-informed behaviour sustains or undermines the viability of our planet and varied life that it supports (Donahue and Johnston 1998). In too many instances we see the development and management of water resources, and the related environmental degradation and displacement of previously cohesive communities, as a significant threat to the means that sustain communities and culturally-distinct ways of life.

Water Cultures and Waterscapes

Everything is culturally mediated, in all societies. Economic activities; politics; the way we think about and interact with the material environment. And every social group and every actor in society has a cultural engagement with water. Some of this human/water engagement is manifested in the form of *water culture*: the knowledge, traditional customs, and behaviour that support the development and reproduction of a stewardship ethic, or the political organization of societies to manage and maintain water resources. Other aspects of the human/water engagement are manifest in the form of *waterscapes* – where human imagination and endeavour produces a material form which contains, shapes, moves, disperses, reveres and celebrates the essence of water (Strang 2009). Such waterscapes may involve considerable human agency in the direction of water: for example through the construction of dams, reservoirs and irrigation channels, water pipes and taps. Or they may be low-key, producing only temporary fish weirs and small settlements around resource-rich wetlands. They may involve celebrating aesthetic or spiritual ideas about water, through fountains, fonts and sacred sites, or striving for population health through the provision of sanitation services. In whatever form, waterscapes are concerned with the ongoing materialization and perpetuation of particular cultural lifeways (Strang 2004). Deepening individual and collective understandings of how water is understood and valued, and the consequences of that valuation for how flows and cycles of water are used, is a crucial step in laying the foundations for sustainability.

Integrated Water Resources Management

In terms of the practicalities of resource management, the most common approach to water in recent decades has been integrated water resources management (IWRM). IWRM takes an ecosystem perspective of water together with its human uses; encourages broad stakeholder participation; and stresses that water, in all its competing uses, must be valued as an economic good. The overall goal is to manage water resources in ways that sustain the place and society in both immediate and long-term timeframes. This requires involving and addressing the views and concerns of diverse resource users; consciously recognizing the relationship between water development and the health and productivity of the economy, including varied agricultural, industrial, and energy uses of water; creating a sustainable drinking water supply and improving quality of water in rivers and other sources to ensure a healthy population; and recognizing the principal role of water for ecosystems and ensuring release of water to sustain and restore ecosystems and the biodiversity sustained by such systems. Practitioners have found IWRM a powerful tool to recognize and value ecosystem services associated with river systems and develop management strategies that accommodate human and environment needs, but its use has also generated considerable criticism over its failure to incorporate lesser-quantifiable values.

Environmental Flows, Cultural Flows

A core goal of IWRM is to balance human and ecosystem needs of water. To achieve this goal, the varied uses of water are identified and assigned an economic value, acknowledging the human need for water supply, recreation, and aesthetic values, as well as the role of the free flow of water in providing ecosystem services. The minimum flow required to sustain this basic human/ecosystemic integrity is called *environmental flows*.

A healthy flow of water will sustain human and environmental needs. However, flows for the environment and for cultural purposes are not one and the same. For example, the connections held by Indigenous people to water, and to particular features or dimensions of the waterscape, may have been part of their existence for thousands of years, yet such values are not always reflected in environmental flow allocations. In Australia, this issue has generated the need for a new concept of *cultural flows* to further expand the IWRM model to manage water resources in a more holistic, equitable, and sustainable fashion. *Cultural flows* refers to the dynamic ways in which water sustains diverse cultural beliefs, values, and ways of life, as well as the ways in which groups value, care for, and sustain the health of aquatic systems. This approach involves not only managing flows to sustain ecological processes, but also ensuring that the water flows allocated to non-consumptive or environmental purposes are sufficient to recognize, respect, and sustain cultural ways of life.

The concept of *cultural flows* offers a way to further expand the IWRM model to manage water resources in a holistic, equitable, and sustainable fashion. How are we to do this? One strategy has been to document qualitative descriptions of the varied meanings that waterways have for people, and the knowledge and relationships that people have with these waterways. Such efforts produce a set of social and cultural indicators that are then used to structure and monitor management goals. A more complex strategy is to document the relationship between cultural values and *ecosystem services*. Examining values in customary fisheries, for example, and the knowledge of them held by Indigenous people, provides a useful starting point in the quantification of indigenous values for water planning. Both strategies recognize that indigenous communities and other cultural groups whose long-standing ties to the land and customary ways of life constitute a *rights-holder* status.

Water Rights and Human Rights

Despite its overall adaptive potential, the cultural diversity described here is also vulnerable to erosion, particularly in situations where major power imbalances exist between local actors and those advocating particular kinds of development and/or structural change. Essays and case studies in this book argue that sustainability is achievable if collective laws, policies and actions:

- Encourage global recognition of the relationship between biodiversity and cultural diversity and the sustaining role of water resources;
- Acknowledge and nurture the varied roles played by cultural values, knowledge, and traditional resource management systems in protecting water quality, regulating access, and ensuring sustainable use of freshwater resources;
- Recognize and address the challenges that local communities confront as large-scale solutions to energy and water scarcity are proposed and implemented;
- Develop and implement the legal framework and social safeguard mechanisms to protect fundamental rights to water, livelihoods, and culturally diverse ways of life; and
- Work to incorporate cultural diversity into watershed management and water resource development.

International human rights law recognizes that all peoples have the right to self-determination and to the pursuit of economic, social and cultural development. While states have sovereign rights to protect, preserve and develop their resources in ways that benefit the nation as a whole, all of their citizens have the right to know about proposed development projects, such as dams, mines, logging and other large infrastructure projects, which may affect them in some way. All project-affected communities have the right to participate in project planning and decision-making processes; the right to just compensation for any damages and losses associated with development; and the right to benefit materially from such development. Indigenous communities have additional rights, recognised in the United Nations Declaration on the Rights of Indigenous Peoples (Article 32[2]). They have the right 'to give or withhold their *free, prior and informed consent* to actions that affect their lands, territories and natural resource'.

Claiming these rights is often difficult and contentious, especially when proposed developments involve significant upheaval and the loss of critical resources that sustain cultural diversity. Some states and their transnational development partners have laws and regulatory procedures designed to respect these rights and allow citizens to participate fully in development planning. However, in many instances, these rights are not recognized, respected, or implemented. Contributions to this book therefore argue that the right to free and prior informed consent not only reflects international law, but more importantly, offers an opportunity to protect biocultural diversity and manage water resources in a truly sustainable fashion. Claiming such rights requires:

- Timely access to information about the developers and the positive and negative effects on host communities and ecosystems;
- Discussion of the proposal within the affected communities;
- Consultation over the scope, scale, and terms with project developers;
- Access to legal assistance and independent advice;
- Decision-making processes to determine if consent will be given; and
- Should the project proceed, formal processes to implement the terms of any compensation, resettlement, and other benefits be negotiated by project-affected communities.

Structure

The book is divided into five parts. Although their themes, interests, and geographic locations intersect and overlap, each of the parts has a particular orientation.

Part I explores water's fundamental place in life, flowing through all organic processes, shaping all of the earth's environments, and coursing through humanity's traditions, values, meanings, politics, economics, art and engineering – i.e., culture in its broadest sense. The case studies and brief vignettes, conceptual essays and graphic imagery of this Part articulate the "culture of water" and the environmental consequences of human relationships with water; they speak to the breadth and depth of cultural diversity in practices around water. Part I is a thick description of lived actualities and possibilities, with a central theme being the necessary foregrounding of cultural diversity in water policy.

Part II considers the "culture of water" through an explicit focus on traditional ecological knowledge and water resource management: approaches that have historically served to sustain the lifeways of indigenous groups and ethnic minorities. These contributions reflect the input of scientists, scholars, and advocates whose work with diverse cultural groups considers two key questions: how are traditional ways of life threatened by the degradation or loss of water resources, and, how might traditional knowledges contribute to future water security?

Part III examines current patterns of water resource management in various ecoregions and geopolitical contexts. It considers problematic contexts where water resource development and management have undermined the viability of culturally diverse groups (asking what are the lessons learned?), as well as cases where water resource management has achieved sustainable societal goals by strengthening biocultural viability (what best practices might be identified?). Case studies explore problems such as access to clean water and environmental health, and the role of culture and power in shaping, and addressing, vulnerability.

Noting current trends in global environmental change and state-sponsored efforts to ensure future water and energy security, **Part IV** considers the changing and possible future dynamics of intersections between water, biodiversity, and cultural diversity, with a critical focus on the lessons learned from the past several decades of hydrodevelopment. The rivers that run through mountain valleys and plains, and empty into major deltas sustain many of the world's most biodiverse areas. These resource rich areas also support many of the world's remaining indigenous nations, minority cultural groups, and place-based peoples. Hydroelectric energy and water resource development plans are shaped with an emphasis on project-specific proposals and funding, and on impact assessment frameworks that are place-specific, rarely allowing consideration of the cumulative effects such developments may have on regional or global patterns of biocultural diversity. Emerging trends include a massive increase in hydrodevelopment, with the associated displacements and ecosystemic changes posing a significant threat to the world's remaining biocultural diversity. Thus this section asks 'what lessons can be gleaned from past experience, and how might these transform understandings of the cumulative effect and synergistic forces at work in hydrodevelopment?'

Part V sketches out alternative scenarios for the future, arguing that a sustainable approach to water resource development must, first and foremost, be one that sustains the cultural and biological diversity of life. Contributions expand upon the case-study lessons with strategic recommendations for incorporating sociocultural perspectives into water resource management systems; and, with a more inclusive notion of sustainability, they address rights and entitlements to water for all human groups and all species, as well as reaffirming established stewardship principles and responsibilities.

A Final Note

Is it possible to chart a way forward through the current crises that ensures that human needs and desires are met and water resource, cultural diversity and biodiversity concerns are addressed in a peaceful and sustainable fashion? Authors in this collection of essays and case-specific vignettes argue that such outcomes are indeed feasible, especially when the actions taken recognize and respect the role that water plays as the lifeblood of the planet, the universality of culture in human life, and the critical contribution that cultural diversity plays in the ongoing adaptive potential of human beings. Meeting the manifold challenges of human population growth and patterns of resource use and an increasingly stressed global environment requires truly holistic strategies. The increasing vulnerability of water resources calls for diverse approaches to water management to be recognized and supported. Indigenous and local communities are invaluable partners in this regard, representing a vast array of adaptive knowledges and practices which have not only sustained engagements with water for millennia, but also offer much potential for creativity and innovation.

However, local water practices do not exist in a vacuum. They are embedded in increasingly large socio-political systems governed by national and sometimes international organizational mechanisms. When water resources are brought under centralized, bureaucratic control, the resilience of local forms of governance is often diminished. Long-term sustainability may require a re-embedding of social and political institutions at a local level, to enable sustainable water resource governance.

Fully inclusive collaboration, both horizontally and vertically, is required to create a more workable balance between national and international aims, and the health and well-being of local communities and ecosystems. Meeting current and future human-environmental challenges depends on the active involvement of all water users.

Barbara Rose Johnston
Marcus Barber
Veronica Strang
Irene Klaver
Lisa Hiwasaki
Ameyali Ramos Castillo

References

Donahue, John, and Barbara Rose Johnston. 1998. *Water, culture & power – local struggles in a global context.* Washington, DC: Island Press.

Dugan, P., A. Delaporte, N. Andre, M. O'Keefe, and R. Welcomme. 2010. Blue harvest: Inland fisheries as an ecosystem service. United Nations Environment Programme and WorldFish Center, Penang. http://www.worldfishcenter.org/resource_centre/Blue_Harvest.pdf. Accessed 5 Nov 2010.

Hanson, T., and T.M. Brooks, D.A. Fonseca, A. Gustavo, M. Hoffmann, J.F Lamoreux, G. Machlis, C.G. Mittermeier, R.A. Mittermeier, J.D. Pilgrim. 2009. Warfare in Biodiversity Hotspots. *Conservation Biology* 23(3 June):578–587.

Palmer, Margaret A. 2010. Water resources: Beyond infrastructure. *Nature* 467(Sept 30): 534–535.

Strang, Veronica. 2004. *The meaning of water.* London: Berg.

Strang, Veronica. 2009. *Gardening the world: Agency, identity, and the ownership of water.* Oxford/New York: Berghahn Publishers.

United Nations Convention on Biological Diversity. 2010. Nagoya Protocol. http://www.cbd.int/nagoya/outcomes/

United Nations Declaration on the Rights of Indigenous Peoples. Adopted by General Assembly Resolution 61/295 on 13 September 2007. http://www.un.org/esa/socdev/unpfii/en/drip.html. Accessed 5 Nov 2010.

United Nations Educational, Scientific and Cultural Organization-International Hydrological Programme (UNESCO-IHP). 2009. Policy Brief on Mainstreaming cultural diversity in water resources management. http://www.unesco.org/water/ihp/pdf/CDW_FINAL_WEB.pdf. Accessed 5 Nov 2010.

Vörösmarty, C.J., P.B. McIntyre, M.O. Gessner, D. Dudgeon, A. Prusevich, P. Green, S. Glidden, S.E. Bunn, C. A Sullivan, C. Reidy Liermann, and P. M. Davies. 2010. Global threats to human water security and river biodiversity. *Nature* 467: 555–561.

Acknowledgements

This book was developed out of a project titled Water and Cultural Diversity by UNESCO's International Hydrological Programme (UNESCO-IHP). It benefited greatly from partnership with the Traditional Knowledge Initiative of the United Nations University-Institute for Advanced Studies (UNU-IAS TKI) and the Center for Political Ecology (USA).

The vision, direction and content of the book were originally developed by members of the expert advisory group for the Water and Cultural Diversity project, and refined by an editorial board consisting of Barbara Rose Johnston, Lisa Hiwasaki, Irene Klaver, Ameyali Castillo Ramos, and Veronica Strang. Section editors not on the editorial board were Marcus Barber and Daniel Niles. In addition to his role as section editor, Barber made significant contributions to the book, most notably to the preface. Besides those on the editorial board, members of the expert advisory group, which met four times during 2008–2009, are: Fenda Akiwumi, Elena Benitez, Michael Blackstock, Monica Morgan, Prachanda Pradhan, Jie Qian, and James Webster. Barbara Rose Johnston provided exceptional leadership throughout the entire process of preparing and finalizing this publication. Her work was supported in part by the UNESCO-IHP and in part by a grant from the Global Biocultural Initiative program of The Christensen Fund.

The Research Institute for Humanity and Nature is gratefully acknowledged for their work in hosting the International Symposium "Water, Cultural Diversity and Global Environmental Change: Emerging Trends, Sustainable Futures?" in October 2009, which brought together many of the contributing authors and other experts for a 3-day meeting.

Illustrations in this book include imagery supplied by contributing authors and their organizations; photos obtained through the creative commons; and, images generously provided by International Rivers. Drawings created by children and youth in Canada, Colombia, Iran and Cuba are included thanks to an initiative on youth art supported by the UNESCO-IHP with contributions from the UNESCO National Commissions in these and other countries. Michael Blackstock, Lisa Yamaoka of South Kamloops Secondary School, B.C., Canada, and teachers at el Colegio de Education Basica Luis Manuel Parra Caro, Sativanorte, Boyaca, Columbia should be specially acknowledged for their involvement. Special gratitude goes to Dennis Erwin, Matt Story, Joseph Mayfield and Megan Rush, students of University of North Texas' Philosophy of Water Project, who helped edit parts of Part I and produced the world and case-specific maps under the direction of Irene Klaver. The Dixon Water Foundation is thanked for its generous funding of the Philosophy of Water Project.

A special note of acknowledgement is addressed to Catherine Cocks whose editorial and copy edit skills were instrumental in improving a diverse set of contributions from around the world.

UNESCO Office, Jakarta, led by Hubert Gijzen and his staff, in particular Estradivari, provided the administrative assistance needed to bring this publication into fruition.

Data Reference for Ecosystem Water Bodies noted in all maps:

World Resource Institute, *Earthtrends*, Water Resources and Freshwater Ecosystems database. All data from 1992–1993. Retrieved July 2010. http://earthtrends.wri.org/searchable_db/index.php?action=select_variable&theme=2

Cited Source: Loveland, T.R., Reed, B.C., Brown, J.F., Ohlen, J.F., D.O., Zhu, Z., Yang, L., and Merchant, J. 2000. *Global land cover characteristics database (GLCCD) Version 2.0*. Available online at: http://edcdaac.usgs.gov/glcc/globdoc2_0.html

Contents

Foreword v

Preface xi

Acknowledgements xxiii

About the Authors xxxiii

Acronyms xlv

Part I Water and Cultural Diversity

1.0 Introduction: Water and Cultural Diversity 3
Irene J. Klaver

1.1 Placing Water and Culture 9
Irene J. Klaver

Box 1.1a Waters and culture in the Doce River Basin of southeast Brazil: Mobilization and construction of a new perspective 21
Claudio Bueno Guerra and Fabiane Torres

1.2 The Paradigm Shift in India's River Policies: From Sacred to Transferable Waters 31
Kelly D. Alley

Box 1.2a Isoso Guarani culture and livelihoods shaped by, and dependent on, the pulsing Parapeti River in the arid Gran Chaco of Bolivia 39
Janis B. Alcorn, Elio Ortiz, Antonio Mendez, and Alejo Zarzycki

1.3 Rethinking the Role of Humans in Water Management: Toward a New Model of Decision-Making 49
Marcela Brugnach and Helen Ingram

Box 1.3a ‘Space for the River,’ space for diversity? 54
Jeroen Warner

1.4 **Local Water Management in the Andes: Interplay of Domination, Power and Collective Participation** 65
Rutgerd Boelens

1.5 **The Power of a Disappearance: Water in the Jerid Region of Tunisia** 77
Vincent Battesti

Box 1.5a Climate change, water, and development in daily life in Tuvalu 88
Heather Lazrus

1.6 **Diverting Water: Cultural Plurality and Public Water Features in an Urban Environment** 97
Veronica Strang

Part II Culture and Water in Diverse Environments

2.0 **Introduction: Culture and Water in Diverse Environments** 119
Marcus Barber and Ameyali Ramos Castillo

2.1 **Watersheds and Marinescapes: Understanding and Maintaining Cultural Diversity Among Southeast Alaska Natives** 123
Thomas F. Thornton

Box 2.1a River, sea, and sky: Indigenous water cosmology and coastal ownership amongst the Yolngu people in Australia 124
Marcus Barber

Box 2.1b Blue ecology and the unifying potential of water 132
Michael D. Blackstock

2.2 **The Influence of Westernization on Water Resources Use and Conservation Among the Maasai People of Kenya** 137
George M. Ogendi, Rose K. Morara, and Nicholas Olekaikai

Box 2.2a Sand dams and plant indicators: Indigenous knowledge in water resource use and management among the Kamba People, Kenya 141
Cyrus M. Kilonzi, George M. Ogendi, and Rose K. Morara

2.3 **Groundwater and Qanats in Syria: Leadership, Ownership, and Abandonment** 149
Joshka Wessels

2.4 Case Studies from the Americas ... 163

Box 2.4a Water, livelihoods, and morality around the Panama Canal ... 163
Ashley Carse

Box 2.4b Saving the Pilcomayo River in Argentina: Traditional *criollo* ranchers resist destruction of an ecosystem and a way of life ... 166
Janis B. Alcorn and Luis María de la Cruz

Box 2.4c Working on water: Cultural survival in the 'tri-state water wars' ... 168
Becky Blanchard, David McLain, and Linda Raffield

2.5 Nourishing Diversity in Water Governance: The Case of San Cristobal de las Casas, Chiapas, Mexico ... 171
Ameyali Ramos Castillo

2.6 Water Knowledge, Use, and Governance: Tibetan Participatory Development Along the Mekong (Langcangjiang) River, in Yunnan, China ... 185
Yin Lun

Box 2.6a Water culture traditions: Pond myths ... 189
Anwar Islam and Shireen Akhter

Box 2.6b Water and culture in the land of the gods ... 192
Tashi Tsering

Box 2.6c Development, disaster and myth: A case study ... 200
Anwar Islam

2.7 Ecological Change and the Sociocultural Consequences of the Ganges River's Decline ... 203
Georgina Drew

Box 2.7a Mitigating the impacts of Himalayan glacial melt: Trends and policy recommendations ... 205
Georgina Drew

Box 2.7b Perceptions of chaotic water regimes in northeastern Siberia ... 214
Susan Crate

Part III Water Value, Access, Use, and Control: Sociocultural Contexts of Water Scarcity

3.0 Introduction: Water Value, Access, Use, and Control: Sociocultural Contexts of Water Scarcity ... 221
Daniel Niles

3.1 Culture, Gender, and Vulnerability in a Vietnamese Refugee Community: Aftermath of Hurricane Katrina 225
Gennie Thi Nguyen

Box 3.1a Water in a Hungarian Roma community 230
Judit Bari and Krista Harper

3.2 Water, Culture, and Gender: An Analysis from Bangladesh 237
Farhana Sultana

Box 3.2a Managing domestic water in Bangladesh after the arsenic crisis: A case study 241
Kazi Rozana Akhter and Suzanne Hanchett

Box 3.2b The purity problem and access to water in Bangladesh 246
Shireen Akhter

Box 3.2c A new type of 'social problem' – Comments of some male and female union council members in Ramganj Sub-district, Laksmipur District, Bangladesh.................................. 249
From Suzanne Hanchett's field notes, 2006

3.3 Privatization and Collective Stewardship of Water Resources: Case Studies 253

Box 3.3a Zimbabwe's water crises: The importance of environmental governance and a right to water 253
Bill Derman and Emmanuel Manzungu

Box 3.3b Extremely private: Water trading and the commoditization of water in Australia................................. 257
Veronica Strang

Box 3.3c Acequia communities and the politics of water 260
Sylvia Rodríguez

3.4 Manufacturing Water Scarcity, Generating Environmental Inequity.. 265
Barbara Rose Johnston

Box 3.4a Water, indigenous 267
Marcus Barber and Sue Jackson

Box 3.4b Legacies and challenges in integrating cultural diversity in the Sierra Leone water sector 270
Fenda A. Akiwumi

Box 3.4c The vulnerabilities of native peoples in the Mackenzie and Athabasca drainage systems: Tar sands, gas, and uranium 277
Alexander M. Ervin

Part IV Hydrodevelopment, Cultural Diversity and Sustainability

4.0 Introduction: Hydrodevelopment, Cultural Diversity, and Sustainability 291
Barbara Rose Johnston

4.1 Water, Culture, Power: Hydrodevelopment Dynamics 295
Barbara Rose Johnston

Box 4.1a Consequential damages of dam development: Maya A'chi experiences in Guatemala 298
Barbara Rose Johnston

Box 4.1b Schistosomiasis 303
Barbara Rose Johnston

Box 4.1c One person, one vote – The voice of communities in development decisions 307
Monti Aguirre

Box 4.1d Power struggles – dams on Brazil's Xingú River 311
Terence Turner

4.2 The Lesotho Highlands Water Project: Water, Culture, and Environmental Change 319
Robert K. Hitchcock

Box 4.2a The fight over fish in the Lesotho Highlands 328
John A. Ledger and Robert K. Hitchcock

Box 4.2b Dams, diversions and transboundary river management 330
Barbara Rose Johnston and Robert K. Hitchcock

4.3 Not All Dams in Africa Are Developmental: Advocacy Perspectives from the African Rivers Network 339
Robert Kugonza Akiiki

Box 4.3a People-centred development 341
Robert Kugonza Akiiki

Box 4.3b Ethiopia's Gibe 3 dam endangers Kenya's lake Turkana: An activist's view 344
Ikal Angelei

4.4 **Drowning Under Progress: Water, Culture, and Development in the Greater Mekong Subregion** 349
Nathanial Matthews

Box 4.4a Rasi Salai, Thailand 350
Nathanial Matthews

Box 4.4b Myanmar 353
Nathanial Matthews

4.5 **Damming China's Angry River: Vulnerability in a Culturally and Biologically Diverse Watershed** 367
Bryan Tilt

Box 4.5a Lessons from Tibet's only successful anti-dam campaign 375
Tashi Tsering

4.6 **Cultural Survival, Tribal Sovereignty and River Restoration on the Central Northwest Coast, North America** 387
Colleen E. Boyd and John B. Boyd

Box 4.6a Declining salmon, large dams, and the capitalist world-economy: A case study of biocultural diversity in the Nez Perce homeland 388
Benedict J. Colombi

Part V The Ways Forward

5.0 **Introduction: The Ways Forward** 405
Lisa Hiwasaki

5.1 **Managing 'Water Traditions' in Uttarakhand, India: Lessons Learned and Steps Towards the Future** 411
Amitangshu Acharya

Box 5.1a Path to sustainable development: Water parliament in India 412
Aradhana Parmar

5.2 **'El Agua es Vida/Water Is Life': Community Watershed Reserves in Intag, Ecuador, and Emerging Ecological Identities** 433
Linda D'Amico

Box 5.2a Resurgence of campesino community associations to reforest and calm an angry river, Upriver Parapeti River, Sub-Andean Gran Chaco, Bolivia 443
Janis B. Alcorn and Alejo Zarzycki

Box 5.2b The sustainability of a customary Andean water institution among the urban poor of Cochabamba, Bolivia 446
Amber Wutich

5.3 Cultural Flows: Asserting Indigenous Rights and Interests in the Waters of the Murray-Darling River System, Australia 453
Monica Morgan

Box 5.3a Environmental flows 458
Angela H. Arthington

5.4 Environmental Flow Assessments: A Participatory Process Enabling Maori Cultural Values to Inform Flow Regime Setting 467
Gail Tipa and Kyle Nelson

Box 5.4a Creating a cultural health index for assessing stream health 486
Gail Tipa

5.5 Droplets of Hope: Searching for Sustainability and Common Ground in the Arab/Israeli Conflict 493
Rosina Hassoun

Box 5.5a Participatory action research: water in a Hungarian Roma community 502
Krista Harper and Judit Bari

5.6 'Water for Life'… Water for Whose Life? Water, Cultural Diversity and Sustainable Development in the United Nations 509
Lisa Hiwasaki

Box 5.6a The influence of culture on interactions between stakeholders of rural water, sanitation, and hygiene promotion programmes: Lessons from Uganda 515
James Webster

Erratum E1

Appendices

Water, Cultural Diversity and Biodiversity: United Nations Statements, Resolutions and Treaties 533

Glossary 537

Index 557

About the Authors

Amitangshu Acharya is a development analyst by profession. He has been engaged with grassroots environment and development issues in India and has written extensively on these issues in various national newspapers and magazines. He worked as Project Officer at Arghyam Trust, an Indian non-profit where he spearheaded a programme on Participatory Ground Water Management. Presently, he is pursuing an MSc in Environment, Culture and Society at the University of Edinburgh, U.K. amitangshu.acharya@gmail.com

Monti Aguirre is the Latin America Program Coordinator at International Rivers. She supports local movements for the protection of rivers; identifies new dams and helps to examine their impacts; and designs strategies to counteract their effects. She has worked for more than a decade in support of Amazon indigenous peoples' rights and is co-producer of Amazonia: Voices From the Rainforest, a film on the fight of grassroots groups to defend their lives and their land. She has also been a tireless supporter of the people affected by Chixoy Dam in Guatemala. monti@internationalrivers.org

Kazi Rozana Akhter has Master of Science (demography) and Master of Arts (philosophy) degrees. She has more than 15 years of experience in community-based fisheries management projects, participatory planning studies, NGO-government partnerships, disaster response, and evaluation research in water-related and sanitation projects and others She has published a study of Bangladeshi women's responses to floods. She is an Associate of Planning Alternatives for Change LLC. akrozana@gmail.com

Shireen Akhter has Master of Social Science (sociology) and Master of Science (anthropology) degrees and more than 14 years of experience with training, social research, program planning and program implementation in Bangladesh. Her specialties include water and sanitation, gender and development, and poverty. She has published an evaluation study of a water and sanitation program in urban slums. She is an Associate of Planning Alternatives for Change LLC. akshireen@yahoo.com

Fenda A. Akiwumi is an environmental geographer-hydrogeologist. She is assistant professor, Department of Geography, University of South Florida, USA and a member of UNESCO-IHP's expert advisory group on water and cultural diversity. Her research takes a holistic approach to resource management including conflict, cultural perception, environmental history, and socio-culturally sensitive and sustainable resource policies. Her research areas of expertise include water and mining in Sierra Leone. fakiwumi@usf.edu

Janis B. Alcorn is an independent consultant with 30 years experience in conservation and development, including field experience in Asia, Africa and Latin America. She is Adjunct Professor at the University of Manitoba, Canada, a Fellow of Rights and Resources Initiative in Washington, D.C., and Fundacion Urundei in Salta, Argentina, and co-chair of the Theme on Governance, Equity and Rights of the IUCN Commission on Environmental, Economic and Social Policy. janisalcorn@yahoo.com

Kelly D. Alley is Alumni Professor of Anthropology at Auburn University, USA. She has worked in the Ganga river basin of India for 18 years, studying multiple interpretations of this sacred river and problems with river pollution and wastewater management. She is the author of a book and numerous book chapters and articles on religion and ecology and environmental law and justice in India. She is now working on water governance in the larger Ganga-Brahmaputra-Meghna basin that spans India, Bangladesh, China, Nepal and Bhutan. alleykd@auburn.edu

Ikal Angelei is a co-founder of Friends of Lake Turkana, an association that works on the conservation and protection of Lake Turkana, its ecosystem and its peoples. A graduate from Stony Brook University with a Master's degree in Public Policy, Ikal's focus is on the governance of natural resources: land, water and extractive resources looking into policies that govern the conservation protection, management and access to the natural resources by communities. ikal@friendsoflaketurkana.org

Angela Arthington is a professor, freshwater ecologist and Senior Fellow in the Australian Rivers Institute at Griffith University (Queensland, Australia) and a member of the Scientific Committee of the DIVERSITAS (IUCN) Freshwater Cross-Cutting Network. She has pioneered ecosystem approaches to environmental flow assessments for rivers, served as advisor to international water management and research agencies, and is currently advising the Australian Commonwealth Government on water allocations in the Murray-Darling Basin. a.arthington@griffith.edu.au

Marcus Barber is an anthropologist based at the Australian government research agency, the CSIRO. His research focuses on Indigenous people and water issues incorporating sea rights actions and coastal co-management as well as freshwater allocations and river planning. Marcus is an Associate Lecturer at James Cook University, Australia and from 2007 to 2009 was a member of the UNESCO-IHP expert advisory group on water and cultural diversity. Marcus.Barber@csiro.au

Judit Bari is a community development expert and human rights activist. She is working as an independent consultant for civic and academic initiatives on minority, gender and environmental issues mostly in relation to Roma communities. She was one of the first activists fostering environmental justice movement in minority and other vulnerable communities in Hungary. She is the leader of a community and environmental development CBO and member of Pakiv European Network. jutka.bari@gmail.com

Vincent Battesti is a social anthropologist, researcher for the CNRS (French National Scientific Research Centre) at the Muséum national d'histoire naturelle (Paris, France), Lab. Ecoanthropology & Ethnobiology (UMR 7206). His research in environmental anthropology focuses mainly on North Africa (Algeria, Tunisia, and Morocco) and the Middle East (Egypt, Yemen, and Sudan), and mainly on the relationships between local communities and oasian environment, analyzing the ways to conceive and practice irrigated agriculture in desert locations. Battesti@vbat.org

Michael D. Blackstock is an independent scholar from the Gitxsan First Nation, Canada. He is also a painter, carver, poet, forester and chartered mediator. His research focuses on indigenous perspectives on water. He believes in hope and he works to build a reconciliation foundation for younger generations to survive and thrive from. Michael lives with his family in Kamloops, BC, Canada. pioneer@telus.net

Becky Blanchard is a PhD Candidate in anthropology and NSF IGERT Trainee in the Adaptive Management of Water, Wetlands, and Watersheds program at the University of Florida, USA. She conducts research on environmental knowledge and policy-making networks in Florida's Apalachicola Bay estuary. bblanch@ufl.edu

Rutgerd Boelens is Associate Professor of Environmental Sciences at Wageningen University, The Netherlands, and Visiting Professor of Social Sciences and Law at the Catholic University of Peru. He coordinated the international WALIR Water Law and Indigenous Rights program; currently he directs the research program "Struggling for Water Security in the Andean countries" and the international action-research alliance "Justicia Hídrica/Water Justice" on water accumulation, power, culture, conflicts, and resistance. Rutgerd.Boelens@wur.nl

Colleen E. Boyd is an environmental anthropologist and ethno-historian. She is an associate professor of anthropology at Ball State University, Indiana, USA and has conducted research regarding the effects of colonialism and hydroelectric development on indigenous cultures and natural resources in the Pacific Northwest. ceboyd@bsu.edu

John B. Boyd (Lower Elwha Klallam/Arrow Lakes Band) is a writer, artist and storyteller. He teaches for the Writing Program at Ball State University, Indiana, USA. In addition to writing and speaking about his homeland in the Pacific Northwest and other contemporary indigenous issues, he is a carver, canoe-builder and salmon fisherman. jboyd@bsu.edu

Marcela Brugnach is Assistant Professor at the University of Twente (The Netherlands). She has an interdisciplinary background that combines social sciences, ecology, engineering and modelling. She specializes in collective decision-making processes and has written extensively on the themes of uncertainty, ambiguity and decision making in water management. m.brugnach@utwente.nl

Ashley Carse is a Ph.D. candidate in the Department of Anthropology at the University of North Carolina, Chapel Hill, USA. He is currently completing a dissertation project that examines the changing social, technological, and ecological relationships that have bound the Panama Canal and its rural hinterlands over the past century. The project foregrounds issues of water and watershed management. acarse@email.unc.edu

Benedict J. Colombi is Assistant Professor of American Indian Studies, Anthropology, and Natural Resources at the University of Arizona, USA. His research examines relationships between Indigenous resource management, cultural and environmental sustainability, and globalization, and he is writing several articles and developing a book about large dams, Pacific salmon, and the world economy in the Columbia River Basin. He also sustains interests in expanding his research to include Southwestern watersheds and combining field sites in Russia, Canada, Norway, Tasmania, Chile, and the United States. bcolombi@email.arizona.edu

Susan Crate is an Associate Professor of Anthropology in Environmental Science & Policy at George Mason University (Virginia, USA). She specializes in environmental and cognitive anthropology focusing on human-environment interactions. Her recent work focuses on perceptions, understandings and adaptations to local effects of global climate change in arctic and domestic settings. She is senior editor of Anthropology & Climate Change: From Encounters to Actions (2009, Left Coast Press). Scrate1@gmu.edu

Linda D'Amico is a cultural anthropologist and Associate Professor at of Global Studies and Women's and Gender Studies at Winona State University in Minnesota, USA. She lived in Latin America for two decades and worked for UNESCO, the Ecuadorian Ministry of Education and Culture, indigenous, women's and environmental groups. She recently completed a book on Otavalan women and globalization, and is currently writing one about ecological identities in Intag, Ecuador. LDamico@winona.edu

Luis María de la Cruz is an anthropologist and hydrologist with over 30 years experience in rural development in northern Argentina; Fundación para la Gestión e Investigación Regional (FUNGIR), Formosa, Argentina. fundacionfungir@gmail.com

Bill Derman is an environmental anthropologist and Professor of International Environment and Development Studies at the Norwegian University of the Life Sciences. His current research is on water and land reform in South Africa and human rights NGOs in Zimbabwe. Bill.derman@umb.no

Georgina Drew is a PhD Candidate in the anthropology department at the University of North Carolina, Chapel Hill, USA. Her work addresses water management and development challenges in South Asia and elsewhere. For her dissertation research, she examined the socio-cultural implications of hydroelectric development and climate change along the Himalayan stretch of the Ganges River. Georgina is involved in several initiatives to promote gender sensitive and culturally appropriate water policies. gdrew@unc.edu

Alexander M. Ervin, Professor of Anthropology at the University of Saskatchewan, Canada, is an applied and environmental anthropologist focusing on cultural change and globalization issues. He did his first fieldwork in the Mackenzie Delta of Canada's High Arctic and is now engaged in community-based, political ecological based issues surrounding energy and water in the Canadian West. He is the author of Applied Anthropology: Tools and Perspective for Contemporary Practice (2005). a.ervin@usask.ca

Claudio Bueno Guerra is an environmental engineer who has been working for 20 years at Doce River Basin, mainly involved with community participation on waters matters. He currently woks as environmental consultant for river basin committees, NGOs, universities, governments, and the National Water Agency. He received his MS at UNESCO-IHE Institute for Water Education, in the Netherlands. cbguerra@oi.com.br

Suzanne Hanchett Ph.D., a social anthropologist, is a Partner in the consulting firm Planning Alternatives for Change LLC. She has done basic and applied research on social structure, gender, and poverty related issues in Bangladesh, India, and several other countries. For the past 12 years she has been studying the ways in which communities respond to water and sanitation programs. Suzanne@planningalternatives.com

Krista Harper is a cultural anthropologist at the Department of Anthropology and the Center for Public Policy (University of Massachusetts, USA). She has studied environmental and social justice movements in Central and Eastern Europe since the mid-1990s. She is author of the 2006 book, Wild Capitalism: Environmental Activism and Postsocialist Political Ecology in Hungary. kharper@anthro.umass.edu

Rosina Hassoun is an anthropologist and Assistant Professor in the Program for Public Health, teaching courses on Poverty, Water, and Health, and courses on the Middle East and Arab Americans in Integrative Arts and Humanities at Michigan State University. She presented with the UNESCO-IHP project on water and cultural diversity at the International Symposium in 2009 in Kyoto, Japan, as well as at the Fifth World Water Forum in Istanbul, Turkey. hassoun@msu.edu

Robert K. Hitchcock is an anthropologist and Professor of Geography and the Center for Global Change and Earth Observations (CGCEO) at Michigan State University, Michigan, USA. He is also a member of the Panel of Environmental Experts (POE) for the Lesotho Highlands Water Project. He focuses on human rights, development, resettlement, and the environment, with particular reference to indigenous peoples, refugees, and farmers in southern Africa. hitchc16@msu.edu

Lisa Hiwasaki is an environmental anthropologist who launched the UNESCO-IHP project on Water and Cultural Diversity in 2007. Since April 2010 she has been working as Programme Specialist for Small Islands and Indigenous Knowledge at UNESCO's Regional Science Bureau for Asia and the Pacific in Jakarta, Indonesia. l.hiwasaki@unesco.org

Helen Ingram is a Research Fellow at The Southwest Center at the University of Arizona, USA. She is the Warmington Chair Emerita at the University of California at Irvine. She has written widely on topics of politics and policy related to water resources and climate change. hingram@uci.edu

Anwar Islam, who has a Bachelor of Science degree in sociology, is a researcher with more than 15 years of experience in numerous studies, programs, and projects, including several relating to water and sanitation. He is a senior researcher in a PAC Water and Culture project, doing ethnographic and folklore studies in his home area of Delduar, Tangail. He is an Associate of Planning Alternatives for Change LLC (PAC). anwar_escherbd@yahoo.com

Barbara Rose Johnston is an environmental anthropologist and senior research scholar at the Center for Political Ecology (Santa Cruz, California USA) and a member of UNESCO-IHP's expert advisory group on water and cultural diversity. A leading scholar on political ecology, environmental health, and human rights, she has served as an advisor to the World Commission on Dams, the Marshall Islands Nuclear Claims Tribunal, and governments and dam-affected communities in Guatemala and Chile. bjohnston@igc.org

Irene J. Klaver is Associate Professor in Philosophy at the University of North Texas, USA, and Founding Director of the Philosophy of Water Project www.water.unt.edu. She focuses on social-political and cultural dimensions of water and has directed and produced water documentary films and imaging projects. She is a member of UNESCO-IHP's expert advisory group on Water and Cultural Diversity and Co-Director of the International Association for Environmental Philosophy. Klaver@unt.edu

Robert Kugonza Akiiki, an environmentalist by profession, is among the pioneers of environmental education in Uganda. Since 1995, Robert has worked tirelessly at different levels and with different initiatives locally and internationally to promote environmentally friendly practices and conservation. Currently, Robert coordinates African Rivers Network (ARN) a network of CSOs in Africa that works with and advocates for dam-affected communities' rights. Robert is also a rights activist. robertk@nape.or.ug

Heather Lazrus is an environmental anthropologist engaged in research on perceptions and responses to weather and climate impacts and other hazards. She is a postdoctoral research associate and the deputy director of the Social Science Woven into Meteorology (SSWIM) program jointly funded by the National Oceanic and Atmospheric Administration at the University of Oklahoma, USA. lazrus@ou.edu

John A. Ledger is an environmentalist and energy specialist and Associate Professor of Energy Studies in the Department of Geography, Environmental Management and Energy Studies at the University of Johannesburg in South Africa. He is a member of the Panel of Environmental Experts (POE) for the Lesotho Highlands Water Project. He focuses on environment and heritage, endangered species and biodiversity, with particular reference to threatened ecosystems, dwindling natural and energy resources and population stabilization in southern Africa. John.Ledger@wol.co.za

Yin Lun is associate professor of the Yunnan Academy of Social Science, China. He received his MA in Social Anthropology Studies at the Ecole Des Hautes Etudes en Science Sociale in Paris (2001). Currently he continues his PhD study in Ethno-Ecology at Minzu University of China. He is also the Deputy Director of Center for Biodiversity and Indigenous Knowledge. His major field of interests is the Indigenous Knowledge among the mountain ethnic groups in Yunnan and Eastern Himalaya Mountain, China, which links watershed management, environment, biodiversity, and climate change. lun.yin@gmail.com

Emmanuel Manzungu is a well-published authority on water policies and reform in southern Africa. He is Professor of Soil Science and Agricultural Engineering at the University of Zimbabwe. He is currently engaged in several research projects including how to build participatory watershed management. manzungu@mweb.co.zw

Nathanial Matthews has a background in political science, development studies and environmental science. During the production of this book he was working as the Education and Research Officer for the International Water Centre (Queensland, Australia). He is currently undertaking PhD research at Kings College London, UK. nathanial.matthews@kcl.ac.uk

David McLain is a retired Army Colonel who has dedicated the last 10 years to the protection and preservation of the Apalachicola River and Estuary and the communities they support. He is currently serving as the Coordinator for the Apalachicola Sub-Basin Caucus on the Governing Board of the Apalachicola –Chattahoochee-Flint (ACF) Stakeholders. www.acfstakeholders.org

Richard A. Meganck is a part-time Professor of International Water Policy at Oregon State University. He retired from the United Nations in 2009 after 34 years working in international development. His most recent position was that of Rector of the UNESCO-IHE Institute for Water Education located in The Netherlands. Dr. Meganck has authored some 100 scientific and technical articles and 5 books on a wide variety of environmental topics. rameganck@gmail.com

Antonio Mendez is an Isoso Guarani anthropologist with Fundacion Yeporaka, Camiri, Bolivia. funda_yeporaka@yahoo.es

Rose K. Morara is a textiles, clothing and interior design Lecturer at the Department of Applied Community and Development Studies, Egerton University, Kenya. Her research interests are in the area of African cultures and their influence on development and resource allocation, use and conservation. She is actively involved in the preservation of material culture of various communities in East and Central Africa. rosekim75ke@yahoo.com

Monica Morgan is a Yorta Yorta woman who has 35 years experience advocating for her people's rights. A leading advocate for Yorta Yorta Nation in the 10-year campaign of the landmark Yorta Yorta Native Title claim, she co-founded and facilitated the Murray Lower Darling Rivers Indigenous Nations and developed the Indigenous Action plan for engagement of Indigenous peoples in management of water and cultural heritage for the Murray Darling Basin Commission. Monica is currently Manager of Demand Dignity & Indigenous Rights with Australia's Amnesty International, working on our global priority campaign and other related Indigenous rights projects in Australia. monica-morgan@hotmail.com

Kyle Nelson (Ngai Tahu, New Zealand) affiliates to Te Runanga o Moeraki, one of the papatipu runanga of Ngai Tahu (the iwi in the lower part of the South Island of New Zealand). Since graduating from University of Otago with a degree in political science and economics, Kyle has been working as a Research Associate for Tipa and Associates Ltd principally in the areas of water allocation and aquatic restoration

Gennie Thi Nguyen is a PhD student at the University of Oregon, USA, in the Department of Anthropology. Her research focuses on issues of environmental justice and race along the U.S. Gulf Coast specifically among Vietnamese American communities. nguyen.gennie@gmail.com

Daniel Niles is assistant professor of geography at the Research Institute for Humanity and Nature, a national research institute located in Japan. He is interested in human-environmental interactions and whether and how contemporary societies may actually make a "transition to sustainability". He has written on the concept and goals of sustainable agriculture, the challenges facing island nations, and landscape change in Japan. dniles@chikyu.ac.jp

George M. Ogendi is an ELP Senior Fellow and Lecturer of Environmental Sciences at the Department of Environmental Sciences in Egerton University, Kenya. A registered Environmental Impact Assessment Lead Expert in Kenya, he advises the Kenyan government on the impacts of proposed projects on human health. George does research on water quality as well as the role of indigenous knowledge in water management and conservation and is keen on improving access to potable water for low-income communities. george1971ke@yahoo.com

Nicholas Olekaikai is a Technologist at the Department of Natural Resources, Egerton University, Kenya. He has extensive knowledge on rangeland resources of Kenya, particularly Maasai indigenous knowledge that relates to management and conservation of animals, plants and water. He is active human rights activist for the indigenous communities in Kenya where advocates for equal access to natural resources and sharing of benefits that accrue from their utilization. nolekaikai@yahoo.com

Elio Ortiz is an Isoso Guarani indigenous anthropologist with Fundacion Yeporaka, Camiri, Bolivia, and is currently serving in the Bolivian government. funda_yeporaka@yahoo.es

Aradhana Parmar is an Associate Professor of the Development Studies Program, Department of Communication and Culture, University of Calgary, Canada. Currently, Dr. Parmar is working on her ongoing research on Water, Women and Empowerment issues in India funded by Social Science and Humanity Research Council, Canada. aparmar@ucalgary.ca

Linda Raffield is a community organizer, writer, and photographer in Apalachicola, Florida.

Ameyali Ramos Castillo is an adjunct research fellow at United Nations University - Institute of Advanced Studies (UNU-IAS). Ameyali's research focuses on strengthening linkages between Indigenous Peoples and international policy processes, especially on issues relating to climate change and water. ramos@ias.unu.edu

Sylvia Rodríguez is Professor Emerita of Anthropology and former Director of the Ortiz Center for Intercultural Studies at the University of New Mexico, USA. She has conducted ethnographic as well as collaborative and participatory research on traditional acequia or irrigation communities in northern New Mexico. Her most recent book is Acequia: Water Sharing, Sanctity, and Place (School for Advanced Research Press, 2006). sylrodri@unm.edu

Veronica Strang is an environmental anthropologist at the University of Auckland, and is internationally recognised for her work on water issues. She has participated in steering and advisory groups for UNESCO-IHP's programmes in Ecohydrology and in Water and Cultural Diversity. In 2007 she was named an international prize "Les Lumières de L'Eau" at the Cannes International Water Symposium. Her most recent book is Gardening the World: agency, identity and the ownership of water (2009). v.strang@auckland.ac.nz

Farhana Sultana is Assistant Professor of Geography at Syracuse University, USA. Previously she taught at King's College London (UK) and was a Programme Officer at the UNDP (Bangladesh). Farhana's research interests and publications are in water governance, development contradictions, political ecology, and gender dynamics of climate change. She has extensive research experience in South Asia, especially in gender and water. Her forthcoming book is titled 'The Right to Water' (2011, Earthscan). sultanaf@syr.edu

Thomas F. Thornton is Senior Research Fellow at the Environmental Change Institute, School of Geography and the Environment, University of Oxford, UK. An anthropologist, he has written widely on human ecology, adaptation, local and traditional ecological knowledge, conservation, coastal and marine environments, conceptualizations of space and place, and the political ecology of resource management among the indigenous peoples of the Pacific Northwest. His most recent book is Being and Place among the Tlingit (University of Washington Press, 2008). thomas.thornton@ouce.ox.ac.uk

Bryan Tilt is an environmental anthropologist whose research focuses on economic development and environmental protection in China. He has conducted fieldwork and published widely on pollution control, sustainable development, and water management in China, including a recent book entitled The Struggle for Sustainability in Rural China: Environmental Values and Civil Society. He currently teaches in the Anthropology Department at Oregon State University, USA. Bryan.Tilt@oregonstate.edu.

Gail Tipa (Ngai Tahu, New Zealand) affiliates to Te Runanga o Moeraki, one of the papatipu runanga of Ngai Tahu (the iwi in the lower part of the South Island of New Zealand). For the last 14 years a focus for Gail's work has been projects of benefit to Iwi. Tackling freshwater issues has been to the fore. A focus has been working with others to develop new water management tools for Maori to use. More recently, Gail has been working in aquatic restoration and allocative decision making, specifically the development of new tools to help Maori determine appropriate stream flows. gtipa@xtra.co.nz

Fabiane Torres is a journalist who has been working for the NGO "Projeto Aguas do Rio Doce" as redactor of the regional magazine "Águas do Rio Doce" and the website of the same name. She has been involved with social communication towards social mobilization for waters, with focus on the strengthening of the management of water resources in the regional society, particularly valuing the cultural identity and the sense of belonging of the population. fabianetorres@oi.com.br.

Tashi Tsering is a PhD candidate in Resource Management and Environmental Studies program at the University of British Columbia

Terence Turner, Ph.D. Harvard 1965, Professor of Anthropology University of Chicago, USA (Emeritus) and Cornell University (Adjunct), has worked with the Kayapo of Central Brazil since 1962. He has served on the Ethics Committee and the Committee for Human Rights of the AAA. He has published widely on Kayapo social organization and culture and has worked as an activist with several Kayapo, Brazilian, European and American NGOs seeking to defend indigenous territories, societies, and the Amazonian environment, focusing on resistance to the planned Xingu River dams. tst3@cornell.edu

Jeroen Warner, a political scientist by training, is currently Assistant Professor of Disaster Studies with Wageningen University, the Netherlands and Senior Researcher with the Public Administration department at Erasmus University Rotterdam. His PhD dissertation, 'The Politics of Flood Insecurity' investigates the politics and framing of risk in flood management interventions. Different cultural frames play an important part in explaining conflicts over river interventions. jeroen.warner@wur.nl; jeroenwarner@gmail.com

James Webster is an anthropological engineer with more than 20 years of broad and diverse overseas experience in rural water, sanitation and hygiene promotion programmes in low income countries, with a particular interest in the influence of culture within these programmes. He is Director of his own consultancy firm, is a member of UNESCO-IHP's expert advisory group on water and cultural diversity, and was formerly a Lecturer at Cranfield University. Jameswebster1@btinternet.com

Joshka Wessels is a Dutch geographer/anthropologist, communication specialist and environmental filmmaker. During her PhD research she specialized on Traditional Water Technology in the Middle East. Joshka runs her own environmental media company Sapiens Productions and is associate trainer in media for development for the Centre for International Development and Training (CIDT), University of Wolverhampton (UK) and associate media consultant for the IRC-International Water and Sanitation Centre in Delft, The Netherlands. Joshka.wessels@btinternet.com

Amber Wutich is a cultural anthropologist on the faculty of the School of Human Evolution and Social Change at Arizona State University, USA. Her research examines human ecology in water and food-scarce environments, with a particular emphasis on cultural and institutional adaptations. In cooperation with local communities and stakeholders, she conducts long-term ethnographic fieldwork in Bolivia, Paraguay, and the southwest United States. amber.wutich@asu.edu

Alejo Zarzycki is a forestry engineer with over 30 years experience in rural development in the Gran Chaco of Bolivia and Argentina; Fundación Yangareko, Santa Cruz de la Sierra, Bolivia, and Fundacion Urundei, Salta, Argentina. urundei@gmail.com

Acronyms

AC	Asociación Civil (Mexico)
ACF	Apalachicola-Chattahoochee-Flint Basin (USA)
ADB	Asian Development Bank
AfDB	African Development Bank
AIC	Association d'intérêt collectif (Collective Interest Association) (Tunisia)
ANA	National Waters Agency
ANC	African National Congress (South Africa)
AMNI	Area under Integrated Management (Bolivia)
ARN	African Rivers Network
AWM	adaptive water management
BNHUOP	Banco Nacional Hipotecario Urbano y de Obras Públicas (Mexico)
C&D	construction and demolition waste
CBD	Convention on Biological Diversity
CEH	United Nations-sponsored Commission on Historical Clarification
CEPL	Central European University's Centre for Environment and Policy
CGRB	Bañado Watershed Management Committee (Bolivia)
CHI	cultural health index (New Zealand)
CHMPs	Cultural Heritage Management Plans (Australia)
CIMI	Missionary Council of the Church (Brazil)
COMA	cultural opportunity mapping and assessment (New Zealand)
CRDA	Regional Centre for Agricultural Development (Tunisia)
CSAs	customer satisfaction assessments
CSOs	civil society organizations
DECOIN	Defensa y Conservación Ecológica de Intag (Ecuador)
EC	European Community
EFA	environmental flow assessment
EGAT	Electricity Generating Authority of Thailand
EIA	environmental impact assessment
EO	Ecological Ordinance (Intag, Ecuador)

EPA	Environmental Protection Agency (Republic of Marshall Islands United States)
ESIA	Environmental and Social Impact Assessment (Sierra Leone)
EU	European Union
EZLN	Zapatista National Liberation Army
FRAG	Flood Risk Action Groups (England)
FUNAI	National Indian Foundation
GAP	Ganges Action Plan (India)
GMS	Greater Mekong Subregion
GRBA	Ganges River Basin Authority (India)
GWP	Global Water Partnership
HHW	household hazardous waste
HSS	Himalaya Seva Sangh (India)
HWRMT	Hongpo Water Resource Management Team (Deqin County, Diqing Province, China)
HYPHEN	Himalayan and Peninsular Hydro-Ecological Network
IADB	Inter-American Development Bank
IBAMA	Brazilian Institute for the Environment
IBRD	International Bank for Reconstruction and Development
IFC	International Finance Corporation
IHP	International Hydrological Programme
IIRSA	South American Regional Infrastructure Integration Initiative
IMF	International Monetary Fund
INGO	international non-governmental organization
IPCC	Intergovernmental Panel on Climate Change
IPP	Indigenous Partnerships Program (Murray Darling River Basin, Australia)
IR	International Rivers
IRN	International Rivers Network
ISA	Socioenvironmental Institute (Brazil)
ITSN	International Tibet Support Network
IUCN	International Union for Conservation of Nature
IWM	integrated water management
IWRM	integrated water-resources management
JPA	Joint Planning Approach
KFF	Katse Fresh Fish (South Africa and Lesotho)
KOBWA	Komati River Basin Management Authority (Swaziland and South Africa)
LHDA	Lesotho Highlands Development Authority
LHWP	Lesotho Highlands Water Project
LINKS	Local and Indigenous Knowledge Systems
LLEs	local legal entities (Lesotho)
MAB	Movement of Dam Affected Peoples in Brazil
MCF	Mau Complex Forests (Kenya)
MDGs	Millennium Development Goals

MENA	Middle East and North Africa region
MLDRIN	Murray Lower Darling Rivers Indigenous Nations (Australia)
MLIT	Ministry of Land, Infrastructure, Transportation and Tourism (Japan)
MRC	Mekong River Commission
MRGCD	Middle Rio Grande Conservancy District (New Mexico, USA)
MTAP	Mineral Sector Technical Assistant Project (Sierra Leone)
NAILSMA	Northern Australian Indigenous Land and Sea Management Alliance
NAPE	National Association of Professional Environmentalists (Uganda)
NGO	non-governmental organization
NGRBA	National Ganga River Basin Authority (India)
NMAA	New Mexico Acequia Association (USA) 2.4a
NOAA	National Oceanic and Atmospheric Administration (USA)
NPHD	National Policy on Hydropower Development (India)
NRCD	National River Conservation Directorate (India)
NREGA	National Employment Guarantee Act (India)
NWDA	National Water Development Agency (India)
OECD	Organisation for Economic Co-operation and Development
PAEW	Producers' Association of the Estrella Wetlands (Argentina)
PAR	participatory action research
PDR	Lao People's Democratic Republic
PPA	participatory poverty assessment
PROMAGUA	Program for the Modernization of Water Management Companies (San Cristóbal de las Casas, Chiapas, Mexico)
RDP	Reconstruction and Development Programme (South Africa)
REDLAR	Latin American Network Against Dams and for Rivers, Communities and Water
RIS	Reservoir-Induced Seismicity
RMI	Republic of the Marshall Islands
RMU	Independent Review Mechanism Unit
SAPAM	Sistema de Agua Potable y Alcantarillado Municipal (San Cristóbal de las Casas, Chiapas, Mexico)
SCS	Soil Conservation Service (United States)
SQUIDS	Stormwater Quality Improvement Devices (Brisbane, Australia)
SRP	Sustainable Rivers Project (USA)
STEG	Tunisian power company
TK	traditional knowledge
TRC	Truth and Reconciliation Commission (South Africa)
UN	United Nations
UNCESCR	United Nations Committee on Economic, Social, and Cultural Rights
UNDP	United Nations Development Programme
UNEP	United Nations Environment Programme
UFMG	Federal University of Minas Gerais (Brazil)
UNESCO	United Nations Educational, Scientific, and Cultural Organization
UNIVALE	University of Vale do Rio Doce (Brazil)
UNPFII	United Nations Permanent Forum on Indigenous Issues

UNU-IAS	United Nations University-Institute of Advanced Studies
WASH	water, sanitation, and hygiene
WCD	World Commission on Dams
WFD	Water Framework Directive (European Union)
WHO	World Health Organization
WSSD	World Summit on Sustainable Development
WSUD	water-sensitive urban design
WWAP	World Water Assessment Programme
WWDR	World Water Development Reports
ZINWA	Zimbabwe National Water Authority

Part I
Water and Cultural Diversity

Cover art by Anabel Fundora Medina, 11 years old, Cuba

Chapter 1.0
Introduction: Water and Cultural Diversity

Irene J. Klaver

Truisms about the relation between water and culture abound: Water shapes culture and culture shapes water. Water is crucial for the flourishing of cultures, and vice-versa. A cultural stewardship relation towards water is crucial for the health and viability of the world's water resources. Simple truisms often emerge from immensely complicated and multifaceted human experiences that are in turn, influenced and shaped by conceptual, sociopolitical, and practical realities. In this complex constellation cultural diversity emerges as an indispensable concept, making visible the ways in which human-water relationships reflect natural processes, sociocultural understandings, relationships and norms, and, political and economic interests and institutions. Part I sets out to explore these relations with a sampling of cases and essays that suggest some of the key linkages between cultural diversity and water.

Cultural diversity refers to the variety of human cultures or societies in a given region and across the world. Cultural diversity, as recognized by UNESCO, is a driving force of development, not only in generating economic growth but also as a means of leading a more fulfilling intellectual, emotional, moral and spiritual life. Cultural diversity, like biological diversity, is intimately shaped and sustained by the interconnected realms of ecological, genetic and species diversity. As observed by the International Union for the Conservation of Nature (IUCN), it is no coincidence that areas of linguistic and ethnic diversity are also areas rich in biodiversity. Most of the world's languages are spoken by indigenous and other tribal peoples in countries that harbour great biodiversity. Indeed the biocultural diversity of a given region represents a complex history of human interaction, knowledge, values and stewardship of the environment, and especially of the essential role of water in sustaining life.

The term biocultural diversity is increasingly central to a general discourse of sustainable futures in the face of human-induced environmental change, recognizing

I.J. Klaver (✉)
Department of Philosophy and Religion, University of North Texas, Denton, TX, USA
e-mail: Klaver@unt.edu

B.R. Johnston et al. (eds.), *Water, Cultural Diversity, and Global Environmental Change: Emerging Trends, Sustainable Futures?*, DOI 10.1007/978-94-007-1774-9_1,

both the common struggle *and* the adaptative potential, and facilitating a political, conceptual and regulatory toolbox with nested concepts such as endangered species, cultural values and behaviour, ecological knowledge, and customary rights. In order for us to protect the planet's water and the biological and cultural diversity that water supports, alternative institutional infrastructures need to be nurtured or developed with local knowledge serving as a key component in the decision-making framework (Conca 2006). This safeguarding of sustainable local practices is crucial to achieving equitable and sustainable development.

The notion of biocultural diversity as a core element in shaping the water-culture interface runs throughout this book. In this introductory section – using such concepts as water culture and waterscapes – we consider some of the varied cultural values of water; the importance of the everyday routine and the structures and practices of engagement that create a sense of belonging or water-identity; the integrative nature of local knowledge and the physical manifestations of the water/culture relationship; and, finally, we examine how these notions, behaviour and concrete realities reflect sociopolitical conditions and inequities.

The section begins with a historical–geographical essay by Klaver that considers how water is never simply present (or absent) but produced in a cultural landscape of social, political and infrastructural systems. A mixture of conceptual and material themes surfaces, some of which are further analysed by other authors of this section. A recurrent topic is the homogenizing effect of the nation state on local water practice: state-driven water management, based on highly specialized science, law and engineering, has historically replaced the integrative approach of local knowledge.

Thus, Alley describes the emerging water nationalism in India with the large infrastructure development creating a national water grid through river-linking, diversion and water flow management strategies. This engineering has involved a shift from quality-driven policy to a quantity-driven approach, with a concomitant expansion of water privatization, transforming the notion of water as a sacred power, public utility and human right, to a market resource. Religious practices such as Hindu worship rituals, and other daily practices of bathing, washing clothes, fishing, transportation and small-scale industry have been deeply affected by this ongoing transformation, as public uses are restricted and services increasingly priced out of reach.

In the Jerid region of Tunisia water control has shifted from local farmers to the French colonial state, and again to the Tunisian national government, undermining a complex oasis culture with communal systems of collective hydraulic maintenance. Battesti shows how these transitions take place in multiple domains: shifts in economic and social power, in whose knowledge counts as expertise, and in management from a communally operated to a contract labour structure. Recently, a new economic transition is taking place in the Jerid: a shift from irrigated agriculture to tourism. Ironically, this development allows for older water structures to surface again as signifiers for traditional ways of oasis culture life.

Boelens observes a similar trajectory from colonial to postcolonial state control in irrigation politics in the ethnically diverse societies in the Andes in South America. State authorities are usually from a dominant culture of a ruling ethnic group who exert social control through controlling everyday water management. Analyzing the social political water arrangements that disempower local people, Boelens determines mutually constituting domains of knowledge, power and economy. Where in so-called traditional ways of knowledge these domains are intertwined, in mainstream science and water policies they are separated: hydraulic infrastructure is just a technical and material matter; water laws and rights are simply legal matters; water organization is the work of sociologists and social organizers; economists lead the planning of productive wealth and 'rational economic institutions'. Thus, water rights in an Andean community are embedded in historical, political, economic and cultural relations, closely related to a 'water-identity' of communities. Protecting a water community in the Andes entails a socio-physical creation process in the context of collective hydraulic property: a confluence and interaction of all water control domains.

Brugnach and Ingram point out that the interests of indigenous peoples and small farming communities are often overlooked in integrated water resource management (IWRM) and adaptive management. While these relatively new management models intend to integrate social, environmental and hydrological concerns and to explicitly accommodate participatory and inclusive practices, they regularly reflect traditional power asymmetries. Administrative procedures calling for participation do not guarantee equal participation. Cultural barriers limit the contribution of disadvantaged populations. Open and transparent forums alone do not compensate for differences in resources and skills or for power differentials among participants. Furthermore, expert-driven discourses dominate current decision-making processes, in which nature is externalized and independent from human experience. Brugnach and Ingram contrast this with decision-making models based on a relational concept of knowledge, which is situational, experiential, integrative and invitational for collaboration.

Warner further analyses participatory initiatives in contemporary water management, examining a joint Dutch, German, and French river management project pursued through the European Water Framework Directive which prioritized early public engagement with multiple stakeholders. As illustrated in this case, even the best-planned participatory models can be hijacked by power plays or media attention. Warner concludes that multi-stakeholder planning processes should neither be idealized as the pinnacle of democracy nor dismissed as hegemonic instruments or ineffective smokescreens; rather, they should be seen for what they are: social networks representing a degree of diversity.

Recognizing that the challenges before us not only require integrating cultural concerns on the local scale, but doing so on a global scale, Lazrus asks the reader to the consider water/culture realities confronting residents of the Tuvalu archipelago, one of the places on our planet most vulnerable to climate change. A Polynesian nation of reef islands and atolls in the central Pacific Ocean, Tuvalu is, at its highest point, only 4.5 m above sea level. Lacking significant aquifers, rivers, or lakes, rainwater is the main source of fresh water and the nation is economically and technically dependent on development aid to guarantee enough storage capacity. Without considerable effort from the world's nations to reduce carbon emissions and halt or reverse the profound climate changes that are now underway, even this form of intervention will soon prove insufficient in meeting the threat from rising seas. Water has been acknowledged as important on a global level: yet, integration of local cultural diversity in global conversations and planning has still a long way to go.

Culture is a capacious term. Definitions of culture are plentiful. They all configure relations – relations on multiple scales, among multiple planes, along multiple vectors. Heritage, traditions, habits and customs are usually emphasized, but futurity has a crucial role as well, generating ideals: culture can be seen as a capacity to aspire (Appadurai 2004). It is in a dialogue between traditions and aspirations that engagement or involvement emerges. The effectiveness of cultural diversity is predicated upon the capacity to be involved. Thus, various pieces in Part I speak to the social political infrastructure and the motivational contexts that need to be in place for people to be able to be engaged and to participate or to be interested in being involved.

Klaver ends her opening essay with sketching modes of water engagement in contemporary urban context. She connects restoring rivers and watersheds with restoring a cultural mindset or mentality of re-connecting and re-engaging with water, rivers and watersheds. Cultural re-connecting to a river in combination with restoring its water quality is the explicit theme of the vignette or textbox of Guerra and Torres. They highlight the cultural transformations around the Doce River Basin in southeast Brazil, the major mining area of the country. Serious river pollution led to a mobilization of many groups and the construction of a new water management perspective with an active involvement of local knowledge, cultural activities, industrial clean up practices, with the help of schools and universities.

Alcorn, Ortiz, Mendez and Zarzycki depict a most engaged river relation in the arid Gran Chaco of Bolivia. They describe how Isoso Guarani culture is shaped by, and dependent on, the pulsing Parapeti River: the Isoso people adjust their practices to the 'decisions' of the river, moving their homes when the river moves. A similar versatility or flexibility can be found – in different shape and form – in modern cities if urban planning takes biocultural diversity into consideration. That means planning that gives space to diversity. Various strategies are on their way, such as the development of so-called, green infrastructure and carefully chosen material culture like specific water features in urban design.

Strang concludes Part I with a cultural analysis of the wide variety of urban water features in the city of Brisbane in Australia. Public material culture sustains communities, gives an opportunity to community identity and belonging. Strang emphasizes that water features not only celebrate collectiveness but can also express exclusion and expose a complexity of power relations in material culture. In the wide variety of water features representing the scope of Australian culture, ranging from European historical ideals, natural imagery, cosmopolitan modernist style, to Chinese water gardens, one community is conspicuously absent: the Aboriginal culture.

The case studies and brief vignettes, conceptual essays and graphic imagery of Part I speak to the breadth and depth of cultural diversity in practices around water.

This section is a thick description of lived actualities and possibilities, with its central theme being the necessary foregrounding of cultural diversity in water policy. As such it is also a tribute to potentiality, and emerging promise.

References

Appadurai, Arjun. 2004. The capacity to aspire: Culture and the terms of recognition. In *Culture and public action*, ed. Rao Vijayendra and Walton Michael. Stanford: Stanford University Press.

Conca, Ken. 2006. *Governing water: Contentious transnational politics and global institution building*. Cambridge: MIT Press.

Chapter 1.1
Placing Water and Culture

Irene J. Klaver

Water has seeped into unexpected domains. Water-related advertisements abound, and not only in the bottled water industry or in describing waterfront property. There are fashionable lines of clothing and cosmetic products with 'water,' 'aqua,' or 'H_2O' in their names. In high-end shopping districts of the world's capitals, one can find skin-care stores called *H_2O*, built of sleek aquamarine-coloured glass. Water keeps re-entering the cultural imagination in new shapes and forms, subliminally exerting its aesthetic appeal.

In sculpture, architecture, multimedia, and landscape art, contemporary artists such as Tadao Ando, William Pye, Roni Horn, Basia Irland, and Herbert Dreiseitl—to name just a few—all explore innovative ways of working with water, implicitly articulating a larger cultural interest in the element, while highlighting the fundamental fact: earth is the 'Blue Planet', the water planet.

Fig. 1.1.1 Atelier Dreiseitl (from *left* to *right*): Potsdamer Platz, Berlin, Germany; Water traces, Hannoversch Münden, Germany; Heiner-Metzger-Platz Neu-Ulm, Germany (Photo credit: all are courtesy of http://www.dreiseitl.net, Accessed 13 Aug 2010)

I.J. Klaver (✉)
Department of Philosophy and Religion, University of North Texas, Denton, TX, USA
e-mail: Klaver@unt.edu

B.R. Johnston et al. (eds.), *Water, Cultural Diversity, and Global Environmental Change: Emerging Trends, Sustainable Futures?*, DOI 10.1007/978-94-007-1774-9_2,

Nearly three-quarters of earth's surface is covered by water, mostly by oceans. Water gives rhythm and pulse to life. It pervades everything. Moving through any living entity—our bodies, the land, the atmosphere, and our cultures—water connects, transports, and transforms. Earth has a biosphere because it has a hydrosphere. Earth has a culture sphere because it has a hydrosphere.

But, ubiquitous as water may be, fresh water is limited. 97.5% of the planet's water is saline. More than two-thirds of the 2.5% fresh water is locked in ice at the polar regions or in distant glaciers in mountainous areas. A little less than one third is groundwater, and only 0.3% of the fresh water is surface water, such as rivers, lakes, and reservoirs.

Rivers are the blue ribbons of the earth: relatively small in comparison to landmasses, they tie oceans and mountains together. Rivers mediate between aquatic and terrestrial ecosystems, providing habitat for land and water species alike. Over geological time rivers have shaped the lay of the land through erosion, flooding, and meandering. Carrying and depositing sediments, rivers form intricate landscapes with fertile floodplains at the mouth of their basins.

The amount of water on the planet remains fairly constant over time; however, the quality and quantity of water at any specific time and place are highly variable. Many factors account for this variability. Human culture has become one of the determining factors.

1.1.1 Water in Culture

Water has a presence in every aspect of our daily lives. We drink it, bathe in it, and cook with it. Our food consists largely of water: spinach equals milk in being approximately 90% water, tomatoes 95%, and beef, seemingly so solid, is 61% water. We ourselves are largely composed of water (66% of our body weight), and we need about 3 litres a day to live. We can survive for a month without food, but only a week or less without water.

Given its vital role, water has always had cultural significance. Virtually all cultures developed around water: tribes settled at the shores or on the banks of water bodies, cities originated at confluences of rivers. The first complex societies were irrigation-based cultures with ingenious water management structures—societies as diverse as ancient Rome, China, India, Mesopotamia, pre-Columbian Mexico and Peru. Karl A. Wittfogel called these 'hydraulic civilizations' because they were characterized by a centralized governments and extensive division of labour centred around complex water management. Roman culture thrived with skilful engineering of numerous aqueducts for public water supply using various techniques of hydraulic architecture.

Through the ages, poetry, music, and religion have found a deep well of meaning in water: crosscurrents of meaning, reflections and emotions. Gazing at his reflection in a pond, Narcissus epitomized the rise of self-consciousness.

Fig. 1.1.2 The Pont du Gard is part of a 50 km aqueduct in southern France constructed by the Romans in the first century A.D. It was added to UNESCO's list of World Heritage Sites in 1985 (Photo credit: Armin Kübelbeck, http://commons.wikimedia.org/wiki/File:Pont_du_Gard_01.jpg

The cosmogonic power of water has been a major theme of many ancient accounts of human origin. These poetic sources are precursors to later scientific theories of evolution, which confirmed that all life forms emerged phylogenetically and ontogenetically out of water.

In the *Enuma Elish*, the Mesopotamian-Babylonian creation epic of the third millennium BC, the primordial waters are Apsu, meaning sweet water 'ocean,' 'deep abyss' or 'outermost limit', and Tiamat, the one who 'is too deep to fathom,' the salt sea, whose name refers to 'primeval waters.' Apsu and Tiamat are brought under control by gods (their offspring) in order to create the topography of earth and sky.

We see a similar structure in Genesis, the first book of the Old Testament (written between 1400 and 400 BC). Clearly influenced by the *Enuma Elish*, its opening lines read: 'In the beginning…darkness was upon the face of the deep; and the Spirit of God was moving over the face of the waters.' The Hebrew word *Tehom*, meaning 'deep [waters],' is etymologically related to 'Tiamat.'

The first Greek philosopher, Thales of Miletus (c. 624 BC–545 BC), considered water to be the beginning, an originating and guiding principle or *archê*. There were similarities between Thales's philosophy and the ancient legends and myths, specifically the ones about Oceanus and his consort Tethys, who was both sister and wife of Oceanus and whose name has etymological ties to Tiamat and Tehom. However, Thales rejected the traditional beliefs that the gods organized, shaped, and controlled the cosmos. Thus he was, for Aristotle, no longer a theologian like the old poets, but the founder of natural philosophy—investigating the basic principles of matter and theoretically moving towards a scientific treatment of natural phenomena.

Nearly a century later another Greek philosopher, Heraclitus of Ephesus (c.535 BC–475 BC), found in the *movement* of water a guiding principle: '*panta rhei*' (all things flow). Heraclitus was a proto-phenomenologist, interested in the everyday experience of change: 'Cold things warm up, the hot cools off, wet becomes dry, dry becomes wet.' With these simple examples he revealed how opposites are intrinsically related to each other and continually change into one another. This transformational flux was not random but governed by *Logos*, the first proposed 'law of nature' in Western natural philosophy. The widely contrasting workings of water capture this co-constitutive dynamic between opposites. In ebb and flow and rain and drought, water comes and goes—powerful and vulnerable, calming and dangerous, quiet and violent. It leads Heraclitus to one of his famous sayings: 'one cannot step in the same river twice.'

Fig. 1.1.3 Alhambra, Spain: Patio de los Arrayanes (Photo by Ra-smit); Patio de la Acequia (Photo credit: Andre Dunn)

Similarly, the ancient Chinese text *Tao Te Ching* of Lao Tzu (sixth century BC), invokes the paradoxical powers of water: 'Nothing in the world is as soft and yielding as water. Yet for dissolving the hard and the inflexible, nothing can surpass it.' Water is most powerful, and at the same time unassuming. Therefore: 'Water is the supreme good, which nourishes all things without trying to. It is content with the low places that people disdain. Thus it is like the Tao.' Lao Tzu suggests that water is a metaphor of exemplary leadership, because despite its power it accepts the lowest place: 'All streams flow to the sea because it is lower than they are. Humility gives it its power.... If you want to govern the people, you must place yourself below them. If you want to lead the people, you must learn how to follow them.'

Humility is also an important value in the Hebrew and Christian Bible, thematically often combined with God's punishment of human hubris. Water has a crucial metaphorical place in this. The flood, for example, is the most famous biblical water story, the ultimate expression of God's wrath. In the deluge only Noah, his family, and the pairs of animals on his ark are saved. At that point the imagery moves from water as punishment to water as cleansing, and the water-formed rainbow appears, the symbol of God's covenant with his people, in which humans have the first right to the goods of his creation but in return have to take care of the earth.

There is a similar ethic of stewardship in Islam: the blessings of water come with human responsibility for its proper use. All life forms—including plants and animals—should be supported according to their needs. The Koran observes that the supply of water is fixed and should not be wasted. This respect for water found a highest material expression in the majestic architecture of the Alhambra, the fourteenth-century Muslim palace in southern Spain.

Fountains with running water can be heard from every room, resulting in a cooling effect. At the entrance a stately pool, symbol of power, reflects the magnificent building.

1.1.2 Rivers in Culture

Throughout recorded history humans have shaped rivers for irrigation, navigation, and flood protection. Ancient Sumerian cultures and, later, the Assyrians (2400 BCE) flourished in the Fertile Crescent, an area whose astounding agricultural productivity was due to its main river systems, the Tigris and Euphrates. These societies constructed elaborate irrigation and flood-control projects that enabled the emergence of cities and sophisticated tax-based economic and legal systems. However—and there is an obvious lesson here—like most early irrigation-based societies, their civilizations collapsed because of soil depletion, salinisation, and a high vulnerability to invasion. By the time the Greek historian Herodotus (c. 484–c. 425 BCE) visited the area, the Persians had established their rule. 'Rivers... they revere,' he wrote about the Persians. 'They will neither urinate nor spit nor wash their hands in them, nor let anyone else do so.'

The only irrigation-based ancient society that proved to be sustainable was that of the Egyptians, who maintained their agricultural methods successfully for 5,000 years. Their secret lay in the fact that their method of irrigation made use of the natural rise and fall of the Nile's seasonal fluctuations and hydrology. The river's yearly flooding rejuvenated the land with water and also with a rich layer of sediments. A famous bas-relief from 3100 BCE—showing one of the ancient kings ceremonially holding a hoe for cutting an irrigation ditch—testifies to the existence and importance of irrigation waterworks going at least back to that time. This annual flood-deposition cycle continued until the Aswan High Dam disrupted the natural flow pattern of the Nile in the twentieth century. The flooding of the Nile also

Fig. 1.1.4 Mace head scene of the Scorpion King cutting irrigation ditch (Photo credit: 'duckie-monster' [http://www.flickr.com/photos/duckiemonster/2081142989/])

precipitated the development of mathematics: The world's first numerical system was invented to apportion the land after floods and sediment depositions had obliterated boundaries (geometry literally means 'land measurement') and to determine planting and harvesting schedules. Herodotus rightfully called Egypt 'the gift of the Nile' (Herodotus 1921–1924).

Rivers are revered in many traditions. Along African rivers one finds small offerings such as bowls of fruit or bundles of flowers. For Hindus in India, rivers are goddesses. The doorways of early Hindu temples were decorated with images of the two main rivers: Ganga rode on a crocodile and Yamuna on a tortoise. 'Mother Ganga'—known internationally by its Anglicized name, Ganges—is still the most sacred river in India. Her generative and purifying powers are invoked in various initiation rites and drinking her waters is believed to bestow longevity. Pilgrims visit

Fig. 1.1.5 da Vinci: depiction of human arm's veins; birds-eye view of the Arno river (Photo credits: http://www.universal leonardo.com via the Royal Windsor Library)

her banks for daily ritual bathing, which strengthens the weak, heals the sick, and washes away sins. People give the bodies of the dead to the river or spread their ashes over her, and she carries them in a last transition to the land of their ancestors.

In many cultures rivers represent the capacity for transformation. The most definitive transition is between life and death, symbolized in Greek mythology by the rivers Styx and Lethe, the former providing a journey to the netherworld, and the latter washing away human memories so that the afterlife can be begun anew. In many traditions time is represented as a river. In Dante's *Inferno* ice, not fire, is at the centre of hell, and frozen stasis is its most severe punishment. Small rivulets, trickles thawed out of ice, symbolize the narrow path to renewal and hope.

Rivers have thus always been anchors of civilization and bones of contention (the word *rival* is cognate with *rivulet*). In the early 1500s Leonardo da Vinci and Niccolò Machiavelli conspired to divert the Arno River from the city of Pisa in what is today Italy. This diversion would have deprived the city of water, thereby giving victory to Pisa's rival, Florence, after a 10-year war. For various reasons the plan failed, inspiring Machiavelli to compare fortune to a river—something that is unpredictable, violent, and irresistible. Even though the scheme to divert the Arno did not come to fruition, the underlying paradigm of the control of fortune through a powerful combination of economic practices, engineering, and strategic planning was a precursor of modern river management.

1.1.3 Rivers in the Modern Era

With the industrial revolution, rivers became the focus of energy production and economic activity, and were increasingly confined to the mercantile back regions of towns. They became functional arteries for transport, harbours, and other economic purposes. Often they were exiled underground, or dried, paved over, and turned into roads.

Until relatively recently, Europe's rivers were treated as cheap waste transportation to the sea. Heavily polluted, with both industrial and human waste, they have

Fig. 1.1.6 The Cuyahoga river in Cleveland, Ohio, has caught fire several times—in 1936, in 1952 (shown here), and in 1969 (Photo credit: United Press International)

harmed human health and degraded the quality of coastal and marine waters. The biodiversity of thousands of kilometres of waterways has been affected. The advent of intensive industry brought similar changes in the United States, culminating in a famous incident in 1969 when a thick layer of oily industrial pollutants on the Cuyahoga River in Cleveland, Ohio, caught fire from the sparks of a passing train. Even the revered Ganges and Yamuna rivers turn into open sewers after they pass through Varanasi and Delhi, respectively. Each river enters the city alive and leaves nearly dead, soiled by sewage and other waste, anaerobic with a zero dissolved-oxygen level as gaseous sludge rises from the bottom and floats to the surface. Cities were not the only problem: along the Mississippi River, farming and logging on a massive scale had caused immense erosion by the early twentieth century. Vast amounts of topsoil washed down the river into the Gulf of Mexico, a trend that had a disastrous impact on local agriculture.

The twentieth century saw an unprecedented impact on rivers, with the building of approximately 800,000 dams worldwide, of which 48,000 are 15 m or higher. The collective weight of the harnessed water and trapped sediments has caused a measurable change in the angle of the Earth's axis and the speed of its orbital movement. One-fourth of the world's sediment ends up in reservoirs behind dams instead of nourishing floodplains and estuaries. Silt's colour accounts for the names of the

Fig. 1.1.7 Itaipu dam during release (Photo credit: Angelo Leithold 2005)

Yellow River and the Colorado River (*colorado* is Spanish for 'colored'), but dams have robbed them of the significance of their names: the released water is as clear as glass. And many of the smaller reservoirs behind dams have silted up to form marshy plains.

Dependence on dams is hard to undo: they provide almost 20% of the world's electricity supply. The massive dam at Itaipu (built from 1975 to 1991), for example, on the mighty Paraná River between Brazil and Paraguay, provides energy to São Paolo (a city of 11 million people) as well as Rio de Janeiro (a city of six million), and furnishes 20% of Brazil's electricity and 93% of Paraguay's. The Itaipu Dam shifted the course of the seventh-biggest river in the world. The American Society of Civil Engineers called the dam one of the seven wonders of the modern world, echoing in a secular fashion, Jawaharlal Nehru, India's first Prime minister, who called dams 'temples of modern India' inaugurating modernization in post-Independence India. Nehru perfectly expressed the engineering culture of modernity in which temples are associated with the mystical, traditional and ancient, while dams signify rationality, progress and modernity.

Dams, however, come at high costs. They profoundly disrupt long established ecological, hydrological, and cultural systems. They have also displaced 40–80 million people, either by forced eviction or because of the loss of their traditional livelihoods as a result of dam-induced environmental change. The stagnant waters of dam reservoirs breed mosquitoes and are infested with freshwater snails that carry disease-causing parasites.

It is therefore not surprising that dam building has triggered protests all around the world. Influential advocacy groups, such as International Rivers, question the conventional or modern 'development' model that dams epitomize. They research other ways of meeting people's needs for water, energy, and protection from damaging floods. In 1997 the World Commission on Dams (WCD), an independent panel to review large dams, was created by the International Union for Conservation of Nature (IUCN) and the World Bank. The Commission's 2000 report, *Dams and Development: A New Framework for Decision-Making*, acknowledges the important benefits of dams for human development but also concludes that the price paid, both socially and environmentally, is unacceptable (WCD 2001).

Once widely regarded as a symbol of a culture of progressive technology, dam building is now commonly viewed as a threat to the earth's ecosystems. Nevertheless, major dams are still being built and still being planned. A series of dams are planned for the headwaters of the Ganges River in India, where the Tehri Dam, which began to fill in 2004, already affects the river's flow from the Himalaya Mountains. In April 2008 leading Indian scientist, Professor G. D. Agarwal, former dean of the Indian Institute of Technology at Kanpur, announced a 'fast-unto-death' against damming on the first 125 km stretch of the River Bhagirathi, the Himalayan headwaters of the Ganges, 'to oppose the destruction of this ecological marvel and the epitome of Hindu cultural faith.' (Letter from Prof. Agarwal to the government of India, April 14, 2008; see also Chap. 2.7, this volume). The initial free-running of the river from its source at the glacier is a crucial element of its sacred status. For Nehru the dam was the temple; now the river is once again the centre of religious attention. After the Government of India ensured perennial environmental flows in the pristine stretch of the river, Prof. Agarwal stopped his fast only to recommence half a year later, in January 2009, when work on the hydropower projects had resumed. When Prof. Agarwal came close to dying, the Ministry of Power called off the dam construction. The Government declared the Ganges a National River and set up the National Ganga River Basin Authority (NGRBA). It remains to be seen if these interventions will affect the river quality and its environmental flow.

At the beginning of the twenty-first century, one in ten of the world's major rivers no longer reached the sea for part of the year because of upstream impoundment or diversification of their waters, mainly for irrigation. The Nile, the Yellow, the Indus, the Rio Grande, and the Colorado rivers, along with many others, now regularly end in sand, sometimes hundreds of kilometres before they have a chance to reach their mouths. They no longer can fertilize their deltas to maintain ecologically rich estuaries where sweet and salt water mix into one of the most productive kinds of ecosystem on earth. Because floodwaters are no longer being cleansed by floodplain wetlands, more pollution is reaching inland and coastal seas. Whereas rivers once symbolized transition, they are now themselves in transition. No longer do the periodic floods of untamed rivers shape river channels and redistribute sediment, creating habitats essential to fish and other riverine life.

Human and ecosystem water needs ought to be balanced, according to water specialists Sandra Postel, director of the Global Water Policy Project, and Brian Richter, director of the Freshwater Initiative of the Nature Conservancy (Postel and Richter 2003).

Fig. 1.1.8 Rio Grande Big Bend National Park. (Photo credit: Department of Agricultural Economics and Agricultural Business, New Mexico State University); and, Caption: Cocopa Indian in dry Colorado Delta (Photo courtesy of The Water Encyclopedia.com)

More than 60% of the world's 227 largest rivers have been fragmented by infrastructures such as dams and diversions. Rivers are turned on and off instead of flowing by natural rhythms. Many rivers are thus but shadows of their former selves and the blue lines on the map are often tokens of faded glories.

1.1.4 Legacies of Modernity

In this whole process of damming, draining, engineering, and diverting rivers, water consumption has skyrocketed. A growing world population has led to greater needs for food and industrial production and an explosive rate of urbanization. Given these increasing pressures on water use, the amount of water available for humans and other species keeps declining. The lion's share, about 70% of global fresh water, goes to agriculture, 22% goes to industry, while domestic and municipal use accounts for a mere 8%. The high percentage of water for agriculture is partially due to low water-use efficiency, further aggravated by archaic water laws and irrigation subsidies that do not provide incentives to use water more efficiently. The complexity of agricultural water issues is further highlighted by the United Nations Millennium Development Goal on hunger, which entails doubling food production by 2050.

An important 'invisible' water allocation issue has been brought to light through John Anthony Allan's notion of 'virtual water,' also referred to as 'embedded water', 'embodied water', 'hidden water', or 'water footprint'. It is the water that is used in the production of a good or service. Allan called it virtual because, once a particular good is produced, say, an irrigated crop of wheat is grown, the water used to grow or produce it is no longer actually contained, molecule for molecule, in the end product. Thus, a cotton T-shirt contains 2,000 litres of virtual water, a hamburger 2,400 litres of, and a car 400,000 litres of water. To import goods, grain, and fruits, means to 'virtually' import water.

The uneven distribution of water worldwide is one of the most poignant problems. Roughly 1.2 billion people—one-fifth of the current world population—lack access to potable water. Many African and Asian women and girls spend hours each day walking to get water, severely reducing their participation in other productive activities, including education. People in poorer countries use on average 10 litres of water per person each day, while in Europe the average is 135 litres per person and in the United States, 570 litres.

Water quality is seriously declining, too, polluted by heavy metals and other industrial toxic wastes, intensive agricultural fertilizers, and microbial pathogens and excessive nutrients from untreated sewage. Approximately 2.4 billion people—two-fifths of the earth's population—do not have adequate sanitation services. Water-related diseases cause 80% of all illnesses and deaths in poorer parts of the world.

Fresh water is a critical limiting factor for health, food security, economic growth, biodiversity, and environmental sustainability. Hence, fresh water is a limiting factor for culture. The total volume of water on earth may be sufficient to accommodate human needs on a sustainable basis, but, as has been argued by various United Nations (UN) agencies, non-governmental organizations (NGOs) and other water organizations, attaining sustainability requires a far more serious political commitment than has yet been achieved.

Box 1.1a Waters and culture in the Doce River Basin of southeast Brazil: Mobilization and construction of a new perspective

—Claudio Bueno Guerra and Fabiane Torres

> The territory is the ground and the people, the identity and the facts, the feeling of belonging (Milton Santos 2000).

Box Map 1.1a.1 Map of Doce River Basin, showing the main municipalities as well as the drainage water system of that territory (Credit: Claudio Guerra)

Box Fig. 1.1a.1 Aerial photo of the Rio Doce (Sweet River). Upper part of the river: extracted from the book: O Vale do Rio Doce, VALE company, 2002, pg 71

The territory of the Doce River Basin is like a small country with an intricate interplay of land, nature, people, and work. The area is widely acknowledged as holding a wealth of water and mineral resources, as well as a rich local diversity. Although the region faces enormous and varied economic, social, and environmental problems, the cultural values of its people and their land are strong enough to resist the globalization processes that threaten their cultural foundations and social cement.

The region is a societal melting pot of native Indian populations, mixed with European and African descendants. In the old town of Ouro Preto, the centuries of diverse history, culture, and traditions are well represented. Located in the upper area of the basin, the town was the most important economic and cultural centre of colonial Brazil, heavily supported by the gold mining industry. It preserves the largest set of buildings in the architectural baroque in all of Brazil, and it was recently placed on UNESCO's list of World Heritage Sites.

The Doce river is the primary means of connection between the rich cultural tapestry of regions, histories, peoples, and economies within this territory. Its waters link and influence the economy, public health, fishing, leisure, and most other cultural manifestations.

(continued)

Box 1.1a (continued)

Box Fig. 1.1a.2 Children demonstrating at Santo Antonio river, a main tributary of Doce River: "We want to swim in our river. Pollution is enough" (Credit: Claudio Guerra)

Box Fig. 1.1a.3 Home to the largest steel making complex in Latin America, the Rio Doce drainage is heavily polluted from persistent organic pollutants from sugar cane and coffee cultivation, urban sewage and industrial effluents, and mercury and other toxins from gold and iron mining, including the tailings from the open pit mining at Mina de Gongo Soco (Credit: Mario Carvalho Neto)

1.1a.1 Social mobilization for waters

Since the 1990s, several societal movements have focused on bringing people and institutions together to confront the major water pollution of the river. These mobilizations were part of a larger journey toward the implementation of a new model of water resource management in Brazil, culminating in the Water Act of 1997.

Of all these mobilizations, we mention specifically a series of environmental expeditions by a group of professional engineers, biologists, sociologists, journalists, and others along the Piracicaba, Santo Antonio, and Doce rivers. Lasting 2–3 weeks, these expeditions took the professionals from the springs to the river's mouth. Along the way, they gauged the rivers' environmental health and conducted public hearings about water in the cities they visited. The primary products of these expeditions were video documentaries, slide shows, technical reports, and the forming of a River Basin Committee. Months after the expeditions, these products returned to the basin and became the foundational building blocks of a regional 'Parliament of Waters'.

The organization of the Doce River Basin Committee had the logistic and financial support of ANA (National Waters Agency). A multidisciplinary work group was formed by specialists, local consultants, and volunteers, which started elaborating a Master Plan, in order to develop a programme of social mobilization to improve water resource management. The strategy adopted prioritised environmental education based on the physical immediacy of local reality.

Environmental capacity courses were held in 11 towns, 'regional centres', carefully chosen in the upper, medium and lower part of the river basin, which

(continued)

Box 1.1a (continued)

worked as 'dissemination nuclei' for the committee organization. In 2002, almost 2 years and several preparatory regional meetings later, a significant water conference (involving representatives of governments, the private sector, and civil society) was held, and the Doce Committee was established.

Parallel to this movement, the leading companies of the region (steel, mining and cellulose), motivated initially by pressures from the international markets, signed a Term of Adjustment and Control with the State Secretary of Environment of the Minas Gerais government. This agreement started a process of investments in pollution control and certification, according to ISO 14001 (an internationally acknowledged standard required for organizations wishing to operate in an environmentally sustainable manner).

Box Fig. 1.1a.4 Rio Doce Festival, 2007, pollution dragon (Credit: Fabiane Torres)

As a consequence of these two initiatives, two major projects occurred in the beginning of the twenty first century: In 2004 a local NGO, Projeto Aguas do Rio Doce (PARD), organised the first annual Doce River Waters Forum. Empowered by the support of local governments and leading companies in the region, PARD's main objective is to strengthen water resources management for sustainable development within the basin by encouraging broad based popular participation and cooperation. The third meeting of the Waters Forum, held in March 2007 in the city of Ipatinga, drew an audience of 140,000 in 4 days with an agenda which paired social and cultural issues with science and education by combining a spectacular river festival with scholarly presentations. This forum is already considered one of the most important events of the water sector in Brazil.

In 2008, the Federal University of Minas Gerais (UFMG) and the University of Vale do Rio Doce (UNIVALE) took the initiative to develop an interdisciplinary pilot project called Doce River Culture to encourage people to rediscover and reconnect with the river. With its facilities located at UNIVALE, on the banks of the Doce, this project intends to implement a partnership between the universities and the local communities in order to construct an arena in which culture and the arts meet science and technology through a creative collaboration of academic and non-academic partners. Thus, the Doce River brings together representatives from diverse institutions and communities, as well as contrasting perspectives and perceptions of water and culture within the territory with the purpose of establishing a centre of biological, environmental, cultural, social, and historical studies.

(continued)

Box 1.1a (continued)

1.1a.2 Conclusions

Rivers are the silent witnesses of history. Going downward towards the great meeting with the ocean, they not only carry waters but the dreams, memories, stories, and histories of the peoples living in their basins. An understanding of river reality through a new lens is called for, one that takes these everyday cultural practices seriously, in order to find new paths toward sustainable alternatives. While current management models are still dominated by a paradigm of 'expertise' in terms of engineers, politicians, and water managers, a recognition is rising, that the most effective and sustainable technical solutions turn toward a broad-based cultural expertise and take local knowledge and experience into account. Culture is a complex mixture in which technology, economic capacity-building programmes, social and historical perspectives, art and religious practices flow together into a sophisticated water research management and water science. Local leadership will be vital in addressing the challenges in the decades to come. The Projeto Aguas do Rio Doce, the Doce River Waters Forum and the Doce River Culture Project are first steps in this direction and represent hope for the future of a river, which has been for ages the lifeblood for the region.

Reference cited

Santos, Milton. 2000. *Por uma outra globalização*, 96. São Paulo: Editora Record.

1.1.5 Promise of the Twenty-First Century: Toward a Truly Integrated Sense of Water?

Peter Gleick, a water policy specialist and director of the Pacific Institute, speaks of a 'soft path' that complements the twentieth-century's large-scale, centralized infrastructure with 'lower cost community-scale systems, decentralized and open decision-making, water markets and equitable pricing, application of efficient technology, and environmental protection' (Gleick 1993; Gleick et al 2006).

An example of the beneficial effects of legal change is the implementation in the United States of two major 1970s water laws: the Clean Water Act and the Safe Drinking Water Act, which required industries to clean up their wastewater. These laws made many industries more water efficient, because it turned out to be cheaper not to produce the waste than to clean it up. It now takes, for example, 5 or 6 tons of water to make a ton of steel, while it used to take 200 tons.

Water privatization is presented as another tool of efficiency. Although 90% of the world's water provision and sanitation systems are publicly owned and operated, there has been a shift toward private sector participation in these very basic municipal services. The proponents of privatization argue that it will improve the quality of the services, reduce the costs, and mobilize more financial investment and technical expertise. Opponents counter that privatization leads to poorer services and higher costs because profits are valued over service. There have been intense political campaigns against privatization in Ghana (2001), Uruguay (2004), and, most famously in Bolivia, where the Cochabamba Water Riots of 2000 occurred, resulting in the death of a young man and numerous injuries to protesters. Some countries (for example the Netherlands in 2004) have enacted laws banning the privatization of public water supply.

Progressive water pricing is often mentioned as a way of managing demand and encouraging conservation. In such a system, a daily minimum of water is sold at an affordable price: as a customer's use increases, the price per unit also increases. This runs contrary to the usual approach of markets, in which high-use customers are charged less per unit than low-use customers. From a progressive water-pricing perspective, agricultural water stands out as being seriously underpriced. In reaction to conventional anti-progressive water pricing and water privatization, a growing rights-based movement has emerged for implementing a UN-mandated *human right to water*, which entitles all people to sufficient, safe, accessible and affordable water. This right is supposed to trump all property rights in water.

Fig. 1.1.9 Water protests in Cochabamba, Bolivia, 2000 (Photo credit: Thomas Kruse/1World Production)

Water gets more attention: there are awareness campaigns about the value of water and the importance of conservation to keep more water in the system. Sandra Postel, of the Global Water Policy Project, calls conservation our 'last oasis' (Postel 1992; 1999). Examples include water education programs in schools and municipalities, venues to involve various stakeholders in water management, and citizen-based projects of wetland, riparian, and coastal restoration. Environmental restoration is as much a practice of our *relation* to the land as of restoring the land itself. Watershed restoration involves a broad spectrum of human-water-land relations, varying from ethics, aesthetics, politics, and participatory activities to modes of knowledge as diverse as science, engineering, elder experience, storytelling, and children's imagination. Celebrations such as World Water Day and local river festivals provide playful ways to reconnect and re-engage with water, and to enhance motivation to learn about water quality and quantity.

1.1.6 Resurfacing Rivers

In 2000 the European Parliament adopted a water-policy framework, the European Community Water Framework Directive (WFD), an unprecedented attempt to design a regime for managing water quality and instream flow for each watershed, thus creating a basic geographic unit for resource planning. The WFD aims to manage whole watersheds or river basins in a holistic manner (a strategy referred to as

Fig. 1.1.10 Seoul River Walk (Photo credit: ddol-mang [flickr.com/photos/ysjjhfox/247932125; http://flickr.com/photos/ysjjhfox/247932125])

integrated water-resources management [IWRM]) at the basin level in order to integrate land and water and to encompass upstream and downstream water, surface water, groundwater, and coastal water. A more transparent and participatory transnational governance alternative has thus replaced a politically fragmented approach. In an effort to achieve consensus among multiple stakeholders, the Directive aims to ensure that hydrological and engineering expertise is meshed with ecological knowledge, and that urban, agricultural, industrial, and recreational interests are taken into account. Because water connects all elements of society, an integrative water approach, one that views local problems of water quality and quantity in transregional and global political contexts, becomes essential.

The major water laws were created in an era focused on the political and economic control of rivers. The emerging new paradigm gives greater priority to ecological health. Around the globe, in areas as diverse as Asia, Australia, South Africa, Europe, and the U.S. state of Texas, governments are implementing policies that are intended to allocate water for ecosystem support: mainly minimum instream flows to maintain environmental quality and sustainability.

Fig. 1.1.11 Los Angeles River (Photo credit: Irene J. Klaver)

Rivers are resurfacing in the public imagination as cultural and ecological corridors, creating a cultural rejuvenation around urban renewal projects. Many cities are re-inventing themselves around their rivers with river walk promenades, theatres, cafés, and restaurants and new housing developments, often in the water.

Although some of these projects are driven largely by an economic culture, most stimulate a cultural awareness of the river's ecological and economic wealth and inspire public education about local water issues. Some of them explicitly aim to increase stewardship of the river and engage communities in river-related projects. For example, the Los Angeles River is about to be unlocked from its concrete ditch and restored by means of a riparian-community-based process meant to create a blue ribbon along housing developments, parks, and walkways, thus revitalizing river and community alike, by reintegrating the river into the social life of the community. Many cities celebrate their river with festivals: the Brisbane River Festival in Queensland, Australia; the Hudson River Festival in New York City; London's Thames Festival; Danube Day throughout southeastern Europe, Sail Amsterdam in the Netherlands, the 'One River Mississippi' festival celebrates the Father of Waters (which is what 'Mississippi' means in the Cherokee language). Furthermore, in many rural areas river restoration is underway: the re-meandering of water courses, done by many of the same engineering firms that straightened the waterways 50 or 60 years ago, restores floodplains. New management regimes are seeking to work with, not against, rivers to enhance their flow and allow them the freedom of flood in a natural cycle of renewal and rejuvenation for humans and the entire manifold of life that flows into and out of these vital arteries of the earth. At the same time there are movements that want to minimize any culture of managerialism and propose a more integrated culture of re-learning to live with the floods, and to undertake riparian restoration and nurturing wetlands to take out pollutants, fertilizers, and pesticides before the water drains back into the river.

With the combined application of interdisciplinary engineering, hydrological, and ecological expertise, the twenty-first century promises a renewal of a respect for rivers that echoes the ancients' reverence for waterways. We are relearning the old Taoist and Heraclitean aquatic wisdom: in water's constant flow lies its stability, in its humility its power.

Rivers are archives. They record deep time and shallow time, revealing what happened a million years in the past as well as a moment ago upstream. All cultures are shaped by this ongoing cycling of water. On a planetary level we are all downstream.

References Cited

Agarwal, G.D. 2008. Letter from G. D. Agarwal to the government of India, April 14, 2008. http://tapasya-bhagirathi.blogspot.com/. Accessed 15 Apr 2008, and 3 July 2010.

Gleick, Peter H. (ed.), Alan P. Covich, Malin Falkenmark, Gunnar Lindh, Stephen McCaffrey, Linda Nash, Sandra Postel, and Igor A. Shiklomanov. 1993. *Water in crisis: A guide to the world's fresh water resources*. New York: Oxford University Press.

Gleick, Peter, Heather Cooley, David Katz, and Emiley Lee. 2006. *The world's water 2006–2007: The Biennial report on freshwater resources*. Washington, DC: Island Press.

Herodotus. 1921–1924. *The histories*, ed. A.D. Godley, 4 vols. London/New York: W. Heinemann/ G.P. Putnam's Sons.
Postel, Sandra. 1992. *Last oasis: Facing water scarcity*. New York: W.W. Norton Company.
Postel, Sandra. 1999. *Pillar of sand: Can the irrigation miracle last?* New York: W.W Norton.
Postel, Sandra, and Brian Richter. 2003. *Rivers for life: Managing water for people and nature*. Washington, DC: Island Press.
World Commission on Dams (WCD). 2001. *Dams and development: A new framework for decision-making*. London: Earthscan. Available from http://www.dams.org/report/contents.htm

Chapter 1.2
The Paradigm Shift in India's River Policies: From Sacred to Transferable Waters

Kelly D. Alley

The beliefs and practices of Hinduism are directly tied to many public uses of rivers across India and have been central to the reverence for rivers appearing in religious worship for centuries and nationalist movements and rallies over the last 50 years. There are also a multitude of public uses of rivers apart from widespread worship rituals: washing persons, clothes, and animals; general household consumption; fishing; transportation; and small-scale industry. These combined with small- and large-scale Hindu ritual bathing practices constitute public uses of rivers in India and are distinct from industrial or urban uses in that they are unpriced and sometimes essential to subsistence. Citizens are usually not prevented from using river waters, but the times when and places where they can attain access are sometimes regulated. With the development of private water and hydroelectric projects and the supporting state controls, one might expect that these public uses will be threatened or restricted over time.

It is with this thesis that I would like to draw upon data on Indian river-resources policies that I have collected between 1995 and 2010. I look specifically at references to the role of religious or spiritual uses of river water, because these references are a window, I suggest, into the ways people value rivers socially and politically and define and approve their acceptable uses. More broadly the meanings of rivers in religious exegesis and practice over time reflect in profound ways the eternal connections between humans and water. If these references erode or disappear in policy over time, their absence may signal the growing strength of other uses, such as hydropower and industry that may displace public needs.

I look at river pollution prevention and dam and diversion policies and projects since the 1980s to outline the diminishing policy concerns for water quality and flow that benefit religious practices and public, unpriced uses. This decline in

K.D. Alley (✉)
Anthropology Program, Auburn University, Auburn, AL, USA
e-mail: alleykd@auburn.edu

B.R. Johnston et al. (eds.), *Water, Cultural Diversity, and Global Environmental Change: Emerging Trends, Sustainable Futures?*, DOI 10.1007/978-94-007-1774-9_3,

governmental interest appears to be happening even as public demand for water for religious bathing, washing, and drinking remain high. Dams, diversions, and flows of wastewater have caused low water in specific reaches of rivers across India and the loss of aquatic biodiversity within river basins, limiting the possibilities for fishing and water transportation. What the declining availability of water may mean is less an effective barring or constraining of religious uses than a reduction in the quality and quantity of the water available for public use. In my conclusion, I will outline the implications of this diminishing concern for river basin science and policy, religious worshippers as stakeholders, project evaluation, and the future of public water uses.

1.2.1 Sacred Waters

All the rivers of India are feminine, and many are important goddesses in the Hindu traditions. Two of the most eminent river goddesses are the Ganga and Yamuna rivers, which together water the lands forming the cradle of Hindu and Buddhist pilgrimage culture in India. Some of the most important centres of spiritual learning and healing have thrived for centuries along the banks of these rivers and their tributaries. At these pilgrimage centres and at countless other, smaller sacred spots, pilgrims have worshipped Ganga and Yamuna and carried away their water for use in worship and purification rituals (Alley 2002; Haberman 2006).

According to the Hindu view, the sacred is not detached from the material (*bhautik*) realm of ecology and the built environment, but rather provides a context for understanding the truth of the physical world. Interpretations of sacred space describe intersections and conjunctions between divine power and the physical world. Hindu mythologies and sacred texts weave together understandings of sacred and natural ecology, portraying the intersections of the sacred with material and human culture.

Devotees describe Ganga as a goddess who absolves worldly impurities and rejuvenates the cosmos with her purifying power. She is a mother who cleans away human sin and mess with loving forgiveness. Goddess Yamuna is also a mother who sustains and provides, but she is more concerned with the blessings of this life than the Ganga, whose role is purification and preparation for death (Haberman 2006:60). Ganga and Yamuna's immanent form as water (*jala*) is a central element of ritual practices for Hindus. Ablution or *snan* in the early morning hours is an essential component of daily ritual for residents and pilgrims in sacred cities along both rivers. Hindus also immerse the ashes or bones of the cremated in the Ganga to ensure their safe journey to the realm of the ancestors. Using Gangajala and Yamunajala for *puja* (offerings to deities) affirms the eternal essence of these rivers. Devotees also carry Gangajala in jugs to temples, where they perform *jalabhisek* (pouring of Gangajala over a Siva *linga*, the aniconic representation of the God) and other worship rituals. As Haberman (2006:61) puts it, the Ganga is often associated with the

Fig. 1.2.1 Hindu devotees bathing and performing ablutions in the Ganga river (Photo credit: Kelly Alley)

yogic culture of the ascetics who seek release from this world, whereas the Yamuna is associated with the loving, devotional traditions that seek enjoyment in this life. The popular phrase, 'Bathe in Ganga, drink Yamuna' (Ganga snan Yamuna pan) indicates the central role that both ritual and consumption play in the sacred and very public uses of these rivers.

Devotees recite eulogies to these goddesses from the sacred texts of the Hindu traditions, the *Ramayana*, the *Mahabharata*, the *Puranas*, and the *Mahatmyas.* In sacred places, pilgrims and residents also revere the waters and seek blessings by undertaking the ritual of *arati,* a ceremonial offering of fire. While standing on the riverbank, devotees wave an oil lamp and other sacred objects in front of Ganga and Yamuna. The sounds of bells, gongs, drums, and conch shells play a prominent role. A devotee gains power by chanting *mantras* or formulaic phrases understood as verbal codifications of sacred texts. Singing also accompanies the last rites of the *arati* ceremony.

In the sacred cities of these river basins, master narratives about Ganga's purifying powers and Yamuna's unrequited love, are tenable as personal and family convictions because they are reproduced through occupations (*pesa*) and everyday ritual practices. Because these narratives tend to disregard material, mundane, and contemporary oscillations in favour of the eternal and transcendent, devotees are able to argue that Ganga's purifying power can withstand grave assaults. For them, the eternal and transcendent bring out the true essence of sacred ecology and power. Likewise, Yamuna's love and compassion and seeing the world through her eyes can turn misconduct into worship and bring humanity back from the brink of disaster. Sacred values drawn through religious narratives and everyday devotion affirm the powers of these important river goddesses and bring their worship forward into modern narratives of environmental decline. In the devotee's mind, science reaffirms the power of Ganga's purity and Yamuna's love and of Hindu *dharma* (order and duty) more generally. These two rivers have been very visible in water policy and environmental plans over the last 15 years, so I highlight their significance in this paper as examples of a more general turn in river policy. There are, however, many other sacred rivers of India that claim a central role in the devotional, ecological, agricultural, and everyday lives of citizens.

1.2.2 Water Quality Policies

In addition to earlier projects built over the last 1,000 years, large-scale water projects were developed in the nineteenth century to irrigate land in the Indus, Ganga, and Yamuna river valleys. During that century these projects produced wide-ranging effects in river basin ecosystems, agricultural production, rural power, and relations between farmers and the state. Canal projects developed the colonial state's hydraulic modelling of the environment, as Gilmartin (1995:210-236) has put it, laying the foundation for state control of river water in the post-independence period (after 1948). More recent dam projects affirmed state control and centralised decision-making as they garnered financial sponsorship from international banks and support from technology consultants and engineering firms (Singh 1997). All these projects are part of an emerging water nationalism – sketched out via a national 'water grid' – to unite the nation's water resources conceptually and geopolitically. More recent water policies are also taking a neoliberal turn as they focus on the outsourcing of dam and diversion projects to private companies, some with very little prior experience in the water sector.

These water policies are what I would call 'quantity-driven approaches'. Apart from such projects, the government of India has, since the 1980s, planned and executed river action plans to prevent intrusion of raw wastewater into rivers. These are what I am calling 'quality-driven approaches'. In these, the aim has been to divert wastewater for treatment before routing the treated water back into rivers. The government of India relied on British, Dutch, Japanese, and Australian donors to fund the two initial phases of the first river programme, the Ganga Action Plan. From this sprang other river action plans on the Yamuna and Gomati, with others following, designed to collect municipal, state, and central funds to build and operate wastewater diversion and treatment systems. During this period, the understanding was that public uses of rivers, and in particular Hindu ritual uses, required pollution prevention schemes to improve water quality, especially in religious bathing areas. References to the religious importance of the Ganga and Yamuna rivers, in particular, appeared throughout the policies of the action plans for these bodies of water from the outset, and pictures of religious worship framed many government publications and web sites.

In academic and government research and policy reports on these plans, scientists and government officials invoked the religious meanings of the river to justify the utility of water quality studies and sewage treatment systems. Although officials generally dismissed any relevant connection between ecology as it is known through science and the cosmic theory of Ganga's origin and essence, officials did not cast off religious symbols and narratives as if they had no valence among the wider public (for details see Alley 2002: 67-69). Backing up these policy statements are public narratives of Ganga's unique properties. Many devotees invoke the scientific notion of water quality to support the master narrative of Ganga's sacrality. This master narrative is embedded in the Gangajala story and supported even when devotees invoke the modernist notion of microbial impurity. Many argue that if kept in a jar or glass at home for years and years, Gangajala will never 'spoil' or 'develop bacteria' as mineral water or tap water do.

Haberman describes officials of the Yamuna Action Plan who mixed devotion to Yamuna with exhibits on pollution prevention. Haberman notes that when he interviewed one official, the man sat under a sign in Hindi that read, 'Holy Yamuna is the beloved of all people. Yamuna's water is pure and cool' (Haberman 2006:159). Haberman (2006:159) also describes a celebration for 'loving service' to Yamuna on the first day of the new millennium. At that time, municipal officials called upon the devotion to Yamuna as the Mother of all to motivate efforts to clean up the city and the riverbanks.

The aim of the river pollution prevention schemes in the Ganga and Yamuna Action Plans was to restore water quality to bathing standard (now called Class B status), safe for public access and especially for ritual and multipurpose bathing. By the end of the first phase of these action plans (1985–1996) they were both consolidated under the National River Conservation Directorate (NRCD) in the Ministry of Environment and Forests, and this body carried along the water quality model in its pollution prevention programmes. In the first 5 years (1996-2001), officials in the NRCD had enough funds to contract for the construction of treatment and diversion facilities through state governments. River plans grew in number and covered major waterways across the country. However, as the plans were drawn up and the costs escalated, full funding from local and state coffers became almost impossible to procure. The attempts to cost-share programmes between the central and state governments were not reaping results, and the central government's projects began to starve.

Fig. 1.2.2 United States National Public Radio (NPR) team in boat at Varanasi (Photo credit: Kelly Alley)

In late 2004, I returned to field locations I had been visiting since 1992 to record, with a radio team, the state of the Ganga Action Plan. After almost 20 years of the plan, we found the facilities in a dilapidated state of existence. Most were not running 24 hours a day or even every day of the week, portions of plants were lying dormant and unused, and staff had not been paid for months. The British had completed their projects, government of India asked the Dutch to discontinue as a project donor, and the then Japan Bank for International Cooperation was considering new investment to salvage the facilities. To my memory, that state of affairs was far bleaker than the situation had been at the time that the first large treatment plants were being constructed under the first phase of the Ganga Action Plan (1985-1996). The situation was indeed worse: populations had grown; consumption behaviour had changed to include more use and disposal of water, plastics, paper, and toxic substances; industries, cities, and farms were emitting heavy metals, pesticides and other toxic chemicals into surface waters at an alarming rate; and projects representing big investments lay dysfunctional or lacked key elements (including uninterrupted electricity) to run properly.

Meanwhile, the Ganga Action Plan was passed off as a success as other river pollution prevention plans were developed on paper. Since most projects remained starved for funds, only minimal infrastructure building occurred. Meanwhile, public uses of rivers and ritual practices did continue on a large scale. Citizens

Fig. 1.2.3 Sewage flowing from Kirki nala (Photo credit: Kelly Alley)

were not barred from religious bathing at sacred sites, but upstream diversions, urban and industrial effluents, fertilizer and pesticide runoff, and more affected the physical or chemical quality of the river water they were using. These effluents have been changing the quality and experience of use, even if they have not gone so far as to undermine religious devotion to the river goddess or worship practices. Although the sacred purity of the Ganga may override such pollution – an issue of devotion I do not dispute – public river uses bring citizens into direct contact with untreated effluent and wastewater, contacts that have consequences for human health.

Haberman notes that officials of the Yamuna Action Plan told him that they had successfully tapped wastewater drains and set up functioning sewage treatment plants in and around the city of Mathura in 2001. There is no independently corroborated data, however, to confirm this report. Instead, a number of critiques (see, for example, Ghose and Kaul 2004) of the Yamuna Action Plan claim that the same failures and the continuation of massive wastewater drainage and other runoff into the river have occurred here as on the Ganga.

Intensive uses of river water and riverbeds as effluent channels affect water quality, watershed ecology, and ecosystem services. Peer-reviewed scientific research has documented the rise in levels of faecal coliforms, bacteria, pathogens, and metals in rivers and the deterioration of water quality measured by the parameters of biological oxygen demand and dissolved oxygen. Water quality studies have shown a marked decrease in the oxygen needed to sustain aquatic life, fishing and public uses. Pollution considerations began to appear outdated, as citizens and officials turned their attention to the idea of transferring surplus water from one basin to another. The interest in transference – entailing distribution among agricultural, urban and industrial users – began to overshadow the problem of pollution and the importance of the cultural practice of bathing in a sacred river. The water quality model has been giving way to a water quantity model, as statements about the growing needs of agriculture, industries, and cities eclipse the importance of religious rituals. In the process, national policies have moved from a focus on river basins to a vision of a national water grid connecting water supplies through a network of canals.

1.2.3 River Linking and Surplus Water

So how has this interest in transferable and marketable waters grown to the point that it appears to be bankrupting river pollution prevention schemes? In 2002, the government resurrected with euphoria a river-linking plan based on one slowly dying in the files of the National Water Development Agency (NWDA). In just 2 years it went from a sleepy, fund-starved plan to a symbol of resource nationalism. The central government at the time, the National Democratic Alliance, used sketches of the river-linking plan in its manifesto and election campaigns and made general references to sacred and life-giving rivers as it stirred up the technological courage to move 'surplus' waters around (see Alley 2004). Proponents of river linking rarely mentioned or promised benefits to religious practices and uses.

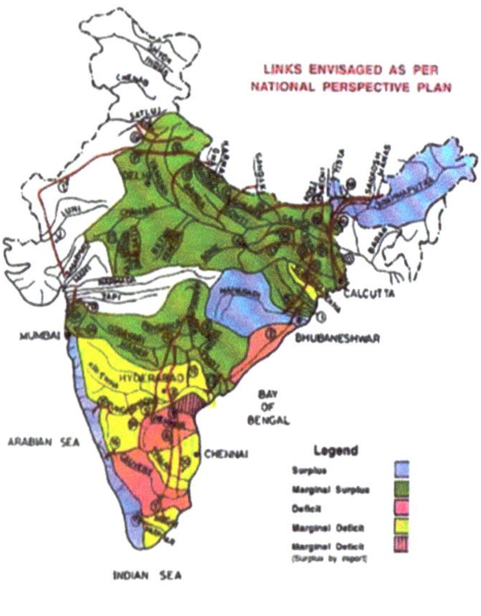

Map 1.2.1 Map of the river linking scheme announced by the Government of India in 2002

After the Supreme Court ordered the creation of an ad hoc Task Force on the Interlinking of Rivers (see Alley 2009), the team announced that the epic and plentiful Brahmaputra and Ganga waters could be transferred south, to meet the needs of farmers in South India. Politicians and members of the NWDA and the task force also promised that conflicts in water-sharing agreements between the states of Tamil Nadu, Karnataka, and Kerala would be allayed, and not complicated, by the flow of additional water from the north (Iyer 2003). When the ad hoc task force dissolved in early 2005, the river-linking data and feasibility reports were transferred to the NWDA, and some of them were then presented on the agency's web site (see http://nwda.gov.in/).

This shift in agency assignments provoked vociferous demands by nongovernmental actors and scientists for access to classified documents. The NWDA responded to the new Right to Information Act of 2005 by posting selected reports and documents on its web site. However, all the significant documents – namely, impact assessments and detailed project reports – remained categorised as classified. A few feasibility studies were later posted on the official web site after NGOs and citizens demanded more information be posted in the public domain. Parliament also demanded the disclosure of more information as members discussed water resources and river linking. In response to the closed-sourcing of government data, scientists, NGOs, and other ecology experts outside government began to engage in an open-sourcing of science, using the information and knowledge generated in the public domain to critique government plans (Alley 2004). This public scrutiny has raised questions about the processes of debate essential to the production of verifiable knowledge and 'best practices' for public assets and services. It has also raised questions about the democratic process and the fundamental rights of citizens guaranteed under the Constitution more broadly.

Box 1.2a Isoso Guarani culture and livelihoods shaped by, and dependent on, the pulsing Parapeti River in the arid Gran Chaco of Bolivia

—Janis B. Alcorn, Elio Ortiz, Antonio Mendez, and Alejo Zarzycki

Isoso Guarani of the inhospitable Gran Chaco region of eastern Bolivia define their world and annual activities in accord with the cyclical changes of the Parapeti River, on which their society depends. The Parapeti originates from multiple small streams running down the humid Andean foothills to the west and flows eastward onto the flat dry forested plains where it sinks into sands, dividing itself into multiple shifting channels until it ultimately ends, burying itself in seasonal wetlands, called Yandeyari. The Yandeyari wetlands are respected and tabooed as the Mother of Life, because this is the place on which depend fish, animals, and birds (those of Isoso as well as those of the neighbouring Kaa Iya National Park, the largest protected area in South America).

The Parapeti annually divides Isoseños' year into halves of abundance and scarcity, yet Isoseños celebrate this full cycle as the beauty and balance of life itself. The word Isoso means 'water that lets itself loose in repetition'. For Isoseños, the word *ɨ* (water) is the root for Guarani words for wind, semen, earth and blood. The river is seen as pulsing like a great heart that keeps life alive. Isoseños have the task to maintain this equilibrium, so that the world will continue to exist.

(continued)

Box 1.2a (continued)

Box Map 1.2a.1 Illustrative map locating Isoso in Guarani territory along the Parapeti River in the Bolivian Chaco region (Reproduced from Ortiz et al. 2008)

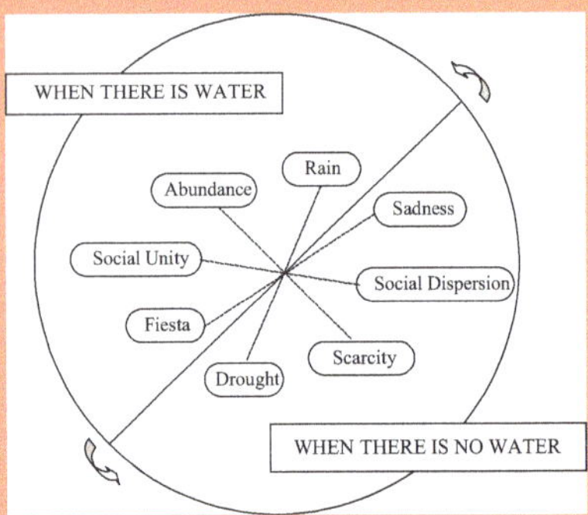

Box Fig. 1.2a.1 Isoso bipolar world shaped by annual cycle of the Parapeti River (Reproduced from Ortiz et al. 2008)

For the 'happy' half of the year, when annual rains fall in the Andean foothills, water becomes abundant in Isoso. The Yandayari wetlands become a nursery for fish while the Parapeti itself becomes is a broad, fast flowing river that provides essential water to irrigate the dry fields of Isoso. As the river drops and the wetlands dry, fields are harvested, maize beer is shared, and the fish swim upriver from the wetlands to be collected by Isoseños. Concentric circles of dancers moving in opposite directions celebrate this axial moment in time. Then follows the 'sad' part of the year, when the Parapeti River becomes a wide horizon of immense sand dunes moving in the fierce dry winds common to this season in the Gran Chaco ecoregion, a time when little water remains in the streambed and wooded wetlands, and hunters harvest game that fattened during the preceding 'green' time.

The river 'makes the decisions', and people adjust to enjoy the benefits from the river and maintain the abundance from nature's balance. The elders and *arakuaa iya* (Keepers of Knowledge) interpret the changes and variations that continually arise, and renew people's relations to the river and the landscape as it changes. Over the years, communities have moved their homes, as the river (water and dunes) have moved. Evidence from older dunes and oral memory of the locations of prior settlements indicate the river has shifted northwestward over time. The annual floods have created higher natural levees on one side of the meander and undercut the opposite side to create lower land along the river's east-west sinuous S-curve as it crosses

(continued)

Box 1.2a (continued)

Box Fig. 1.2a.2 Isoseño fishing ritual uniting oppositions and generations. Isoseños are fishing according to their customs and rules. The river fills with fish returning from the Yandayeri wetlands, swimming back upriver at the time when the waters begin to disappear (Photo credit: CPI Chaco)

Box Fig. 1.2a.3 Isoso dancer wearing a mask representing the Iya and the ancestors during fiestas. After the dance the mask is thrown into the Parapeti river to be carried back to the east from whence the ancestors came (Photo credit: CPI Chaco [Commisión de Los Pueblos y Comunidades Indígenas del Chaco], Santa Cruz, Bolivia)

Box Fig. 1.2a.4 The huge Parapeti River dunes, created and moved by fierce winter winds, are shown below a typical summer sky (Photo credit –CPI Chaco (Commisión de Los Pueblos y Comunidades Indígenas del Chaco), Santa Cruz, Bolivia)

Box Fig. 1.2a.5 People on foot or horseback crossing the Parapeti River. When the river is low the dunes attempt to claim the riverbed, allowing people to walk or ride across it. Vehicles, however, sink into the sand as water flows just below the surface, creating a sort of quicksand. Entire vehicles can disappear here (Photo credit: CPI Chaco)

Isoso. The irrigation canals for agricultural lands are on the low side, and sacred woodlands for hunting are on the high sides of the curve.

Governance is also based on balance around the river, the centre of whose "S" divides Isoso into upper and lower halves. Upper Isoso and Lower Isoso

(continued)

Box 1.2a (continued)

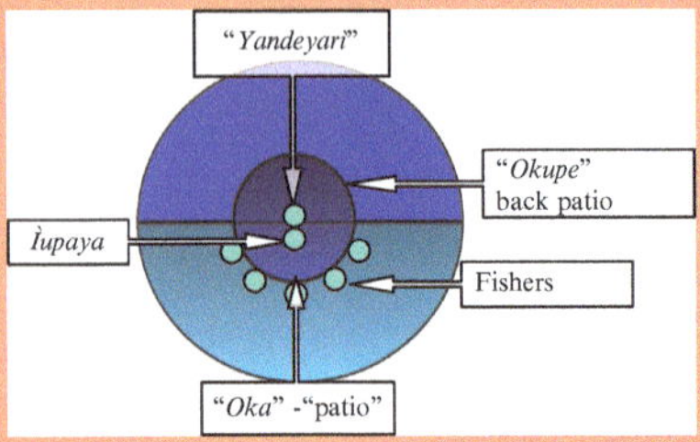

Box Fig. 1.2a.6 Isoseño fishing ritual spatial and symbolic orientation (Photo credit: Reproduced from Ortiz et al. 2008)

governments unite the 22 communities to work together to manage their Indigenous Territory via their traditional government, Capitania of Upper and Lower Isoso (CABI). CABI has also successfully co-managed the Kaa Iya National Park for 10 years, demonstrating the strength of governance built around maintaining nature's balance based on traditional knowledge.

Places and things have their spirit caretakers with whom Isoso Guarani maintain a mutual relation in order to make use of all places and things, including water. Everything is sacred, and must be respected according to guiding principles. A ritual spatial and social arrangement ensures that the fish and the people prosper in an orderly way. The leader of the ritual (*Iupaya*) first locates the Centre (*oka*) and the doorway (*okë*) of Grandmother (*Yarì*), the keeper of the fish. Watched silently by the fishing families, he wades into the water until it reaches his waist. He thus arrives at the symbolic doorway, and there converses with the spirits (*Iya*), vowing to observe all the sacred norms. This fishing ritual is a key social ritual, because it joins families of Upper and Lower Isoso, reminding everyone of the rules, the oppositions, and the origins of Isoso. Only then do young adults fan out into the deeper Okup to drive fish toward the area where elders and children await to catch fish.

Reference

Ortiz, E., A. Mendez, A. Zarzycki, and J.B. Alcorn. 2008. Fox walker on the Parapeti River, Bolivia – How we Guarani live in Ivi. In *Pre-Columbian landscapes of creation and origin*, ed. J.E. Staller, 161–202. New York: Springer.

1.2.4 Hydro-Power in the Himalayas

Since 2005 the vision of river uses has also been transmuting into a vision of hydro-marketable regions as the government drew up plans for a multitude of run-of-the-river schemes along rivers across the Himalayan states of India. After completing several schemes along rivers in the northwestern states, the government shifted more concentrated focus to rivers in the northeastern states, supported by a National Policy of Hydropower Development report that only 1.6% of the 63,257 megawatt (MW) capacity had been tapped there (Menon and Kohli 2005: 22). New incentives for investors included open access and the freedom to sell power on a merchant basis, the transfer of hydrological risks to the public, and the cost-plus approach to tariffs (Dharmadhikary 2008). Across the country, peak power demand in the year 2007-2008 was 108,886 MW, while the peak power supply was only 90,793 MW; the shortfall was 18,093 MW or 16.6% of peak demand. These uses of rivers for hydropower are meant to meet growing needs for power and electricity, but the rapid pace of planning and construction have created wide ranging effects for society and environment, especially in the northwestern Himalayan states where more than fifty large hydropower projects are completed or under construction. This rapid development also falls in line with a shift from public service and publicly funded efforts to private initiatives that build transnational markets in water and power.

Meanwhile, what is rather clear is that river linking, as a grand scheme, will never be feasible as envisioned from 2002 through 2006. Instead, the scheme has drawn many experts and concerned citizen groups into a public debate about water uses and the assumptions, projections, and actual instances of water use. However, as this debate ensues, state governments have been entering into memoranda of understanding to allocate hydropower projects to bidders and arrange for large scale infrastructure for power evacuation and redistribution.

Map 1.2.2 Map of proposed hydro projects in the northeastern state of Arunachal Pradesh

1.2.5 Implications and Recommendations

So what does this diminishing concern for water quality mean for river basin science and policy, religious worshippers as stakeholders, project evaluation, and the future of public water uses in India? Debates about water uses will continue to develop as complex networks of policy analysts, professionals, and scientists evolve across these river basins. Although most groups are equipped with authoritative scientific and professional knowledge and some legal power, only a handful of official agencies have the institutional legitimacy and control to make key decisions. The official agencies (central and state governments and key consultants) work with the World Bank, the Asian Development Bank, and other banks and funding agencies, affiliated scientists, and the judiciary. The unofficial groups work with scientists, professionals, and citizens inside and outside India and with justices and bureaucrats through public interest litigation. The difference in power and capital between government policymakers and nongovernmental experts allows the former to continue closing off the latter through such methods as the classification of data and control over key resource decisions. The judiciary has bridged that divide by taking up citizens' pleas to monitor, correct, or punish governmental practices and to order scientific studies and their dissemination. However, the Supreme Court also legitimises government plans for large-scale resource use projects and decides its own vision of the national good. Inevitably, all remain inextricably tied to the bifurcation of science in India: a closed 'legitimate' sourcing of data and information and an emergent form of open-sourcing in the public domain. The effect is that peer-reviewed research occurs more often outside the agencies that plan and implement dam and link projects. Moreover, this critical discourse on policy does not open the door in any effective way for participation by the country's residents who use unpriced services, especially those engaged in religious bathing.

This failure means that a significant public user group slips into invisibility in water allocation equations and outlines of a river's ecosystem services. Environmental impact assessments may document livelihood uses, but they do not verify that these public users are stakeholders with an allocated share of the project benefits. Stakeholders who use unpriced services such as water for religious bathing appear more and more like liabilities – 'impacts' – and like the villagers who are resettled from a dam catchment site, the government deals with them by moving their practices out of the way.

River-linking policies, in accounting for power, urban, industrial, agricultural, and drinking water needs, do not differ much from the international bank and aid policies that direct funding to national governments and private corporations. The policy of the International Finance Corporation (IFC), for example, places the issue of unpriced uses in the performance standards for social and environmental risks and impacts. The IFC's proposal to dilute the rules on land acquisition and involuntary resettlement to facilitate large dam and link projects helps to internationalise this shifting definition of public uses in water allocation records and schemes (International Finance Corporation 2005). Meanwhile, the wastewater intrusions that affect all public uses and the sustainability of usable surface water more generally have not been sufficiently problematised. The lofty goal of pollution prevention spelled out in the river action plans has been atrophying under the weight of surplus water notions and now under ever more popular hydro-visions.

Finally, the classification of official data and reports, because it blocks the free exchange of information and prevents the participation of a diverse group of experts and citizens, enables opportunism in government contracts for dam and link projects by limiting auditing, accountability, and evaluation. Even if the agencies involved admit the past problems with the technology, maintenance, and the human costs of pollution prevention programmes and dam and link projects, there is no requirement that the system should be modified to correct these problems.

What all this discussion suggests is a point Vandana Shiva (2002) has made very clearly in her book, *Water Wars*: that policy shifts to expand water privatization change the fundamental notion of water from a sacred power (and a public utility or human right) to a market resource. But to be clear, it is only since the constitutional changes of the late 1970s that water has been treated as a basic human right in India as a matter of policy; it has, for much longer, been a living source of sacrality and purity, serving important use values not just for Hindus but for all basin residents. This sacred value guaranteed, in some ways, access to river waters for all residents of a river basin, especially during auspicious days marked on the Hindu calendar. This slow erasure of religious definitions from current river policies can be considered an 'enabling condition' for phasing out public, unpriced uses. The declining government interest in guaranteeing public uses occurs as the interest in rivers for irrigation or hydropower outputs gains more official and public attention than projects to treat wastewater and generally improve water quality, values more critical for small-scale, public uses. This kind of enabling condition has not been cited in the literature. Instead, the tendency is to note that World Bank and Asian Development Bank policies on cost recovery and water pricing 'enable' the development of tradable water rights and a fuller commodification of water (Asian Development Bank 2007; Kijtiwatchakul 2002, cited in Siregar 2003).

The needs to maintain sacred values and prevent pollution of surface waters for unpriced or public uses are needs losing ground in India's early twenty-first-century water politics. Given that surface water is in such high demand for agricultural, industrial and public needs and will continue to be so in the near future, as glacial formations melt and arsenic threatens groundwater in the east, more concentrated effort must be paid to protecting these water sources from pollution and wastewater intrusions. Wastewater treatment is extremely expensive and Indian cities are far behind in treating it, especially before it enters precious surface water bodies. More brainstorming on ways to keep wastewater separate from the surface water used for human consumption and religious observance can help to create policies that avoid expensive and ultimately unworkable treatment schemes. A greater concern for the available water flowing in rivers needs to be combined with quantity concerns to create a better vision of environmental flows. The Brisbane Declaration (2007) is one good example of an effort to create standards for minimum flows in rivers to sustain ecology. Its definitions of environmental flows and the policy models evolving out of these definitions combine the importance of quality and quantity and argue for maintaining river flows for ecosystem functions and human needs and well-being. This model can shift the lens further, beyond the transfer and market models currently in vogue.

In February 2009, the government of India focused again on the River Ganga and created the National Ganga River Basin Authority (NGRBA). The government declared the Ganga a national river and announced a new commitment to pollution prevention. Soon after, the government requested a three billion dollar loan from the World Bank, and the Bank responded with a billion dollar loan and a carefully crafted project appraisal report for another round of river cleanup.

Taking a broader perspective on water quality, minimum ecological flows and sustainable access, the NGRBA is finalizing specific plans in 2011 to reach the goal of treating all 3,000 million litres of sewage dumped daily into the Ganga by 2020. The objective of treating all wastewater running to the Ganga and also the Yamuna river has been pronounced many times before, starting in the early 1980s by government leaders and civil servants. Ten years ago the Supreme Court scolded the government for failing to implement its order to stop all pollution of the Yamuna river (Alley 2009: 811–912). Activists watching this latest iteration of government concern and the emergence of new loans are asking for details and real commitments to public participation and scientific assessment and monitoring (Babu 2009; Singh 2009). Indeed, the Indian nation has a talented cadre of intellectuals, scientists, and technologically skilled personnel, and these resources should be used to create forward-thinking plans for the next iteration of pollution control programmes and for surface water management across the country. It is possible that the political climate in the near future will allow for a greater degree of sharing between governmental and non-governmental experts in water resource management. Data and information sharing can include input from local level organizations and civil society groups. All the available human resources will be needed to fix the enormous problems that now threaten the available water resources needed to meet human and ecosystem needs as well as urban, industrial, and agricultural development. It is hoped that the decision-making table will also expand to include the leaders and citizens who depend upon surface water for their religious and cultural practices. These are practices that, if sustained, can provide a vital link to the ongoing human engagement with living rivers.

Fig. 1.2.4 Resident of Varanasi preparing offerings for the Ganga (Photo credit: Kelly Alley)

References Cited

Alley, K.D. 2002. *On the banks of the Ganga: When wastewater meets a sacred river*. Ann Arbor: University of Michigan press.

Alley, K.D. 2004. The making of a river linking plan in India: Suppressed science and spheres of expert debate. *India Review* 3(3): 210–238.

Alley, K.D. 2009. Legal activism and river pollution in India. *The Georgetown International Environmental Law Review* 21(4): 793–819.

Babu, Suresh. 2009. *Curious case of Ganga*. India Water Portal Org Blog, December 4. http://www.indiawaterportal.org/blog/suresh-babu/9006. Accessed 15 Dec 2009.

The Brisbane Declaration. 2007. Adopted at the 10th international riversymposium and international environmental flows conference. Available online from http://www.riversymposium.com/2004/index.php?element=2007BrisbaneDeclaration241007.

Dharmadhikary, Shripad. 2008. *Mountains of concrete: Dam building in the Himalayas*. Berkely: International Rivers.

Ghose, R.K., and Vijayendra N. Kaul. 2004. *Report on Government of NCT of Delhi of 2005*. Audit report on the Yamuna Action Plan, prepared for Delhi Jal Board and National River Conservation Directorate.

Gilmartin, D. 1995. Models of the hydraulic environment: Colonial irrigation, state power and community in the Indus Basin. In *Nature, culture, imperialism*, ed. D. Arnold and R. Guha, 210–236. New Delhi: Oxford University Press.

Haberman, D.L. 2006. *River of love in an age of pollution: The Yamuna river of northern India*. Berkeley: University of California Press.

International Finance Corporation. 2005. International Finance Corporation's Policy and Performance Standards on Social and Environmental Sustainability. Public Release Draft, September 22

Iyer, R. 2003. *Water perspectives, issues, concerns*. New Delhi: Sage.

Kijtiwatchakul, K. 2002. *ADB's new water policy*. Bangkok: Foundation for Economic Recovery.

Menon, M., and K. Kohli, comps. 2005. *Large dams for hydropower in northeast India: A dossier*. New Delhi: South Asia Network on Dams, Rivers and People and Kalpavriksh.

Shiva, V. 2002. *Water wars: Privatization, pollution, and profit*. Cambridge: South End Press.

Singh, S. 1997. *Taming the water: The political economy of large dams in India*. Delhi: Oxford University Press.

Singh, Binay. 2009. Wake up sons of Gangā. *Times of India* (Vārānasi, page 2), TNN 16 December. http://timesofindia.indiatimes.com/city/Vārānasi/Wake-up-sons-of-Gangā/articleshow/5344961.cms. Accessed 16 Dec 2009.

Siregar, P.R. 2003. World Bank and ADB's Role in privatizing water in Asia, *JubileeSouth* (Indonesia), December 12, 2003. http://www.jubileesouth.org/news/EpZyVyEyylgqGYKXRu.shtml. Accessed 22 June 2009.

Chapter 1.3
Rethinking the Role of Humans in Water Management: Toward a New Model of Decision-Making

Marcela Brugnach and Helen Ingram

During the first decade of the twenty-first century, water availability and distribution have become increasingly important for sustainable development and biodiversity conservation. Issues of water scarcity, quality, and accessibility affect the livelihood of many communities across the globe, as well the sustainability of water systems and associated biodiversity. Although not the only cause, human activities are a major factor in triggering problems of water scarcity and quality. Acknowledging the intrinsic relationship between water and human culture and behaviour has led to a re-evaluation of water resource management (Whiteley et al. 2008; Blatter and Ingram 2001) and the development of new approaches, such as integrated water resource management (IWRM) and adaptive management (Gunderson et al. 1995; Lee 1999; Pahl-Wostl 2007a; Walters 1986). These new models try to integrate social and environmental interests and to facilitate participatory and inclusive practices (Feldman 2007), recognizing that water issues involve multiple equally valid ways of understanding. The underlying rationale is to provide effective solutions through collective actions, accommodating diverse perspectives on water management (Ingram and Lejano 2010; Lejano and Ingram 2009).

Although the benefits of these approaches are very promising, real applications have remained elusive. We argue here that this lack of traction for the new methods is related to the fact that the recent focus on participation has not been accompanied by a concomitant shift in decision-making processes. Despite the great advances in promoting participatory and inclusive practices, the new models still contain antiquated decision-making paradigms, in which the 'natural system' is seen as external to human

M. Brugnach (✉)
Faculty of Engineering Technology, Water Engineering and Management group, University of Twente, Enschede, The Netherlands
e-mail: m.brugnach@utwente.nl

H. Ingram
Southwest Center, University of Arizona and School of Social Ecology, University of California, Irvine

B.R. Johnston et al. (eds.), *Water, Cultural Diversity, and Global Environmental Change: Emerging Trends, Sustainable Futures?*, DOI 10.1007/978-94-007-1774-9_4,

experience and decisions are mainly informed by scientific or expert information, favouring technical and production-oriented solutions and evaluating feasibility mainly in economic terms (Pahl-Wostl 2007b). Furthermore, the new management plans often reflect power asymmetries that ultimately undermine ecological considerations (Ingram and Stern 2007). The interests of indigenous peoples and small farming communities are usually overlooked in the policy processes (Boelens 2008).

Administrative procedures calling for participation are clearly not a sufficient guarantee for equal participation: the determination of who gets to participate and how is ultimately a political decision (Bloomquist and Schlager 2005). Open and transparent forums alone do not compensate for power differentials among participants, or for differences in resources and skills. Cultural barriers limit the participation of disadvantaged populations (Whiteley et al. 2008). Conflicting values and perceptions remain regularly unresolved, transforming participation into an often controversial and futile process (Gray 2003). Because expert-driven discourses dominate current decision-making processes, the experiential and traditional knowledge practices of indigenous populations have little legitimacy in water policy arenas. Cultural and religious values that disadvantaged populations associate with water are mostly discounted or ignored (Rodríguez 2006).

We suggest that decision-making models are called for that are based on a relational concept of knowledge and that are congruent with the new management approaches. Participatory solutions require a reformulation of decision-making models that will take into consideration those who make the decisions and the processes by which decisions are agreed upon, as well as their influence upon ecosystem functioning. We discuss these ideas and offer practical recommendations to support revised knowledge-creation processes within water management.

1.3.1 Why the Assumptions of Current Decision-Making Models Are Not Adequate

Current decision-making models make assumptions that are not always in agreement with the claims of participation and inclusion made by new management frameworks. First of all, 'natural' and 'human' systems are seen as two separate entities, with the environment subservient to the humans who control it (Eldredge 1995). Descriptions and understandings of the natural system are externalised and independent from human experience. Solutions are interventionist and based upon the control of nature as an exterior system. Indigenous (and some rural) populations, on the other hand, see the human–nature relationship as integrated. Because such a frame of reference is not taken seriously in current water management models, these groups are often alienated from decision-making processes.

Contemporary decision-making models also rely heavily on scientific and expert knowledge. Physical sciences form the criteria, overshadowing the insights of anthropology, sociology, and other social sciences. This despite the fact that local realities often vary substantially from the summary statistics upon which many physical scientists base their conclusions. Local, traditional, and indigenous knowledge systems have guided, and continue to guide, on-the-ground practices across the world (Nakashima and Nilsson 2006). These knowledge systems stem from long periods of adaptations to specific socioecological environments (e.g., irrigation systems, the use of sacred lands, or access to water), and they illustrate societies' interactions with the environment, resources, practices, the capacity to adapt and to change, languages, rituals, and spirituality. These local knowledge systems constitute the bases for local decision-making, and as such, should not be ignored (Berkes 1999; Rodríguez 2006).

Sylvia Rodríguez (2006) argues that water in the *acequia*, or ditch system, in northern New Mexico influences identity formation, community, culture, and a sense of place that far outweigh the scientific value of water that hydrologists, economists, or even environmentalists might place on it. The acequia and the collective work of dividing the water and cleaning ditches so that it may flow to members of the community is important for maintaining social relations. She writes:

> All of this adds up to the fact that the New Mexico acequia or irrigation communities involve a moral system, a way of life, a social and cultural identity, and an attachment to place. This is why acequia associations resist the loss or transfer of water rights away from the ditches to nonagricultural use: it threatens the integrity of the whole, by removing not only water from the system, but also labor and participation from the ongoing communal effort to maintain the ditches. (Rodríguez 2006:116)

Rodríguez, among others, argues that profit maximization, as the engine of contemporary decision-making models, can no longer be the foundation for management approaches that include a diversity of stakeholders. Although economic values may represent the interests of many, solutions should be in accordance with the values and beliefs of all participants in order to be effective (Bouwen and Taillieu 2004; Schusler et al. 2003). Various authors have argued strongly that considering water as an economic good opposes human rights. Privatization, they argue, leads to market solutions, preventing the access to and control of water resources by economically disadvantaged communities (Corpuz 2006; Solón 2006). Further, because of the dynamic nature and adapting capacity of socioecological knowledge systems, profit maximization as the foundation of water management systems proves futile.

1.3.2 Toward a Reformulation of Decision-Making Processes

Overcoming the problems found within the currently employed models implies a reformulation of the role humans and their specific ways of knowledge play in decision-making processes. As several authors have suggested, natural and human systems should be conceived as a single dynamic socioecological system, able to adapt and change (Gunderson et al. 1995; Pahl-Wostl 2007b). Thus, human actors have the dual role of modifying a system that they are part of, while at the same time changing the rules they themselves create to govern their behaviour. This acknowledgement has profound implications for how decisions are made, since human actors have the capacity to adapt to new conditions and learn new ways of acting (Brugnach and Pahl-Wostl 2007).

These ideas offer a new understanding of the knowledge-creation processes in water management. What we know about the water system is no longer external to human experience but is bound within a situational, subjective representation and understanding. Furthermore, decision makers are inherently part of a larger social network (Brock and Durlauf 2001). From this perspective, the interpretation of a problem and the type of solutions sought are not independent from the decision maker and the influence of those participating in the decision-making process, any more than they are independent from the responses of the natural system.

Hence, in decision-making processes, the social context of the decision maker can significantly affect the way in which a problem is understood (Wenger 1998). What is known or unknown about a problem, from a water policy standpoint, depends not only on scientific or expert consideration but also depends on the knowledge, views, and preferences of the decision-makers and their relations with other actors (Schusler et al. 2003). 'Knowing' goes beyond understanding content alone, because what is known about a system can be influenced by the interaction between different actors and elements within the system (Brugnach et al. 2008).

Decision-making can be more effective when knowledge production processes are situation-specific. Therefore, we suggest that a relational concept of knowledge is appropriate for participatory practices. Under this rationale, knowledge is understood to have both content and relational aspects (Bouwen 2001). The content refers to what is being understood, including formal and systematic knowledge such as hard and quantifiable data (e.g., scientific knowledge). The relational aspect refers to who is being included and excluded in problem conceptualization.

Central to the relational view is a dynamic conception of knowledge and the recognition that knowledge is continually being formed and enacted into practice. As Wenger (1998:141) explains, 'every practice is in some sense a form of knowledge, and knowing is participating in that practice'. Knowing-through-practice implies that knowledge is specific to a particular situation and, as such, is tied to specific communities. It is through interactions that people make sense of a reality, negotiating the meaning they give to their surroundings and their actions. In this context, knowing involves the coordinated action among actors who engage in some form of collaboration (Bouwen and Taillieu 2004). For decision-making models, this perspective means the ability to support processes of dialogue and negotiation from which knowledge can be generated that is conducive to action. However, collaborations involving stakeholders from different communities and sectors present more than one challenge.

1.3.3 What Are the Challenges of Collaborating?

Collaborative practices aim at developing a shared construction of reality through the understanding of local contexts from the perspective of the involved parties in order to derive knowledge informed by context and equitable solutions. However, doing so is not easy.

One unavoidable problem in multiparty collaborations is differences in interpretation about what the problem is and how it can be solved. The involvement of multiple parties of diverse backgrounds means accommodating a spectrum of opinions, experiences, expectations, values, and forms of knowledge. In such situations, there are often multiple equally valid ways of framing a problem (Dewulf et al. 2005), which may result in ambiguities and conflicting values about the problem domain and its solution. For example, a water shortage can be seen as a problem of 'insufficient water supply' for one actor and one of 'excessive water consumption' for another. This difference is relevant in decision-making processes because formulating a problem in a different way elicits distinct preferences and points towards different solutions. As argued in the section above, framing water as an economic good has favoured market-based solutions, which are at odds with solutions produced by framing water as a human right. Although these differences are unavoidable, coping with them requires the capacity to resolve conflicts and disagreement constructively.

Yet another challenging problem is that the stakeholders involved may come from very different knowledge systems (e.g., indigenous communities and expert advisors). Dealing with this discrepancy requires an ability to integrate disparate knowledge systems. There are contextual, methodological, and substantive differences among knowledge systems, each representing a particular way of making sense of reality, providing understandings of, and applying meanings to a situation (Agrawal 1995). Thus, a knowledge system not only carries factual information but also reflects the way in which knowledge holders interact with the environment. Even when people share facts, they may disagree about the meaning and implications of the shared information (O'Flaherty et al. 2008). As a consequence, the problem of knowledge integration cannot be reduced to translating content from one system to another; rather, the role that knowledge holders have in decision-making processes needs to be reformulated.

Additionally, including diverse stakeholders also means handling disparities in power and resources, since stakeholders belong to communities that engage in different practices (Dewulf et al. 2005; Craps et al. 2004; Wegner 1998). Underlying the generation and use of knowledge are intrinsic assumptions of power and control that are anchored in institutions and values. Ignoring such assumptions leads to discrimination against those knowledge systems that are not commensurate with what the *status quo* accepts as valid.

Perceptions concerning sources of information, including credibility and legitimacy, are situational, and different stakeholders are likely to hold very different views (Cash et al. 2002, 2003). Water agency officials are likely to hold different perspectives than lay people, especially those with little political power. The water sector has historically been highly technical and dominated by longstanding bureaucratic agencies that embrace physical sciences more readily than social sciences and depend almost exclusively upon information from credentialed sources (Conca 2006). In contrast, indigenous people and insular minorities have a basic distrust of 'official' sources, having long experience with agencies' use of data to control decisions. For instance, the inability of Native American and rural Hispanics living in the American Southwest to assert their water rights absent hydrologic data built a legacy of suspicion. Instead of longitudinal monitoring data, indigenous peoples in the American Southwest depended on oral histories and direct experience (Brown and Ingram 1987). 'Official' data lacked credibility for them because it was produced by sources whose motives were suspect. Further, 'official data' contradicted both more trusted sources and immediate direct experience, and this distrust even extended to the courts, since the rules of evidence often privileged the economic over other values (Rodríguez 2006).

Whereas technical and monitoring information depend upon averaging of datum from limited numbers of monitoring sites, direct experience and traditional knowledge are linked to place-based details that are highly relevant to residents. Differences in soil type, slopes, land cover, and other particulars related to place, can result in wide variations from the mean experience in a particular watershed. Potentially, experiential knowledge can provide details that can be very helpful to water management, but despite this potential, many agencies hold a very restricted view of these knowledge systems, believing that valid systems must follow protocols and data analysis handed down to citizens by experts. As a consequence, credibility and legitimacy gaps exacerbate tensions between officials, who hold power, and people who do not.

If solutions ought to accommodate the multiple different interpretations that surround natural resource management problems, as claimed by new paradigms, then issues of knowledge exchange and empowerment should not be avoided. We suggest looking at more innovative solutions based on deliberation and empowerment, which can open a space for dialogue and transformation. We offer some pragmatic recommendations for doing so.

Box 1.3a 'Space for the River,' space for diversity?

—Jeroen Warner

Box Map 1.3a.1 Western Europe

Why think of cultural diversity in planning? Throughout Western Europe, the fear of climate variability has spurred projects to make rivers climate-proof. Rather than building even higher levees to stop overflowing, recent projects have tried to 'loosen the laces'[1] – the banks, dikes, and floodwalls – to allow rivers to meander again, within limits. These so-called 'Space for the River' projects attempt to accomplish many goals at once – safety, shipping, recreation, emergency irrigation, cultural heritage, urban regeneration, wastewater discharge, and natural values.

Moreover, they also need to comply with the European Water Framework Directive, a European directive adopted in 2000 committing EU member states to achieve

(continued)

Box 1.3a (continued)

good qualitative and quantitative status of all water bodies by 2015. As a result the proposed river projects require working across boundaries of many sorts with multiple interests and actors, such as agricultural interests, regional economic development, natural values, water safety and water quality issues, to complete each project.

Safer, scenic, sustainable rivers through cost-covering interventions... this seems like such a golden combination, it might be hard to imagine anyone would oppose such a project. Yet, it turns out that the 'Space for the River' projects encountered quite a bit of local protests. German citizens worried about wet cellars because of higher groundwater levels, Britons complained about the ever-rising project cost, Dutchmen felt they did not have enough say, and the French were unsure whether their safety was guaranteed.

Why not spend the project money on safety alone? If safety were the only issue, less space would be required. Space given to the river is taken away from someone else, usually farmers. But also city-folks, who are in general quite open to change, are not always convinced of the projects: why should they sacrifice their peace and quiet? Municipal authorities and investors usually support the 'Space for the River' engineering initiatives, but, inevitably, planners come across people who come with competing plans or who resist their plans as such because they see them as an invasion of their area.

To deal with this complex and often controversial planning situation a new way has been developed to include a variety of voices through the so-called Joint Planning Approach (JPA).

1.3a.1 One method of spanning boundaries and achieving diversity: Joint planning

The Joint Planning Approach was developed at the Centre for Sustainable Management of Resources of Radboud University in Nijmegen, the Netherlands, in 2006–2008, with Dutch, German, and French partners (see http://www.jointplanning.eu/en). The project sought to enable a diversity of social actors to contribute their knowledge, resources, and capacity, and to find ways of enabling partnerships and participation where they are not always institutionally facilitated. The Joint Planning Approach involves multiple stakeholders in river management projects from the start. It sees public engagement with a multiplicity of perspectives as an opportunity,

(continued)

Box 1.3a (continued)

rather than a threat. According to the Joint Planning Approach, interactive planning should involve the following:

1. Joint Learning
2. Joint Visioning
3. Joint Rule-making/Institution-Building
4. Joint Options
5. Joint Design
6. Joint Implementation

These steps need not be taken sequentially, and some may be skipped along the way, but all appear useful in promoting promising alternatives and forestalling unnecessary conflict. However, these measures do consume time and energy, which may not sit well with public officials under mandate to move quickly while at the same time preventing potential lawsuits against the project.

In Bréhémont, a town on the Loire in central France, the river authority completed a study on widening the river in 2007. Upstream, protests over dams had already alerted the river manager that the citizenry was active in local water politics, and would not just blindly comply with the plans. The river authority agreed to apply a 'joint planning as a pilot, and a host of government agencies were joined by a local NGO. The group looked at every detail of the plan and found an error in the hydrological model that necessitated a remodelling exercise. As a result of this success, the JPA has been adopted by or proposed for other river basin authorities.

1.3a.2 Implementing diversity remains problematic

1.3a.2.1 Veessen-Wapenveld

There are various 'Space for the River projects' along several branches of the Rhine River, but one particularly controversial site is Veessen-Wapenveld on the river IJssel, in the eastern part of the Netherlands. The river meanders through an aesthetically pleasing landscape of sandy soils. A river bypass was planned by the national Public Works department to divert water in case of extreme high-water events. However, the bypass would require five or six operations and move three or four homes. It is also an area that is not so easily evacuated.[2]

When the bypass was first proposed in 2005, the municipality and the water board protested together with angry farmers and citizens, and they hired an engineering consultancy for a second opinion. The Dutch parliament largely ignored these protests. After a controversy, however, the province, Overijssel, initiated 'joint design' sessions with stakeholders to work out the details together.

(continued)

Box 1.3a (continued)

The new consultation process appeared to follow proper Joint Planning procedure. The farmers initially complied with a search strategy for the Environmental Impact Assessment that explored alternative solutions – including a very narrow bypass, a separate wider bypass that included new agricultural opportunities, and a citizen alternative. Some alternatives promoted recreational facilities; others favoured agricultural development in the region.

Yet when the farmers felt they had painted themselves into a corner, they decided to push their favourite option through in a make-or-break session. As it turned out, they did so successfully. Whether this move grew out of a fear of cooptation, or a plain power play, is a moot issue – defending cultural diversity does not prevent power politics on either side of the table.

1.3a.2.2 River Thames

A British 'green bypass channel' along the Thames, the Jubilee River, was actually confronted by a high water event in 2003. The bypass provides the citizens of Windsor, Eton, and Maidenhead with better flood protection. Maidenhead had been permissive to building in the flood plain, so that several thousand properties were at risk from flooding.

The initiating water management agency contracted environmental consultants and worked with environmental and cultural heritage NGOs to ensure cultural and environmental values were tended to. Already at the early planning stages, the neighbours – famous Eton College, Taplow Parish, and Buckinghamshire County – resisted, citing the integrity of cultural and natural heritage.

The Environment Agency, the public body responsible for the project, worked hard to placate the dissenters with amenities, like a bridle path and better traffic facilities. The placation was not as successful with the parishes immediately downstream of Maidenhead. Central government resisted a follow-up project extending the channel downstream because of cost-benefit considerations. The failure to extend the bypass meant that Datchet and Wraysbury would not be protected. Residents of these two parishes feared the embankments protecting upstream Maidenhead would bring even more flooding to their houses downstream. The project was 'called in' by the environment minister and subjected to a public enquiry. At this enquiry, Datchet and Wraysbury were reassured the channel would only raise flood levels marginally.

But in 2003, when houses in Datchet experienced flood damage, people were quick to point their fingers at the bypass channel, where 'natural banks' had failed to settle and had eroded considerably, a situation exacerbated by the malfunctioning of the sluice gates. The owners of the damaged properties demanded a public inquiry.

(continued)

Box 1.3a (continued)

The response of the environment agency was evasive: it refused to accept blame, rejected a public enquiry, and held 'Open Days' to explain rather than listen. When a report by independent consultants showed the project was structurally flawed, the agency sued its main contractors. Then the agency did reach out to local stakeholders. Immediately after the flooding, it established three Flood Risk Action Groups (FRAGs). It continues to have a weak institutional position and must work with stakeholders to be more effective in coping with increased flood risk, especially since 2001, when the Association of British Insurers threatened not to provide coverage to areas that are insufficiently protected (Huber 2004).

1.3a.2.3 The Westerscheldt controversy

The final example crosses national boundaries. The Belgians and Dutch have a long, fraught history of managing the Scheldt River estuary: for centuries the Netherlands blocked the harbour of Antwerp, which depends on the estuary for sea access. Since the last few decades, relations have improved, and Antwerp Harbour is in business, but the Belgians' dependence on the Dutch remains. Allowing even larger containers to travel on the river requires dredging in Dutch territory, and harbour authorities want this work done as soon as possible. Environmentalists, however, fear that dredging will have an adverse impact on the river's fragile ecology and the rare birds living in the estuary.

Dutch and Belgian authorities agreed to a visioning process for the estuary. A key topic was the environmental effect of the proposed deepening. Money was provided for research and discussions among the stakeholders, including public, private, and NGO groups. In the Netherlands, farmers and citizens agitated against 'depoldering' – removing embankments to make space for water in the tidal region. In February 1952, the province of Zeeland had suffered badly from the biggest sea flood in recent history, which claimed 2,000 lives. Between 1950 and 1997 the world famous Dutch Delta Works were built, a complex engineering construction of dams, sluices, locks, dikes, and storm surge barriers, to prevent such a flood from ever happening again; however, environmental consciousness and the decreasing importance of agriculture led to a plan for giving land back to the water. Many people from Zeeland, the main Dutch province in the middle of the estuary, were uncomfortable with this plan.

A Zeeland politician arrived at a compromise but shortly thereafter died, and the carefully constructed alliance fell apart. First to defect were Dutch agriculturalists, who lobbied to reverse the decision. They succeeded in

(continued)

Box 1.3a (continued)

influencing the Dutch prime minister, who came from the region. The environmental groups took the matter to the Administrative Supreme Court and sued for breach of contract. In the end, the court agreed that the Dutch had not conclusively shown that deepening would not harm the local environment.

1.3a.3 Some risks

Facilitating diversity essentially implies risk taking. The examples above demonstrate there are financial and procedural risks. Furthermore, there is the risk that other actors will refuse to play along with those attempting to span geographic, conceptual, political, and other boundaries.

A Joint Planning process can be a fruitful way to achieve a diversity of values and viewpoints at the negotiating table. It can be successful in conflict resolution and constructive planning, yet it can also be thwarted and sidelined by power differentials, as its policy outcomes may be simply ignored by the project initiators or by a new local authority (Faysse et al. 2007). If the joint planning network has only tenuous links with the formal decision framework, even a well-designed participatory process will have a hard time convincing passionate stakeholders. In such a case, Joint Planning takes place in a vacuum, has little effect, or ends in confrontation. Non-participants can have considerable obstructive power by resorting to judicial, extra-parliamentary, and media venues to get a hearing for their concerns.

Multi-stakeholder planning processes should neither be idealized as the pinnacle of democracy nor dismissed as hegemonic instruments or ineffective smokescreens; rather, they should be seen for what they are: social networks representing a degree of diversity.

Reference cited

Huber, Michael. 2004. *Reforming the UK flood insurance regime. The breakdown of a gentleman's agreement*, ESRC/LSE, Discussion paper no. 18. http://www.lse.ac.uk/collections/CARR/pdf/DPs/Disspaper18.pdf. Accessed 5 July 2010.

Notes

1. Natuurmonumenten & Wereld Natuur Fonds, 'Veters Los!' 1997.
2. www.hoogwatergeul.nl

1.3.4 Strategies and Recommendations

1.3.4.1 Dealing with Framing Issues and Ambiguities

Framing differences and ambiguities within and between diverse knowledge systems can be done in myriad ways. Doing so does not necessary imply reaching consensus but rather the creation of a shared problem definition that represents the values and beliefs of the different stakeholders and rightsholders. In practice, this goal can be achieved in different ways, and several strategies based on deliberative approaches for resolving conflictive views are presented below.

Persuasive communication: This approach consists in convincing others of one's own frame of reference, by presenting it as an attractive and worthwhile option (see, e.g., Bouwen and Fry 1991).

Dialogical learning: This approach aims at better understanding each other's frames and perspectives. It is based on open dialogue and encourages learning and discovery for all stakeholders and rightsholders (see, e.g., Argyris and Schön 1978).

Negotiation: This approach aims at reaching a mutually beneficial and integrative agreement that draws from multiple perspectives or frames (see, e.g., Leeuwis 2000). The negotiation can range from an 'integrating' approach, in which actors develop synergetic win–win outcomes, to a 'distributive' one, in which a win–lose position distributes profits and gains in an antagonistic way.

Oppositional modes of action: This approach constitutes one way of dealing with multiple frames when opinions and interest are polarised (see, e.g., Gray 2003). This mode of action can take the form of cold or hot conflict. 'Cold conflict' means that distancing and avoiding each other is a dominant mode of operating. 'Hot conflict' refers to heated opposition and adversarial actions in which parties attempt to impose their frame of reference upon others forcefully. It should not be assumed that conflict necessarily disadvantages the politically powerless: on the contrary, it may serve to generate interest, make issues more salient, and mobilise the disengaged.

1.3.4.2 Creating Opportunities for Participation

Decision-making processes must increase public involvement in water-related issues and greater interaction between the silos of user interests that tend to dominate, such as agriculture, environmentalists, recreationalists, power generators, and others. Additionally, physical scientists – such as hydrologists and engineers – should engage and interact more with ecologists, anthropologists, and other social scientists. Face-to-face engagement has the potential for bringing about trust, shared experiences, empathic understanding, positive relationships, and other community-oriented consequences that will enable people to work toward new knowledge systems that are more amenable to collective solutions (Feldman and Ingram 2009; Innes and Booher 2003; Ostrom 1990). Boundary objects (Star and Griesemer 1989) provide an artifact, such as a model, scenario, template, prototype, compact, and the like, that draw upon necessary but different, ways of knowing and enhance mutual understanding. Boundary organizations, like advisory committees or task forces relying on both scientific and lay members, may perform similar services across organizational boundaries (Guston 2001; Jasanoff 1990). Shared, or boundary, experiences along with critical reflection and dialogue can facilitate collaborative action (Feldman et al. 2006).

Collaborative processes are not magic bullets, and simply changing decision-making processes alone would not reduce problems. However, substantive change is not likely to take place without changes in the decision-making processes, and only through such changes can new techno-scientific ideas emerge that reflect the social context. Only through collaborative processes involving participation, discourse, and communication can effective political strategies be developed (Ingram and Endter-Wada 2009).

1.3.5 Conclusions

In this chapter, we have challenged the assumptions of decision-making models that current water management practices apply. Despite the claims of participation and inclusion that many new management approaches make, their decision-making models still reflect an impaired management paradigm based on controlling the natural system through technical solutions developed with expert information. Although this management may have been appealing in the past, it contains assumptions of power differentiation, control, and cultural dominance that have resulted in solutions that embraced the interests of few. This approach slighted ecological considerations and excluded disadvantaged populations, indigenous peoples, and insular minorities. Failing to address the biological and cultural diversity associated with water problems is no longer suitable. Instead, a cross-cultural approach that encompasses diversity is needed.

From this point of view, problems and solutions need to embrace multiple ways of knowing, taking into consideration what individuals and groups value and believe important. However, doing so is not yet common. One instance of success in taking multiple perspectives into account is the effort of the government of the Northwest Territories (NWT) in Canada. The NWT has a policy respectful of traditional knowledge systems, defined as 'knowledge and values which have been acquired through experience, observation, from the land or from spiritual teachings, and handed down from one generation to another' (Mills 2008). Credentialed science is now augmented with traditional knowledge in official decision-making. It is important to note that indigenous peoples are the majority constituency in the Northwest Territories and are therefore not without great political power.

Multiparty water management must be analyzed as a knowledge creation process whereby knowledge is redefined relationally. Behind any knowledge system are particular assumptions about human–nature relationships, as well as power relationships, which endow the practice and the actors with meaning. This approach to knowledge development opens up the possibility for an improved integration of interests in water policy management. We claim that this type of approach is practicable only in a context in which serious political deliberation can occur.

It would be naïve to think that the influence of political and economic power can be overcome by knowledge alone, however inclusive of multiple perspectives knowledge might be. Further, we recognise that power sometimes distorts knowledge creation processes, including those governed by strict scientific protocols. Full and fair deliberation of diverse perspectives has the potential to create understandings and alliances that challenge the unbridled use of power. Such deliberations might result in the creation of institutional arrangements that provide opportunities for cooperation, learning, and discourse.

Acknowledgments The authors greatly acknowledge the insightful comments of Art Dewulf and the insightful comments of anonymous reviewers.

Resources

Agrawal, A. 1995. Dismantling the divide between indigenous and scientific knowledge. *Development and Change* 26(3): 413–439.

Argyris, C., and D. Schön. 1978. *Organizational learning: A theory of action perspective*. Reading: Addison Wesley.

Berkes, F. 1999. *Sacred ecology: Traditional ecological knowledge and resource management*. Philadelphia: Taylor & Francis.

Blatter, J., and H. Ingram. 2001. *Reflections on water: New approaches to transboundary conflicts and cooperation*. Cambridge: MIT Press.

Bloomquist, W., and E. Schlager. 2005. Political pitfalls of integrated watershed management. *Society and Natural Resources* 18(2): 101–117.

Boelens, R.A. 2008. *The rules of the game and the game of the rules: Normalization and resistance in Andean water control*. Wageningen: Wageningen University.

Bouwen, R. 2001. Developing relational practices for knowledge intensive organizational contexts. *Career Development International* 6(7): 361–369.

Bouwen, R., and R. Fry. 1991. Organizational innovation and learning: Four patterns of dialog between the dominant logic and the new logic. *International studies of management and organization* 21(4): 37–51.

Bouwen, R., and T. Taillieu. 2004. Multi-party collaboration as social learning for interdependence: Developing relational knowing for sustainable natural resource management. *Journal of Community and Applied Social Psychology* 14: 137–153.

Brock, W.A., and S.N. Durlauf. 2001. Discrete choice with social interactions. *Review of Economics Studies* 68(2): 235–260.

Brown, F.L., and H. Ingram. 1987. *Water and poverty in the southwest*. Tucson: University of Arizona Press.

Brugnach, M., and C. Pahl-Wostl. 2007. A broadened view on the role of models in natural resource management: Implications for model development. In *Adaptive and integrated water*

management: Coping with complexity and uncertainty, ed. C. Pahl-Wostl, P. Kabat, and J. Möltgen, 187–203. Berlin: Springer.
Brugnach, M., A. Dewulf, C. Pahl-Wostl, and T. Taillieu. 2008. Toward a relational concept of uncertainty: About knowing too little, knowing too differently, and accepting not to know. *Ecology and Society* 13(2): 30. http://www.ecologyandsociety.org/vol13/iss2/art30/.
Cash, D., W. Clark, F. Alcock, N. Dickson, N. Eckley, and J. Jäger. 2002. *Salience, credibility, legitimacy and boundaries: Linking research, assessment and decision making*. Faculty Research working paper series. November 2002 RWP02–046. Cambridge: John F. Kennedy School of Government, Harvard University.
Cash, D.W., W.C. Clark, F. Alcock, N.M. Dickson, N. Eckley, D.H. Guston, J. Jäger, and R.B.H. Mitchell. 2003. Knowledge systems for sustainable development. *Proceedings of the National Academy of Sciences* 100(14 July 8): 8086–8091.
Conca, K. 2006. *Governing water: Contentious transnational politics and global institution building*. Cambridge: MIT Press.
Corpuz, T.V. 2006. Indigenous peoples and international debates on water: Reflections and challenges. In *Water and indigenous peoples*, Knowledge of nature, vol. 2, ed. R. Boelens, M. Chiva, and D. Nakashima. Paris: UNESCO.
Craps, M., A. Dewulf, M. Mancero, E. Santos, and R. Bouwen. 2004. Constructing common ground and re-creating differences between professional and indigenous communities in the Andes. *Journal of Community and Applied Social Psychology* 14(5): 378–393.
Dewulf, A., M. Craps, R. Bouwen, T. Taillieu, and C. Pahl-Wostl. 2005. Integrated management of natural resources: Dealing with ambiguous issues, multiple actors and diverging frames. *Water Science and Technology* 52(6): 115–124.
Eldredge, N. 1995. *Dominion*. New York: Henry Holt and Co.
Faysse et al. (2007). Chapter 2: Supporting local organisations in peri-urban Cochabamba, Bolivia. In: *Peri-Urban Water Conflicts: Supporting dialogue and negotiation*, eds. J. Butterworth, R. Ducrot, N. Faysse, and S. Janakarajan. IRC International Water and Sanitation Centre, Technical Paper Series No. 50, Delft 128: 25–26.
Feldman, D.L. 2007. *Water policy for sustainable development*. Baltimore: Johns Hopkins University Press.
Feldman, D.L., and H.M. Ingram. 2009. Climate forecast, water management, and knowledge networks: Making science useful to decision makers. *Water, Climate and Society* 1(October 2009): 19–21.
Feldman, M.S., A.M. Khademian, H. Ingram, and A.L. Schneider. 2006. Ways of knowing and inclusive management practices. *Public Administration Review* supplement to 66(December): 89–99.
Gray, B. 2003. Framing of environmental disputes. In *Making sense of intractable environmental conflicts: Concepts and cases*, ed. R.J. Lewicki, B. Gray, and M. Elliott, 11–34. Washington, DC: Island Press.
Gunderson, L.H., C.S. Holling, and S.S. Light (eds.). 1995. *Barriers and bridges to the renewal of ecosystems and institutions*. New York: Columbia University Press.
Guston, D.H. 2001. Boundary organizations in environmental policy and science. *Science, Technology and Human Values* 26(4): 399–408.
Ingram, H., and J. Endter-Wada. 2009. Frames and ways of knowing: Key considerations for policy responses to climate risk and vulnerability. Paper for the 7th International Science Conference on the Human Dimensions of Global Environmental Change, UN Campus, Bonn, Germany, 26–30 Apr 2009.
Ingram, H., and R. Lejano. 2010. Transitions: Transcending multiple ways of knowing water resources in the U.S. In *Water policy entrepreneurs*, ed. D. Huitema and S. Meijerink. Cheltenham: Edward Elgar.
Ingram, H., and P. Stern (eds.). 2007. *Research and networks for decision support in the NOAA sectoral applications research program*. Washington, DC: National Academies Press.
Innes, J.E., and D.E. Booher. 2003. Collaborative policymaking: Governance through dialogue. In *Deliberative policy analysis: Understanding governance in the network society*, ed. Maarten A. Hajer and Wagenaar Hendrik, 33–59. Cambridge: Cambridge University Press.

Jasanoff, S. 1990. *The fifth branch: Science advisors as policy makers*. Cambridge: Harvard University Press.
Lee, K.N. 1999. Appraising adaptive management. *Conservation Ecology* 3(2):3–16. [online] URL: http://www.consecol.org/vol3/iss2/art3/
Leeuwis, C. 2000. Reconceptualizing participation for sustainable rural development: Towards a negotiation approach. *Development and Change* 31: 931–959.
Lejano, R.P., and H. Ingram. 2009. Collaborative networks and new ways of knowing. *Environmental Science and Policy* 12(6 October 2009): 653–662.
Mills, C. 2008. Role of TK in aquatic effects monitoring, Powerpoint presented to aquatic effects monitoring program workshop, Yellowknife, Northwest Territories, Canada, 21–22 Oct 2008.
Nakashima, D., and A. Nilsson. 2006. Linking biological and cultural diversity: Local and indigenous knowledge systems. (LINKS) project. In *Sixty years of science at UNESCO, 1945–2005*, ed. P. Petitjean, V. Zharov, G. Glaser, J. Richardson, B. de Padirac, and G. Archibald, 385–388. Paris: UNESCO.
O'Flaherty, R.M., I.J. Davidson-Hunt, and M. Manseau. 2008. Indigenous knowledge and values in planning for sustainable forestry: Pikangikum first nation and the whitefeather forest initiative. *Ecology and Society* 13(1): 6. http://www.ecologyandsociety.org/vol13/iss1/art6/.
Ostrom, E. 1990. *Governing the commons*. New York: Cambridge University Press.
Pahl-Wostl, C. 2007a. Transition towards adaptive management of water facing climate and global change. *Water Resources Management* 21(1 January): 49–62.
Pahl-Wostl, C. 2007b. The implications of complexity for integrated resource management. *Environmental Modelling and Software* 22(5 May): 561–569.
Rodríguez, S. 2006. *Acequia: Water sharing, sanctity, and place*. Santa Fe: School of American Research Press.
Schusler, T.M., D.J. Decker, and M.J. Pfeffer. 2003. Social learning for collaborative natural resource management. *Society and Natural Resources* 15: 309–326.
Solón, P. 2006. Cultural diversity and privatization of water. In *UNESCO, 2006, water and indigenous peoples*, Knowledge of nature, vol. 2, ed. R. Boelens, M. Chiva, and D. Nakashima. Paris: UNESCO.
Star, S.L., and J. Griesemer. 1989. Institutional ecology, translations and boundary objects: Amateurs and professionals in Berkeley's museum of vertebrate zoology. *Social Studies of Science* 19(3): 387–420.
Walters, C. 1986. *Adaptive management of renewable resources*. New York: McGraw-Hill.
Wenger, E. 1998. *Communities of practice: Learning, meaning and identity*. Cambridge: Cambridge University Press.
Whiteley, J., H. Ingram, and R. Perry. 2008. *Water, place, and equity*. Cambridge: MIT Press.

Chapter 1.4
Local Water Management in the Andes: Interplay of Domination, Power and Collective Participation*

Rutgerd Boelens

Map 1.4.1 Andean region, South America

Water management, rights, and distribution practices manifest themselves simultaneously in water infrastructure and technology, normative arrangements, and organizational frameworks for operating and maintaining water control systems, each embedded in diverse political-economic and cultural-symbolic contexts. This situation implies that technology, organizations, culture, political economy, and ecology fundamentally influence and structure possibilities of water captured in contexts of cultural diversity and environmental change. Water rights analysis requires an interdisciplinary focus – one that allows for analyzing the politically contested nature of water resources and water rights as well as the interacting domains that constitute water control systems. I use the concept of 'domains' of water rights and control not in the sense of 'arenas' or social fields of interaction with territorial and political

*This paper is largely based on elements of my book, *The Rules of the Game and the Game of the Rules. Normalization and Resistance in Andean Water Control* (Wageningen, The Netherlands: Wageningen University, 2008).

R. Boelens (✉)
Environmental Sciences Group, Wageningen University and Departments Social Sciences and Law, Catholic University Peru, Wageningen, The Netherlands
e-mail: Rutgerd.Boelens@wur.nl

B.R. Johnston et al. (eds.), *Water, Cultural Diversity, and Global Environmental Change: Emerging Trends, Sustainable Futures?*, DOI 10.1007/978-94-007-1774-9_5,

boundaries, but as (distinct but interlinked) thematic fields producing knowledge on water control. Irrigation water control studies in the Andean region, like investigations and narratives from other parts in the world, have shown the need for conceptualizations that dynamically interrelate the organizational, technical, and normative as interdependent 'subsystems' of water control that interact with cultural and political-economic forces and structures of their societal context.

The struggles of Andean water collectives give a deeper understanding of how 'mutual bonds of rights and obligations' and a 'sense of belonging' among water users are strengthened, how a common 'hydraulic property' is created and reaffirmed, and how 'water cultural identities' are acquiring their actual substance. As Cohen observes, 'to operate effectively, a group must define its membership and its sphere of operation, by defining its identity and exclusiveness, within the political field in which it operates' (1986:68). This political field has many levels, from household and community to national and global ones, and at each level dominant players try to present the field's members with 'a ready-made blue-print for living, with a design for selfhood' (1986:60). Therefore, to understand the water cultures of the subjugated, it is crucial to focus on the water cultures of those who subjugate, because each stands in direct relation to the other.[1] Ethnicity, identity, and subject-formation stem not just from the Self and self-definition but also, and importantly, from the confrontation with the Other and the ways in which the Self is 'othered' (Fanon 1967; Said 1993).

Gelles (2010) rightly observes that the hierarchical submission of local water control systems to the state takes on an added dimension in ethnically differentiated and culturally plural Andean societies, in which the cultural politics of irrigation bureaucracies, market-mechanism based projects and secular forms of water organization generally reflect and reproduce the legitimacy and authority of a dominant culture of a ruling ethnic group. Historically a deep-seated racism has characterized bureaucratic myopia in the Andean nation-states (Gelles 2002; Boelens et al. 2006).

In this chapter I focus on the ways in which ruling groups have aimed to supplant the diversity of water cultures and rights with the categories and frames of reference of dominant water players in order to get grip on and control over everyday water management and social relations. The authorities have often done so by presenting the rules of the dominant group as the objective, universal schemes of rational water culture, identity, and belonging. The policies of nation-states to recognize local water cultures often serve as a way to 'normalize' deviant groups, rights, and behaviour, and to keep them from 'wrong-doing' in the eyes of dominant powers (see Boelens 2009).

---

[1] The 'social organization of cultural difference' (Barth 2000), based on 'situational facts', historical roots and actual perceptions of group difference as well as on the 'invention of traditions' (Hobsbawm and Ranger 1983) and the construction of 'imagined communities' (Anderson 1983) and 'usable histories', takes place as relational ('inter-subjective') and most of all confrontational processes, within economic and political structures and struggles.

Fig. 1.4.1 Female campesina promotor training her fellows on water design, Ecuador (Photo credit: Rutgerd Boelens)

Fig. 1.4.2 Community water users organization, Ecuador (Photo credit: Rutgerd Boelens)

Fig. 1.4.3 Indigenous woman in collective irrigation labor party, Ecuador (Photo credit: Rutgerd Boelens)

Resisting normalization, local water user groups define their own cultural-political projects in which they refuse to accept selfhood as a mechanical reflection of prevailing power relations, laws, or scientific categories. On the one hand, they dynamically use other normative systems and discourses, appropriating those elements that can legitimate their claims. In this context, ethnicity and Andean or indigenous identity can be considered as a political strategy in the struggle of local communities and supra-local organizations to defend their rights against the colonial and postcolonial state, its agencies, and other powerful water interest groups. On the other hand, Andean communities show specific historical and cultural forms of collective action and resource management, embedded in specific Andean cultures with particular normative repertoires, symbols, meanings, livelihoods, and local economies (Baud 2010; Gelles 2010; Mayer 2002). Their water identities and boundaries are not only 'limitless', 'tactical', 'fluid', and 'disposable', but they also dynamically relate to each other in varying combinations, requiring an understanding of the Andean highlands '... as a place of synthetic and shifting identities that have grown out of the multi-layered interactions of the local, the regional, and the global since pre-Columbian times' (Starn 1994:20).

> As Gelles argues, "Andean culture is best viewed as having been created from a hybrid mix of local mores with the political forms and ideological forces of hegemonic States, both indigenous, Iberian and others. Some native institutions are with us today because they were appropriated and used as a means of extracting goods and labor by Spanish colonial authorities and republican States after Independence; others were used to resist colonial and post-colonial regimes" (2002:12).

It is not just dominant economic groups and policy institutes but also dominant science that have powerfully influenced the conventional ways of seeing

Fig. 1.4.4 Collective irrigation working party, Ecuador (Photo credit: Rutgerd Boelens)

Fig. 1.4.5 Women irrigating Canterones, Ecuador (Photo credit: Rutgerd Boelens)

Fig. 1.4.6 Opening the irrigation gate, Ecuador, Chimborazo province (Photo credit: Rutgerd Boelens)

water and of shaping water subjects, categories and identities. For example, mainstream science and water policies usually treat hydraulic infrastructure as just a technical and material matter; commonly consider water laws and rights as simply legal matters; restrict water organization to the work of sociologists and social organizers; and assume that economists lead the planning of productive wealth and 'rational economic institutions'. They see water culture as nonmaterial and merely 'soft', 'social', and rooted in static tradition and folklore – or as a useful opening for convincing water user groups of the need to engage in so-called integrated and participatory projects with objectives and rules largely established by outside agencies (see Chap. 1.3, this volume).

These issues lead me to question the universalistic conceptualization and mono-dimensional categorization and compartmentalization typical of irrigation and water management studies and to come with an alternative framework for understanding. Let us look, for example, at 'water rights': these tend to be seen – even in interdisciplinary water control frameworks – as 'the legal component', with only a juridical, normative, or socio-legal dimension. Moreover, despite the enormous variety of space- and time-specific properties and definitions of 'water rights', most approaches (and policies) typically analyze the concept of 'water rights' as a 'black box' with universal definitions, rather than as contextually composed, complex bundles. However, a misplaced conceptual understanding of the categories of water rights is not the main reason why an alternative analysis is hard to accomplish. The main difficulty lies in the boundaries that divide the categories and obscure trans-boundary linkages and a more 'realistic' or adequate perception. In

Fig. 1.4.7 People at headwaters of the watershed, Ecuador, Chimborazo province (Photo credit: Rutgerd Boelens)

modern science, water 'domains' have been 'dominated' (demarcated and encroached) by scientific disciplines that separated them, in order to produce water truth claims, backed by the disciplines' own system of valuing, codes of esteem and norms of correctness, and sustained by each discipline's methods of categorization, comparison and judgment. Vernacular combinations of these water control domains, at the other hand, can be analyzed as 'non-dominant' ways of water knowledge and truth production. Although obviously bound by power mechanisms and societal interaction, judging their soundness does not rely on the approval by dominant knowledge systems. Therefore, in order to understand local water visions, management and practices, water rights must be made clear and specific and at the same time revelatory of the fact that they encompass multiple domains – interlinked, thematic fields of knowledge, conceptualization, and interpretation, mutually constituting each other. While the conceptualization of water rights in these domains depends clearly on context and the situated perspectives of each subject, it is useful to describe the domains and inter – domain linkages to facilitate an alternative approach of analytical understanding. I distinguish the following domains.

The *socio-legal domain* focuses particularly on the contextualized constructs of water rights and property relations: definitions of the contents of water rights and the associated privileges, obligations, sanctions, operational norms and rules; the accepted mechanisms for acquiring, materializing, and upholding rights; agreements on procedures to formulate new rules and rights and to establish authorities, etc. Important here is that water rights express an agreement about the legitimacy of the right holders' claim to water and water-related decision-making. This agreement is intimately linked to social relations of authority and power and can be based on a variety of grounds. Since water rights in Andean irrigation systems exist in conditions of legal pluralism in which rules and principles of different origin and legitimization coexist in the same locality, constructions of water rights are often hybrid.

The *technical and biophysical domain* of water rights relates to the need to have or acquire adequate means to take water from a source and transport it, while considering the location-specific physical, climatic, and ecological opportunities and constraints. Adapted to the ecological and hydrological setting, infrastructure (such as intakes, canals, outlets, division boxes, and biophysical protection structures) enables or blocks the actual use of water rights. Having the legal possibility (and social power) to take water is in itself meaningless without the necessary irrigation facilities, artifacts, and technical and agro-productive skills. As I will show later, infrastructure also 'embeds' water rights and norms in multiple ways.

The *organizational domain* centres around the need to manage the planning, organization, and monitoring of water and the operation of infrastructure, as well as around mobilizing resources and organizing decision-making. To materialize water rights requires adequate organization of labour and resources to operate and maintain the technology, to distribute the water, to direct and organize water users' behaviour, implement collectively required rules and rights, keep records of contributions, and penalize non-compliance. Therefore, the collective capacity to access and mobilize inter-human and inter-institutional relationships is central. In Andean community systems, having a right to water often accompanies the right-holders' opportunity to participate in system operation and management and to fulfil a number of duties and obligations established collectively by users.

The *political-economic domain* of water rights provides insight into the allocation and distribution of this powerful resource and sheds light on the distribution of decision-making power in water control: bundles of water rights include both water access rights and decision-making rights. The distribution of water rights (allocation) reflects economic, political, and discursive power structures and is at the same time an important foundation for reproducing this power, because power structures contribute significantly to shaping the specific content and modes of acquisition of water rights. To what degree are actors able to structure water control and define, allocate, and enforce water rights according to their needs and objectives?

The *cultural-metaphysical domain* focuses on how the rules, rights, and duties attached to water are closely linked to cultural systems of meanings, symbols, and values. In many user-controlled systems, water allocation and distribution are deeply embedded in local, historical institutions and networks of both human and supernatural actors and powers that (are seen to) influence and define water control. Supernatural authority often reinforces the legitimacy of particular human authorities and actions, and the symbolic and metaphysical power to control water and regulate water users' behaviour may be mobilized both unconsciously and with clear practical (e.g., agro-productive or political) purposes. Often, metaphysical water rights concepts include all kinds of non-water related rights and duties and inform locally specific forms of organization.

The cultural and political domains of water rights significantly address the question of the legitimacy of actors' inclusion in and exclusion from irrigation water use and decision-making processes – in the eyes of 'insiders' and 'outsiders'. In this sense water rights clearly embody socio-technical power relations. They are intimately linked to the existing social and cultural organization and relations of authority and power, and the process of defining, imposing, defending, disputing, or internalizing 'water norms' involves technical tools and managerial capacities – a process in which both technology and human labour organization are shaped or 'moralized'.

The domains, as outlined above, are intimately related, always in locally specific ways. Water infrastructure, rights, and organization neatly and functionally entwine, while political-economic and cultural context steer their interrelation. In the following, I give an example how this intertwinement works.

Within collectively managed irrigation systems, water users need to formulate and materialize the fundamental building blocks of the system to enable day-to-day

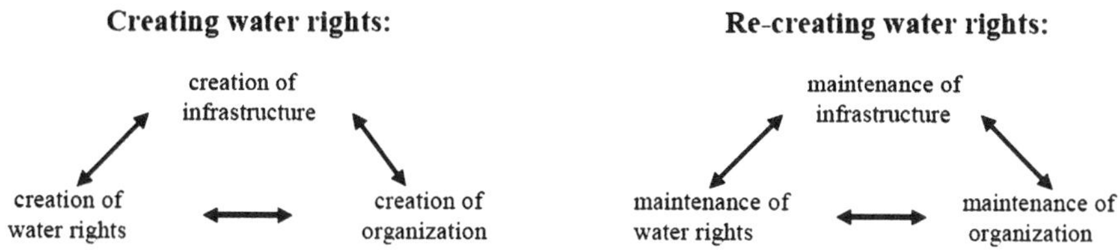

Fig. 1.4.8 'Driving force' behind collective action in user-controlled systems

water management. Without a clear (although contested) definition of these basics, a water use system cannot work. The multiple domains provide the contents and context in which these elements and their interaction are embedded. Action research, that is, participatory based research, in Ecuador, Bolivia, and Peru – as well as in places ranging from Nepal to Switzerland and Tanzania – shows that the key conditions for the generation and regeneration of local, sustainable irrigation systems, include coherent interaction, especially among technological, normative, and organizational drivers of an irrigation system. Sustainable user-managed systems have a firm interrelationship among three parallel processes:

- construction and rehabilitation of the infrastructure,
- creation and reconfirmation of water rights, and
- creation and strengthening of the organization.

In other words, the infrastructural system, normative system, and organizational system are fundamentally interdependent. Changes in one of them correspond – or ought to correspond – to changes in the others. Rules, rights, and obligations shape and are shaped by collective action and social organization around collective ownership of the irrigation infrastructure. Users attempt, consciously or unconsciously, to synchronize and harmonize these main aspects.

For example, when local users have identified the possibility of obtaining irrigation water from a given source, they begin defining the initial *norms* (especially preliminary rights and obligations of each participant) for water distribution. They also know that they need to (re)create a suitable *organization* for the physical and social work to be done, and they will make this organization functional first for the process of lobbying and negotiation, later for the construction activity. When construction of *infrastructure* begins, water users know that they not only generate their canals and structures but also their rights.

And vice versa, irrigation infrastructure is adapted during the construction process to the agreed rights: the canal network, its layout and conduction capacity, the division facilities, and so on, directly reflect the norms agreed upon for water distribution. In other words, infrastructure must enable the users to concretize their rights. Simultaneously, the channel network must coincide with the location of user-groups that distribute the water among themselves in an organized manner.

In fact, the development of irrigation infrastructure simultaneously establishes property relationships among the system creators, as Coward (1983) argued. By investing in the facilities, users create their collectively owned 'hydraulic property', which is the factor binding irrigators together and driving their collective action. This collective property forms the foundation that guarantees realization of the various operation and management activities required by a user-managed system. The mechanism also guarantees that peasant and indigenous communities, as collective bodies, will have effective control over the development and application of their own norms for managing their system (Boelens and Hoogendam 2002). 'Investments to create irrigation facilities always create or rearrange property relationships with regard to those new facilities. In other words, one cannot build facilities without establishing property.... The creation of irrigation works establishes among the creators property relations, which become the social basis for their collective action in performing various irrigation tasks' (Coward 1983:12).

So the appropriation of families' individual rights directly corresponds with the appropriation of the group's collective rights, and these water access and control rights are directly connected to the collective infrastructure and underlie collective system management. In many traditional systems, families obtain irrigation rights not only through their own contemporary investment in building the collective facilities, but also by inheriting the investment made by their ancestors. Such rights are also a loan granted by the deities, which is confirmed through rituals rooted in irrigation practice.

After generating rights, users must consolidate them. They do so by fulfilling their obligations within the irrigation system, which generally takes the form of user investment. In this phase, the organization is adapted again, now to the need to operate and maintain the infrastructure (suited to the technology-specific usage requirements) and to oversee compliance with established rights. Maintenance activities reconfirm rights, conserve the infrastructure, and strengthen the organization itself. Thus, interaction is also *continuous* in later phases of the system. At each point in time when they are rehabilitating, maintaining, or changing the irrigation infrastructure, local users tend to reason on the basis of this dynamic unity: rights-organization-infrastructure. If one changes, then the other elements must also change and 'harmonize' for the system to work properly. Seen in this light, the system is dynamic and 'living'. The precise agreements about how to relate the creation (and upkeep) of water rights with the building (and maintenance) of infrastructure and with the setup (and consolidation) of water organization depend on the historical, cultural, and political-economic context of each community.

By contrast, many of the systems (co-)constructed by intervening agencies lack this correspondence and synchronization among the normative, organizational, and infrastructural systems. This leads often to serious problems in operation and maintenance. Irrigation interventions often break this unity – often unintended – by establishing rigid, separate planning of infrastructure, organization, and operational and distributive norms, solely based on institutional timetables. Engineers and economists tend to intervene in infrastructure and planning without considering the direct impacts on water rights; social promoters work on 'water organization' without understanding its intrinsic relation to infrastructure; lawyers and legal promoters aim to train people in dealing with water rights but neglect how the water rights of individuals and collective users are profoundly related to (and 'embedded in') infrastructure and community agreements about investment.

Usually agencies do not understand the relationship between prior and new user investments and collective hydraulic property creation. 'Participatory' projects call for local contributions but without stating that these inputs are individual investments in a co-owned system (which *by definition* requires prior clarification about the relationship between contributions and benefits for each user, as well as a common understanding about the ownership of the system and the collective water rights). When project is over, the users have no solid foundations on which to organize water distribution, much less to maintain their channels as collective property (Beccar et al. 2002).

Upsetting the existing property relations lies at the heart of many intervention failures. That is why, when one visits a 'post-project' irrigation system, it is not unusual to find the users' group still bitterly arguing about conditions for access to water and the corresponding maintenance obligations. Particularly problematic are the many cases in which external agencies intervene and invest in existing irrigation facilities that were founded on local rights agreements and based on prior user investment. These new, agency-led investments often destroy existing collective and individual property rights and, thereby, the necessary collective action to sustain the system (Beccar et al. 2002).

Water techniques and water flows are thus shaped and structured by societal context, its power structures, cultures and human interactions, and in turn, the norms embedded in and the emergent properties of irrigation techniques co-structure organizational, legal, cultural, and political relations of water control.

Finally, throughout Andean history including today, ruling groups have sought to articulate local water management values, norms, and practices – like the ones described above – for their own interests. Andean communities have their own strategies and approaches to counteract this domination and achieve greater autonomy. Continuously and dynamically, they reconstruct their own representation systems, with a remodeled interaction between fundamental elements of the water control domains. They do not attempt to 'restore traditional values, norms and practices' (although historical norms do inform their strategies), but rather they interweave many contemporary elements to build strong alliances between the old and the new, between their 'own' and 'outside' factors. Thus, local water rights embody particular combinations of an array of normative sources, both official and unofficial. When inserted into local norms, the elements of each of these legal sources are 're-construed' to occupy their own places in the locally dominant mixture manifested as 'local water management'.

Informal organizational agreements and water rights 'in action' (Boelens and Zwarteveen 2005) play a major role in this kind of situations. Such agreements and practices go beyond referential rules of the state and the formalized rights and associations of the users themselves – even communities' own regulations. Therefore, to analyze water rights in practice, a merely functionalist or instrumentalist approach is not enough. Even if it does not underestimate the truly significant functions of local rights and rules, a merely instrumental approach will necessarily misinterpret local law because the analysis of cultural and metaphysical water domains is often neglected.

As with the present-day functional power of political discourses, the dominant classes in the region have always understood the great power of myths and beliefs in the Andes. Because of this power local communities actively continue to practice (and redevelop) their own myths, beliefs, and metaphysics. When localized and rooted, the diverse myths, legends, narratives, rituals, and beliefs of communities and water user collectives are the fertile soil and medium for common morals, values, and rules. They provide part of the ideological background for local distinctiveness and identification with their territories, water sources, and fellow water users. As such, they sustain collective action to defend or re-conquer these sources to resist intrusion and reinforce local rule-making autonomy. Communities and peoples who lose their myths lose their heart, particularly when they lose their myths of origin, constitution, and future evolution. In many cases, Andean communities produce an enormous variety of localization strategies 'to bring metaphysical control back home'.

Usually this self-defence does not manifest itself through violent conflicts and large-scale mobilizations. Hidden strategies, covert actions, concealed social and political spaces, and encoded practices are fundamental elements of and complements to public protests. Within their territories, local water user communities actively use, extend, or aim to create these relatively autonomous spaces in which they materially practice and extend their own water rights and discursively construct their counter-narratives. In these covert social spaces and concealed 'water rights territories', far beyond 'just' the metaphysical or symbolic contestation strategies, there is a confluence of and dynamic interaction among all water control domains to defend local rights and contest encroachment, surveillance, and repression.

Thus, water rights are embedded in historical, political, economic and cultural relationships that determine the nature, value, and function of water. They are closely related to a 'water identity' of user communities. Community embeddedness and creation of water rights, grounded in mutual dependence and intrinsic obligations for intensive cooperation among users in contexts of adverse power and agro-physical conditions, reinforce context-specific cultural and socio-territorial water bonds. The logic of the defence and reproduction of 'water community' in the harsh Andean context, far from being just an ideological construct, relates to the material creation of hydraulic property linking individual action and property rights to the collective water property owners group. Jointly with the historical struggle for water, the collective defence of community authority, and the development of the community's own rules and customs, this socio-physical creation process is at the heart of collective action in water control.

References Cited

Anderson, Benedict. 1983. Imagined communities: Reflections on the origin and spread of nationalism. London: Verso.

Barth, Fredrik. 2000. 'Boundaries and Connections'. *In Signifying Identities. Anthropological Perspectives on Boundaries and Contested Values*, ed. P. Cohen, Anthony. London and New York: Routledge.

Baud, Michiel. 2010. Identity politics and indigenous movements in andean history. In *Out of the mainstream: Water rights, politics and identity*, ed. R. Boelens, D. Getches, and A. Guevara, 99–118. London: Earthscan.

Beccar, Lily, Rutgerd Boelens, and Paul Hoogendam. 2002. Water rights and collective action in community irrigation. In *Water rights and empowerment*, ed. R. Boelens and P. Hoogendam, 1–21. Assen: Van Gorcum.

Boelens, R. 2009. The politics of disciplining water rights. *Development and Change* 40(2): 307–331.

Boelens, Rutgerd, and Margreet Zwarteveen. 2005. Prices and politics in andean water reforms. *Development and Change* 36(4): 735–758.

Boelens, Rutgerd, and Hoogendam Paul (eds.). 2002. *Water rights and empowerment*. Assen: Van Gorcum.

Boelens, Rutgerd, Chiba Moe, and Nakashima Douglas (eds.). 2006. *Water and indigenous peoples*. Paris: UNESCO.

Cohen, Abner. 1986. *Two dimensional man: An essay on the anthropology of power and symbolism in complex society*. Berkeley/Los Angeles: University of California Press.

Coward Jr., E. Walter. 1983. *Property in action: Alternatives for irrigation investment*. Document for the workshop on Water Management and Policy, University of Khon Kaen, Khon Kaen.

Fanon, Frantz. 1967[1952]. *Black skin, white masks*. New York: Grove.

Gelles, Paul. 2002. Cultural politics and local resistance in highland irrigation development. In *Water rights and empowerment*, ed. R. Boelens and P. Hoogendam, 22–35. Assen: Van Gorcum.

Gelles, Paul. 2010. Cultural identity and indigenous water rights in the andean highlands. In *Out of the mainstream. Water rights, politics and identity*, ed. R. Boelens, D. Getches, and A. Guevara, 119–144. London: Earthscan.

Hobsbawm, Eric and Terence Ranger (eds.). 1983. *The invention of tradition*. Cambridge: Cambridge University Press.

Mayer, Enrique. 2002. *The articulated peasant: Household economies in the Andes*. Boulder: Westview Press.

Said, Edward. 1993. *Culture and imperialism*. New York: Knopf.

Starn, Orin. 1994. Rethinking the politics of anthropology. The case of the Andes. *Current Anthropology* 35(1): 13–38.

Chapter 1.5
The Power of a Disappearance: Water in the Jerid Region of Tunisia

Vincent Battesti

In arid North Africa, control over water has long been and continues to be central to political and economic power for a wide variety of actors. The recent history of Tunisia, and particularly that of the oasis gardens of the Jerid region, demonstrates the social and environmental consequences of the transfer of control over water from local to colonial and then national elites. The oasis is the scene of conflicts over fresh water resources; local people are constantly adapting to new situations, learning to diversify their practices and discourses.

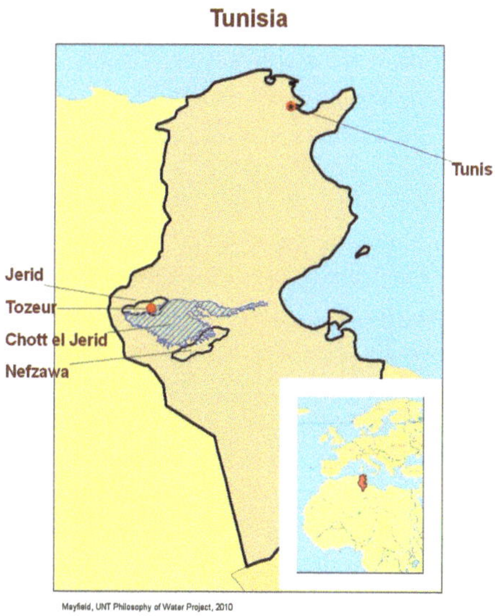

Map 1.5.1 Tunisia

1.5.1 History of Water Control in Jerid

Just as absence of water generally defines a desert, both conceptually and ecologically, the presence of water is the essential principle of oases. However, water alone does not assure their existence. An oasis is the association of a town or village and a cultivated area, often a palm grove (of *Phœnix dactylifera* L.), in an arid or

V. Battesti (✉)
Muséum national d'histoire naturelle. Lab. Ecoanthropology & Ethnobiology (UMR 7206), CNRS (French National Scientific Research Centre), Paris, France
e-mail: Battesti@vbat.org

B.R. Johnston et al. (eds.), *Water, Cultural Diversity, and Global Environmental Change: Emerging Trends, Sustainable Futures?*, DOI 10.1007/978-94-007-1774-9_6,

desert environment. Such groves are artificial and anthropogenic places that rely upon irrigated agriculture, classically intensive and characterised by mixed farming.

Not all deserts in the world have seen the creation of oasis ecosystems, because they require the presence of people with the proper technological culture, knowledge and socioeconomic tools to build elaborate hydraulic structures such as wells, channels, small dams, and so on. Labour force and suitable land are other crucial elements. Water remains the essential, life-sustaining element of an oasis.

The technological changes of the twentieth century have transformed the immediate significance and the role of water in the oasis gardens of Jerid, in southwest Tunisia: all the springs and some '*wed*' (artificial rivers collecting the water from the springs) have disappeared and deep wells from which water must be pumped have taken their place as well as irrigation systems buried underground to avoid evaporation. These shifts in technology have marked shifts in power, as the control over water moved from local farmers to the French colonial state and then to the Tunisian national government.

The joint history of Jerid and water is ancient. Whoever wants to control local society must control the water. For settlers, nomads, or conquerors, control of the oases was always a key element enabling the domination of the surrounding areas of the desert. Controlling the oasis springs means controlling the vast territories surrounding them. For instance, the French colonization of the Sahara progressed incrementally, conquering (or losing) the oases one by one. Many palm groves owe their existence to their status as 'stopovers' on trans-Saharan routes. Jerid was one of the 'desert's doors', and its strikingly intricate water distribution systems, built over several centuries, makes visible and materialize intricate negotiations around water resources.

Prior to French colonization, collective duties concerning water by the local communities were vitally important for the survival of the oasis, particularly the annual maintenance of the hydraulic systems. Landowners sent their *khammes* (tenant farmers) to clear up the springs and dredge the drainage system. The *shorfa*, the group of families of supposedly prophetic origin with religious, economic, and social prestige, were the largest landowners in the Jerid palm groves. The wealthiest families formed local councils (*jamâ`a al-mâ´* or *miyâd*), with significant local influence – beyond the strict space of agriculture – and took charge of the running of the oasis. In Tozeur, this council was known as the '*`ashra kbâr*' (the 'ten largest' landowners, each with 5–10 ha).[1] Today, the intervention of external authorities and 'modern' management methods have changed the *shorfa*'s role in managing the oasis, which decreased its religious, economic, and social status.

Although the history of the oases is a series of struggles for control over water (Bou Ali 1982), of which French colonization in the nineteenth century is just one episode, the French intervention brought unprecedented innovations in water control that took the resource out of the hands of Jerid landowners. The colonial

[1] In fact, in Tozeur, a decree of the 8 February 1913 already substitutes for the former council '*miyâd*' (the *Qa´id* with some close friends) a new 'Union Association of the Landowners of the Oasis of Tozeur' (Attya 1957).

administration was not satisfied with the existing palm groves because they did not reflect the French conception of productive agriculture, and were too deeply rooted in local 'negotiations' that produced gardens organised in ways that were not very intuitive to the new rulers. Therefore French colonials preferred to create their own palm plantations in the desert: they cultivated new land, drilled water wells, and implemented a new model of labour that relied upon a novel personnel and labour management – the salaried staff. In doing so, the colonial rulers freed themselves from the complex local politics surrounding the control of water, land and the workforce, the three essential elements of an oasis.

The more ambitious local colonial authorities, however, also attempted to understand the local hydraulic system in order to control it. The work of Penet (1912), for example, a name the people of Jerid still remember, in deciphering and analyzing the system at the beginning of the twentieth century was not the innocent pastime of a '*contrôleur civil*' (director of the local colonial civil service). One of the first measures enacted after the installation of the French 'Protectorate' in Tunisia was the nationalization of all the springs, including those that irrigated the gardens of Jerid.

Under the French colonial regime, unprecedented technical means were deployed to transform these 'patches of greenery kept by the stubborn digging of men in spite of the dunes, the wind and the sun', to improve these 'flocks driven as at the biblical times by nomads in search of sparse pastures' (Jacques Soustelle, former French minister of the Sahara, quoted by Gaudio 1960:104). Colonization programmes required new tools that visibly transformed the environment as well as the local thinking and practice. Quite consciously, the French settlers attempted to demonstrate the ability of the Western civilization to inscribe itself on this desert *tabula rasa* thanks to their technological prowess. 'To give life to deserts, thanks to the presence of trees and greenery, to break up with the sorrowful monotony of a bare ground, to populate those mournful and silent sands, this could be the oeuvre of the artesian wells... Nothing resists our powerful tools, whilst a thick layer of stone is, for those indigenous well-diggers, an insurmountable barrier' (Tissandier 1867:363).

Even though the French settlers and the natives cultivated more or less the same date palm, the French formalised methods, technical means, and workforce distinguished the new settlers from their indigenous counterparts (Battesti 2005). According to the French modernistic approach towards the environment and later taken over by the Tunisian engineers, nature is primarily understood as an object of technical exploitation. Unlike the Jeridi tradition, palm groves were reduced to solely productive areas, no longer 'recreational' ones. Farmers now had to consider the productivity of monoculture, the focused cultivation of a single crop, rather than the production of a multiple crops. The so-called 'salvation' of the local agriculture was bound with the importation of technical modern support (drilling, plastic coverings for protecting dates bunches, tractors *versus* hand labour) and groundwater mining, which soon turned into overexploitation. In this ideology nature is to be colonised, the desert to be fertilised: the subject stands external the environment, and this externality legitimises and authorises the scientific approach, presumed to be the only way of rational production. Nowadays, the Regional Centre for Agricultural Development (CRDA), located in Tozeur, transmits scientific information, application practices

Cliché de l'auteur, février 1900.

Fig. 1.5.1 Boring of an artesian well, '*Forage d'un puits artésien près de Touggourt*' in Feb. 1900, photo by J. Brunhes. "Every year, cultivation wins on the desert, thanks to this instrument of civilization *par excellence*, the artesian drill. Officers run the boring workshop, but this workshop works both for the military authorities and for individuals" (Brunhes 1902: 268)

(mostly through physical installations) and governmental policy to the farmers through extension services (*morshed*), to guide agricultural methods in this 'rational' way.

In the late nineteenth and early twentieth centuries,

> the French took…an interest in the cultivation of the date palm in the Saharan region and so we have seen mainly in the South of Constantine the birth of many *magnificent new oases, established,* so to speak, *from scratch* right in the middle of a desert region. (*Les produits algériens, Les dattes*, n.d. [circa 1922], emphasis added).

This trend toward removing control of oasis water from the local authorities continued after Tunisia became independent from France in 1956. Scientific knowledge and engineering became the foundation of state authority, and the state considered its intervention necessary for development. Civil engineers of the Tunisian state thus proceeded to press for 'modern progress' in Jerid, while the local peasant culture was seen as an obstacle obstructing the path of modernization. The state attached little to no importance to the local knowledge and aspirations of oasis people.

However, the transfer of power from the *shorfa* to Tunisian state officials did not occur without difficulties and regret from members of the largest landowners council. To give an example, a particular *sherîf* (member of the *shorfa*) could not understand why the local research centre for date palms[2] did not seek his opinion about oasis issues (interviews in 1996, in Tozeur). Strengthened by the modern transfer to the scientific field of the legitimacy of knowledge, the research centre no longer required the *shorfa*. The *sherîf* argued in these interviews that the tradition they embodied already empirically knew the data and models to which scientists lay claim today (for instance, about the number of litres of water a date palm needs). In this way, he attempted to demonstrate and justify their former authority while using the rival's argument.

The result of intensifying outside intervention is that springs no longer naturally irrigate the oasis gardens in Jerid. About fifty deep wells have replaced them, the first dug initially by the French colonial farmers and administrators, and later by the Tunisian Ministry of Agriculture to create new irrigated perimeters and to offset the decline of the hydrostatic level that was, ironically, caused by the construction of deep wells in the first place.

1.5.2 Recent Moves to Control Water

The replacement of springs by deep wells over the past 50 years has had serious repercussions for local lifestyle, the specific cultivation methods and environmental practices.

Before the deep wells were dug and the springs withered, the local community of landowners was responsible for managing common water resources, maintaining and preserving the oasis' agricultural activity. The disappearance of natural springs corresponds to the end of this collective management and the beginning of state control.

Water lost its local social significance: although it was still needed, water was no longer under local control. Today, collective work in agriculture is confined to the mutual aid between independent farmers concerning periods of intensive labour (harvest, pollination, and so on) or the employment of temporary workers. The CRDA has the responsibility for maintaining the drains and wells. Local gardeners ('gardener' and 'farmer', in this context, are interchangeable) only manage water inside their garden during their 'water turn' (the *nûba*).[3]

A brief study on the commonly used word '*jar*' (palm grove) shows that local etymologies suggest different social meanings of water management. The word '*jar*' in Jerid points out the palm grove or one of its parts. In the Nefzawa (bordering oasian region, on the other side of the Shott el-Jerid), Bédoucha (1987:399, 355,

[2] CRPh, *Centre de recherche phœnicicole*, now the CRRAO, *Centre de régional de recherche en agriculture oasienne*, in Degache, near Tozeur.

[3] By area of palm grove, water is supplied in each garden for a limited length of time of the 'water turn'. This amount of water nowadays is *pro rata* of the surface (the ratio may vary from a palm grove to another) while it used to be considered as a distinct property.

336) gave similar meanings; however, she added the following sense: 'the time it takes for water to go from one point to another'. Her definition reminds us of the intimate link between land and water, the fragile dependence of each upon the other. According to Bédoucha, the word *'jar'* could derive from the verb 'run' and, when applied specifically to water, means 'to flow'. Jerid people provide another etymology: the word could come from *'jâr'*, neighbour. This sense of the word emphasises the composition of the palm grove as a jigsaw puzzle of neighbouring gardens; to go through the gardens, from one to another, is *'min al-jâr lil-jâr'* (Battesti and Puig 1999). Note that this etymology does not specifically refer to water. Sometimes water is invisible, but is still audible within the local discourse. If one speaks with a gardener, no matter the topic, the subject of water surfaces.

1.5.3 Diversification of Farming Practices in Extracting Water

In general, workers in Jerid's palm groves trust the common water supply system: they use it as their main water provider. However, they increasingly also use private wells dug (*bîr*) inside their gardens (thus extracting water from the groundwater aquifer) as a means to control risk. These small private wells are equipped with a small power-driven water pump (probably the most successful mechanised input within the palm groves Jerid has experienced). This technology, however, does not eliminate difficult manual pumping, but creates new maintenance issues. The private pump offsets the insufficiency of the water supplied by the *nûba* with supplementary irrigations (see Fig. 1.5.2). The use of such small wells marks the difference between the 'classic' gardens and these illegal 'extension plots' surrounding the old palm groves.

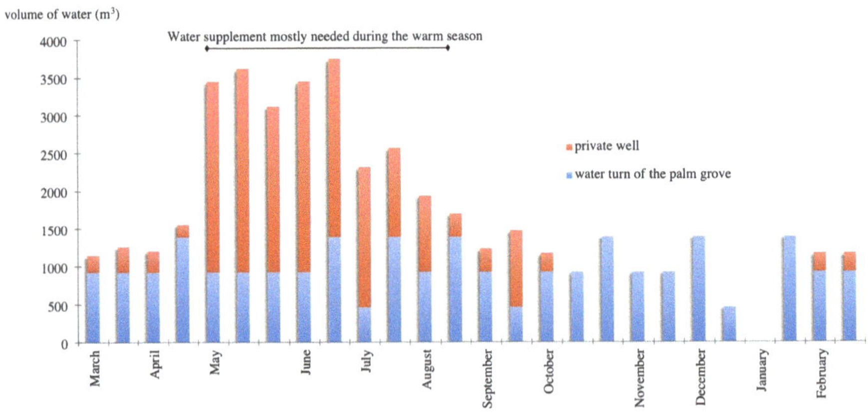

Fig. 1.5.2 Annual evolution of the water volume for irrigation in a classic garden of Nefta palm grove of 1.30 ha (years 1995–1996) (Battesti 1997)

Despite the latter's illegality,[4] they nevertheless conform to the modernistic spirit of agriculture, as determined by the state.

In extension gardens, the private well and its power-driven pump provide all the water because illegal plots cannot benefit from the *nûba*. Called in Jerid (at least in Tozeur) *maziûd* or *sênya* (plural *swânî*),[5] these kinds of gardens are the work of landowners wishing to create *ex nihilo* a garden for production purposes rather than renovating an old one. The illegal plots may also be the work of tenant farmers (*khammesa*) who thus become landowners (*mâlek*) and raise their social status and standard of living. These illegal gardens enable farmers to increase their income because these gardens were not inherited but created from desert lands: local people give to these lands an agricultural and social 'quality' that allow these lands to host cash crop (see Battesti 2005).

Even though the extension plots tend to look very similar, always equipped with a small well and its water pump, their conception and organization differ greatly from those of the 'classic' gardens inside the palm grove. The extension plots are organised according to the colonial, and now the state pattern, supposedly 'rational' and 'modern'. Here date palms are aligned following a grid of low density of plantation (100–150 plants/ha, in contrast to 300/ha in 'classic' gardens), and the ratio of the *deglet nûr* cultivar, the best-selling date, to other types is very high – sometimes 100%. Moreover, these new plots display an open vegetable structure and do not display the three vegetation levels typical of the oasis ecosystem. Although they do not reach the scale common in the Nefzawa (east shore of the Shott el-Jerid, see Brochier-Puig 2004), agricultural extensions in Jerid are quite numerous, constituting 18% of irrigated lands in Tozeur (CRDA Statistics department in 1996).

These extension gardens are evidence of a radical diversification of farmers' practices and conception of agriculture and their relation to the environment. Farmers extricate themselves from the 'classic' and complex system of water supply to depend on their private one. The use of a private well thus redefines the working time as the farmer chooses the moment for irrigation instead of being subject to the common *nûba*, which turns between neighbours on a round-the-clock schedule, all around the year. After the state seizure of the common irrigation system, this disconnection from a centuries-old system of collective agricultural management prompted the next step – individualization.

This shift means that the farmer has to deal with a different quality of water, for when water is pumped in the low-depth groundwater aquifer,[6] it is less fresh than water from the deep aquifers. The individual gardener also has to manage the choice

---

[4] By official decree: this ban aims precisely to hold back and avoid an expansion of the lands under cultivation in old palm grove, according to the view that the volume of supplied water is not extensible. Just about the same time, the state was creating new large irrigated perimeters.

[5] In Jerid old palm groves, the word used for garden is in general '*ghâba*'.

[6] It is irrigation water infiltrated through the soil. This water has a high level of salinity (considered dangerous for crops in the region at a rate higher than 6 g/l). Unfortunately, water reaches its higher rate of salinity in the periphery of the palm groves (where are extensions plots).

of different varieties and cultivars to plant (some are more resistant to the salinity than others) and different rotation of crops. This management requires in depth knowledge of ecosystem's patterns and preferences.

These changes have diversified the way that gardeners think about water and water use in the Jerid. First, there is a diversification of thoughts and practices between different actors (Battesti 2005) within the same 'class' of population, as when some Jeridi gardeners continue to cultivate their 'classic' gardens, while others start extension plots, assimilating the modern pattern. Second, individual gardeners diversify their practices; the owners of extension plots often also have classic gardens in the old palm grove. The diversification occurs both geographically and in the way that land, water, energy, and biological resources are used and conceptualised.

1.5.4 Water: A Rare Means of Free Expression

Technologically speaking, the people working on palm groves no longer have to worry about digging or draining the canals, cleaning up springs, or distributing water between gardens. The state, through its public departments, deals with these issues nowadays.

However, water remained one of the rare subjects on which people could express themselves freely. In Tunisia before the January 14th Revolution (2011), discussion of power and politics was forbidden; the presidency and government did not allow for divergences from the official policy. Discussions about water have become a venue for expressing one's opinions about politics and relations between local communities and the national power in North Tunisia. Hence, when speaking with a local gardener, the topic of water invariably springs forth in the conversation, many times in the form of implicit social or political protest.

This tight relationship between water and politics is especially evident in the newest palm groves built by the national government within the last 30 years. These

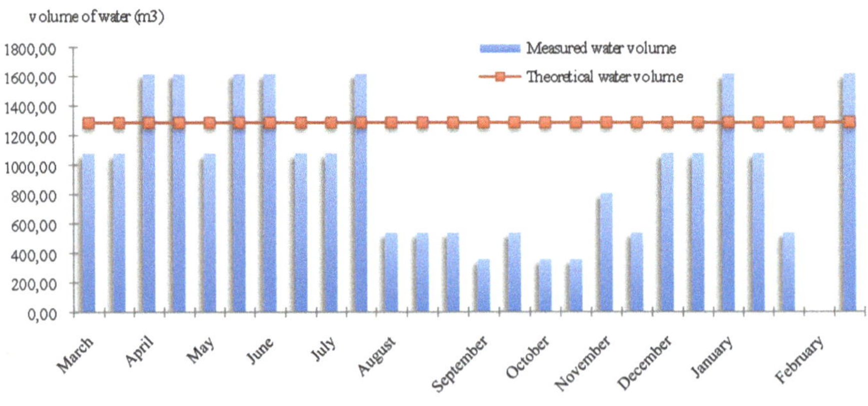

Fig. 1.5.3 Annual evolution of the water volume for irrigation in a plot of 2 ha in Ibn Chabbat palm plantation (years 1995–1996) (Battesti 1997)

newly irrigated areas were conceived and constructed by non-local engineers and technicians, with a non-local specialised technical skill set, without involving local farmers. The farmers are under an obligation to maintain the structure of *deglet nûr* date palms in line (one every 10 m) with a prescribed monoculture. Where farmers in the old palm groves do not express dissatisfaction for poor results, in the new palm plantations returns on investments are inevitably expected. Water here is seen as an investment, while in the old palm grove, water is seen as a means of reproduction. Local farmers appeared to have participated in the creation of the new areas only through paying their water bills – or, as in the case of the Ibn Chabbat palm grove, their refusal to pay the bills.

In the mid-1990s, the first settled farmers on these newly irrigated lands began to refuse to pay their water bill because the expected yields and financial income from the state modernization programme took longer than expected, causing politically charged language to form around the issue of water (see Fig 1.5.3: an evident lack of water regarding the volume promised by the administration, and still not itself sufficient.)

When the government cut off the water supply, the farmers' response (February 1996) was virulent: 'they treat us as if we're not Tunisian: they don't help us, they cut off water' and 'Even Israel doesn't do that [to Palestinian people]'. Water became a bone of great contention between the state and the farmers. The very same farmers who refused to pay their bills stated angrily: 'we're tired of this project. There are two solutions: either to burn up Ibn Chabbat, or for the State to work with the farmers.' Water issues opened the door to a more profound political sentiment: people were so desperate that they commended the 'time of the French whose 40-year-old wells still work, unlike here' (August 1995).

The state reacted in 1996 by creating an AIC (*Association d'intérêt collectif*), a Common Interest Association officially aiming at managing water resources. AICs were already present in the other palm plantations (several coexist in larger palm groves such as El-Hamma, Tozeur, and Nefta). Now, AICs deal with the STEG (the Tunisian electricity and gas company), which supplies water-pumping plants with electricity. Since the natural springs have dried up, irrigation requires wells and pumps that must have the electricity supplied by STEG. STEG signs a contract with the AIC, and if the AIC does not pay, STEG may legitimately cut off the electricity (and so the water). AIC collects money from the gardens owners to pay STEG. Thus disengaged, the agriculture services are confined to the maintenance of drilling, and STEG is no more than a service provider. The coercion formerly exerted by the CRDA has now to be performed among farmers, through a "water union" – the AIC.[7] There is no real gain of control for the local farmers through this process: they lost non-payment as a leverage strategy and still have not resolved the problem of electricity pricing.

[7] These AIC became GIC (*Groupements d'intérêt collectif*, common interest organizations), and then in 2003/2004, the GDA (*Groupement de développement agricole*, agricultural development organization). The policy remained the same, heading towards liberalization of the sector under the state's firm rule. More and more, the GDA have also had to take care of the maintenance of the drilling itself as well.

As this history shows, water is not a neutral element. Local people have lost control over it, but as a subject of conversation, political discourses in the guise of water discourses had been a kind of safety valve. Since the criticism was not frontal it provided citizens a voice within a broader crisis while protecting them from violent repercussion. Discussions on water could pass as purely technical, locally and nationally. While it was impossible to have a real debate on issues such as democratic decision-making processes, there could be debates over water.

1.5.5 Long-Term Consequences

The most significant consequences of the colonial and later Tunisian policies are undoubtedly yet to come. Drillings from the Complex Terminal, a relatively deep aquifer system – about 500 m – has dried up the natural springs. This depletion meant also that drilling became increasingly less effective while demanding greater amounts of energy to draw the same quantity of water that 50 years ago naturally spouted out. It became necessary to call upon the even deeper fossil aquifers reserves of the *Continental Intercalaire.* Consequently, springs that flowed naturally from that aquifer basin in the districts of Timimoun and Touat in Algeria are dry today (Richter 1995). At first, the drilling provided an increase of profitable surface area for new agricultural activity, but now the upkeep of these plots, combined with the maintenance of the old ones, is increasingly difficult. Deeper modern drillings have sucked up the water that sustained the natural springs of the old oases. It is important to recognise that there was not true profit, but just an expensive dislocation of the same water: first from springs, then to artesian wells, then to wells requiring large pumps (see Fig. 1.5.4). Since 2002, more than 90% of the volume of water has had to be

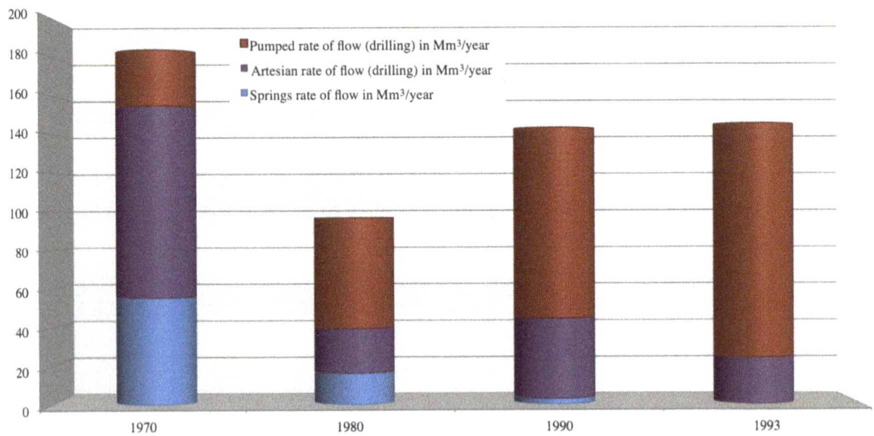

Fig. 1.5.4 Evolution of the exploitation of the deep aquifers for Tozeur Governorate (Data from Mamou 1995)

pumped in the Jerid region, and less than 10% is artesian. A tube well in Jerid is called a *bîr*, the same word used to name a private garden well. The only difference is that the tube well provided 'state water' you have to pay for, says the farmers.

People of Jerid refer to the time when water was 'free' as their golden age. However, water was never really free of charge. According to local discourse, the state simply took the water to redistribute it in return for payment. Landowners in the past 'paid' for their access to water in kind, by their participation in the collective maintenance (usually the time of work of their *khammes* invested in the upkeep of the irrigation and drain networks, and the cleaning-out of water sources). The allocation of water resources was very uneven, quite separate from property rights, which came with owning land. Those who did not need the entirety of their water could theoretically sell it to others. Today, the sale of water is prohibited since the distribution of water now depends on the surface rights.

One could say that the Tunisian Ministry of Agriculture's management of water resources is equivalent to a mining exploitation. The water exploitation – or, better, overexploitation – was made possible by adoption of methods used in mineral mining and later in oil extraction – with use of beating drilling and then of rotation drilling (Penet 1912) – and, furthermore, as in mining, the water resources exploited are not particularly renewable. Nobody knows if the volume of water reserve will last 15 or 25 years, or what its quality will be. The aquifers of the '*Complex Terminal*' and the fossil water of the '*Continental Intercalaire*' are both severely under-recharged (Mamou 1995). Nearly 70 years ago, the situation was already discouraging: 'In spite of the miracle of the multiplied water by the European technique, Algerian oases remain centres of weak life and bad production' (Capot-Rey 1944). One can seriously wonder why those oasis zones should be developed at all. It seems to make more sense to focus agriculture in the area of the '*Maghreb utile*' ('useful Maghreb'), the narrow coastal strip of North Africa, excluding mountains and deserts.[8] However, the reasons for development are political, including an effort to delineate Tunisian territory on its borders with Algeria. Oases, cultivations and human settlements mark the national territory to the border: desert territories are too open to have any significance for the neighbouring state. Furthermore, the government hopes that drilling will help to maintain, if not increase, national agricultural production.

Preserving the oases also aids two other projects to which the government is committed: keeping a balance between urban and rural populations by preventing or managing migration out of the countryside and promoting tourism (Battesti 2009b). Since the tourism opportunities on the coast are already quite saturated, the state is looking at developing Saharan tourism. The state's interest in preserving the oases meets the tourism project, and indicates that the state is open to the new sensibilities crossing the planet, namely, the worldwide environmentalist movement in favour of sustainable development. While some state engineers generate plans to restore and

[8] They take part, on the contrary, in the shadow zones (*zones d'ombre*) of Tunisia, in accordance with the presidential terminology.

Box 1.5a Climate change, water, and development in daily life in Tuvalu

—Heather Lazrus

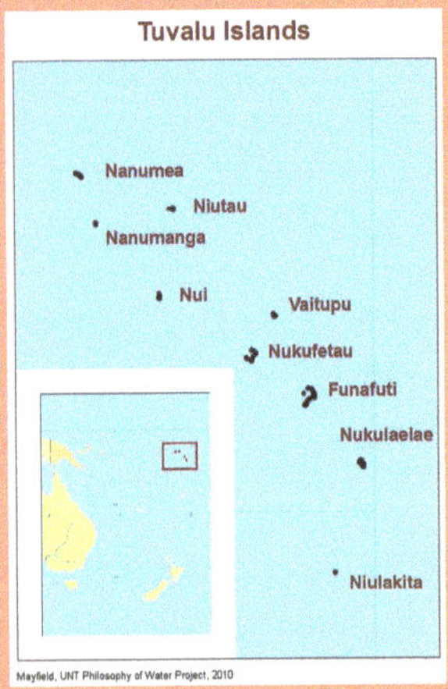

Box Map 1.5a.1 Tuvalu

The nine islands that comprise the Tuvaluan archipelago arc across a stretch of the central Pacific Ocean between 5° and 10° south and 176° and 179° east. Running from north to south the islands are Nanumea, Niutao, Nanumaga, Nui, Vaitupu, Nukufetau, Funafuti, Nukulaelae, and Niulakita. Together, the islands cover a total land area of less than 26 km^2 and support a population of approximately 11,000 people. Settlement may have occurred as long ago as 2,000 years before present when sea levels fell during a small ice age to expose the atolls (Nunn 2007:117–125). Some archaeological evidence indicates that settlement of the islands occurred at least about 1,000 years ago. Tuvaluans have thus been adapting to atoll life over several generations.

Box Fig. 1.5a.1 Causeway: A narrow-causeway has been constructed to connect two parts of Funafuti Atoll. On one side the narrow strip is buffeted by the energy of the Pacific Ocean, protecting the calm lagoon on the other side. Sea level rise will threaten the causeway used to access Funafuti's municipal rubbish dump (Credit: Heather Lazrus)

The low-lying atolls average just three metres above sea level and have only limited supplies of naturally occurring fresh water. Rising sea levels, changing precipitation patterns, and extreme events including storms and droughts are among the challenges that Tuvalu's water supply faces as anthropogenic influences transform the nature of our global climate. The Intergovernmental Panel on Climate Change (IPCC) has warned of the unequivocal threats to islands and island-dwelling communities based on their shared characteristics of 'physical size, proneness to natural disasters and climate extremes, extreme openness of their economies, and low adaptive capacity' (Mimura et al. 2007:687–716). Tuvalu epitomises these characteristics, a fact that has led to its being described as one of the most vulnerable places on the planet to climate change.

(continued)

Box 1.5a (continued)

Supplies of fresh water are one of the most essential and uneven resources for atoll life. Rain water is the primary source of fresh water, but highly variable levels of precipitation makes management of freshwater supplies difficult and renders atoll inhabitants deeply susceptible to drought. The islands receive varying amounts of rainfall according to latitude. Although the population numbers on the outer islands are relatively uniform, the southern islands may receive around 3,500 mm per year and this amount decreases to 2,700 mm per year in the northern islands (Sharp and Henson 1997). Funafuti, the national capital, receives slightly more rain on average (3,800 mm) but also houses roughly a third of the national population. On all islands, water availability is limited by the technologies available to harvest rain water.

Because much of the actual precipitation is lost through high evaporation and high soil porosity and permeability, the islands of the Tuvaluan archipelago are characterised scientifically as 'dry islands' (Mclean et al. 1991). Rainfall percolating through the porous material of an atoll replenishes the freshwater lens that floats on top of the denser salt water permeating the atolls at the level of the surrounding ocean. Rising sea levels and increased storm activity can damage the fragile freshwater lens by mixing it with salt water. Coastal mining for construction material also destroys the natural coastal buffer and may make the freshwater lens even more susceptible to storms and sea level rise.

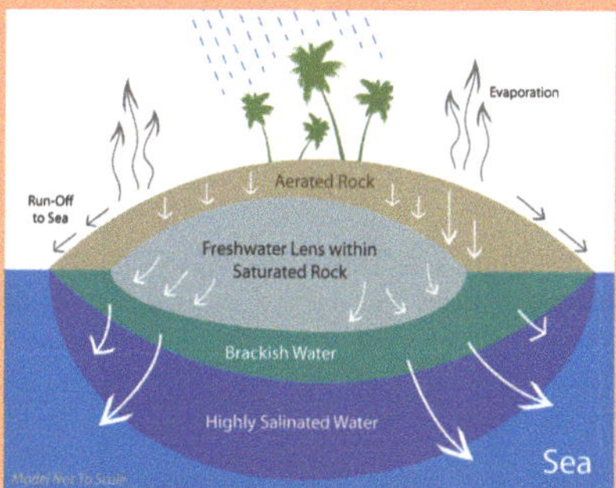

Box Fig. 1.5a.2 Graphic of freshwater lens by Matt Story, Philosophy of Water Project, University of North Texas (Credit: Matt Story, Philosophy of Water Project, University of North Texas)

With the construction of the first communal cement rainwater catchment cisterns in the early decades of the 1900s, Tuvaluans no longer had to rely exclusively on the freshwater lens (Chambers 1984). Prior to the introduction of catchment tanks people used local knowledge to dig wells carefully where the freshwater lens was at its thickest, and they sometimes had to travel across the atoll's lagoon to bring water home in canoes for the family. They also collected water in intricate palm frond containers. Today, almost every public building and family household has a catchment tank that collects the runoff from corrugated iron roofs. Most tanks were constructed by Save the Children, a humanitarian NGO, in the late 1980s and the United Nations Development Programme, in the early 1990s. These

(continued)

Box 1.5a (continued)

Box Fig. 1.5a.3 A young boy bathes in a tub left on Nanumea Atoll by American soldiers after World War II. In the background are the household water tank and a faucet with a buoy that, with its top cut out, is now used for hauling water (Photo credit: Heather Lazrus)

Box Fig. 1.5a.4 Broken tank: Rainwater catchment tanks are prone to cracking and breaking when not maintained and kept full of water. A broken tank on Funafuti Atoll collects organic material and makes a fertile home for new palm growth (Photo credit: Heather Lazrus)

ferro-cement tanks are nearing the end of their expected 25- to 30-year life span. The government of Tuvalu also sells green plastic tanks, but these are very expensive. Desalination technology exists on the islands, but is often out of commission due to lack of spare parts and highly technical expertise needed for its maintenance.

Water, like other resources, plays a cultural role and is socially and economically mediated. The introduced technology changes how people use and perceive water. In the words of one Nanumean elder: 'we used to have one coconut shell of water in the morning, and one in the evening. We were satisfied. The kids used to be in charge of bringing in the water, and we would always be careful not to play around the bucket because if it accidentally knocked over and spilled… But now people spend all day sitting around drinking tea. They are always thirsty'.

Taste preferences have shifted away from the brackish well water used before cement tanks, and needs have increased with more clothes and dishes to wash. Indoor plumbing and flush toilets are replacing beaches and long-drops, but also increase the need for water. The improved toilets often have

(continued)

Box 1.5a (continued)

water-sealed sewage, which is less effective during droughts. Freshwater supplies are prone to pollution from ineffective wastewater management, hazardous chemical by-products called Persistent Organic Pollutants, and contamination of tanks and cisterns by dead insects and animals.

Climate change will have a significant effect on fresh water on Nanumea Atoll in different and compounding ways. In times of drought, the water from household and community catchment tanks is rationed and people also begin to make more use of the naturally occurring freshwater lens. However, when there are prolonged periods of no rain, the tanks and cisterns run dry and the freshwater lens shrinks. Changing precipitation patterns are likely to produce more intense but less frequent rain (Mimura et al. 2007:687–716). The storage capacity on the islands is not great enough to cope with this increasing variability. Unforeseen events, including funerals that bring large families together and stress water supplies, multiply the ecological impacts.

Rising sea levels and changing precipitation patterns will threaten freshwater supplies, and storm surges and prolonged drought periods will further erode the resilience of the freshwater lens. Simultaneously, higher household incomes,

Box Fig. 1.5a.5 Nanumea well: A farmer collects water from a well on his land for his pigs on Nanumea Atoll. Few wells that access the fragile freshwater lens are still maintained on Nanumea Atoll (Photo credit: Heather Lazrus)

Box Fig. 1.5a.6 Nanumea cistern: A woman collects water from a community cistern on Nanumea Atoll when her household tank has run dry after higher than usual water consumption during a family funeral (Photo credit: Heather Lazrus)

(continued)

Box 1.5a (continued)

largely from remittances earned by family members working overseas, and the growing desire for Western-style homes and lifestyles are manifesting in more intensive uses of local resource. Although Tuvalu's population remains relatively steady, uses for water are increasing and put pressure on the limited supply of harvested rain water while cement buildings are constructed of coral and sand mined from the coastal areas that protect the freshwater lens.

To a large extent, the availability and use of technology to provide water are more important limiting factors than environmental changes. In a complex interplay, compounding factors increase the urgency of enhancing Tuvalu's water supply: the catchment systems are nearing the end of their anticipated life expectancy; the socially driven needs for water are increasing, and the local effects of climate change are intensifying. The residents of Tuvalu are highly dependent on development aid for freshwater storage, yet international investment is tricky to procure for a small country facing ecological devastation from the effects of global climate change. The ability to adapt to climate change will likely revolve around the availability of adequate fresh water to support the population of Tuvaluans inhabiting these nine islands.

References

Chambers, A. 1984. Atoll economy: Social change in Kiribati and Tuvalu, No. 6. In *Nanumea*, 2nd ed. Canberra: Australian National University, Development Studies Centre.

Mclean, R.F., P.F. Holthus, P.L. Hosking, and C.D. Woodroffe. 1991. *Tuvalu land resources survey. Nanumea*. Auckland: Department of Geography, University of Auckland.

Mimura, N., L. Nurse, R. McLean, J. Agard, L. Briguglio, P. Lefale, R. Payet, and G. Sem. 2007. Small islands. In *Climate change 2007: Impacts, adaptation and vulnerability. Contribution of working group II to the Fourth Assessment Report of the Intergovernmental Panel on Climate Change*, ed. M.L. Parry, O.F. Canziani, P.J. Palutikof, P.J. van der Linden, and C.E. Hanson, 687–716. Cambridge: Cambridge University Press.

Nunn, P. 2007. Holocene sea-level change and human response in Pacific Islands. *Transactions of the Royal Society of Edinburgh: Earth and Environmental Sciences* 98: 117–125.

Sharp, R., and B. Henson (eds.). 1997. *National environmental management strategy*. Apia: South Pacific Regional Environment Programme.

modernise the oases to increase domestic production and agricultural exports, others argue that the oases should be preserved as 'tourist Edens'. In the best-case scenario, as Caratini (1994:124) reports of the Touat in Algeria: 'Order was given to developers to develop without destroying'. A mission report of that project[9] offers a remarkable summary:

> Taking into account the historical and cultural importance of these *foggara* [which water the palm plantations], we will ensure the protection of around one thousand hectares of traditional gardens in connection with the genetic conservation of the varieties of date palm and other original species, within the framework of tourist and cultural activities.

This programme claims to support the scientific cause (which will also take advantage of the creation of 10,000 ha of new palm plantations and 25,000 ha devoted to vast cereal fields) and at the same time the environmentalist cause (by safeguarding the oases and promoting tourism). In the near future, Jerid may appoint 'traditional' gardeners to maintain the 'natural/tradition capital' of the old oases. This capital is also at the root of the tourist industry – a movement which could be called 'patrimonialization' (Battesti 2009a).

The environmentalist or protectionist concern emerges as a third way of practicing and conceptualizing water and its use. Young people who refuse to work or even to set foot in their father's garden will defend the palm grove, or rather will advocate for the safeguarding of its aesthetics. For example, when the agricultural administration tried to line the main *wêd* (wadi) bed of Tozeur with concrete to reduce losses by infiltration (February–March 1996), the *wild sûq*[10] criticised the project, saying: 'the landscape will be spoiled', and 'they should do this only in the hidden places of the oasis', that is, places where tourists do not go. Criticisms of the ugliness of recent construction merge with concerns about the loss of the 'authentic character'. The *wild sûq* defended the traditional 'landscape' object since it draws tourists, but they also expressed genuine feeling for safeguarding heritage. This was made possible thanks to their perception of the oasis from an outside perspective, that is, imagining first the foreigner's representation of water and environment rather than the farmer's.

In response to the growing interest in tourism and the preservation of heritage, municipalities of Jerid oases might begin to treat agricultural land as 'patrimony'. To give an example of this process: drilling has begun (with cooperation from Monaco, 2006–2009) in the area of Nefta's palm grove where the springs used to be, before they dried up during the 1980s, with the intent of channelling water into the former wadi beds. In the 1970s, a German project had buried all these traditional irrigation systems to avoid evaporation. Despite the shortage of water for agriculture, the goal of these new drillings is not primarily to water the gardens. It is rather to create a 'visual and sound re-enchantment' of this small part of the palm grove to attract tourists to the '*corbeille de Nefta*' (former springs location), overhung by the famous Sahara Palace, which is now a hotel zone. The authorities wanted to restore the picturesque attributes of an oasis palm grove with a wadi like the one that still exists at Tozeur. Nefta people laugh at the new concrete and oval water tank built on

[9] About Algerian French project of agricultural planning for the *wilaya* of Adrar, region of Touat. Mission report of D. Dubost quoted by Caratini (1994:124).

[10] 'Sons of the centre town' is a term to designate idle young persons who spend their days there, looking for business opportunities with tourists rather than working in the agricultural sector.

Fig. 1.5.5 (**a**) at the *top*, *Nefta, l'oasis, la Corbeille*, 1920 ca, LL. (Lehnert & Landrock) Lévy & Neurdein réunis, Paris, édité spécialement pour les hôtels "Transatlantique" (coll. V. Battesti) and *below*; (**b**) the new *Corbeille de Nefta*, November 2008 (V. Battesti)

the *corbeille*, calling it the *besin lahthem* (the egg pool). The project, now completed, includes a waterfall, a promenade, and '*aménagement de bon goût*' (a tasteful layout),[11] but does not primarily serve agricultural production.

However odd the Nefta project may seem, there is actually an even more 'eccentric' project: a golf course. Experts working closely with the Minister of Tourism offered a 50 ha golf course in the desert (150 ha planned), opened in November 2006 on the edge of the palm grove, also near the former springs area (now dry). Along with thalassotherapy (seawater therapy) and cultural tourism, golf was part of the Tunisian government's new strategy to diversify the tourism product. On the Tozeur Golf website,[12] the company boasts that the greens are 'irrigated with recycled water to preserve the water table' (A New Golf 2009). This was less than true at first, but even if completely realised, it comes to the mind of people that this recycled water could still be used for agriculture. Although tourist guidebooks still say that the Jerid has 'over 200 springs' (Lonely Planet Guide, Hole 2007:256), no more springs naturally irrigate the oasis gardens here. In this region, where farmers complain about lacking water, projects such as the *corbeille* and the golf course are good examples of the competition between agriculture and tourism for water resources.

In the desert Southwest of Tunisia, rainfall is not sufficient to maintain cultivation, not even to grow date palms: local people and officials have to manage underground water and irrigation. Jerid is a perfect example of a hydraulic society. Water shapes the palm groves and the design of the gardens. The complex geometry of the old palm grove garden with the numerous and overlapping beds of cultivation requires a technical *savoir-faire*, a distinctive feature of Oasians.

During the last century, means and scales of intervention on the water resources have changed considerably, from a local control to national control. A complex mixture of shifts in water availability, changes in social practices, and conceptions of 'nature', and ecological, political, economical, or geographical situations and pressures are all intricately interwoven to create an ongoing dynamic landscape in the desert.

Resources

A New Golf Course in the Heart of Sahara. Tozeur Oasis Golf. Société Golf Oasis. Website http://www.tozeuroasisgolf.com/en/. Accessed July 2009.

Attya, H. 1957. L'organisation de l'oasis. *Les cahiers de Tunisie* 17–18: 39–43.

Battesti, V. 1997. *Les oasis du Jérid: des révolutions permanentes?* Montpellier: CIRAD-SAR GRIDAO; CRPh, INRAT.

Battesti, V. 2005. *Jardins au désert, Évolution des pratiques et savoirs oasiens. Jérid tunisien.* Paris: Éditions IRD.

[11] Quoting a tourist. See (with pictures): http://www.nachoua.com/Nefta-avr-2009/Corbeille.htm (accessed July 2009).

[12] See http://www.tozeuroasisgolf.com/en/ (accessed July 2009).

Battesti, V. 2009a. De Siwa au Caire, la fabrique du patrimoine se nourrit du désir des autres. *Égypte/Monde arabe, 'Pratiques du patrimoine en Égypte et au Soudan'* 5–6: 69–101. Le Caire: CEDEJ.

Battesti, V. 2009b. Tourisme d'oasis, Les mirages naturels et culturels d'une rencontre? *Cahiers d'études africaines*, 'Tourismes: La quête de soi par la pratique des autres' XLIX(1–2): 551–582. Paris: Éditions EHESS.

Battesti, V., and N. Puig. 1999. Le sens des lieux, Espaces et pratiques dans les palmeraies du Jérid (Sud-Ouest tunisien). *JATBA, Revue d'ethnobiologie* XLI(2): 19–44.

Bédoucha, G. 1987. *L'eau, l'amie du puissant, el-Mâ' sâhib al-sultân, Une communauté oasienne du Sud-tunisien.* Paris: Édition des Archives contemporaines.

Bou Ali, S. 1982. Le Djérid (Tunisie): Un essai d'analyse régionale. Thesis Géographie, Université de Toulouse.

Brochier-Puig, J. 2004. *Les usages de l'eau dans les oasis en voie d'urbanisation. Le cas du Nefzaoua (Sud tunisien).* Montpellier: Publication Montpellier III, CNRS, coll. Territoire en mutation, vol. 11.

Brunhes, J. 1902. *L'irrigation, ses conditions géographiques, ses modes et son organisation dans la Péninsule ibérique et dans l'Afrique du Nord, Thèse présentée à la Faculté des lettres de l'Université de Paris.* Paris: C. Naud, Études de géographie humaine.

Capot-Rey, R. 1944. *Problèmes des oasis algériennes.* Alger: Publications du CNRS.

Caratini, S. 1994. Dialogues sahariens. In *Le métis culturel, Internationale de l'imaginaire. vol. 1, nouvelle série*, 107–126. Paris: Éd. Maison des cultures du monde, Babel.

Gaudio, A. 1960. *Le Sahara des Africains.* Paris: Editions René Julliard, Collection d'Histoire et Voyages.

Gouvernement général de l'Algérie, Direction des Territoires du Sud. N.d. (circa 1922). *Les produits algériens, Les dattes.* Alger: Imprimerie algérienne.

Hole, A., M. Grosberg, and D. Robinson (eds.). 2007. *Tunisia.* Hawthorn: Lonely Planet Publications Ltd.

Mamou, A. 1995. Développement des zones sahariennes en Tunisie et son incidence sur les ressources en eau. *Les oasis au Maghreb, Mise en valeur et développement*, Actes du séminaire Gabès 12 (Nov 1994): 71–86. Tunis: Cahier du C.E.R.E.S.

Penet, P. 1912. *Tableau d'eau de Tozeur. Au septième siècle de l'Hégire (XIIIe de l'ère chrétienne) Ibn Chabbat, Imam de Tozeur et historien, a fondé le règlement de répartition des eaux de l'oasis de Tozeur. Ce règlement s'est perpétué par tradition orale. De 1911 à 1912 (1329 à 1330 de l'Hégire), sur l'initiative et par les soins de P. Penet, Contrôleur civil, assisté de Mahmoud Snadly, interprète de contrôle civil, la tradition orale a été relevée et fixée par écrit.* Tunis: Régence de Tunis, Protectorat français, Direction générale de l'intérieur.

Richter, M. 1995. Les oasis du Maghreb: typologie et problèmes agro-écologiques. *Les oasis au Maghreb, Mise en valeur et développement*, Actes du séminaire Gabès 12 (Nov 1994): 29–56. Tunis: Cahier du C.E.R.E.S.

Tissandier, G. 1867. *L'eau.* Paris: Louis Hachette et Cie.

Chapter 1.6
Diverting Water: Cultural Plurality and Public Water Features in an Urban Environment

Veronica Strang

1.6.1 Introduction

Map 1.6.1 Brisbane

There are many diverse cultural engagements with water: different spiritual and secular beliefs and understandings; multiple views about how resources should be owned, used and managed; varied interactions with water in all its forms; and a wide spectrum of ideas about how to achieve sustainability and engage with environmental change.

Flowing underneath these highly diverse cultural engagements is a layer of cross-cultural commonality: powerful themes that have persisted for centuries, crossing temporal, spatial, and cultural boundaries. These larger cross-cultural themes cast water as the elemental basis of productivity and wealth generation. There are complex ties between these meanings and ideas about water as 'nature' or as an

V. Strang (✉)
Department of Anthropology, University of Auckland, Auckland, New Zealand
e-mail: v.strang@auckland.ac.nz

B.R. Johnston et al. (eds.), *Water, Cultural Diversity, and Global Environmental Change: Emerging Trends, Sustainable Futures?*, DOI 10.1007/978-94-007-1774-9_7,

'essence' of being. As the substance of which humans and all organic things are composed, water serves as an image of spiritual essence, social identity, and belonging. It defines particular communities and creates a literally and conceptually fluid connection between them and the places that they inhabit (see Strang 2005; Strang and Garner 2006).

Every cultural group engages with these broad cross-cultural themes in a relationship with water that is framed by a particular time and place. Each specific cultural relationship is expressed in material terms: in artefacts, in methods of water use and management, and in representations of water (Schama 1996). An examination of material culture is useful in elucidating both commonalities and diversities in people's relationships with water.

Water features are a widespread form of material culture; they can be found in all urban environments. There are deep historical roots for this phenomenon in early settlements; the village well was a practical necessity and a focus of social interaction. Despite the diminishing utilitarian need for centrally accessible water supplies, cities have maintained water features in central public spaces, and they have generally enlarged them and made them into more decorative objects. The ubiquity of public water features suggests some substantial reasons for their enduring popularity. Water features in public spaces lend themselves to statements of community identity and belonging, effectively representing the vitality, generative potency, and power held by particular social groups (Stewart and Strathern 2003; Tilley 1994; Tilley and Bennett 2004). Water features also have a range of practical and aesthetic purposes; they improve the quality of life, providing cooler, greener, and more aesthetically and physically enjoyable spaces (Anderson and Tabb 2002).

Every public water feature has the potential to assert the physical presence and authority of the group that it represents. A statement of social and political potency is not something to be taken lightly. The majority of capital cities have centrally placed fountains that express a powerful vision of national identity and cultural vitality. Examples include the massive Jet d'Eau in the lake at Geneva, the Apollo fountain at Versailles, the Trevi fountain in Rome, and the Samson fountain at Peterhof in Saint Petersburg.

The design of smaller scale water features in cities suggest that they perform a broadly similar role, making local statements of collective identity, community, and belonging. They function to 'sustain' communities. By examining urban water features, this chapter points to a critical relationship between cultural diversity and biodiversity and underlines the interdependence of social and ecological sustainability.

As with any kind of material culture, form is critical. Urban water features can suggest that their inhabitants share large-scale national or cosmopolitan identities, or they can define more particular cultural identities. Water is subject to powerful and emotive ideas about self and other, purity and pollution (Strang 2004). Ideas of race, ethnic, and cultural identity are expressed through fluid metaphors. People can be seen as sharing a common substance with each other and particular places, or as alien and thus potentially 'polluting'. Water features not only celebrate collectiveness and sociality but can also express separateness and exclusion, or imply ideas of dilution, absorption, and assimilation of 'the other'.

Fig. 1.6.1 Brisbane City (Photo by Veronica Strang)

Recent ethnographic research in Queensland, Australia,[1] demonstrates how water features in and around Brisbane illustrate these realities. It adopts a classic anthropological approach to material culture, considering the form of the various features as expressions of particular ideas about cultural identity and belonging (Hodder 1991; Miller 1998). It considers how this public material culture supports practices that are similarly directed towards establishing or maintaining community identities and celebrating social cohesion.

1.6.2 Featuring Water in Brisbane

Brisbane's public water features can be loosely classified into four main types: those focused on the celebration of natural forms; those that emphasise European historical traditions; those that are culturally distinctive in referring to other societies in their form and intent; and those that attempt to suggest modern, cosmopolitan visions of social being. There is considerable overlap, but each type channels these cultural meanings in particular directions, with varying degrees of cultural specificity.

[1] This work was part of a long-term anthropological research interest in water in Australia, under the aegis of a 5-year project on two river catchments in Queensland, funded by the Australian Research Council (2003–2008).

There are numerous water features in the first category, employing organic forms to bring an idea of nature into cityscapes. For example, the University of Queensland has several water features that attempt to suggest plants, springs, ponds, and waterfalls.

Fig. 1.6.2 'Waterfall' at the University of Queensland (Photo by Veronica Strang)

There are numerous 'waterfalls' in the city, for example on the river frontage and on the South Bank.

Some water features, such as the pebbled pond at the Art Gallery, try to evoke natural pools and streams. This organic ideal is exemplified by the large 'lagoon' and 'beach' on the South Bank, which make an explicit effort to bring elements of nature into the cityscape.

Fig. 1.6.3 'Waterfall' into the river at Brisbane's ferry dock (Photo by Veronica Strang)

Fig. 1.6.4 'Waterfall' on the South Bank (Photo by Veronica Strang)

Fig. 1.6.5 Pebbled 'pond' in Brisbane art gallery (Photo by Veronica Strang)

Fig. 1.6.6 South Bank lagoon (Photo by Veronica Strang)

Fig. 1.6.7 South Bank beach (Photo by Veronica Strang)

In the 'colonial history' category, water features draw on European art and architecture from earlier periods of history. One such example is located in the centre of the downtown business district.

This feature is readily recognised as speaking to a long tradition of landscape art in Europe, albeit with a more contemporary design. It retains a recurring European mythological theme defining water as nature/female and women as 'water carriers'. It is classical in its shape and style, echoing the Italianate Renaissance villa fountains that greatly influenced the design of water features throughout Europe. It therefore affirms the traditions and identities carried to Australia in the colonial period.[2]

The imagery carries with it a complex of religious and secular ideas about watermelding Greco-Roman and Judao-Christian cosmologies. Though reflecting a

[2] There are numerous material echoes of this in other colonised spaces: across the Tasman Sea, for example, there is a similar classical water carrier on the fountain in Albert Park in Auckland, which (having been cast by the same company) is a virtual replica of the one that can be seen at the Raffles Hotel in Singapore.

Fig. 1.6.8 Fountain in Brisbane's business district (Photo by Veronica Strang)

fondness for classical Greek and Roman mythology, colonial style water features also refer to a longstanding Biblical vision of water as the substance of the Holy Spirit. Both the Old and New Testaments envision the Holy Spirit as the rain sent by God to enliven all things on earth and as a sacred flow of knowledge and wisdom 'from the temple'. The Holy Spirit is thus a religious framing of water as a powerful generative force, a source of wealth and health, and as a substance carrying knowledge and social identity (Schmemann 1976). Contemporary water features of this kind therefore draw on a deep well of European ideas, images, and social identities, serving to uphold these in 'the New World'.

Although European forms of material culture predominate in Australia, there are other distinctive cultural statements of identity. Like most major cities in Oceania, Brisbane has a large Asian population. This community is a longstanding part of Australia's history: Chinese explorers charted the shores of the continent well before Europeans 'discovered' it, and Chinese workers were an important part of the earliest settlements in Queensland, participating in the gold rushes of the 1800s. Brisbane therefore contains a small Chinatown, where a central mall was created in 1987 and refurbished in the 1990s. Ornate Chinese gates bound the area and pairs of stone lions, gifts from the People's Republic of China, spiritually guard the entrances.

Fig. 1.6.9 Fountain in Albert park (Photo by Veronica Strang)

A large water feature was placed at the heart of the neighbourhood, in the Chinatown Mall.[3] Like the European colonial fountains, this one replicated a specific cultural tradition in its form and intent, drawing on Chinese architecture and art. It consisted of a pagoda on red columns with a curving, layered roof of terracotta

[3] At the time of writing, this whole area was about to undergo further redevelopment, designed by architects from Brisbane's sister city, Shenzhen, in southern China, and arranged according to the principles of feng shui. A new water feature will relate an ancient Chinese myth about a Yellow River carp.

Fig. 1.6.10 Fountain in Singapore (Photo by Veronica Strang)

tiles, decorated with figurines of mythical beings. Bright ceramic paintings and tiles further enlivened the pagoda, which stood alongside a fountain and pond, landscaped in the style of a traditional Chinese water garden. The water feature encapsulated specifically Asian ideas about water, wealth, and good fortune, and it affirmed the Chinese community's identity as a distinct cultural group within the wider city population. By affirming this social identity and signifying its vitality and generative power, it also served to delineate as 'Chinese' a specific spatial area within the city.

Brisbane has continued to develop and enlarge its economic ties with Asia, and the city was keen to celebrate these connections when it hosted the World Expo in 1988. Thus, in the late 1980s, as part of the 'rejuvenation'[4] of what was then a largely derelict area, it constructed water features to celebrate the city's transnational links with various Asian communities, homelands, and cultural landscapes.

[4] The fact that water features are so often integral to exercises in 'rejuvenation' underlines its core meanings as powerful, generative substance.

Fig. 1.6.11 Chinese fountain (Photo by Veronica Strang)

One of the key water features remaining from this endeavour is the Nepalese Peace Pagoda. Like the Chinese fountain, this structure incorporates traditional Asian motifs. There are intricate hand carvings, a tea house on the second level, a Buddhist *stupa*, stone lions at the entrance, and a decorative pond alongside. The Association to Preserve Asian Culture provided the pagoda for the Expo site. The building continues to house traditional Buddhist and Hindu rituals, as well as providing a place for wedding celebrations and other social events.

Again, these 'culturally specific' water features make use of traditional imagery to materialise a particular cultural identity and negotiate its presence and agency in a particular space.

Fig. 1.6.12 South Bank Peace Pagoda (Photo by Veronica Strang)

1.6.3 Cosmopolitan Pools

When the occupancy of space is contested – as seems inevitable in a postcolonial environment – material statements of identity have important social and political dimensions. Groups that do not share these cultural traditions must either cede space to these identities or be assimilated into them.

In the early part of Australia's settlement by Europeans, the settlers generally assumed that whatever remnants of its indigenous population remained would (and should) be assimilated into the new European-Australian society. The emergence of the Aboriginal rights movement in the 1960s and 1970s challenged these assumptions and brought to the surface ongoing conflicts about identity and belonging in Australia (see Dodson 1996). The growing diversity in the groups immigrating into the country and a concurrent trend towards urbanisation and more cosmopolitan ways of life have complicated these struggles. European or Asian statements of identity in public water features may be acceptable in particular neighbourhoods, but publicly funded and spatially central water features now have to encompass greater cultural diversity.

From an indigenous perspective, 'multiculturalism' is a further assimilative force which denies the potential for bicultural acknowledgment of the special status of indigenes. However, public discourses tend to stress the democratic and egalitarian 'inclusiveness' of multiculturalism, and this prevailing view is evident in the design of water features in many town centres. Thus in Brisbane's central public spaces, water features tend toward a cosmopolitan and modernistic style. For example, in King George Square, in front of the city hall, a determinedly futuristic water feature replaced the conventional round pool and fountain built in a more comfortably Eurocentric period of development.

Fig. 1.6.13 Water feature, King George square (Photo by Veronica Strang)

This suggested little in the way of a geographically or ethnically specific cultural tradition, only a more abstract vision of 'modernity' and a cosmopolitan, future-oriented identity into which all city dwellers can be incorporated and all subcultural identities potentially subsumed. In response to the major droughts 2005–2007, the water in the pool was 'temporarily' replaced by a 'Watersense' garden composed of drought-resistant plants.[5] Intended to encourage the local

[5] Like Chinatown, the central city area, including King George Square, is being redeveloped at the time of writing.

population to eschew heavy water use, this transformation of a fountain into a dry garden also raises both practical and symbolic questions about water security and the potential productivity of southeast Queensland.

Fig. 1.6.14 South Bank water feature (Photo by Photo by Veronica Strang)

On the South Bank are similarly unspecific features in which monolithic forms suggest urbanity and modernity but offer few culturally specific clues. The water feature at Mount Coot-tha, near one of the city's earliest reservoirs, combines vaguely modernistic shapes with the idea of a natural spring and waterfall. The same cosmopolitan style appears with increasing regularity in contemporary corporate architecture, in the atria of the skyscrapers along the banks of the central business district. These private monuments often favour monolithic visual forms: dark slabs of granite, marble or glass, with water flowing over the surfaces.

There are many such public water features in Brisbane. All express the cross-cultural meanings encoded in water, in particular ideas about vitality and generative power. They also manifest ideas about social identity and provide a focus for activities that celebrate and affirm communities. Some valorise Australia's specifically European history and cultural identity, whereas others demonstrate that quite different cultural communities can maintain distinct social and physical spaces and identities within the larger cityscape. Many – in particular those associated with

Fig. 1.6.15 City Centre water feature (Photo by Veronica Strang)

local governance or situated within the spaces of multinational companies – try to communicate more cosmopolitan visions of urban identity. And there are some important absences, omitted or subsumed in a determined assertion of plurality.

1.6.4 Water Margins

Prior to the colonial invasion a dense population of indigenous people lived around the Brisbane River estuary. Each clan had important sacred sites and camping, fishing, and hunting places, all of which were named in detail. Some of the original names remain in the modern city, but there is otherwise little material reference to the precolonial cultural landscape. Although Brisbane still has a small indigenous community,[6] few references in its contemporary water features or in its more general architectural design speak to or about Aboriginal culture. Indeed, offers to introduce

[6] Indigenous people represent only about 2.4% of the Australian population as a whole, though this percentage is perhaps slightly higher in Brisbane, as Aboriginal people are more heavily concentrated in the northern parts of the country.

Fig. 1.6.16 Traditional indigenous water container (coolamon) (Photo by Veronica Strang)

such imagery have been actively resisted.[7] If Aboriginal people are represented in Brisbane's public water features at all, it would appear to be at a larger, assimilative cosmopolitan scale or as part of 'nature in the city', rather than as a distinct and self-directing cultural community that perceives itself as the original owner of that landscape and as the other half of a bicultural colonial history.

In a country where conflicts over land and water remain central to debates, the erasure of an indigenous cultural landscape remains integral to non-Aboriginal assumptions of ownership. Water features (or other artefacts) that explicitly materialise indigenous cultural identities and histories potentially challenge this view. It is therefore not surprising that such expressions are generally confined to certain locations: central spaces in remote indigenous communities or – in more populated areas – places that are specifically devoted to indigenous affairs (for example, centres for indigenous health or education).

There are other potential reasons for their omission in Brisbane. For example, there is a long-standing anthropological notion that highly overt 'statements' of identity

[7] For example, in 1986, there was an offer by Japanese fundraisers to donate, to the Expo site, a large bronze sculpture of an Aboriginal man called Pemulwy, beheaded by settlers in Sydney in 1802. This offer was firmly rejected by the Queensland State Premier, Joh Bjelk-Peterson. The curator of sculpture at the Queensland Art Gallery reportedly said 'I hope they put it as far away from the Art Gallery as possible, preferably in Sydney' (Monaghan 1986).

often reflect a lack of certainty in this regard. Where place-based cultural groups, such as indigenous communities, have been securely associated with particular landscapes for many millennia, explicit material statements of their relationships with particular places may have seemed superfluous prior to the colonial invasion. No doubt local groups felt some need to differentiate themselves and their rights from those of others, but such statements could be quite subtle and ephemeral. It is only when identity and rights to land and resources are seriously challenged that the need for explicit material statements of identity and belonging becomes more pressing, though this is also, of course, when such statements are most likely to be excluded.

Thus the creation of water features in Brisbane that speak primarily to non-indigenous groups can be seen in two ways: as an appropriative assertion of non-Aboriginal ownership, and as a form of place-making that allows a settler population to deal with the ambiguities around 'belonging' associated with colonial movements into new spaces.

There are other factors: a persistent colonial vision associating Aboriginal people with 'nature' and an understanding of their traditions as being embedded in an ancestral landscape seem to have segued into an idea that they are culturally located in remote, rural environments, rather than in the city. To some extent indigenous communities also feel that their particular cultural landscapes have been more disturbed and subsumed in heavily urbanised spaces. So there are reasons why both Aboriginal and non-Aboriginal groups have some difficulty in seeing indigenous culture as being located in modern cityscapes.

The exclusion of Australia's indigenous people from fluid representations of urban identity is clearly politically and socially important. That exclusion presents a view of 'authentic' Aboriginal life as necessarily outside the city, in remote 'natural' areas, and so undermines the sense of identity of indigenous communities that do live in urban areas. Although Brisbane has an Aboriginal Council of Elders whom authorities must consult on a variety of issues, indigenous people's involvement in local water management is minimal.

1.6.5 Engaging with Water

Just as the material form of public water features is telling, so too are the collective activities that take place around them. Rituals of social belonging often involve water and water features. In Brisbane, these include ANZAC Day[8]; River Festival events; rituals of governance and investiture; and public gatherings for entertainment, sport, or even protest. It is also clear that water features are integral to the

[8] ANZAC Day is Australia and New Zealand's annual day of remembrance for those who have died or been injured in each country's military actions. It was founded originally to honour the members of the Australia and New Zealand Army Corps who fought at Gallipoli in WWI.

Fig. 1.6.17 Shopping centre fountain, Brisbane City centre (Photo by Veronica Strang)

active creation of collective social space in an increasingly privatised urban landscape, enabling the city to maintain and materialise ideas about shared social and cultural being and 'imagined community' (Anderson 1991).

Critically, water features also provide a focus for recreation: people relax, picnic, and socialise around them. As the aesthetic centrepieces of recreational spaces, they encourage affective responses and engender emotional attachment to places. Recreational engagement has been shown to have an important influence on the extent to which people develop concern for the well-being of local environments and ecosystems (see Milton 2002; Strang 1997). As engagements with land in Australia have shifted from rural, agricultural relationships to more urban, non-commercial forms of interaction, there has been a concurrent rise in interest in environmental issues. In the last two decades many local catchment management groups have emerged, and Brisbane now has nearly 40 small groups involved in taking care of the myriad tributaries to the Brisbane River and encouraging the wider population to be more conservative in its water use. These local catchment groups complement the larger, more formal regional organisations established by the federal government, in theory to provide democratic management of Australia's rivers by all water 'stakeholders' (see Strang 2009).

The recent rise in support for environmental care has significantly influenced local water management. The city has initiated a number of schemes aiming to create more sustainable water usage. As well as supporting the catchment groups in their activities and initiating its own 'water wise' educational schemes, the city council has made major moves to improve water treatment (and thus river water quality) and to increase water recycling, primarily for industrial and agricultural use. It has also invested in new kinds of water features, building a number of Stormwater Quality Improvement Devices (SQIDS) in local parks. These are new small-scale stormwater treatment structures that, by filtering water through vegetated swales, create green spaces in areas that would otherwise revert to dry grass and dust.

A discernible shift away from traditional 'decorative' water features to those that consciously encapsulate communities' aims for more sustainable environmental engagements is evident. This move reflects an international trend in the design of urban water features toward water-sensitive urban design (WSUD) that encapsulates environmental concerns, reuses water, and seeks to treat water without using chemicals that are damaging to local ecosystems (see Dreiseitl 2005). The aim is to create material engagements with water that reduce the environmental costs of human activities. In this sense, water features in urban environments are part of larger adaptive process, reflecting responses to social and ecological pressures. With an international exchange of knowledge, these new features combine global and local technologies, aspiring to balance human needs with diverse cultural and ecological realities.

1.6.6 Conclusion

When situated within this larger frame, Brisbane's water features can be seen as material expressions of both global and local processes. At a local level, the city's fountains and pools demonstrate particular ideas about social and political identity and power, offering celebratory affirmations of collective identities and values and expressing the tensions and conflicts pertinent to the city's urban demographics. Some features underline the cultural history of the European settler population and others celebrate and uphold the identity of specific immigrant communities. They show how the material formation of the urban landscape contributes to a dynamic process of place-making and identity construction in which cultural strategies are employed to create, support, and sometimes to suppress particular visions of social being.

Resources

Anderson, B. 1991. *Imagined communities: reflections on the origins and spread of nationalism*. New York: Verso.

Anderson, S., and B. Tabb (eds.). 2002. *Water, leisure and culture: European historical perspectives*. Oxford: Berg.

Dodson, M. 1996. *Assimilation versus self-determination: No contest*. Darwin: Northern Australia Research Unit, Australian National University.

Dreiseitl, H. (ed.). 2005. *New waterscapes: Planning, building and designing with water*. Basel: Birkhäuser.
Hodder, I. 1991. *The meaning of things: Material culture and symbolic expression*. London: Harper-Collins Academic.
Miller, D. (ed.). 1998. *Material cultures: Why some things matter*. Chicago: University of Chicago Press.
Milton, Kay. 2002. *Loving nature: Towards an ecology of emotion*. London: Routledge.
Monaghan, D. 1986. Brisbane Baulks at the Big Bronze Aboriginal. *The Sydney Morning Herald*, 9 April, P2.
Schama, S. 1996. *Landscape and memory*. London: Fontana Press.
Schmemann, A. 1976. *Of water and the spirit: A liturgical study of baptism*. London: Society for Promoting Christian Knowledge.
Stewart, P., and A. Strathern (eds.). 2003. *Landscape, memory, and history*. Cambridge: Cambridge University Press.
Strang, V. 1997. *Uncommon ground: Cultural landscapes and environmental values*. Oxford: Berg.
Strang, V. 2004. *The meaning of water*. Oxford: Berg.
Strang, V. 2005. Common senses: Water, sensory experience and the generation of meaning. *Journal of Material Culture* 10(1): 93–121.
Strang, V. 2009. *Gardening the world: Agency, identity and the ownership of water*. Oxford: Berghahn.
Strang, V., and A. Garner (eds.). 2006. *Special issue of Fluidscapes: Water, identity and the senses. Worldviews 10:2*. Leiden/Boston/Tokyo: Brill Academic Publishers.
Tilley, C. 1994. *A phenomenology of landscape: Places, paths and monuments*. Oxford: Berg.
Tilley, C., and W. Bennett. 2004. *The materiality of stone: Explorations in landscape phenomenology*. Oxford: Berg.

Part II
Culture and Water in Diverse Environments

Cover art by Andrea Sanders, Canada

Chapter 2.0
Introduction: Culture and Water in Diverse Environments

Marcus Barber and Ameyali Ramos Castillo

This section delves further into the concept of biocultural diversity and the key sustaining role that water plays with examples from across the globe of the way human communities interact with water and of how these complex interactions maintain life in diverse environments. Collectively these contributions emphasise the role that local/traditional/Indigenous peoples and knowledges have in the sustainable management of water and they describe some of the challenges that such peoples face in having their knowledge recognised and acted upon by others.

In assembling the case studies and vignettes, we chose to begin with a 'geographic' logic, grouping the examples in terms of continents or related cultural contexts. The section begins with two examples of Indigenous peoples living within developed capitalist democracies: Thornton's discussion of Southeast Alaskan Natives, complemented by Barber's description of water cosmology in tropical Aboriginal Australia. This grouping according to related circumstances is followed by two complementary stories from Africa about the Maasai (Ogendi et al.) and the Kamba (Kilonzi et al.) respectively. Wessels then moves us northward to review the status of the ancient Syrian qanat groundwater systems. From the Middle East the focus shifts to the Americas, beginning with Carse's description of the huge freshwater demands of the Panama Canal and their consequences for the changing livelihoods of local residents. Alcorn reminds us of the potential strength of local resistance with her account of ranchers along the Pilcomayo River in Argentina. And, Blanchard et al., in a case from the southeastern United States, demonstrate the synergistic links

M. Barber (✉)
Ecosystem Sciences, the Commonwealth Scientific and Industrial Research Organisation (CSIRO), Darwin, Australia
e-mail: Marcus.Barber@csiro.au

A. Ramos Castillo
Traditional Knowledge Initiative, United Nations University, Institute of Advanced Studies (UNU-IAS), Yokohama, Japan
e-mail: ramos@ias.unu.edu

B.R. Johnston et al. (eds.), *Water, Cultural Diversity, and Global Environmental Change: Emerging Trends, Sustainable Futures?*, DOI 10.1007/978-94-007-1774-9_8,

between environmental health, biodiversity and cultural diversity in their description of a downstream community campaign to demonstrate the human environmental effects of upstream water diversions. The stories from the Americas conclude with Ramos Castillo's look at the importance of nourishing diversity in urban water supply through an analysis of water in the city of San Cristobal in Mexico.

The geographic focus then shifts to the Himalayas, which supplies the freshwater that sustains and feeds a huge proportion of the world's growing population. Yin Lun demonstrates the value of appropriate participatory water resource development in villages in the upper reaches of the Mekong (Langcangjiang) River in Tibet, while Tsering describes the significance of Tibetan water cosmologies in organising local property regimes and motivating determined political resistance to the many large dams planned for the region. Drew explores the implications of melting Himalayan glaciers for a downstream community in India, introducing the current and future impact of climate change. A similar focus and concern is found in the final contribution, Crate's account of northeastern Siberian cattle herders and the challenges they face in maintaining their way of life in a warming world. The geographic spread of these case studies spans the globe, demonstrating the cultural and environmental diversity that is so crucial to human interactions with water and to its ongoing management and sustainable use of water.

From this picture of global diversity emerge certain themes, commonalities, and contrasts worthy of note. These themes lead us on a path through the section, a path that at times weaves back on itself or suggests alternative routes going in different directions. The two case studies of coastally located 'Fourth World' Indigenous peoples demonstrate the kind of holism characteristic of their cosmologies; rivers and seas form interconnected aquatic wholes that are as significant for what they imply for peoples' sense of individual and communal identity as they are for their consequences for ongoing legal, property and management arrangements.

Like that of Native Alaskans and Australians, the story of the Maasai in Africa is a story of colonisation and the difficulties of maintaining indigenous ways of life in the face of the changes wrought by the colonisers, in this case even after those colonisers have departed. Ogendi and Kilonzi also supply us with a counter to the usual story of colonisation and loss with an account of the Kamba people's successful adaptation of indigenous plant knowledge to a new context and their successful adoption of a new kind of small-scale sustainable sand dam. The effectiveness of traditional water technologies in the arid deserts of Syria and the overuse and mismanagement of recently introduced technologies are the subject of Wessels' analysis.

Opening the examples from the Americas, Carse notes that local resistance to efforts to implement sustainable water regimes is not a consequence of 'backward' ancient traditions but emerges from a recent history of regional development that complicates the attempt to turn smallhold farmers in Panama into foresters. Carse's challenge to those who would champion local action in all its forms as the unproblematic key to sustainability is met in turn by Alcorn's account of local criollo ranchers in Argentina, who determinedly resist dam developments sponsored by large international agencies in order to protect the Pilcomayo wetlands that are so important to their way of life. An insistence on the value of the local requires engagement with local variation and complexity, as Carse and Alcorn's examples demonstrate.

The final examples from the Americas underscore the importance of local engagement in designing and implementing sustainable water use systems. Blanchard et al., demonstrate the vulnerabilities that local downstream users, here the people of the Apalachicola Bay in Florida, can face when they are not in control of water planning and consumption decisions further upstream. In putting a human face to the issues, Florida oystermen were able to successfully argue for cultural flow allocations in the tri-state river basin, improving near shore habitat and reducing coastal community vulnerabilities to upstream water management decisions. Ramos Castillo demonstrates the autonomy, flexibility, and sense of community that can be generated by local respect for and control of the freshwater supply, even in circumstances of rapid population growth in an urban context, in this case the Cuxtitali barrio of Mexico City.

The focus then moves to Asia and perhaps the most pressing water issues faced by humanity. The Tibetan Plateau is the world's 'Third Pole' in terms of the amount of freshwater it holds, and it is the 'water tower' of Asia, indirectly feeding some 20% of the world's population. Those downstream users are often the primary emphasis in contemporary discussions, but Yin Lun and Tashi Tsering instead focus on the inhabitants of the Tibetan highlands to show that local water beliefs and water management practices must be taken into account in the use of this critical resource. After describing the intimate connections between local water beliefs and sustainable land and water management practices, Yin emphasises the success of participatory development projects that take local knowledge seriously. Tsering widens the focus, arguing for the direct connection between Tibetan water beliefs, landholding regimes, and Tibetan resistance to the massive dam developments proposed for the region by the Chinese government.

Drew remains in the Himalayas but takes us downstream to India, reviewing the controversies surrounding the dam and hydroelectric developments planned for the Bhagirathi River. This essay also mentions the changes likely to be wrought by climate change and the altered levels of glacial meltwater flowing into the river. Crate's essay amplifies this theme, describing the difficulties faced by horse and cattle breeders in northeast Siberia and the altered lifeways they are already being obliged to adopt because of climate change.

These two elements, the Himalayan 'Water Tower' that feeds the vast population of Asia and the potential impacts of future climate change, more than anything else should focus our collective attention on the need for alternative water strategies. As the Cuxtitali example demonstrates, rapid urban growth does not inevitably lead to unsustainable overconsumption, and such growth need not always be met by massive new water storage capacity that often undermines local, bioculturally diverse environments. It is precisely the diversity of practices and contexts that we wish to emphasise: clearly, this array of human interactions with water and the human needs such interactions imply cannot all be solved by large scale, technocratic solutions. Monocultures – biological, technological, or sociocultural – are inherently more vulnerable to changing conditions because they lack the diversity that underpins true adaptive capacity. In meeting the challenges of a warming world, we must be open to the full range of human possibilities and the knowledges such possibilities have created.

Chapter 2.1
Watersheds and Marinescapes: Understanding and Maintaining Cultural Diversity Among Southeast Alaska Natives

Thomas F. Thornton

2.1.1 Watersheds, Marinescapes and Cultural Diversity

Map 2.1.1 Pacific Northwest

Among the Tlingit, Haida, and neighbouring peoples of the northern Northwest Coast of North America, key watersheds not only define regional dwelling spaces but were owned and managed by lineages (matrilineal clans and house groups), which controlled access and enhanced their productivity in a variety of ways to ensure sustainability. These indigenous peoples also derived critical aspects of their identity and livelihoods from the unique features of these waterways, the differences of which were celebrated in a variety of contexts, including naming, visual art, dance, and rituals such as the potlatch ceremony. Among the Tlingit especially, the relationship between watersheds and marinescape

T.F. Thornton (✉)
Environmental Change Institute, School of Geography and the Environment, University of Oxford, Oxford, UK
e-mail: thomas.thornton@ouce.ox.ac.uk

B.R. Johnston et al. (eds.), *Water, Cultural Diversity, and Global Environmental Change: Emerging Trends, Sustainable Futures?*, DOI 10.1007/978-94-007-1774-9_9,

Box 2.1a River, sea, and sky: Indigenous water cosmology and coastal ownership amongst the Yolngu people in Australia

—Marcus Barber

Yolngu people have dwelt for millennia along the coastlines of Arnhem Land and have a deep understanding of the water flows and cycles of their country. Land, rivers, estuaries, saltwater, and seabed are all created and linked by the ancestral beings who journeyed across the country in the creation period and who still dwell there. Yolngu song cycles describe these journeys and also describe the flows and transformations of water that further unify land, river, sea, and sky into a coherent whole. In the songs, the clouds form at sea on the horizon, are brought in by the wind, and drop their rain on the hills. This rain flows into the rivers and the underground aquifers, giving life to the country and ultimately flowing into the estuaries or bubbling out through beach and subtidal springs, mixing with salt water and flowing out to sea to form new clouds, beginning the cycle once again. The mixture of salt water and fresh water is a symbol of fertility, a source of power and meaning rather than confusion, and the dynamism of water expresses the productive flows of Yolngu social relationships.

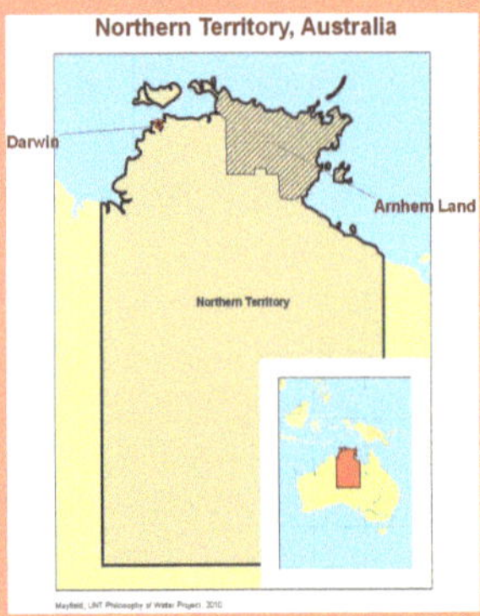

Box Map 2.1a.1 Australia

Australian law has given Yolngu people control over their land, but gaining control over their waters, particularly the sea, has been an ongoing struggle for them. Like indigenous peoples across Australia, Yolngu people have been vocal and vigilant in trying to protect the places that matter to them from harm and in engaging with new forums and new agendas for the control and management of those places. For Yolngu people, owning country means owning land and sea, salt water and fresh. It also means being owned by those places as much as owning them. Co-management becomes more than sharing control among people; it becomes a sense of how the country and people mutually manage one another.

Box Fig. 2.1a.1 Where the clouds stand (Photo credit: Marcus Barber)

Recent environmental changes pose new challenges to the subtle balance of life in Yolngu country. A people attuned to the ebbs and flows of its world is noticing changes to local places: swamps and waterholes disturbed by introduced buffalo and pigs, fresh water receding further on

(continued)

Box 2.1a (continued)

Box Fig. 2.1a.2 Sea rights in Northeast Arnhem Land are partially constituted by and through the circulation of water – an integration of land, sea and sky that is articulated in Yolngu coastal cosmology (Photo credit: Marcus Barber)

the coastal floodplains, and the wind, clouds, and seasons behaving in uncharacteristic ways. Climate change is a new and worrying element in this country, yet a people who have survived 50,000 years on the continent can also draw on the confidence that comes from their past survival. Recent archaeology in Blue Mud Bay explored the middens of sea shells found many kilometres inland on the floodplains. The ancestors lived here when the sea level was once very different, and their legacy lives on in structures in the landscape and the names of the places. Climate change means that many more changes are to come, and if that is the future, it is perhaps even more important to listen to, and learn from, those who survived great changes in the past.

Acknowledgments My sincere acknowledgement and deepest thanks for my Yolngu mentors, teachers, and friends in Blue Mud Bay.

explains critical biological and cultural diversity within the region. For example, sockeye or red salmon (*Oncorhynchus nerka*) streams were highly valued, as were fall dog (chum) salmon (*Oncorhynchus keta*) and coho (silver) salmon (*Oncorhynchus kisutch*) runs because of their temporal 'stretching' of the salmon harvest season. Similarly, marinescapes invisible from the surface, such as Pacific halibut (*Hippoglossus stenolepis*) banks might be defined by a set of relational characteristics between observable surface features, as in the name of one fishing bank, Geesh K'ishuwanyee ('Just on the Edge of the Base of the Kelp'). Such unique and diverse water features, though often not dominant in the physiography, were celebrated as markers of regional identity and culture. The implications of this intracultural diversity are evaluated against current water policy and fisheries management that typically ignores indigenous hydrological units in favour of commercial zoning.

2.1.2 A Tlingit Watershed

Auke Bay is today the busiest harbour in Juneau, the capital city of Alaska, the largest and northernmost state of the United States. Auke is derived from the indigenous Tlingit name for the place, Áak'w, or 'little lake', which references the beautiful freshwater lake that sits just a few 100 yards above the bay in the coniferous rainforest. An arterial stream known as G̲aat Héeni, or 'Sockeye Salmon Stream', connects the lake and the bay. The lake, the stream, and its estuary in the bay together form the watershed, which Tlingits sometimes call Áak'w Héeni, water possessed by Auke Lake. Sockeye or red salmon (*Oncorhynchus nerka*) are the premier species among the five Pacific salmon, valued for their rich taste, high oil content, extended and predictable runs, and harvestability. All of the major rivers that cut through the high coastal mountains separating southeastern Alaska from British Columbia – the Stikine, the Taku, the Chilkat, the Alsek, and even the mighty Nass in Tsimishan country to which the Tlingit trace their origins – support sockeye and two, three, or four other species of salmon, and all at one time supported major Native villages that were inhabited the year round.

Tlingit groups often took their names from important watersheds. Thus the Áak'w K̲wáan, who claimed the Auke Bay area as their home, were known collectively as 'those dwelling around Little Lake'. The k̲wáan, sometimes considered analogous to a tribe though it held few governing powers, consisted of several matrilineal clans organised under two sides, or moieties, and subdivided into house groups (sublineages), each of which claimed and managed its own watersheds, or in the case of joint occupancy of major watersheds, a portion of the larger river system. The house group leader, or *hít saat'i*, organised his kinfolks' labour to take advantage of the short but intense runs of salmon such that they not only caught great quantities but also butchered and dried them, typically in smokehouses. The resulting 'dryfish' could be stored for up to a year and also carried in quantity for trading. It is this unique organization of knowledge and labour around the temporary abundance of northern Pacific watersheds that allowed the peoples of the Northwest Coast culture area to evolve into the wealthiest, most complex hunter-gatherers in the world. 'We never bothered to plant too many crops' (save for a few potatoes), a Tlingit man once told me: 'Why should we, when we can harvest the river's bounty without seeding or weeding…or toil in the soil?'

2.1.3 Cultivating Watersheds and Marinescapes

However, from another perspective the Tlingits *did* cultivate the watersheds and their communities of beings to maintain the river's bounty. Native people conceptualised salmon as 'tribes' with their own undersea social organization and reciprocal relations with humans. Animism, the belief in the spiritual agency of fish, wildlife, plants, and other vital elements of the environment, governed the moral economy of respect and exchange between human and nonhuman persons. First Salmon ceremonies, common throughout the northern Pacific cultural areas, were performed to emphasise this reciprocity. After indigenous people welcomed and honoured the first salmon and ritually consumed its flesh, they reassembled its bones and returned them to the river with a gift and prayer to the spirit of the salmon (which survives the fish's death in disembodied form) so that it would be renewed and encourage its relatives to find their way back to their natal stream.

Fig. 2.1.1 Tlingit Dog Salmon (L'einedí) Clan leader Rosa Miller (center) leads a 1997 ceremony at the site of her clan's aboriginal village with the Auke Cape marinescape visible in the background (Photo credit: Thomas Thornton)

Similarly, people cultivated streams to provide quality habitat for salmon, providing fish rests, egg-laying substrate, upstream pathways, and shade. These ritual and practical techniques were not merely symbolic but essential interventions and exchanges for the maintenance of the ecosystem, in the Tlingits' view. If people did not perform them, the salmon would feel disrespected and would withdraw from the watershed and cease to present themselves to fishermen. The salmon tribe would punish transgressions against their members by withdrawing support for human communities. Contemporary Tlingits still believe strongly that respectful fishing enhances fisheries and habitat rather degrading them. And when fisheries are 'closed' by state managers for extended periods, Tlingits fear that the fishery will atrophy and die because the ecological covenant between humans and salmon has been broken.

The 'Salmon Boy' story, versions of which can be found among indigenous people up and down the northwest coast of North America (e.g., Swanton 1909; de Laguna 1972; Langdon 2007; Thornton 2008), exemplifies the covenant for respectful relations with salmon.

In the Tlingit version, set in a fish camp, a boy insults the salmon by derisively casting aside a piece of dryfish offered him by his mother because it is mouldy on one end. The salmon respond to this insult by 'kidnapping' the boy and taking him to their underwater world, where he learns the ways of the salmon tribe and how they view their human counterparts. Eventually, the salmon 'chief' calls his people to board their 'canoes' and return to their natal spawning grounds. Upon reaching the

Fig. 2.1.2 Cliff Edenshaw laying Western Hemlock branches for herring to spawn upon, near where the Tlingit Salmon Boy encountered herring in his underwater journey (Photo credit: Thomas Thornton)

Tlingit fishing camp, the boy 'stands up' (i.e., jumps) to see the smokehouses and proceeds excitedly up the river to the eddy where his mother is processing fish. The mother admires the beautiful salmon and directs her husband to spear it. He does so, and when he presents the fish to the mother for processing, she discovers that it sports a copper necklace just like the one her son wore. After consulting a shaman about what to do, the parents place the salmon on a large plank at the top of the house near the smoke hole. By the next morning, the salmon boy has transformed back to human form. He then instructs his people in the ways of the salmon and how to treat them respectfully so that they will return each year. Eventually the boy becomes a powerful shaman himself. Some versions of this story add that the boy instructed the people in how to carry out the First Salmon ceremony.

Failure to maintain healthy relations could bring serious ecological consequences. One example of an adverse result occurs when stream weirs block salmon pathways. 'A salmon knows its river', Tlingits say, and will find its way back to its home stream to spawn. When the fish's way is blocked by a weir or other man-made objects for a significant period of time, it may become insulted and abandon the stream altogether. Tlingits, who have possessed weir technologies for thousands of years, sometimes learned this lesson the hard way. One oral history tells of the fate of the 'little sockeye' (red) salmon, or *dagák'*, which today is found only in the Necker Bay watershed near the Alaskan town of Sitka. These rare little sockeye, considered a separate species in Tlingit taxonomy, used to be present in other streams around Sitka, but when these watersheds were blocked or disturbed by human activity, the *dagák'* abandoned them, insulted. Tlingits acknowledge the insult that they perpetrated on *dagák'* and tell the story to avoid further offending this rare but cherished little salmon. 'Those little sockeye get offended if you don't leave them a hole in your [fish] weir; they won't come back if it [the stream] is all blocked off', says elder Herman Kitka, Sr. 'Sitka people have a saying about it', he notes,

> *Tleil dagák' ahawateeni yík'*. Literally, it means, 'Don't stomp off [insulted] like those little sockeyes'. In human contexts, such as at a ceremonial potlatch or memorial party, this adage is spoken to guests whom the hosts want to return for another gathering. As with salmon, the hosts know if they want the guests to return they must cultivate proper, respectful relations with them. (Thornton 2008)

As this story and the Tlingit system of watershed management illustrate, respect (*at yáa awuné*) is not only ideological but practical. Tlingits use conscious, deliberate cultivation of social relations and material conditions to sustain and even enhance the salmon runs upon which their livelihoods depend. This cultivation requires intimate local knowledge and engagement with salmon and their habitats; it also requires controlling salmon fishing according to abundance or scarcity. The Tlingit even went so far as to transplant salmon to different streams to enhance or extend runs (Thornton 2008), a practice similarly found among indigenous peoples of the Northwest coast in the realm of plants (Deur and Turner 2006). Tlingit clan leaders refer to their relationship with salmon streams as 'taking care of it' (*at daat kuyawusitaak*), or stewardship. Taking care of a stream involves cultivating streams to create better habitat, limiting harvests to ensure that sufficient numbers of salmon escape to spawn, culling predators, and honouring salmon through respectful conduct and gifts. The Tlingit 'master of the [salmon] stream' (*héen s'aatí*) is analogous to the master of a lineage house (*hít s'aatí*): he ensures the sustainability of the resources by managing the relations between the various inhabitants (fish and human) of the watershed (Thornton 2008). Thus, for Tlingits managing salmon is similar to managing their own family relations, and managing stream habitats is not unlike managing a lineal house. Violations of sacred management prescriptions or protocols are considered not only disrespectful acts but violations of natural law; indeed, the Tlingit term for taboo is most commonly translated as 'against nature' (*ligáas*).

Protocols varied by watershed because the habitats themselves were different. Large, high-volume mainland rivers, such as the Stikine (from the Tlingit Shtax' Héen, 'water biting itself'), typically contained all major species of salmon and encompassed multiple Native communities and multiple tributaries, some fed by mountain runoff and others by glaciers. In contrast, a small island stream with as few as two species of salmon might be owned by a single house group and managed in its entirety from its headwaters (*héen shaak*) to its mouth (*héen wat*), including its estuary. Sockeye streams, such as the one at Auke Bay (*Gaat Héeni*) or that of the exceptional little sockeye (*Dagák' Héeni*), were especially precious both because of their rarity and because of the particular qualities of the fish. Full size sockeye are lake dwellers that are high in oil content, making them particularly tasty and long-lasting, in that they can be harvested even after they have spawned (whereas other salmon may have deteriorated too much). Watersheds with sockeye also typically contain two or three other species of salmon that run after the sockeye spawn, thus lengthening the harvest period. It is no exaggeration to suggest that Tlingit have co-evolved with sockeye, for wherever we find a sizeable sockeye stream in Southeast Alaska we also find evidence of ancient Tlingit settlements, typically of significant size. Similarly, people celebrated the fall runs of coho and dog salmon species because they blessed communities with a fresh supply of fish just before the long winter, when few fish were available. Salmon harvesting methods traditionally varied according to these constraints and other dynamics of particular watersheds (Langdon 2006).

2.1.4 The Power of Estuaries and Lakes

Estuaries, the interfaces between land and sea, fresh water and salt water, stream flow and tide, were the focal point of Tlingit culture. Here was where the real bounty lay for the 'People of the Tides', as the Tlingit are known. In this coastal zone dwelled myriad fish, invertebrate, and plant resources on which the people relied. But it was a dynamic environment, shifting with the changing geomorphology of the region (including, at times, the highest rate of terrestrial uplift in the world because of retreating glaciers), as well as the seasonal cycles of the herring, salmon, crab, clams, seaweeds, wild rice, and other critical foods. Estuaries are the places where anadromous salmon pause to transform themselves into sustenance for Tlingits and the myriad birds and wildlife that prey on them. The Salmon Boy story nicely illustrates the dynamic and transformative nature of estuaries, as the boy who insults the salmon is initially given the name 'Mouldy End' (Shanyaak'utlaa<u>x</u>) by the salmon people but earns the more honourable name of 'Alive in the Lagoon' (Aak'wtatseen) upon his return and rebirth in the estuarine lagoon of his home stream. Similarly, to maintain his shamanic strength and vision Aak'wtatseen bathes and drums for power in a place called *Xijaa.eix'i* ('Beating Time Slough'), which also lies at the nexus of the saltwater-to-freshwater marinescape.

Tlingits recognised the symmetry and connectivity between saltwater lagoons and freshwater little lakes by using the same linguistic term for the two features: *áak'w*. Both were places of wealth, but whereas the richness of shallow estuarine lagoons was visible, the wealth of freshwater lakes was often hidden and thus had to be gained through intermediaries. The most important of these was a mythical creature known as Tlanaxeeda<u>k</u>w, or 'Wealth-Bringing-Woman', and her offspring, known as Aatayádi, or 'Child of the Lake'. Tlanaxeeda<u>k</u>w was said to have dwelled at Auke Bay, where she and her children would appear to hunters in various forms, including that of an amphibious frog. If a hunter made a gift to her, she would bring him wealth, but if he failed to do so, she would punish him.

As one story, set near Auke Lake, goes:

> A man of the Wolf clan named Heavy Wings (KîtcîdA'lq!) was out hunting and heard a child cry somewhere in the woods. He ran toward the sound very rapidly, but, although the child's voice seemed to be very close to him, he could not see what caused it. Then he stopped by the side of a creek, tore his clothes off, and bathed in the cold water, rubbing himself down with sand. Afterward he felt very light and, although the voice had gotten

Fig. 2.1.3 Auke Lake (Áak'w, "Little Lake"), home of the "Wealth-Bringing-Woman" and multiple species of prized salmon (Photo credit: Thomas Thornton)

> some distance away, he reached it, and saw a woman with an infant on her back. He pulled the child off and started to run away with it, but he did not escape before the woman had given him a severe scratch upon his back with her long copper finger nails. By and by he came to a tree that hung out over the edge of a high cliff and ran out to the end of it with the child in his arms. Then the woman begged very hard for her baby saying, 'Give me my baby'. As she spoke she put her hand inside of her blanket and handed him a copper. When he still refused to give her the child she handed him another. Then he gave the child back, and she said, 'That scratch I made on your back will be a long time in healing. If you give a scab from it to any one of your people who is poor, he will become very rich. Do not give it to anybody but your very near relations'. (Swanton 1909:175).

And so in fact it turned out. The sore did not heal for a long time, not even after he had become very rich. Everything that he put his hand to prospered, and the relations to whom he had given scabs became the richest ones next to him (Swanton 1909).

Perhaps more than anything, the Wealth-Bringing-Woman of the little lake represents the fragile yet powerful nature of watersheds and small lake systems, social relations, and the natural resources that people reply upon. If properly cultivated, all will keep on giving. If not treated with respect, however, each can be scarred and, in turn, can scar one's own bodily landscape, as it were. Such stories instil a strong moral-ecological consciousness in connection with lakes and watersheds.

Box 2.1b Blue ecology and the unifying potential of water

—Michael D. Blackstock

As the rivers of human knowledge flow across the world's landscapes, a diversity of cultural tributaries interweave, streaming with their own enhancing qualities of clarity, flow, and experiences. Each stream has its own history, flavour, and voice, and yet it has the potential to form a larger whole. The concept of Blue Ecology, drawn from indigenous Canadian perspectives and western science concepts, offers a useful framing. Water has the potential to *be the bridge* between cultures as they come to understand how each other thinks about water and thus realise we all share the fear of a dry throat – thirst. Acknowledging water's central functional and spiritual roles in our world is the unifying bridge between indigenous and science-based ways of knowing. An ecological philosophy, acknowledging salt and fresh water's essential, rhythmic life-spirit and central functional role in generating, sustaining, receiving, and ultimately unifying life, in Blue Ecology the living and elemental, fresh and salt water are unified: Earth Mother is an interconnected whole. Blue Ecology celebrates the inherent value and meaning of water.

The five principles of Blue ecology are:

Box Fig. 2.1b.1 Principles of Blue Ecology (Credit: Michael Blackstock)

(a) Spirit: water is a living spirit.
(b) Harmony: harmonious sustainability in a functional rhythm engenders healthy bodies and ecosystems.
(c) Respect: water through ceremony, education, and giving back, else Earth Mother will retaliate by taking water away.
(d) Unity: water has the ability to connect and unify humans because of our common reliance on this basic unit of existence.
(e) Balance: restrained and measured water withdrawals in combination with and giving back (i.e., restoration, monitoring, or ceremony) to watersheds and water).

Such principles suggest a practical outcome. The vision of Blue Ecology is to embrace a water-first approach to planning human interventions in the environment. This means that planned development should not impede the functional delivery of quality water to ecosystems in a healthy rhythm. Implementing this vision demands recognition of the key role each nation play

(continued)

Box 2.1b (continued)

as caretaker of the water that flows through its borders. The opportunity before us, flowing past us, is to inspire collaboration and to mediate disputes in ways that ensure the basic human need for water is met in truly sustainable ways.

---

*This textbox is based on my paper titled "Blue ecology and climate change: Interweaving cultural perspectives on water, an indigenous case study", presented at a symposium held on the island of Capri, Italy, on October 2008 and subsequently included in the proceedings of this symposium edited by Liebscher Hans-Jurgen et al., published by IAHS Press.

2.1.5 Marinescapes as Traditional Cultural Properties

Tlingits also considered the marinescapes adjacent to watersheds to be alive and possessed by animate forces. Making use of them meant confronting what is largely unseen and often ensconced in perilous waters. But Tlingit are fortunate to have Raven, the trickster-demiurge, as their guide in approaching these environs. It was Raven who helped calm the waters around rocky reefs so canoes could land there. Such landing places are referred to as Raven's Wing Canoe Road (*Yéil Kílji Yakwdeiyí*) because the waves form even patterns of small ripples, like feathers on a raven's wing. Similarly, Raven found a way to drag in the great house of fish that lies offshore, to trick king salmon into coming to shore so the trickster could eat him, to find halibut (*Hippoglossus stenolepis*) in the fish holes (*eet*) under the sea by using surface features like kelp beds locate them, and to employ special animated hooks (bestowed with their own names) and bait to catch them. And it was Raven who freed the tides from the grasp of the 'Old Woman Underneath' so that people could take advantage of their ebb and flow to harvest intertidal resources. Hungry, he climbed down the mighty bull kelp to find out what bounty of underwater resources lay deep on the sea bottom. Thirsty, he stole fresh water from the hoarding Petrel's closely guarded spring (blackening his body in the process of escaping out the smoke hole) to create the myriad stream systems that define the Alaskan coast. Raven's watershed and marinescape work is everywhere, leaving permanent traces on the land for inhabitants to see and pathways to follow. Other ancestors' work and pathways serve as similar guideposts on the land and sea.

The ideal Tlingit living space, then, constituted a networked series of meaningful and productive places, centred on a multispecies salmon watershed and linking land and seascapes that encompassed shelter from the open sea, weather, and enemies; bountiful nutrients in the form of fish, wildlife, and plant resources; accessible pathways for travel; and beauty rendered meaningful through story and song. In these dynamic and productive watery environments, Tlingits evolved one of the most complex social organizations and highest population densities of any hunting and

Map 2.1.2 Land of the Aukquwon, map of 'Aukquwon' (Áak'w Kwáan) from Philip Joseph's (1967) historical account its founding

gathering people in the world. The greater Auke Bay land- and marinescape was such an ideal place when the Áak'w K̲wáan came to live there centuries ago.

As L'einedí (Dog Salmon Clan) elder Philip Joseph (1967) relates,

> They came by Outer Point and came to Auk Bay. The Chief then told his people where they would make their new settlement. They landed in Fairhaven [Indian Cove, or X'unáx̲i] and started building. They put up big houses, huts, and smokehouses [and later a fort, Áak'w Noow, nearby]. At the same time most of the people explored the whole bay. They soon find Auk Lake. And they find out the creek [G̲aat Héeni] [that] runs from the lake is a good sockeye creek. They also f[i]nd out the herring spawns in the spring. There were all kinds of berries, game, and shellfish food. The name 'Auquwon' [Áak'w K̲wáan] comes from the lake. In Thlinget, lake means 'auk' [áak'w] and 'quwon' [k̲wáan] means the people. That's how the people who go there were named Aukquwon. The name of Auk Bay in Tlingit is 'Auk-ta' [Áak'w Ta]....They saw ducks of all kinds, many animals like bears and mountain goats. This place suited them and they went right back to report to the Chief. He came and looked the place over. He told his people they will make their settlement in Auk Village to live in winter time. After so many years they start building again for their main village. They built their Dipper House [of the L'einedí clan] at Fairhaven. They worked for years, then they moved to Auk Village. Why the chief took pains to find a village site was that they needed a sandy beach because they use canoes and also had to have a shelter like a boat harbour....After the Dipper House was moved to [downtown] Juneau it has been rebuilt four times. So the Auk Village should be [at least] four hundred years old.

Today, Auke Bay, teeming with boat traffic and the residential and commercial development of Alaska's capital city, no longer supports a Tlingit village or the rich diversity and quantity of species that attracted the aboriginal people to this supreme landscape. The land and sea are co-managed by the State of Alaska and the federal government, which have a fisheries school and laboratory on the shores of the bay. Their collective research and governance efforts have not stemmed the decline of the herring, salmon, invertebrates, and other wildlife that once supported a vibrant Native community in this watershed. The Áak'w Tlingit, meanwhile, have been dispersed, and their village exists only in commemorative form as part of recreation area. Yet members of the L'einedí clan and Áak'w Kwáan still claim ties to this watershed and its associated marinescapes, and recently a part of the their traditional territory on federal land at Auke Cape was found eligible as a Traditional Cultural Property – that is, a landscape of national historical significance because of its association with 'cultural practices or beliefs of a living community that (a) are rooted in the community's history, and (b) are important in maintaining the continuing cultural identity of a community' – under the U.S. National Historic Preservation programme (Thornton 2009). In itself, this designation will not revitalise Auke Bay, however; rather, it must be coupled with a comprehensive watershed management programme that will rehabilitate parts of the bay that have been harmed by development, pollution, and overfishing.

2.1.6 Conclusions

Although Auke Bay is unlikely to return to its aboriginal form in the near future, other watersheds in Tlingit country are still in relatively good health but subject to potentially degrading forces, including human neglect. Meanwhile, the Tlingit themselves have been deprived of their opportunity to be watershed managers, the settlement of their land claims by the 1971 Alaska Native Claims Settlement Act having turned them into for-profit landowning corporations with no aboriginal water, fishing, hunting, or gathering rights. Those seeking to continue to manage watersheds must do so through this attenuated structure or by private allotment or new agreements with state and federal agencies. These new forms of collaborative governance could prove mutually beneficial, especially for watersheds and marinescapes lacking state resources or expertise to manage them. In other parts of Native North America, where aboriginal watershed sovereignty is stronger, Native American tribes and First Nations have forged successful co-management agreements that enable them exercise significant autonomy over their lands and waters. A recent example of this trend is Haida Gwaii in British Columbia, Canada, where cultural rights, livelihoods, and stewardship have been strongly linked to sustainable watershed and marinescape management as well as biodiversity conservation through new environmental governance arrangements. Such participatory governance frameworks have proven beneficial in maintaining and restoring watersheds through both improved monitoring and data collection and respect for indigenous ethics and 'ecosystem justice' (Jones and Williams-Davidson 2000). Just as important, they have served to reground riverine and maritime peoples in the rich watershed and coastal habitats that are the ultimate sources of their unity, diversity, and well-being.

The fisheries scientist Daniel Pauly (1997) has advocated for 'Putting Fisheries Management Back into Places'. He writes,

> Though sometimes tempted by pessimism, I believe that we humans will, in the next millennium, find ways to match our numbers and our demands with what our planet can provide (this is not so for the time being). This will require that we abandon rape and pillage

> as our major mode of interaction with natural resource. For fisheries, it will require rediscovering places for fisheries management. (1997:126)

In the case of the northwest coast of North America, a major step toward this end will be rediscovering indigenous watershed ecosystems as places, like the rich coastal rivers and estuaries that Tlingits, Haidas, and other indigenous coastal peoples have named, celebrated, adapted to, evolved with, and cosustained over millennia. From this case study, it is easy to see how traditional knowledge and practices may contribute to future water security, but only if the integrity of these watershed ecosystems is linked to the integrity of the 'cultures of water' that historically have been their inhabitants and guardians.

Resources

de Laguna, Frederica. 1972. *Under Mount Saint Elias: The history and culture of the Yakutat Tlingít*, Vol. 7, Smithsonian contributions to anthropology. Washington, DC: United States Government Printing Office.

Deur, D., and N. Turner. 2006. *Keeping it living: Traditions of plant use and cultivation on the northwest coast of North America*. Seattle: University of Washington Press.

Jones, R. Russ, and Terri-Lynn Williams-Davidson. 2000. Applying Haida ethics in today's fishery. In *Celebration 2000: Restoring the balance through culture*, ed. S. Fair and R. Worl, 87–94. Juneau: Sealaska Heritage Foundation.

Joseph, Phillip. 1967. The history of the Aukquwon. In *New Alaskan* (Dec).

Langdon, Steve J. 2006. Tidal pulse fishing: Selective traditional salmon fishing techniques of the west coast of Prince of Wales Island. In *Traditional ecological knowledge and natural resource management*, ed. C.R. Menzies, 21–46. Lincoln: University of Nebraska Press.

Langdon, Stephen J. 2007. Sustaining a relationship: Inquiry into the emergence of a logic of engagement with salmon among the southern Tlingit. In *Native Americans and the environment*, ed. M.E. Harkin and D.R. Lewis, 233–276. Lincoln: University of Nebraska Press.

Pauly, Daniel. 1997. Putting fisheries management back in places. *Reviews in Fish Biology and Fisheries* 7: 125–127.

Swanton, John R. 1909. *Tlingit myths and texts*, Bureau of American Ethnology, Bulletin 39. Washington, DC: Government Printing Office.

Thornton, Thomas F. 2008. *Being and place among the Tlingit*. Seattle: University of Washington Press.

Thornton, Thomas F. 2009. Anatomy of a traditional cultural property: The saga of Auke Cape. *George Wright Forum* 26(1): 64–75.

Chapter 2.2
The Influence of Westernization on Water Resources Use and Conservation Among the Maasai People of Kenya

George M. Ogendi, Rose K. Morara, and Nicholas Olekaikai

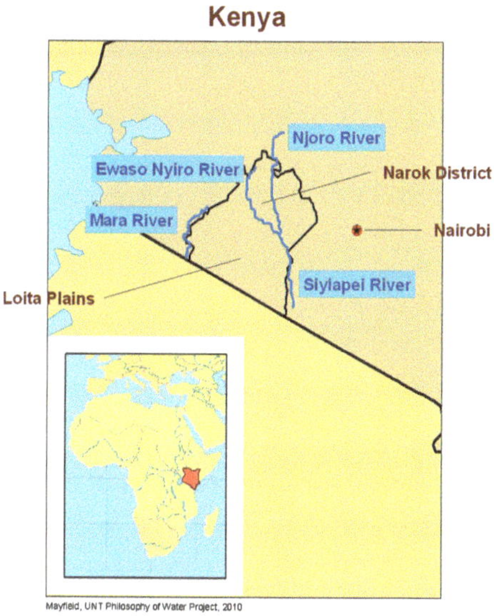

Map 2.2.1 Kenya

Maasai cattle herders of East Africa have adapted to and lived in water-scarce environments for centuries. Prior to the arrival of European settlers in Kenya in the early 1900s, the Maasai occupied large tracts of land in the expansive Rift Valley highlands and the savannah grasslands extending from Lake Rudolf (now Lake Turkana) to Lorochi Plateau in northern Tanzania. However, after more than a century of colonisation and loss of land, they currently live in the savannah grasslands of the Loita plains of southern Kenya and areas adjacent to the foothills of Mt. Kilimanjaro in northern Tanzania. Water scarcity in the region has been worsened by the ever-increasing human population coupled with environmental degradation.

The Maasai believe that rain is driven by the spirits but that the spirits respond to human action and mediation. The spirits show their intentions through signs in the

G.M. Ogendi (✉)
Department of Environmental Sciences, Egerton University, Nakuru, Kenya
e-mail: george1971ke@yahoo.com

R.K. Morara
Department of Applied Community and Development Studies, Egerton University, Kenya

N. Olekaikai
Department of Natural Resources, Egerton University, Kenya

B.R. Johnston et al. (eds.), *Water, Cultural Diversity, and Global Environmental Change: Emerging Trends, Sustainable Futures?*, DOI 10.1007/978-94-007-1774-9_10,

landscape, and the Maasai are able to read these signs. The signs inform Maasai knowledge of rainfall patterns and the cycles of seasons, which in turn inform their decisions about routes of migration and other pastoral activities. The Maasai have considerable understanding about how different pastures survive long spells of little or no rain, enabling the people to identify appropriate areas for sustainable grazing for their livestock. Livestock, including cattle, sheep, and goats, are the Maasai's main source of food and social status. A person's status is measured in terms of the number of animals that he or she possesses.

Over the past century, much Maasai land has been taken over for purposes other than pasture, creating ongoing conflicts as well as substantial changes in Maasai ways of life. Yet the Maasai still depend on reading the signs and their traditional Maasai knowledge for survival. This knowledge of seasonal, hydrological, and ecological cycles must be central to generating sustainable livelihoods in Maasai territory.

2.2.1 The Maasai Age-Sets and Livestock

The Maasai speak the *Maa* language, and the community is organised into different age-sets with specific roles; *iltasat* are the retired elders; *ilmoruak* are active senior elders who allocate resources and settle conflicts; *olaiguenani* are the chiefs, the most important of whom is the *oloiboni* (the spiritual leader and rainmaker); and the *ilmurran* are the warriors and protectors The Maasai are united at the core by their passage through age-sets and the performance of four major ceremonies, each of which initiates a new life state (Olekaikai 2008).

Keeping livestock is fundamental to Maasai existence and Maasai livelihoods. When herding, their diet is based on sour cow milk combined with plant-based preservatives. Cattle are kept mainly for milk and are only slaughtered for meat during circumcision, initiation, and other ceremonies (Spear and Waller 1993). Women play an important role in keeping livestock progeny and production records. Wherever possible, Maasai graze their animals in the plains during wet seasons and move them to higher grounds in search of pasture and water during dry periods. The migratory grazing routes are strategically designed based on a reading of the landscape and the seasonal and inter-annual variations in the quantity and quality of pasture and water to ensure a steady supply of water throughout the nomadic movement.

2.2.2 Ownership and Management of Scarce Pasture and Water

Social, linguistic, and cultural ties are crucial to sustainable Maasai livelihood strategies. In Maasai society, local clans have primary rights to water and pastures in their customary area of residence. People understand natural resources to be exhaustible and therefore they must be shared, particularly during times of scarcity.

Fig. 2.2.1 Nicholas Olekaikai (*centre*), one of the article's authors, with morans (warriors) in full gear during the infamous 1992 ethnic conflict over the Mau Forest watershed (Source: Photo by authors)

For instance, during dry periods of the year, Maasai clans migrate to areas with some water and grazing. Members of visiting clans acquire secondary user rights to clan resources only with permission from the host community. Nevertheless, every member of the Maasai community has a basic right to access resources. Access to water and pasture is strictly controlled during the dry season when resources are scarce, and during times of scarcity, elders from the worst affected Maasai lands will rely on their social ties and connections for access to more plentiful areas. Another adaptation to water and pasture scarcity is herd splitting, in which people loan animals to relatives and friends.

Further, the Maasai reserve particular areas within their community for selective grazing and watering. Areas known as *lokeri* are set aside for calves or weak and old animals and are used during dry times to ensure the survival of vulnerable animals. There are also areas reserved for traditional ritual uses such as initiation ceremonies. Maasai consider such sites sacred and do not allow grazing animals therein without permission from the council of elders. The council only grants permission to graze on these reserved areas during harsh dry weather conditions. These sites (such as those containing *Enamina enkiyoo* or Loita Indigenous Forest) are presently important biodiversity conservation and water catchment zones, further enhancing the sustainability of local livelihoods. In the rainy season, the Maasai avoid settling and grazing in or near riverine vegetation or swamps and floodplains because disease-causing organisms such as tsetse flies, flukes, and mosquitoes are known to inhabit these environments. Grazing in and around the riparian vegetation is limited only to dry seasons, when the elders grant limited access on a first-come, first-served basis. This practice indirectly contributes to catchment protection and water resource conservation by minimizing the destructive activities of humans and livestock in the riparian corridors.

In the Maasai society, water rights normally determine who has access to rangelands, creating a strong connection between water rights and range management for the sustainable use of pasture and peaceful coexistence of different communities (Brehonny et al. 2004). Consequently, conservation measures for water resources and rangelands overlap. The elders set rules to govern the accessibility and use of water resources, including limits on the number of livestock per household, grazing area, and watering point. These regulations ensure equity, as no one individual or group is allowed to exploit water resources to the detriment of the whole community, while also enhancing social cohesion and minimising conflicts.

When water and pasture conflicts do occur, Maasai elders will meet under a sacred tree, preferably a fig, acacia, or baobab, and deliberate over the causes of the conflict, the culpability of the parties involved, and the necessary punishment. The guilty pay fines in form of livestock equivalents or community services, such as fencing a water hole to protect it from destruction by animals. Meeting under a sacred tree makes the decisions binding on all parties, and the offenders and their families will be cursed if they disobey. Elders emphasise sharing, mutual respect, and tolerance among the different clans while utilizing water and pasture.

2.2.3 Colonisation and Loss of Maasai Land

Major changes in land use in Kenya began with the establishment of plantations by British settlers in the early twentieth century. The colonial government forced indigenous peoples into reserves to free large tracts of land to accommodate the plantations. Between 1904 and 1913, more than 70% of Maasai land was taken away for this purpose (Miaron 2002), and further losses occurred with the passage of the National Parks Ordinance in 1945. The creation of national parks,

Box 2.2a Sand dams and plant indicators: Indigenous knowledge in water resource use and management among the Kamba People, Kenya

—Cyrus M. Kilonzi, George M. Ogendi, and Rose K. Morara

Introduction

The Kamba people of Kenya have knowledge that has enabled them to live for centuries in a water-scarce region, and this knowledge has been appreciated and incorporated in water resource development projects by the Kenyan government and its development partners.

Box Map 2.2a.1 Kenya

The Kamba are the third largest ethnic group in Kenya, and are mainly crop and animal farmers. In Ukambani (Kamba land), the sources of water for domestic use are mainly shallow wells, earth dams and pans, rivers, scoop holes and roof catchments. Climate change and other anthropogenic factors have exacerbated the water crisis in Ukambani, and droughts have increased in frequency and magnitude. This has led to the search for more reliable water sources, including wells and sand dams.

A sand dam is a barrier made of locally available materials such as masonry mortar and sandstone, which holds sand and water on the upstream side of an ephemeral river (Mutiso 2002). They increase the water storage capacity of the river and reduce evaporation by storing water within the sand and coarse gravel particles accumulating against the walls of the dam. Water in sand dams is accessed from wells dug into the gravel-sand water storage material. Sand dams are ecologically friendly, cost-effective, and low maintenance, and are estimated to have a lifespan of more than 100 years (Changemakers.net 2009).

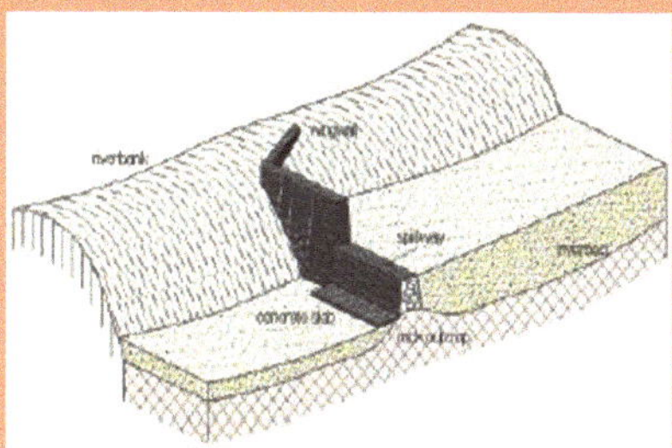

Box Fig. 2.2a.1 Schematic cross-section of sand dam (Source: Borst and Haas 2006)

(continued)

Box 2.2a (continued)

Box Fig. 2.2a.2 A sand dam during the dry season (Source: NWP 2007)

Box Fig. 2.2a.3 A sand dam during the wet season (Source: NWP 2007)

Box Fig. 2.2a.4 Community water well located near *Acacia xanthophloea* (Photo courtesy of the authors)

The reliability of sand dams depends on among others, the number of wells dug on the riverbed, sediment characteristics and the length and frequency of dry seasons.

Kamba have assisted water and environmental engineers in generating new water sources by using their knowledge of the local vegetation, soils, and sediments to determine the location for the construction of san dams. Kamba have known for centuries about plant species which indicate a high water table, including *Acacia xanthophloea*, *Ficus sycomorus, Bridelia microcantha* and *Syzygium cordatum.* Most hand-dug wells for domestic Kamba water supply are located near these species. Engineers constructing sand dams have used Kamba knowledge to appropriately locate these new developments in water rich areas.

Sand dams have greatly improved water accessibility in Ukambani and have transformed livelihoods in Ukambani enabling residents to grow tomatoes, kale, onions, mangoes, bananas, sugarcane, and fruit trees and many other tree seedlings that are known to thrive in high rainfall areas. The year-round availability of water makes small scale irrigation and food production possible, and sales from farm produce have improved family incomes significantly (Changemakers.net 2009). Food and financial security in the region have both been improved by these small-scale and low cost collaborative developments.

(continued)

Box 2.2a (continued)

Box Fig. 2.2a.5 Some of food crops (kales, paw-paws and sugarcane) irrigated by water from a sand dam (Photo courtesy of the authors)

Kamba Indigenous Knowledge has been productively combined with knowledge from elsewhere to improve local livelihoods in arid and semi-arid environments. This case study demonstrates the enormous potential of Indigenous Knowledge in facilitating appropriate local-scale solutions in developing nations.

Literature cited

Borst, L., and S.A. Haas. 2006. Hydrology of sand storage dams, a case study in the Kiindu catchment, Kitui District, Kenya. Master thesis, Vrije Universiteit, Amsterdam.

Changemakers.net. 2009. Social entrepreneurship worldwide. Website accessed on 18 June 2009.

Mutiso, S. 2002. The significance of subsurface water storage in Kenya. In *Management of aquifer recharge and subsurface storage NNC-IAH publication #4*. Utrecht, The Netherlands.

Netherlands Water Partnership (NWP). 2007. *Smart water harvesting solutions. Examples of innovation, low cost technologies for rainwater, fog, runoff water and groundwater*, Amsterdam: KIT Publishers, 41–47.

Amboseli and Tsavo, and the national reserves, Maasai Mara, Kitengela, and Samburu, on Maasai land, further complicated the Maasai's nomadic way of life. Then, after Kenyan independence in 1963 the government gave Maasai land to landless people from other ethnic tribes. The latest episode in the process of dispossession has been the allocation of Mau Complex Forests (MCF) to Kenyans of other ethnicities in the late 1990s. Because of this programme, the Maasai lost most of the dry-season grazing areas, and their socioeconomic activities have been confined to a restricted range. Having no regard for or understanding of Maasai needs and livelihoods, the perpetrators of the recent excision have cleared the forest for human settlements, agriculture, and other developments. Little or no consultation with the Maasai took place before any of the aforementioned land reforms and seizures.

2.2.4 Environmental and Social Effects of Changes in Land Tenure

When the Maasai were removed from what came to be known as the white highlands – Mau Escarpment, Ngong Hills, Laikipia Plateau, and Chylu Hills – for white settlers, these areas had ample water and pasture even in dry seasons. Since the displacement of the Maasai, receding water tables, declining aquifer recharge, and drying rivers have become common occurrences in this region. Currently, Kenyans are demanding efforts to save the MCF as an important water catchment for rivers supplying human settlements. The Maasai expressed discontentment over the destruction of the MCF before any scientific studies were conducted, and findings from scientific studies have since corroborated their fears. Now the Kenyan government is frantically trying to save the forest from total collapse. Unlike the MCF, areas that continue to be sustainably managed by the Maasai people are rich in biodiversity and water. A good example is the Naimina Enkiyoo Loita Indigenous Forest, an intact ecosystem thanks to the strong leadership and indigenous knowledge of the Maasai Oloiboni and the Loita community elders.

2.2.5 Introduced Water Technologies

Although water is necessary for Maasai livelihoods, increasing its availability without regard to its sustainability is detrimental and counter to Maasai knowledge. The loss of territory and the introduction of alternative land tenure systems has compelled the abandonment of other aspects of land use including the conservation of water and pasture, because that knowledge cannot be applied to small or fragmented pieces of land. The transfer of land from communal to individual ownership (Miaron 2002) and the government's provision of boreholes have decreased nomadism. Most Maasai people have been obliged or encouraged to move into sedentary agropastoralism or agriculture, and this sedentarization correlates with an exponential rise in population, from under 30,000 in 1883 to over 500,000 in 1999 (Government of Kenya 2002). These factors, combined with the rapid deterioration of Maasai traditional knowledge and practice, have placed severe pressure on limited land and water resources in Maasai areas. Over the last 100 years, rivers and springs have dried up, forests have been cleared, and biodiversity and soil fertility have decreased. Water levels in rivers such as the Mara, Narok, Njoro, Ewaso Nyiro, and Siyiapei have declined steadily, and conflicts between the Maasai and the neighbouring Samburus, Turkanas, and Kikuyus over water and pasture have escalated in frequency and magnitude. Governmental attempts to ensure water availability have been unsuccessful; boreholes and dams have silted up beyond rehabilitation or collapsed.

While Maasai populations have increased, quality of life has not, and the ever-larger dry areas are forcing people to travel longer distances in search of water and pasture. In some areas, women and girls now have to walk 25 km to a water source

Fig. 2.2.2 Water resource degradation and water transport on donkeys from River Njoro, Kenya (Source: Photo by authors)

and return carrying 20–60 litres of water for their domestic needs. Pastoral development projects geared toward sedentarizing the Maasai have resulted in unsustainable use of previously healthy and sustainable land. Findings from a range of pastoral studies have identified decreased food security and water quality in Maasai communities following sedentarization (Kariuki 1994; Tadingar 1994). Sedentarization and the associated population growth have transformed Maasai culture and negatively affected Maasai livelihoods (Lane and Moorehead 1994).

2.2.6 Conclusion

The Maasai retain extensive knowledge of their ecosystems, including weather patterns, edible plants, location of water, hydrological and seasonal cycles, and many other aspects. Prior to the colonial period, they lived sustainably in these arid and semi-arid environments. Any future development technologies and interventions must recognise that the ecosystems of which the Maasai are a part have thrived because of the beliefs, norms, values, and practices of the Maasai community. Maasai knowledge and practices offer numerous practical, cost-effective and socioculturally acceptable solutions to specific environmental challenges, including techniques to deal with water scarcity in arid and semi-arid lands. The value of dispersed, small-scale, and temporary water sources along pastoral migration routes is already clear (Tadingar 1994); their use increases the radius of available rangeland and reduces overgrazing around water sources.

Nomadic pastoralism remains the best option for these areas of Kenya where resource distribution is patchy and seasonal, and Maasai pastoral knowledge is fundamental to the success of life in those regions. The government of Kenya and its development partners have promoted inappropriate technologies and interventions in spite of strong opposition from the local people, resulting in negative social and environmental consequences. The accurate documentation of local Maasai knowledge and its incorporation into future development programmes are crucial to successful and sustainable futures.

It would be impossible to restore the Maasai way of life, given the socioeconomic and political landscape in present-day Kenya and the future predictions of the impact of climate change on the region. However, the Maasai remain committed to sustainable futures in their own country and to using the knowledge they already possess, as well as new knowledge incorporated in appropriate ways. Maasai culture

Fig. 2.2.3 Maasai herdsmen with their cattle on a heavily degraded grazing land in Kenya (Source: Photo by authors)

is now providing a resource in other ways; public/private partnerships have enabled the Maasai to build cultural and ecotourism villages called *manyattas* (traditional Maasai homesteads) and enabled the Maasai to acquire patents for products developed from their knowledge of medicinal plants. In their new modes of life, the Maasai recognise the crucial role of formal education beyond the initiation ceremony in generating and sustaining environmental awareness, particularly amongst young people (Ogendi and Ong'oa 2009). Increased environmental awareness is necessary at all levels of society: development partners, religious organizations, government, and community-based organizations such as cooperative societies and livestock and food markets. One part of the knowledge that underlies improved environmental awareness must come from those who lived sustainably on the lands of East Africa for many centuries and who continue to live there.

Resources

Brehonny, E., A. Mattee, and B. Nangoro. 2004. Inclusion and opportunity: A way to reduce marginalisation and vulnerability. A study on the development of policy component for Ereto II (commissioned by Ereto II).

Government of Kenya. 2002. *Kenya 1999 population and housing census*. Nairobi: Central Bureau of Statistics, Ministry of Finance and Planning.

Kariuki, G.K. 1994. Socio-economic and phytoecological environment of nomads in smallholder irrigation schemes in Isiolo district, Kenya. MSc thesis, University of Nairobi.

Lane, C., and R. Moorehead. 1994. *Who should own the range? New thinking on pastoral resource tenure in dryland Africa. Pastoral Land Tenure Series no. 3*. London: International Institute for Environment and Development.

Miaron, J. 2002. *Indigenous peoples of eastern Africa and sustainable development*. Nairobi: Indigenous Information Network and Trocaire.

Ogendi, G.M., and I.M. Ong'oa. 2009. A comparative study of water accessibility, use, and conservation among children in selected sites of Kenya and the United States of America. In *Young people, education, and sustainable development: Exploring principles, perspectives, and praxis*, ed. P. Corcoran and P.M. Osano, 215–221. Wageningen: Wageningen Academic Publishers.

Olekaikai, N.K.W. 2008. Resources condition and their influence on the well being of communities in Marigat. Division of Baringo District, Kenya. MSc thesis, Egerton University.

Spear, T., and R. Waller. 1993. *Being Maasai: ethnicity and identity in east Africa*. London: James Currey.

Tadingar, T. 1994. *Pastoral development in Sub-Saharan Africa: An integration of modern and indigenous technical knowledge. The African Pastoral Forum, Working papers series no. 2*. Nairobi: PINEP, University of Nairobi.

Chapter 2.3
Groundwater and Qanats in Syria: Leadership, Ownership, and Abandonment

Joshka Wessels

Groundwater management is among the most important challenges facing the Middle East and North Africa (MENA) region (World Bank 2000). It is the world's most arid, with 1% of the global renewable fresh water available to its population. By the 1990s, eight countries in the Middle East (Kuwait, Qatar, Bahrain, Saudi Arabia, United Arab Emirates, Jordan, Yemen, and Israel) had crossed the red line of 'absolute water scarcity' (Engelman and LeRoy 1993; Swain 1998). The population of nearly 300 million has doubled in the last three decades and is expected to double again by 2025 (Blanche 2001). This increase will mean a massive pressure on the already scarce water resources, yet newer technologies – especially groundwater pumping devices – cannot for long keep pace with rising water demand, as they are already operating at unsustainable rates. Instead, these technologies may further deplete currently viable aquifers and preclude their use even for lower-impact sustainable traditional irrigation systems, such as qanats (Lightfoot 1996).

Qanats are a water management system used to provide a reliable supply of water to human settlements and to irrigate fields in hot, arid, and semi-arid climates. Qanats consist of subterranean wells connected by gently sloping tunnels. Qanats can efficiently deliver significant and sustainable quantities of water to the surface without the need for pumping, as they rely on natural seepage into the tunnel to source the water and on gravity to deliver it. The closed tunnel systems prevent evaporation in hot, dry climates, even when water is transported long distances. The technique is thought to originate in Old Persia (present-day Iran) around 3,000 years ago and has been transported to a diverse array of locations across the globe. Qanats have been used in Syria, the focus of this case study, since at least 64 BC.

J. Wessels (✉)
Department of Political Science, Lund University, Sweden
e-mail: joshka.wessels@svet.lu.se; joshka.wessels@btinternet.com

B.R. Johnston et al. (eds.), *Water, Cultural Diversity, and Global Environmental Change: Emerging Trends, Sustainable Futures?*, DOI 10.1007/978-94-007-1774-9_11,

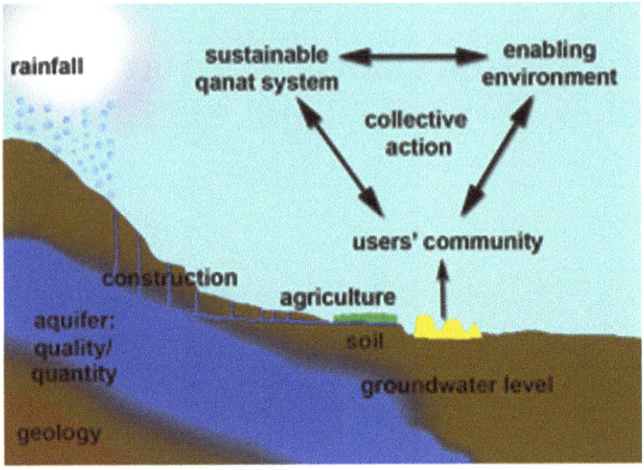

Fig. 2.3.1 Qanats as human ecosystem

Despite their long history of use, qanats are increasingly drying up and being abandoned throughout the MENA region. Technological development, population growth, rising consumption levels, and changing socioeconomic conditions have forced former qanat users to change their practices from agricultural self-sufficiency based on traditional irrigation to pump-driven irrigation or nonagricultural income sources. Qanats, if not maintained properly, can silt up and stop flowing, something that occurs when collective or community structures for qanat maintenance break down. Overexploitation by mechanical pumps has lowered water tables throughout the MENA region, which badly affects qanats. Qanats constitute complex human ecosystems affected by a range of exogenous and endogenous factors. This study focuses on the underlying factors inside a qanat community that influence collective action for the traditional irrigation system, both in terms of protection for its catchment and the maintenance of the water delivery system itself. Contextualisation of such collective action and its breakdown is necessary to better understand how qanats can provide one option for achieving sustainable irrigation in the future in increasingly water-scarce environments.

2.3.1 Short Historical Background of Syrian Qanats

Syria is located at the eastern part of the Mediterranean Sea between Lebanon and Turkey. Water is the main limiting factor for agriculture in the country, and the historical development of qanats greatly influenced settlement patterns. The sites of villages and towns as well as the amount of irrigated land were historically determined by qanat construction, and the need to create a regime capable of digging and maintaining qanats profoundly shaped local politics. In Greek and Roman times, the qanats in Syria played a vital role in the expansion of empires and the development of thriving desert cities like Palmyra. In northern Syria, settlements became centres of religion, science, philosophy and arts, and most of them relied on water from qanats or hand-dug wells.

Having overcome the Byzantine forces at Yarmuk in 636 AD, the Arabs took over the country, along with Iran and Egypt. The Omayyads (661–750) made Syria and Damascus the centre of the Muslim empire (Cotillon 1993). The Omayyad state was highly centralised, and with qanats inherited from previous empires, Syria became the most prosperous province of any Islamic caliphate, with flourishing agricultural production (Lightfoot 1996; Kobori 1990; Goblot 1979). The further diffusion of qanats after the Islamic Empire probably resulted from continued Persian contact and the spread of Islam (Lightfoot 1996).

Qanats are dynamic and must be constantly repaired to maintain waterflow, so local populations have continuously altered their shape and construction. In some cases they lay abandoned for years and were renovated and reused by new settlers. In the twentieth century, qanat building slowed down considerably, except for some occasional attempts in the ghouta (a green agricultural belt that surrounds the city) of Damascus (Goblot 1979; Lightfoot 1996), and since the 1960s the advent of diesel-driven pumps and private wells has resulted in a much faster rate of decline.

2.3.2 Ecology, Characteristics, and Types of Qanats

Rainfall, evapotranspiration, topography, morphology, and hydrogeology determine the locations and original discharge of the qanats in Syria (Lightfoot 1996). All of them are found at or below the 500 mm rainfall isohyet and almost 90% within 25 km of uplands (Lightfoot 1996). Most qanats can be found in arid, alluvial plains at the margins of highland areas or rivers (Lightfoot 1996). Alluvial aquifers are the shallowest and most widespread water bodies in the Middle East. Because of topographical and geological diversity, the Syrian qanats vary considerably in type and construction. We can distinguish three main types:

1. Spring-based – dug in calcareous karstic solid rock near mountainous areas;
2. Infiltration-based – dug in alluvial plains with a relatively flat surface;
3. River-based – dug in the alluvial plain of a major river.

Qanats vary a great deal in Syria. Whereas the qanats in hilly and relatively wet northwest Syria are short and often spring-based, those in central Syria, a dry steppe with a gradual slope and sandy soil, are long and rely on infiltration. Southwest Syria contains river-based qanats mainly around the city of Damascus, where the land is mountainous, as well as spring- and infiltration-based qanats. The yearly snowfall in both the Qalamun and Anti-Lebanon mountains in southwest Syria ensures a good supply of recharge into both the alluvial plains and the underlying geology. This

Fig. 2.3.2 Open channel in qanat community in North West Syria (Photo credit: Joshka Wessels)

reliable water supply is most probably why the region has such a high concentration of qanats. In the eastern part of southwest Syria, the boundary of the steppe runs along the mountain range. Here, longer infiltration-based qanats, such as the ones in Dmayr, can be found.

The locations of qanats in Syria, as in other countries with this system of irrigation (Bonine 1989; Wilkinson 1977), greatly influenced settlement patterns and traditional field systems. Qanat construction determined the village and town sites, as well as the amount of irrigated land. This last was designated at the time of construction based on the estimated flow, cycles, local morphology, and soil types. The size of the land to be irrigated also varies by season, with the smallest surface watered during summer time when the qanats provide a base-flow (also called drought flow, groundwater recession flow, low flow, and sustained or fair-weather runoff, is the portion of streamflow that comes from 'the sum of deep subsurface flow and delayed shallow subsurface flow'. Not to be confused with groundwater flow. Minimum flow is the minimum amount needed to service and sustain the ecosystem). In some sites, such as Dmayr, users have agreed to special summer and winter irrigation cycles. The effect of global climate change on qanats has not been investigated properly, but farmers do report a decrease in rainfall over the past decades, which they blame for the drying of qanats. However, this phenomenon must be understood in relation to the increase in water demand because of population growth, the introduction of pump-driven wells and subsequent overexploitation of the aquifer, resulting in lower groundwater tables.

2.3.3 Ownership and Regulation of Qanats

Different types of qanat ownership systems exist. Property rights in Syria are characterised by a coexistence of the customary law (*'urf*), largely based on Islamic principles on property and land tenure stemming from nineteenth-century Ottoman practices (Métral et al. 1981; Metral 1982, 1987), and the formal legal system of the state (*qanun*) (Rae et al. 2002). All qanats in Syria are either *mulk*, full private ownership, or *amlak ed-dawlah*, state-owned land.

Full ownership in Syria consists of the land where the underground tunnel is situated, along with the rights to use the qanat's water and the land irrigated by that water. This form of ownership ownership (*mulk*) encompasses owning the land and its resources and having the right to use and transfer them, by giving them away, settling them as an endowment (*waqf*), or passing them down to relatives or descendants (cf. Lewis 1987, 2000; Vincent 1995).

Within this broad distinction between privately owned and state-owned, one can categorise ownership as (1) *private*, where one family or tribal group owns the land and water rights; (2) *communal*, where a community consisting of various extended, unrelated families and other groups owns the land and water rights; and (3) *state*, where the government has ownership and responsibility for the qanat and its land and water rights *(amlak ed-dawlah).* Private and communal *ownership* are not necessarily connected to community *leadership.* In many cases, although the qanat is private or communal, people expect the state to keep up the qanat and settle any disputes.

Qanat irrigation systems have a complex organisation of cycles and rights based on time, volume, and discharge (Lightfoot 1996; Beaumont et al. 1989; Wilkinson 1977). In Syria much of this organisation remains partially intact at flowing qanat sites. Three main use rights of qanat water can be distinguished: the right to drink, the right to use the water domestically, and the right to irrigate. The right to drink and use water to wash before prayer is free to everyone who visits the qanat outlet. This right is based on the so-called Islamic Law of Thirst that all people and animals must be allowed to drink the water. The right to use the water for domestic purposes is free but generally confined to those living close to the qanat outlet. Irrigation rights are not free and are confined to certain families and groups. These distinctions are also shaped by Islamic law, under which water cannot be owned unless it is stored and measured (Vincent 1995; Bulloch and Darwish 1993; Metral 1982; Caponera 1973). Moreover, it distinguishes between rivers, springs, and wells (Bulloch and Darwish 1993); the spring category applies to qanats. Natural springs are free for all to use. Those who 'uncovered a spring and caused it to flow' own the water source in common. A spring uncovered on someone's private land belongs to the landowner, but Islamic law obliges him or her to offer water free of charge to others (Vincent 1995; Bulloch and Darwish 1993; Metral 1982). It is difficult to find out who actually uncovered the spring at qanat sites in Syria, but those families and groups who have the right to use the qanat water for irrigation are essentially the community that owns the qanat.

The irrigation rights are bound to a complex system of timeshares and cycles. Each qanat user has the right to irrigate for a certain period at a designated time. His or her water share is measured in time and sometimes volume when the qanat has an irrigation reservoir. For each qanat, the total number of timeshares is arranged in a rotation cycle or period of irrigation called *addan*. The length of the *addan* is rarely changed. The lengths of the timeshares, however, do change and are subject to inheritance, division, and sale.

Another measurement that regulates qanat use is the use of irrigation reservoirs or *birkeh*. If a qanat gives sufficient water, the flow and thus irrigation continues for 24 hours a day. However, the flow decreases and there is not enough pressure to reach the last lands along the qanat, the users build a collection reservoir to regulate the flow. The reservoir is then opened for a specific number of hours per day. The water collected in it has enough pressure to reach the outer borders of the irrigated land. The *birkeh* is often guarded by a *natur*, who opens and controls the valve, directs the flow towards the users at the designated times, and measures the given volume using a pole placed in the birkeh. In most cases, the users pay the fee of the *natur* by pooling money in a system called *sunduq* (literally 'box'). In this system, the users pay a certain fixed amount per year for their irrigation shares to a qanat committee consisting of the biggest shareholders. From this common fund, repairs, fees for the *natur*, and other costs are paid.

Regulations and laws protect the tunnels of the qanat system. The main traditional law to prevent qanats from drying out is the *harim* principle, which mandates that no water well can be sunk within a specified distance from another water project (Vincent 1995, Beaumont et al. 1989; Caponera 1973). In Syria, *harim* is a defined boundary around the qanat tunnel and the source area or outlet within which it is forbidden to drill wells or dig another qanat. The exact boundaries vary from site to site, but the range is between 1 and 5 km, depending on the local conditions. In Syria specifically, the *harim* rule is barely implemented on the existing qanat sites.

2.3.4 Qanat Abandonment: Regulation, Migration, and Collective Action

Why have many qanats been abandoned? Overuse of the aquifer and subsequent falling groundwater tables, high migration rates, the erosion of traditional regimes of water and land rights and specialised local qanat knowledge are all likely factors contributing to the abandonment of qanats. The fact that there are more flowing

Map 2.3.1 Distribution of Qanats in Syria. A snapshot of 2001: Flowing qanats throughout Syria. A 2001 survey undertaken by the author found 17 qanat sites in Middle Syria; 11 were dry, three had static water, and three (Arak, Taybeh, and Breij) were flowing. The most active site is Breij, where the users regularly clean the open sections of the tunnel. In southwest Syria there are 17 qanat sites; six were dry, one flowed only in winter, and ten were flowing (Credit: courtesy of Dr. Dale Lightfoot 1996)

qanats in the southwest region is likely the consequence of favourable biophysical and environmental factors, a long, uninterrupted history of social organisation to tend the qanats and, a policy of regional government support in the Awaj/Barada basin.

At a national level, we can say more about the multidimensional relationship between ownership, leadership, and abandonment of qanats. Water policy is a sensitive issue in Syria, whose national policies on water and irrigated agriculture have always been based on a strong desire for food self-sufficiency. But the current water crisis means that this policy simply will not succeed in the future (Elhadj 2004; Salman 2004). Syria is now water-poor and yet remains heavily on agricultural resources: as a dominant economic sector, it counts for 32% of the GDP and 60% of the non-oil exports (Salman and Mualla 2003; Bazza and Najib 2003; Le Moigne et al. 1992). Irrigation constitutes the main demand for groundwater making up more than 90% of the total water use. The result is the mushrooming of pumped wells and a serious countrywide drop in groundwater levels, except for the coastal areas. Agriculture is the main cause of overexploitation and degradation of water quality in the Arab region (UNDP 2005). Apart from the customary laws and rules, the national water legislation does not refer to qanats in particular. Most of the qanats are regarded as springs underneath private or community lands, a status that implies private or community ownership or control by the Directorate of Antiquities. The local or regional governments do not have specific responsibility or support for qanats or qanat maintenance except in the southwest. In this region, the government has allocated a certain sum on yearly basis to maintain qanats where necessary.

There is a discrepancy between the official legislation on water management and the support and maintenance needs of qanats. Whereas the qanats were subject to the rules and regulations of a central government in the past, today the central government does not regard the systems. The priorities for the government today are bigger groundwater projects that promise a high, albeit short-term, return. This lack of an institutional context and enabling environment for collective action to maintain the qanats could prove to be detrimental for their longevity.

The competition between diesel-powered pumps and qanats for water resources has been evident ever since the introduction of the new pumped well technology to the Middle East in the 1960s. Ehlers and Saidi (1989) reported that 'The change from qanat irrigation to water exploitation by means of motor pumps and large dams has had tremendous effects on the fragile ecology of the arid highlands of Iran'. Between the 1960s and the 1980s, qanats in Iran symbolised backwardness in opposition to the promising future of diesel pumps (Molle et al. 2003; Ehlers and Saidi 1989). Currently, most of Syria's agricultural groundwater supply by far is provided by pumps. Apart from the geophysical consequences of falling groundwater tables, the introduction of pumped wells also had considerable social implications. Formerly cohesive qanat communities disintegrated when the users started to drill pumped wells. Investment costs in construction were high compared to the maintenance costs of qanats, but the growing demand for irrigated agriculture and the short-term high return from pumps led many well-off farmers to opt for drilling a private well. This decision created a discrepancy, as poorer farmers could not afford to invest in modern technology. In northern Syria we found several villages where families killed each other because someone had drilled a pumped well too close to his neighbour. The rapid increase of privately owned pumped wells, or the 'pump revolution', as Molle et al. (2003) calls it, has widespread hydrological, sociopolitical, and economic implications throughout arid countries in the Middle East and Asia (Molle et al. 2003). It is also one of the main causes for the abandonment of qanats. In Syria, migration away from those qanat sites that dried up since the introduction of pumped wells were very high. At the sites where pumps did not destroy the qanats, the social organization stayed relatively intact and water kept flowing through the tunnels.

2.3.5 Collective Action

Falling groundwater tables because of industrialised overexploitation, land reform, and migration have often been identified as main reasons for the abandonment of qanats (Hoogesteger and Vincent 2006; Lightfoot 1996; Vincent 1995; Beaumont 1971, 1993; Safadi 1990; Birks 1984; Kobori et al. 1982). Qanats must be regularly cleaned and maintained to ensure water flow (Lightfoot 1996; Kobori 1990; Beaumont et al. 1989; Goblot 1979; Wilkinson 1977). 'Flowing qanats' can be used as an indicator for collective action; a lack of collective maintenance is strongly related to the dryness of qanats. Types of ownership and leadership have a profound impact on the whether collective action emerges. *Communal* ownership seems to be beneficial for maintenance of qanats, but *state* leadership appears to initiate and support the collective action crucial to maintain regular upkeep. These conclusions seem contradictory, but if we look at investment attitudes, we see that many of the users blame non-human phenomena like God or a lack of rainfall for the dryness of qanats, hence they do not feel inclined to invest in maintaining the system. Culture and religion play a great role in how people perceive and determine cause and effect (McCay 2002; Douglas 2002). In this case, God (by sending little rainfall) and not humans have caused the qanats to dry. Therefore, the responsibility to make the qanat flow lies with God and not humans. In 'cultural theoretical' terms, we can say that in those qanat communities unwilling to invest, the people rely heavily on state authorities (hierarchy) and leave the maintenance to fate (fatalism). Together with the political history of the state apparatus in Syria, the result is a heavy reliance on state and fate to repair and clean qanats.

Urban water supply is an increasing problem in Syria, as the population of cities has grown exponentially in Syria, just as they have in many other countries in the Middle East. The major cities of Damascus and Aleppo regulate the drinking water supply during the summer. The main policy issue in the water sector is to increase water use efficiency on the demand side. Since 2003 there has been a real push for more efficient technologies, like drip irrigation, and institutional reform, such as the establishment of coordination mechanisms and possibly Water Users Associations (Bazza and Naijb 2003).

2.3.6 Conclusion

Qanats in Syria are ancient tunnel systems tapping the groundwater resources using only gravity. Many have currently been abandoned, but in some areas they are still in active use. This abstraction system is the most sustainable way to bring water to the desert. But since the introduction of diesel-powered pump wells in the 1960s, the qanats have been rapidly drying up. The pumps caused an overexploitation of the groundwater resources. Combined with a higher demand for irrigated agriculture because of the growing population, the overexploitation depleted the resources and caused the groundwater tables to lower at an exponential rate.

The abandonment of qanats are warning signs that the groundwater resources in Syria are increasingly being exploited in an unsustainable manner. Although the qanats have been flowing for tens of hundreds of years, it took only the last 30 years for them to dry up at a large scale. This is cause for serious environmental concern. These developments should not be taken lightly; with the increasing impact of climate change in the region, they may forebode major humanitarian disasters ahead. Besides the loss of qanats, the traditional knowledge about qanat maintenance has been lost for the next generations. This valuable human resource could be important for sustainable groundwater management in the future. By rehabilitating the qanat system in combination with efficient groundwater management measures, rural migration to the urban areas can be slowed down and the Syrian water resources saved for future generations.

Fig. 2.3.3 Users renovating and cleaning their qanat in South West Syria (Photo credit: Joshka Wessels)

At the moment, the Syrian national government has identified the increasing problems with the groundwater resources, but illegal and uncontrolled diesel-powered pump well are still being drilled. The crisis in urban water supplies and continuing migration to the cities are also critical issues. Demand management and alternative options like the reuse of waste water could help lower this pressure. Next to that, qanat rehabilitation, conservation of traditional knowledge, agricultural water demand management, and control of pumping activities can all be part of an integrated water management approach at the local, regional, and national levels.

Resources

Agrawal, A. 2001. Common resources and institutional sustainability. In *Drama of the commons*, ed. E. Ostrom, 41–85. Washington, DC: National Academy Press.

Allan, J.A. 2001. Virtual water – Economically invisible and politically silent – A way to solve strategic water problems. *International Water and Irrigation* 21(4): 39–41.

Allan, J.A. 2002. *The Middle East water question, hydropolitics and the global economy*. London: Tauris.

Bazza, M., and R. Najib. 2003. Towards improved water demand management in agriculture in the Syrian Arab Republic. Paper presented at the first international symposium on management and rationalization of water resources use in agriculture, University of Damascus, FAO, Syria, 28–29 Apr 28–29 2003. http://www.fao.org/docrep/008/af981e/af981e00.htm. Accessed 29 July 2010.

Beaumont, P. 1971. Qanats systems in Iran. *Bulletin of the International Association of Scientific Hydrology* 16(1 March): 35–43.

Beaumont, P. 1993. *Drylands, environmental management and development*. London: Routledge.

Beaumont, P., M. Bonine, and K. McLachlan (eds.). 1989. *Qanat, Kariz and Khattara: Traditional water systems in the Middle East and North Africa*. Wisbech: Middle East and North African Studies (Menas) Press, Ltd.

Birks, J.S. 1984. The falaj: Modern problems and some possible solutions. *Water Lines* 2(4): 28–31.

Blanche, E. 2001. Mid-East water crisis: Time is running out. *The Middle East* 311(April): 19–22.

Bonine, M. 1989. Qanats, fieldsystems, and morphology: rectangularity on the Iranian Plateau. In Qanat, Karez and Khattara. Traditional water systems in the Middle East and North Africa, eds. P. Beaumont, M. Bonine, and K. Mc Lachlan. London: Middle East and North African Studies Press ltd.

Briant, P. 2001. *Irrigation et drainage dans l'Antiquité, qanats et canalisation souterraine en Iran, en Égypte et en Grèce, séminaire tenue au Collège de France sous la direction de Pierre Briant*. Paris: Collège de France, Thotm éditions.

Bulloch, J., and A. Darwish. 1993. *Coming conflicts: Water wars in the Middle East*. London: St. Edmundsbury Press.

Caponera, D.A. 1973. *Water laws in Moslem countries*. FAO Irrigation and drainage paper. Rome: FAO.

Cotillon, J. 1993. *Water from dams in Syria*. Paris: ICOLD.

Douglas, M. 2002. Traditional culture, let's hear no more about it. In *Culture and public action*, ed. V. Rao and M. Walton, 85–109. Washington, DC: World Bank.

Ehlers, E., and A. Saidi. 1989. Qanats and pumped wells: The case of Assad'abad, Hamadan. In *Qanat, Kariz and Khattara: Traditional water systems in the Middle East and North Africa*, ed. P. Beaumont, M. Bonine, and K. McLachlan, 89–112. Wisbech: Middle East and North African Studies (Menas) Press, Ltd.

Elhadj, E. 2004. The domestic water crisis in Syria's Greater Damascus region. Occasional Paper no. 47, Water Issues Study Group, School of Oriental and African Studies/King's College, University of London.

Engelman, R., and P. LeRoy. 1993. *Sustaining water: Population and the future of renewable water supplies*. Washington, DC: Population and Environment programme, Population Action International.

Goblot, H. 1979. *Les Qanats, une technique d'acquisition de l'eau*. Paris: École des Hautes Études en Sciences Sociales, Centre de Recherches Historiques.

Hakimian, H. 2003. *Water scarcity and food imports: An empirical investigation of the 'virtual water' hypothesis in the MENA region*. London: Cass Business School, City University.

Hamdy, A. 2001. Water resources management in the Mediterranean countries: Priority actions. *MEDIT* 2(6): 24–29.

Hoogesteger, J., and L. Vincent. 2006. Institutional change and farmer participation in the Abshar irrigation system, Iran. In Proceedings of the CERES Summer School 2006, 358–376. Amsterdam: CERES, KIT.

Hunt, C. 2004. *Thirsty planet: Strategies for sustainable water management*. London: Zed Books.
Kabour, Y. 1992. Planning and management of water resources in Syria. In *Country experiences with water resources management: Economic, institutional, technological and environmental issues*, Technical Paper no. 175, ed. G. Le Moigne, G. Barghouti, S. Feder, G. Garbus, and M. Xie. Washington, DC: The World Bank.
Kobori, I. 1964. Human geography and methods of irrigation in the Central Andes. In *Land use in semi-arid Mediterranean climates, proceedings of the UNESCO/International Geographical Union Symposium, Iraklion (Greece)*, 19–26 Sept 1962. *Arid Zone Research* XXVI, 135–137. Paris: UNESCO.
Kobori, I. 1969. Le Systeme d'irrigation dans le sahara Central-Tidikelt. *Bulletin of the Department of Geography (University of Tokyo)* 1: 1–32.
Kobori, I. 1973. Some notes on diffusion of qanat. *Orient* IX: 43–66.
Kobori, I. 1976. Notes on foggara in the Algerian Sahara. *Bulletin of the Department of Geography (University of Tokyo)* 8: 41–55.
Kobori, I. 1989. Comparative studies on the formation of qanat water system. *Bulletin of the Institute of Social Sciences (Meiji University)* 12: 1–40.
Kobori, I. 1990. Les qanat en Syrie. In *Techniques et pratiques hydro-agricoles traditionnelles en domaine irrigue*, vol. 2, ed. B. Geyer, 321–328. Paris: Librairie Orientaliste Paul Geuthner.
Kobori, I., et al. 1980. *Qanawat Romani of Taibe Oasis*. Tokyo University scientific mission for the comparative study of the foggara oasis in the arid zone of the old continent, Report no. 1. Tokyo: Tokyo University.
Kobori, I. et al. 1982. Case studies of foggara oases in the Algerian Sahara and Syria. Tokyo University scientific mission for the comparative study of the foggara oasis in the arid zone of the old continent, Report no. 2. Tokyo: Tokyo University.
Lambton, A.K.S. 1989. The origin, diffusion and functioning of the qanat. In *Qanat, Kariz and Khattara: Traditional water systems in the Middle East and North Africa*, ed. P. Beaumont, M. Bonine, and K. McLachlan, 5–12. Wisbech: Middle East and North African Studies (Menas) Press, Ltd.
Lassoe, J. 1953. Reflections on modern and ancient waterworks. *Journal of Cuneiform Studies* 7: 5–26.
Le Moigne, Guy, et al. 1992. Country experiences with water resources management: Economic, institutional, technological and environmental issues. Technical Paper no. 175. Washington, DC: The World Bank.
Lewis, N. 1987. *Nomads and settlers in Syria and Jordan, 1800–1980*. Cambridge: Cambridge University Press.
Lewis, N. 2000. Travellers at Tayyibah: The northern desert of Syria, 1600–1980. In *Desert travellers from Herodotus to T. E. Lawrence*, ed. J. Starkey and O. El Daly. Durham: ASTENE, University of Durham.
Lightfoot, D.R. 1996. Syrian qanat Romani: History, ecology, abandonment. *Journal of Arid Environments* 33: 321–336.
Lightfoot, D.R. 1997. Qanats in the Levant: Hydraulic technology at the periphery of early empires. *Technology and Culture* 38(2): 432–451.
Lightfoot, D.R. 2000. The origin and diffusion of qanats in Arabia: New evidence from the Northern and Southern Peninsula. *The Geographical Journal* 166(3): 215–226.
Lightfoot, D.R. 2003. Traditional wells as phreatic barometers: A view from qanats and tube wells in developing arid lands. In Proceedings of the UCOWR Conference: Water Security in the Twenty-first Century, Washington, DC, July 2003.
Lonergan, S. 1996. Water resources and conflicts: Examples from the Middle East. Paper presented at the NATO Advanced Research Workshop, Bolkesjo, Norway, 11–16 June.
Louis, P., and J. Métral. 1987. *L'homme et l'eau en Méditerranée et au Proche Orient*, vol. IV, *L'eau dans l'agriculture*. Lyon: GIS-Maison de L'Orient, Presse Universitaire de Lyon.
McCay, B. 2002. Emergence of institutions for the commons: Contexts, situations and events. In *The drama of the commons*, ed. E. Ostrom et al. Washington, DC: Committee on the Human Dimensions of Global Change, National Academy of Sciences.

Metral, F. 1982. Le droit de l'eau dans le code civil Ottoman de 1869 et la notion de domaine public. In *L'homme et l'eau en Méditerranée et au proche orient*, vol. II: *Aménagements hydrauliques, état et législation*, ed. J. Metral, P. Sanlaville, and F. Metral, 125–142. Lyon: GIS-Maison de L'Orient, Presse Universitaire de Lyon.

Métral, J., P. Sanlaville, and F. Métral. 1981. *L'Homme et l'eau en Méditerranée et au Proche Orient*, 2 vols. Lyon: GIS-Maison de l'Orient, Presse Universitaire de Lyon.

Metral, J. et al. 1987. *L'homme et l'eau en Medtierranee et au Proche Orient IV*. GIS Maison de L'Orient. France: Presse Universitaire de Lyon.

Molle, F., T. Shah, and R. Barker. 2003. The groundswell of pumps: Multilevel impacts of a silent revolution. In Proceedings of the ICID-Asia Meeting, 1–18. Taiwan, Nov.

Naito, M. 1985. Water resource management in the oasis of Damascus: Man's role in changing phases of desertification. *Stuttgarter Geographischen Studien* 105: 103–104.

Noel, E. 1944. Qanats. *Journal of the Royal Central Asian Society* 31: 191–202.

Obaid, T. 2002. Demographic dynamics in the Middle East: A view from the United Nations. *Policy Watch* no. 621. Washington, DC: The Washington Institute for Near East Policy.

Rabo, Annika. 1984. *Great expectations: Perception on development in North East Syria*. Stockholm: University of Stockholm.

Rabo, Annika. 1986. *Change on the Euphrates: Villagers, townsmen and employees in Northeast Syria*. Stockholm Studies in Social Anthropology, No. 15. Stockholm: University of Stockholm.

Rae, J., G. Arab, T. Nordblom, K. Jani, and G. Gintzburger. 2002. Tribes, state and technology adoption in arid land management in Syria. In *Innovation in natural resource management: The role of property rights and collective action in developing countries*, ed. R. Meinzen-Dick, A. Knox, F. Place, and B. Swallow, chap. 7. Baltimore: Johns Hopkins University Press.

Safadi, C. 1990. La Foggara, systeme hydraulique antique, serait-elle toujours concevable dans la mise en valeur des eaux souterraines en Syrie? In *Techniques et practiques hydro-agricoles traditionelles en domaine irrigue, approche pluridisciplinaire des modes de culture avant la motorisation en Syrie*, vol. 2, ed. B. Geyer. Paris/Beyrouth-Damas-Amman: Librairie orientaliste Paul Geuthner/IFAPO.

Salih, A. 2006. Qanats, a unique groundwater management tool in arid regions: The case of Bam Region in Iran. Paper presented at the International Symposium on Groundwater Sustainability (ISGWAS), Spain, Jan. http://aguas.igme.es/igme/ISGWAS/Ponencias%20ISGWAS/6-Salih.pdf. Accessed 30 July 2010.

Salman, M. 2004. Institutional reform for irrigation and drainage in Syria: Diagnosis of key elements. Paper presented at the Food and Agriculture Organization (FAO) Syrian expatriates conference, Damascus, Syria. ftp://ftp.fao.org/agl/iptrid/conf_syria_04.pdf. Accessed 30 July 2010.

Salman, M., and W. Mualla. 2003. The utilization of water resources for agriculture in Syria: Analysis of current situation and future challenges. Paper presented at the Erice International Seminars on Planetary Emergencies, Sharm El-Sheikh, Egypt, December 2003. Seminar Proceedings, 18–26. Rome: FAO. ftp://ftp.fao.org/agl/iptrid/conf_egypt_03.pdf. Accessed 30 July 2010.

Soussan, J. 1998. Water/irrigation and sustainable rural livelihoods. In *Sustainable rural livelihoods, what contribution can we make?* ed. D. Carney. London: Department for International Development.

Sutton, S. 1984. The falaj: A traditional cooperative system of water management. *Waterlines* 2(3): 8–12.

Swain, A. 1998. A new challenge: Water scarcity in the Arab World. *Arab Studies Quarterly* 20(1): 1–11.

UNDP. 2005. *The millennium development goals in the Arab region (summary)*. New York: United Nations.

UN-Habitat. 2005. Islamic land tenures and reform. Islam, land, and property research series, paper 2. Nairobi: United Nations Human Settlements Programme. http://www.unhabitat.org/downloads/docs3546_86904_ILP%202.doc. Accessed 30 July 2010.

Vincent, L. 1995. *Hill irrigation water and development in mountain agriculture*. London: Intermediate Technology Publications.

Wakil, M. 1993. Analysis of future water needs of different sectors in Syria. *Water International* 18: 18–22.
Weisgerber, G. 2003. The impact of the dynamics of qanats and aflaj on oases in Oman: Comparisons with Iran and Bahrain. In Proceedings of the International Frontinus Symposium, Walferdange, Luxemburg, 2–5 Oct.
Wessels, J.I., G. Arab, and A. Aw-Hassan. 2001. Participatory survey on institutions and groundwater use, Aleppo Province, Syria. Unpublished project report for NRMP, ICARDA, Syria.
Wilkinson, J.C. 1977. *Water and tribal settlement in South-East Arabia: A study of the Aflaj of Oman*. Oxford: Clarendon.
World Bank. 2002. *Reaching the rural poor: A rural development strategy for the Middle East and North Africa Region*. Departmental Working Paper 24531. Rural Development, Water, Environment and Social Group, Middle East and North Africa Region, the World Bank. Washington, DC: The World Bank.
World Bank and Swiss Agency for Development and Cooperation (SDC), Yemen. 2000. Summary Report. Middle East and North Africa Regional Water Initiative workshop on sustainable groundwater management, Sana'a, Republic of Yemen, 25–28 June.
Wutmann, M., T. Gonon, and C. Thiers. 2000. The Qanats of Ayn-Manawir (Kharga Oasis, Egypt). In Proceedings First International Conference on Qanats, Yazd, Iran, May.

Chapter 2.4
Case Studies from the Americas

Box 2.4a Water, livelihoods, and morality around the Panama Canal

—Ashley Carse

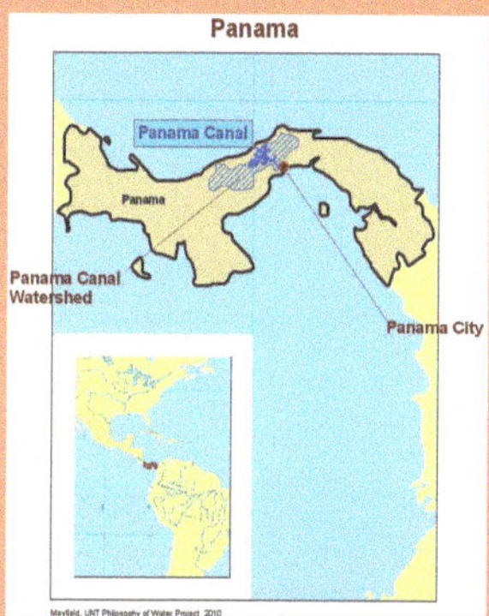

Box Map 2.4a.1 Panama

The Panama Canal is far more than a simple channel between the Atlantic and Pacific Oceans. It is a network of hydroecological systems and engineering technologies that circulate billions of litres of fresh water through rivers, reservoirs, and locks. This network depends upon people and their labour as well as on water, for without the engineers, ship captains, and mechanics – to name just a few – ships and cargoes making up a substantial part of world trade would be unable to pass from sea to sea. Yet despite its obvious importance, the canal does not provide subsistence for everyone living near it. These lands and waters have historically supported diverse rural livelihoods, including farming, ranching, fishing, rubber tapping, and gold panning. Since the late 1970s, however, water and canal managers have understood the forests of the Panama Canal watershed as critical to protecting the canal's water supply, sometimes placing them at odds with local rural people. Cultural difference – particularly historical experience and moral economy – has become crucial in managing the watershed for this key shipping lane.

Each of the 13,000-14,000 ships that transits the canal annually requires 237 million litres of water (Panama Canal Authority 2008). The majority of this water is delivered via six major rivers that drain into the canal's water

(continued)

B.R. Johnston et al. (eds.), *Water, Cultural Diversity, and Global Environmental Change: Emerging Trends, Sustainable Futures?*, DOI 10.1007/978-94-007-1774-9_12,

Box 2.4a (continued)

storage reservoirs. Stored water is released through a system of locks, lifting ships up and across the isthmus and then lowering them back to sea level at the opposite terminus. Since the late 1970s, foresters and hydrologists have warned that watershed deforestation poses two key threats to canal function: increased siltation of the canal and reduced dry season water flows (Wadsworth 1978). Concerns about the watershed coincided with the transfer of control over the canal from the United States to Panama between 1977 and 1999, a move that precipitated the creation of new Panamanian governmental institutions focused on environmental management and the implementation of new watershed initiatives. Canal administrators, politicians, conservationists, and shipping companies came together around the shared problem of watershed deforestation. The solutions that emerged assumed two forms: (1) legal restrictions on *roza*, or slash-and-burn, agriculture and cutting down trees within the watershed; and (2) development projects designed to reform small-holder agriculture in a manner amenable to watershed management.

The history of the Panama Canal watershed narrated in policy documents and media reports fuses oversimplified forest hydrology science with claims of rural ecological ignorance, scripting small-holder land use practices as 'unsustainable', while avoiding the roles played by non-local – particularly state – actors in rural development. Between 1903 and 1979, the watershed was politically bisected, with Panama controlling most of the upper part and the United States most of the lower part. The two nations governed their territories differently. The rural areas of the lower watershed (the canal zone and the US part of the watershed) were depopulated during canal construction and for the most part,[1] maintained in that condition thereafter. By contrast, the Panamanian state, with the financial support of international development organizations, implemented agrarian reform and rural development projects to encourage landless *campesinos* (peasants) from other parts of the country to settle in the upper watershed. Rural development accelerated during the 1950s, and the policy of encouraging campesinos to farm persisted until the 1977 Torrijos-Carter Treaty. This treaty initiated a period of US-Panamanian co-management and guaranteed Panama control of the canal in 2000. During the co-management era, regional development emphasis was replaced by watershed management, which assigned rural landscapes a new function (water provision) and their inhabitants a new responsibility (forest conservation). Although this rapid reversal is rarely officially acknowledged, it is inscribed

[1] The United States experimented with an agricultural land lease programme in the Canal Zone designed to for West Indians and Panamanians during the 1920s and 1930s, but the programme was discontinued because of social welfare and tropical disease concerns.

(continued)

Box 2.4a (continued)

Box Fig. 2.4a.1 Panamanian farmer on the banks of the Boqueron River (Photo credit: Ashley Carse)

on regional consciousness and landscapes, complicating efforts to make watershed management cooperative.

Despite well intentioned sustainable development projects, smallholders often perceive watershed management as an illegitimate strategy to consolidate wealth around the canal at the expense of the poor. These farmers see sustainable development projects as short-lived tokens rather than solutions. Many smallholders say that the moral thing to do, given regional history, would be for the Panama Canal to support viable livelihood alternatives to slash-and-burn agriculture. But what alternative arrangements are recognizable and acceptable? The answer to that question is still under negotiation. A local man stood on the bank of the Boqueron River, watching its muddy water flow toward the canal and ultimately out to sea. 'You don't want me to cut firewood to cook, give me a stove', he commented as the water moved by.

Resources

Panama Canal Authority. 2008. Panama Canal maritime statistics. http://www.pancanal.com/eng/maritime/statisti.html. Accessed 20 Oct 2009.

Wadsworth, F. 1978. Deforestation: Death to the Panama Canal. *Proceedings of the U.S. Strategy Conference on Tropical Deforestation*, 22–24. Washington, DC: U.S. Department of State and U.S. Agency for International Development.

Box 2.4b Saving the Pilcomayo River in Argentina: Traditional *criollo* ranchers resist destruction of an ecosystem and a way of life

—Janis B. Alcorn and Luis María de la Cruz

Box Map 2.4b.1 Gran Chaco of South America

The Pilcomayo River runs along the border between Bolivia, Paraguay, and Argentina, in the Gran Chaco ecological region of South America, and it is part of one the largest river basins in the world - the La Plata. Rainfall here is highly variable (300-1,200 mm per year) but sufficient to create one of the world's largest seasonal wetlands. During the flood season, the waters spread over the flatlands under cacti and dry forest, changing the shape of the land. Over countless millennia, the river has slowly moved northwards, creating undulating areas of silt left by prior floods. The Pilcomayo area is extremely biologically and culturally diverse, and the annual floods sustain this diversity. The *criollo* (creole) descendants of Spanish nomadic herders and the Indigenous communities value the annual flooding because it renews moisture in the soil, sustains the forest on which Indigenous communities depend, and regenerates the wetlands that support the centuries-old criollo herding system. Long-term ecological and economic sustainability requires a multiscalar institutional mechanism that acknowledges the biological and cultural diversity of the region. Effective decision-making processes that integrate local knowledge and ideas about the river must be included in any development plans.

Box Fig. 2.4b.1 Indigenous Toba community in Formosa, Argentina, being flooded by Pilcomayo River because of the diversion of water by IIRSA-related infrastructure development, 2008. Men are standing on an earthen dam rapidly dug to avert the flooding of houses, watching boats bringing refugees from flooded villages, while children look on (Photograph by Janis B. Alcorn)

A current development initiative by the European Union, World Bank, Inter-American Development Bank (IADB), private capital, and national governments threatens to integrate the Pilcomayo and Paraná rivers into the South American Regional Infrastructure Integration Initiative (IIRSA). The IIRSA outlines plans for the development of small dams to fuel speculative land sales for soy farming investments that will displace criollo and

(continued)

Box 2.4b (continued)

Box Fig. 2.4b.2 Criollo on horseback in area not normally flooded in Formosa, Argentina, after new infrastructural developments supported under IIRSA blueprint for development (Photograph by Fundacion FUNGIR)

NOTAS

...NCA DEL PILCOMAYO ECOSISTEMA Y GEOPOLITICA (2)

...librio ambiental puede ser roto

Interrupción de desembocadura del río Pilcomayo en el río Paraguay

Box Fig. 2.4b.3 Diagram in Villamontes newspaper, October 2005, showing how the infrastructure on Pilcomayo in Argentina is blocking upriver migration of fish on which local industry and Weenhayek indigenous people depend in Bolivia (Photograph by Janis Alcorn)

Indigenous communities, forcing them to abandon the river along which they have lived for centuries.

Communities are fighting back to protect the river ecosystem that they depend on. In 2007, a 53-year-old widow with 11 children, a criollo leader in the struggle for the river, organised a 40-day roadblock to protest the flooding created by a road dam funded by the IADB. This successful action led to the consolidation of the Producers' Association of the Estrella Wetlands (PAEW), a grassroots group that represents regional communities. The PAEW filed papers with the government for a judicial inquiry into all the planned dams. The government tried to divide the group by offering relocation to distant titled lands. Since most people in the region lack formal title to their current lands, the offer of title was a powerful incentive to move, and newspapers published false stories of people's accepting relocation to create further dissension. Nonetheless, communities have held their ground and as of mid-2009, a second dam had been put on indefinite hold as a result of their protests. With the help of sympathetic non-governmental organizations, local participants are gathering independent information about changes in the river basin and contributing it to the Grassroots to Satellite Imagery Environmental Monitoring System of the Pilcomayo. As the widowed leader of the protests says:

> We live here and we defend our forest and our wetlands because we like them, they are beautiful. We do not want them destroyed. This land gives food to our animals, and we live from them. We do not have much, but we are rich because we enjoy this beauty that we inherited from our fathers. This is why we defend nature.

Box 2.4c Working on water: Cultural survival in the 'tri-state water wars'
—Becky Blanchard, David McLain, and Linda Raffield

Box Map 2.4c.1 Southeast United States

The Apalachicola-Chattahoochee-Flint (ACF) Basin drains 5,000 ha from the headwaters near Atlanta – one of the fastest growing metropolitan regions in the United States – to one of the nation's most productive estuaries, Apalachicola Bay in the state of Florida. The bay supplies 10% of the entire commercial oyster catch in the United States and these oysters are collected by fishers who have often been involved in oyster harvesting for generations. The oysters need a steady supply of fresh water from the Apalachicola River flowing into the bay to keep salinity at tolerable levels. Higher salinity increases disease mortality in the shellfish, threatening the oysters and the livelihoods of those who depend on the resource for their sustenance and survival.

The biggest factor contributing to dangerous salinity levels is the overconsumption of water by upstream users. At the heart of the 30-year 'tri-state water wars' between the city of Atlanta and the states of Alabama, Florida, and Georgia are the competing demands of municipal, industrial, and recreational water use in metropolitan Atlanta; irrigation; navigation; and environmental flows to sustain both endangered species and the oyster industry. Downstream stakeholders object to Atlanta's unlimited and growing demands on the Chattahoochee River, and there is no basin-wide mechanism to coordinate withdrawals of water, for each state has sovereign jurisdiction over water in its territory. Attempts to resolve this conflict have included federal litigation, a never-completed comprehensive study, and an unsuccessful interstate compact. The interstate conflicts and environmental consequences are further complicated by the fact that the U.S. Army Corps of Engineers (a federal agency) manages dams and reservoirs, using its own standards to set river flows. In the recent 2007-2008 drought, flow reductions brought river levels lower than anything previously recorded and led to a 30-50% increase in oyster disease mortality.

During the drought, oyster fishers' catch per unit effort declined. Historical experience from other bays suggests that such declines may push oyster fishers out of the industry and that when the oysters recover, the culture of oystering does not always recover in the same way. Apalachicola Bay oystermen have a unique place-based culture built upon generations of work on the water. Their intimate knowledge of the bay and the relationship between river

(continued)

Box 2.4c (continued)

flows and oyster harvests cannot be transferred elsewhere, and they inhabit a social landscape characterised by strong extended family ties. The networks of kinship and community assist oyster harvesters in coping with the environmental and economic variability of the industry, but this support system is stretched to the breaking point when all are affected by regional crises such as drought. The hardships endured by the oyster harvesters affect the rest of the region, which relies heavily on the seafood industry and coastal tourism. The waters of the Apalachicola-Chattahoochee-Flint Basin are important to all and affect all.

The conflict over these waters has been protracted and highly technocratic, limiting oystermen's willingness and ability to participate directly. Many voice their concerns through an advocacy alliance that has emerged around downstream interests, and this loose coalition includes organizations representing oyster harvesters and dealers, municipalities, businesses, and environmentalists. It includes staff with legal and scientific training who can attend distant meetings and review management documents. Each group has its own goals and constituencies; however, they have made progress on shared concerns by leveraging each other's strengths. The state of Florida draws upon the reports and testimony of these groups to support its litigation efforts.

The coalition has worked to change public discourse by challenging the upstream framing of the water wars as a 'people-versus-mussels' issue (endangered mussels are one justification for environmental flows). They put a human face on 'environmental' flows by showing their importance for oystermen's cultural survival and the economic survival of the region. These groups have hosted public officials, academics, and journalists from around the world, taking them on the water to meet with harvesters.

Box Fig. 2.4c.1 French journalists interview John Richards, Franklin County Seafood Workers Association President and 50-year oysterman, aboard his boat during the 2007–2008 drought (Photo credit: Linda Raffield)

This strategy has generated both support and unanticipated challenges. Press coverage of oystermen's plight during the 2007–2008 drought may have decreased demand for Apalachicola Bay oysters, perhaps due to perceptions of lower quality. As these groups address ongoing water conflict, they hope to convince water managers to make use of oystermen's knowledge of the

(continued)

Box 2.4c (continued)

rivers and bays and the complex ecology that those waters sustain. Oyster harvesters understand their dependence on the river and their need to respect it. The river and the oysters are a major part of the resilience of these communities in hard times; electricity bills may go unpaid, but as long as people have access to the estuary's bounty, no one goes hungry in Franklin County. The oyster harvesters take pride in caring for their own, and they want to take care of the river that sustains them.

Chapter 2.5
Nourishing Diversity in Water Governance: The Case of San Cristobal de las Casas, Chiapas, Mexico

Ameyali Ramos Castillo

It is May 3, 2009. The mist is receding into the blue-green mountains of San Cristóbal de las Casas, ushered away by the morning sun, as the people of the *barrio* (neighbourhood) of Cuxtitali gather around a sacred water spring. Fresh water from ancient underground aquifers rushes past the flowers, candles, and crosses adorning the spring and, infused with symbolic and cultural meanings, flows into the metal pipes of modern urban infrastructure. Gravity compels the water through 8 km of semi-urban developments and into the homes of 1,100 Cuxtitali residents in a communally agreed rotation.

Less than 5 km away in the Colonia 5 de Marzo, residents are gathered around a public hydrant discussing an alternative and autonomous water development proposal for their community with students from the University of Santa Barbara - California. Driven by solidarity economics and infused with notions of autonomy and self-governance, water flows from the lived reality of public hydrants and rainwater harvesting systems through 2-inch rubber pipes and into the gradually solidifying dream of autonomy for the 250 families in resistance.

Meanwhile, on the slopes of Moxviquil, a mountain on the northern edge of the city, exiled Indigenous Chamulas in La Hormiga barrio prepare their annual contribution for the May 3rd ritual to honour the Water Gods. In exchange for their contribution, Chamula leaders in El Pinar grant La Hormiga access to Chamula water sources for water provision of approximately 25,000 urban Chamula families in La Hormiga.

From a birds-eye perspective these systems of water provision are not that different from the municipal system of water provision in that they all share the common mandate of supplying urban residents with water, they all depend on water from local springs, they all use piped systems for water distribution, and they all deliver

A. Ramos Castillo (✉)
United Nations University, Institute of Advanced Studies (UNU-IAS), Traditional Knowledge Initiative, Yokohama, Japan
e-mail: ramos@ias.unu.edu

B.R. Johnston et al. (eds.), *Water, Cultural Diversity, and Global Environmental Change: Emerging Trends, Sustainable Futures?*, DOI 10.1007/978-94-007-1774-9_13,

water to urban dwellings. Yet, upon closer inspection, the flows of water in these systems are metabolized very differently: each of these systems is a unique and complex temporal and system of water governance that shapes the flows, distribution, and 'relationships to water' of each system.

At a time when climate change and increasing socio-environmental complexities inhibit access to water, it is all the more important to recognize and analyze the role that different systems of water governance play in shaping adaptive, responsive, and resilient 'relationships to water.' Examining the flows of water through cities like San Cristóbal de las Casas allows us to consider how water governance systems and relationships to water evolve, deal with challenges, and suggest alternatives for sustainable urban water governance.

2.5.1 San Cristobal: A Historical Perspective

The city of San Cristóbal de las Casas is located in the small, bowl-like Jovel Valley in the heart of the state of Chiapas in southern Mexico. In 1528 the first Spaniards arrived in San Cristóbal bringing with them a distinct 'relationship to water' that would remain in the circulation of water in the city until the present day. The Spanish regarded water as a passive material resource that had to be managed and controlled. Indigenous slave labour was used to build wells, dam rivers and streams, and exploit streams in order to 'civilize' the wild flows of San Cristóbal. This relationship to water based on notions of control, exploitation, and possession permanently changed the waterscape of the valley.

Map 2.5.1 Mexico

The Spanish aspired to racially segregated cities but economic factors, demand for readily available labor, and the Christianizing agenda worked to redefine the Spanish ideal (Markman 1984). In San Cristobal, colonial authorities settled Indigenous peoples in segregated barrios (neighbourhoods) "outside the official town plan" (Markman 1984) but close enough to provide the necessary services required by Spanish citizens. The distance between the 'official' city and Indigenous barrios was meant to geographically enforce a racially segregated and racially hierarchical society.

This racialized rationality was incorporated into the design and construction of urban infrastructure including the first system of water provision in San Cristobal. In the 1820s, urged by the modernization dream and technological advances, Spanish authorities in San Cristobal built a series of dams and aqueducts that diverted water from the Amarillo and Fogotico Rivers and various springs through ceramic pipes to public water fountains (in some Indigenous barrios) and into cisterns in Spanish homes. From its design, the aqueduct system was intended to provide water only to 'civilized' Spanish citizens. Indigenous populations, who were legally defined as *niños con barbas* or 'children with beards' were expected to continue to rely on water from 'natural' sources or from public fountains that the Spanish purposefully placed in the plaza of the local church. Situating the public fountains in the shadow of the church was seen as a means of advancing the Christianizing agenda.

This fragmentation of access to water infrastructure based on a racialized rationality caused the beginning of differentiated water governances and relationships to water in the city: modern/Spanish/wealthy citizens relied on the 'modern' system of water provision while residents, whose ethnicity and/or socio-economic status precluded them from access to the system, developed their own systems of water provision and governance. This differentiated access to water provision eventually led to the formation of a diversity of water governance systems in San Cristobal among them Chupactic, Cinco de Marzo and La Hormiga.

2.5.2 Chupactic: A Community Water Governance Alternative

Chupactic, a community-based water governance system, emerged in the Cuxtitali barrio of San Cristobal in response to racially differentiated systems of water provision in the city. Earliest archival information suggests that the barrio of Cuxtitali was populated by a mixture of Indigenous Peoples – including Nahuatl and Quiche – who were taken to San Cristobal by the Spanish as slaves as early as 1528. From its inception, the barrio of Cuxtitali was considered an independent municipality despite its proximity to downtown San Cristobal (3 km).

Until the 1800s, Cuxtitali residents depended on surface water for their daily water needs. Women would transport water from the Amarillo River or from the Peje de Oro spring in ceramic containers to their homes. In the 1820s, the dam and aqueduct system introduced the public fountain system and Cuxtitali residents relied on a mixture of public fountains and surface water for their water needs.

In the 1930s the government of San Cristobal began building a modern, centralized, networked system of water provision in the city. In line with previous systems of water provision, however, the networked system of water provision only provided water to Spanish neighbourhoods and excluded Indigenous and mestizo barrios like Cuxtitali.

It was not until 1974 that Cuxtitali residents began the construction of the first piped system of water provision in Cuxtitali. The piped system would later be named Chupactic and was the material product of the fusion of cultural beliefs, local knowledge, engineering innovations, and the urbanization dream. The Chupactic system was completed in 1976 and provided water for approximately 1,500 families in Cuxtitali.

Since the inception of Cuxtitali, the belief of water as a spiritual co-essence and a communal resource had survived and evolved. Indeed, the ritual expressions of water as a living entity were seen as tools of resistance to the spiritual conquest of the region by Spanish culture and Catholic beliefs and were woven into the construction and design of the Chupactic system. When the suggestion of piped water was first introduced, residents in Cuxtitali did not want to have the springs covered and tapped with pipes because they considered water springs to be inhabited by Gods and because water ceremonies to honour the water gods took place at the springs. After much discussion, Cuxtitali residents decided that the San Luis Chupactic and El Pinar water springs, the sources of Chupactic's water, would be left as untouched as possible and would be adorned with ritual and sacred symbolism and that gravity, rather than electricity, would be the preferred means for transporting water from the springs to Cuxtitali.

In Cuxtitali, shared struggles of resistance and a communal memory of oppression also firmly established water as a communal resource. This was reflected in the design and construction of the networked system of water provision: every resident contributed to the construction of Chupactic with both financial resources and labour. The exercise of building the system – communally and with their own hands – meant that residents rather than engineers or hired workers were responsible for deciding where water would flow and where it would not. Thus, every resident was aware of where the pipes were, how much water they could transport, who had laid it and who it would deliver water to. The construction of the Chupactic system not only strengthened the relationship between people & water (by connecting residents to the water source both materially and spiritually) but also between people & people (by promoting community cooperation, collaboration, and participatory decision making).

The governance of Chupactic was also deeply rooted in the communal relationship to water. Cuxtitali residents decided to form an Asociacion Civil (AC) – a Mexican legal term for a non-profit organization – to govern Chupactic. Every resident was recognized as a member of the AC and was granted the right to vote and participate in decision-making. The AC hired two workers to collect contributions and to maintain the system but all the decision-making power remained in the AC ensuring that residents would be the guiding force in the circulation of water in the barrio.

The process of deciding on operating rules for Chupactic was a negotiated and flexible process. From the onset, members decided that the operating rules should reflect cultural traditions specifically the tradition of supplying water to households during wakes. Members also decided that they would continue the tradition of the annual ceremony to honour the Water gods. According to Indigenous cosmovisions, this ritual ceremony ensures that water does not leave the house it inhabits or, put literally, that water continues to flow. In Cuxtitali, residents believed that these rituals would ensure the continued flow of water and mandated that the AC be responsible for collecting donations and organizing the May 3 ceremony.

In order to maintain the system, the AC requested 'monetary contributions' – rather than monthly water fees – to be used for the repair and maintenance of the system, the salaries of the two hired workers, and the costs of the May 3 ceremony to honour the water Gods. Regardless of their monetary contributions, all Chupactic members were expected to participate in the annual ceremony to honour the Water Spring, to attend the AC's meetings, and to help with repairs or other infrastructure needs.

In Cuxtitali the sense of entitlement to water stemmed not from giving monetary contributions but from sustaining an integrated relationship to water, an active interest in the Chupactic system, and a commitment to each other. For a community that had been historically marginalized and socio-politically oppressed, Chupactic represented an alternative development opportunity – one which gave Cuxtitali residents the power to manage their water independently from the municipal government.

Between 1974 and 1992, the Chupactic system thrived. Members reveled in the modern convenience of getting water delivered to their homes and took great pride in caring for and maintaining the Chupactic system. In less than 20 years, Chupactic had succeeded not only in supplying members with water but also in successfully creating and encouraging a water governance model that interwove culture, autonomy, and a unique relationship between people and water.

By 1990, however, the Cuxtitali barrio had practically disappeared under the urban sprawl of San Cristobal. By 2000, the number of registered members of Chupactic doubled to 1,020 registered households and approximately 330 clandestine connections. Despite these challenges, however, today Chupactic continues to thrive even under conditions of continued urbanization and population growth. In San Cristobal, Chupactic has emerged as a symbol of 'autonomy' and 'freedom' and thus, a powerful political model.

The success of Chupactic, however, cannot be solely attributed to the community-based *governance* model. The underlying relationship between Chupactic residents and water has shaped the Chupactic water governance and has been instrumental to Chupactic's success. Historically, specific understandings and relationships to water have shaped the circulation of water in the barrio. And indeed, Chupactic continues to exist because of the unique relationship between people & water and people & people in Cuxtitali. In the following section I will explore the emergence of another unique water governance system in San Cristobal: Cinco de Marzo.

2.5.3 Cinco de Marzo: Water Flows in Resistance in the 'Slums' of the City

> Autonomy is to do things ourselves, with our own ideas, and from our own traditions as Indigenous Peoples.
>
> -Zapatista quoted in Mora 2008

Cinco de Marzo is an informal settlement located in the heart of San Cristobal that was settled by 450 Indigenous families that were forced off their lands by the confrontation between the Mexican National Army and the Zapatista Liberation Army in the 1990s. Eventually these families settled on lands in the outskirts of San Cristobal de las Casas, Chiapas in a barrio that is known as the Colonia Cinco de Marzo. Landless, homeless, and poor, the families erected makeshift camps of wooden shacks with plastic sheeting and, like migrants in informal settlements all around the world, began the difficult challenge of securing access to vital services including the provision of water. But unlike other informal settlements, from the onset Cinco de Marzo declared itself materially, culturally and politically as an autonomous Zapatista community.

The Zapatista movement began in 1994 with the uprising of the Zapatista National Liberation Army (EZLN). At its core the Zapatista movement is a struggle to create "new ways of being political, of doing politics: a new politics and a new political ethic" (Sub Comandante Marcos 2/5/95). The Zapatista's proposal for "new ways of doing politics" is guided by several governing principles including Lead by Obeying, Propose Don't Impose, Represent, Don't Replace, Convince Don't Defeat, Construct Don't Destroy, etc.

Zapatista communities like Cinco de Marzo are commonly referred to as communities in resistance because they tend to refuse government aid and because they develop their own ways of doing and being. Being in resistance challenges communities to be self-reliant, self-sufficient and sustainable and to incorporate communal-decision making processes that are essential for Zapatista governance models.

As a community in resistance, Cinco de Marzo followed the Zapatista policy of rejecting government aid including the provision of water. Thus even before the Municipal government was able to deny Cinco de Marzo water provision, Cinco de Marzo residents began exploring alternative means of water supply. For several years, water from the Navajuelos River was the community's only source of water: residents would carry water from the river to their homes for their daily water needs, women used the river to wash clothes, animals would drink water from the river, etc. Indeed, residents in urban Cinco de Marzo practiced a relationship to water similar to the one they had in their rural communities: water was considered sacred, water sources were sites of curing rituals and ceremonies, and cultural norms were used to govern access to water.

By 1997, however, the population in Cinco de Marzo had grown to 250 families increasing the urgency of having a more stable system of water provision. In May 2000 the community voted to create a Potable Water Council and elected five members to sit on the council. The Council was the community's way to ensure that all residents would participate and feel part of the water provision process.

After discussing the various possibilities, Cinco de Marzo residents agreed via consensus to clandestinely tap into the municipal system to supply water to three public hydrants. Like other Zapatista communities, all residents of Colonia Cinco de Marzo were expected to contribute financially (to purchase materials) and with labour (to construct the system) for the system of water provision. The public hydrant system capable of supplying water to approximately 300 families was completed in October 2000.

Fig. 2.5.1 A Zapatista public hydrant in Cinco de Marzo. Translation of sign: This service was commissioned by Council of Good Government in Oventic, Chiapas, Mexico. The truth sets us free. (Photo credit: Ameyali Ramos Castillo)

Although the public hydrants system facilitated access to water, it did not ensure constant or regular water supply. Water availability in the hydrants fluctuated depending on the amount of pressure in the municipal system and numerous times residents were left without water and had to resort to carrying water from the Navajuelos River.

Perhaps because of this instability, Cinco de Marzo began exploring alternative means of water provision. Cinco de Marzo's affiliation as a Zapatista community granted them unique access to a diverse network of NGOs, civil society organizations, international financing, and activists operating in diverse locations and on multiple spatial scales. By mobilizing these multi-spatial and multi-scalar networks, Cinco de Marzo residents were able to extend their political influence and access unique collaborative opportunities.

Fig. 2.5.2 Rainwater harvesting system in Cinco de Marzo (Photo credit: Ameyali Ramos Castillo)

Between 2007 and 2010, Cinco de Marzo residents, in collaboration with students from the University of California - Santa Barbara, designed and constructed appropriate technologies to help ensure adequate water supply in Cinco de Marzo, among them were: composting latrines, an EcoLavadero (ecologically friendly clothes washing station), and a rainwater capture and collection system.

The projects provided the necessary knowledge and skills to build sustainable, self-sufficient systems of water provision without undermining the Zapatista struggle or destroying the fabric of the Indigenous relationship to water. The projects, besides being easy to build, operate, and maintain, were also seen as a means of achieving self-sufficiency and autonomy in the community.

Although the projects were successful in reducing the community's reliance on water from the public hydrant system, continued population growth has forced Cinco de Marzo to explore complimentary means of water provision. Despite this however, Cinco de Marzo residents continue to exercise and demand their right to access, manage, and govern water according to their community governance models and their Indigenous cultural traditions.

And indeed, in San Cristobal, Cinco de Marzo is not the only community working toward alternative and autonomous water governance. In the following section I explore La Hormiga, an Indigenous water governance system on the outskirts of the city.

2.5.4 La Hormiga: An Indigenous Water Governance System

Water occupies an important position within Indigenous Mayan cultures in Chiapas. In much of the Mayan highlands, cultural histories are intimately entwined with the flows of water and the birth and demise of water springs. The symbolic and ritual importance of water has, for centuries, provided the structural foundation for Indigenous water governance systems in the Mayan highlands, including La Hormiga.

In order to understand the La Hormiga system it is necessary to first briefly explore Chamula governance and water management strategies as La Hormiga is of Chamula origin. The Chamula are an Indigenous Mayan group who speak Tzotzil and live in the San Juan Chamula municipality to the northwest of San Cristobal. Water in Chamula has and continues to be managed according to complex Chamula understandings and relationships to water (Burgete 2000). Both the social and judicial systems in San Juan Chamula are intricately linked with water and, in particular, with the birth and death of water springs. Chamulas believe that the appearance (or birth) and disappearance (or death) of water springs are not the result of physical processes but rather a manifestation of the supernatural and spiritual realms. According to Chamula cosmology, water springs are home to Chamula Gods and for this reason the Chamula consider springs to be sacred. Chamula tradition dictates that those who use water from water springs are obliged to take care of the spring and to express their gratitude through ceremonies and offerings to the Gods that inhabit those springs. In this way, in Chamula, water rights are granted by the Gods that inhabit the water springs rather than by an official or spiritual leader – i.e., if the Gods are pleased and satisfied by the ceremonies and offerings, the water spring will continue to give water (i.e., live) but, if, on the contrary a God is displeased the water spring will dry up (i.e., die).

This understanding of water as sacred and the 'reciprocal' relationship between Chamulas and the Gods who live in the springs form the basis of the conceptual and legal framework for the management of water in Chamula. For instance, all Chamulas – by virtue of their ancestry – have a right to use and access water resources on Chamula lands. Ceremonies, rituals and obligations to these gods are what legitimize water use rights in Chamula. These principles form part of a larger social and symbolic universe that is central to the Chamula identity. And indeed, these the Chamula understandings and relationships to water continue to exert a strong hold even over exiled the Chamulas living in urban San Cristobal barrios like La Hormiga.

The barrio of La Hormiga is located on the northern slope of the city of San Cristobal. It was settled in 1974 by Indigenous Chamula people – although other ethnicities have since added themselves to the mix – that were expelled from their the Chamula communities for adopting Evangelical beliefs. From its inception, residents in La Hormiga relied on public water sources or rainwater collection for their water needs.

In 1988 La Hormiga residents formed a Potable Water Committee that was tasked with petitioning the San Cristobal Municipal government for water provision. Despite

Fig. 2.5.3 Rainwater collection containers outside a home in La Hormiga (Photo credit: Ameyali Ramos Castillo)

numerous appeals, however, by 1994 La Hormiga had yet to receive a formal response from the Municipal government. Municipal authorities justified the lack of response and lack of service provision on the pretence that La Hormiga was an 'informal' settlement and that providing them with public services would encourage more migration from rural areas – an outcome they considered highly undesirable.

In 1995 La Hormiga, faced with the lack of response from the Municipal government, began explore alternative means of water provision. Geographically, the nearest water source outside the San Cristobal municipal system of water provision was El Pinar, a spring located on Chamula land near the peak of the mountain on which La Hormiga is located. But obtaining permission to access water from El Pinar required negotiations between the exiled Evangelical Chamulas and the 'traditionalist' Chamulas responsible for exiling the Evangelical Chamulas.

Throughout 1995, the La Hormiga Water Committee hosted a series of consultations and meetings to discuss the possible benefits and consequences of seeking permission from the 'traditionalists' Chamulas to access water from El Pinar spring. After numerous consultations and active participation from women in La Hormiga, in November 1995 La Hormiga leaders decided to approach the Chamula 'traditionalists' for permission to use water from El Pinar spring.

La Hormiga residents elected representatives to accompany the Water Committee members to El Pinar to seek permission to access water. As per the Chamula tradition of reciprocity, the La Hormiga Committee offered El Pinar community various gifts before broaching the topic of water access (Burgete 2000). After several days of tense negotiations, the traditionalist Chamulas in El Pinar agreed to grant La Hormiga residents access to El Pinar spring because of their common ancestry. But access to the spring was conditioned on La Hormiga's agreement to pay an annual contribution for the May 3rd ritual to honour the Water Gods that inhabit El Pinar spring.

At first La Hormiga residents were against agreeing to the annual contribution believing it was a 'pagan' tradition that undermined their Evangelical principles. But women in La Hormiga convinced La Hormiga residents and leaders that contributing to the May 3rd ritual was more of a 'cultural' tradition than a religious act. This conceptual modification resulted in the strengthening of a cultural Chamula identity among the exiled Chamulas in urban San Cristobal and, in early 1996, resulted in La Hormiga leaders agreeing to the terms proposed by El Pinar.

At its core the agreement between La Hormiga and El Pinar served to strengthen a Chamula identity and to recognize Chamula norms around water. La Hormiga women's re-conceptualization of Chamula understandings and relationships to water as 'cultural traditions' rather than religious practices caused Chamula exiles to re-embrace their Chamulas willingness to recognize and abide by their cultural norms around water that ensured the success of the agreement between El Pinar and La Hormiga.

The agreement also served to seek new forms of Chamula identity and authenticity among the exiled Evangelican Chamulas in San Cristobal. The agreement represented a renewed interest in the preservation and resuscitation of an Indigenous relationship to water albeit as a 'cultural tradition' rather than a religious ritual. This re-acceptance of Chamula customs and norms among Evangelical Chamulas increased their political and social potential especially in negotiations with El Pinar. At its core, the agreement legitimated a new form of 'doing politics' in La Hormiga – one in which residents where simultaneously Chamula and Evangelical rather than one or the other.

The agreement between La Hormiga and El Pinar also marked a new relationship between La Hormiga and Municipal authorities. By using their Chamula ancestry, La Hormiga had successfully solved their water problems without intervention from the Municipal government and had, furthermore, made clear their ability to survive 'independently' and 'outside' state control.

Once the agreement was finalized, La Hormiga residents contributed – financially and with labour – to the construction of a piped system of water provision that delivered water from El Pinar spring to La Hormiga households.

But the agreement between La Hormiga and El Pinar was not flawless and months after the La Hormiga system was inaugurated the El Pinar community disconnected the pipes. The negotiations to reconnect the supply of water revolved around water rights and an Indigenous relationship to water. El Pinar argued that, as 'traditional' Chamulas, the Gods in the water spring had granted them the right to manage and distribute water as was evidenced by the fact that water in the spring was still flowing. La Hormiga residents contested that according to Mexican law, water was the

Fig. 2.5.4 Pipes carrying water from El Pinar to La Hormiga (Photo credit: Ameyali Ramos Castillo)

property of the nation and as Mexican citizens they had the right to access water. When this argument failed, La Hormiga residents argued that by virtue of their Chamula ancestry they too had the right to access water from El Pinar spring. Once again, it was La Hormiga's willingness to recognize and negotiate within Chamula norms, rights and customary practices that convinced El Pinar to liberate the Water Committee members and to reconnect La Hormiga's pipes.

Tensions between El Pinar and La Hormiga continued into the 2000s and this, coupled with a population explosion (estimated 38,000 in 2000), made La Hormiga seek alternative means of water provision. In 2002, La Hormiga leaders once again approached the municipal system of water provision requesting the extension of the municipal system of water provision into La Hormiga on the basis that residents had 'regularized' or formalized their property rights and so were no longer an 'informal settlement.' The municipal system of water provision once against denied their request, this time citing insufficient water availability as the reason for the denial. Finally in 2005, the Municipal government approved La Hormiga's application for water provision.

The municipal system of water provision in La Hormiga was inaugurated in 2005. La Hormiga residents agreed to pay the municipal system a monthly tariff of 45 pesos/month for water provision but also continued giving El Pinar their monthly contributions. Both the El Pinar and the municipal system operated successfully throughout 2005. But by mid-2006, however, the electrical costs of running the pump for the new well exceeded the amount that La Hormiga residents were willing to pay the municipal system and conflicts ensued. The municipal system insisted on

La Hormiga residents paying a higher tariff but La Hormiga residents refused. Conflicts between the municipal system and La Hormiga over monthly tariffs and the quality of service continued throughout 2007 and into 2008 leaving many residents without water.

The strategic alliance between El Pinar and La Hormiga is what has ensured that residents in La Hormiga obtain access to water under circumstances where access is usually denied. La Hormiga's ability to navigate through two entirely different normative frameworks – Chamula customs and traditions and Mexican laws and regulations – as well as their ability to internally organize, has kept La Hormiga hydrated and allowed La Hormiga residents to gradually develop into more or less recognized political citizens in urban San Cristobal, albeit through 'out-side-the state' and sometimes violent means. But La Hormiga's struggle to secure water rights and access to water has also had distinct and unexpected cultural implications, namely the emergence of a hybrid 'Chamula' identity that is being readily embraced by exiled Chamulas. This re-assertion of a cultural "Chamula" identity has served to re-establish networks that link Chamula urban families with rural communities and natural resources – networks that may prove especially important as water in urban San Cristobal becomes increasingly scarce.

2.5.5 Ways Forward: Nourishing Diversity

As documented in this chapter, San Cristobal de las Casas has a diverse array of water governances that operate outside the municipal system throughout the city including community based systems (Chupactic), autonomous systems (Cinco de Marzo), and Indigenous systems (La Hormiga). The exploration of these distinct systems of water governance was meant to illustrate the complexity of water governance in cities in the global South and to interrogate the relevance of conventional frameworks that continue to pursue fully networked systems for urban water governance and distribution.

Despite the rich diversity of water governances in San Cristobal, on the whole the San Cristobal government's approach has been to treat this diversity as a 'problem' and to try to assimilate the diversity into a single, unified, official water governance model. In the eyes of government officials, the diversity of water governances – their alternative and informal origins and their integration with local 'relationships to water' and community structures – presents a problem because their models do not match the government's model nor are their systems of water provision linked with the 'official' state-sponsored networked system (SAPAM). However, notwithstanding the ineffectiveness of a single water governance system for San Cristobal, the Federal and municipal governments continue to embrace and to actively pursue a formal and unified water governance for the city.

This attitude toward a diversity of water governances in the city is not unique to Mexico. Across the global South, national and international water policies have failed to recognize the complexity of urban water governance systems. Urban water governance policies are based on the assumption and aspiration of a fully networked, universal, and connected urban systems of water provision (Kooy and Bakker 2008). But the reality of cities in the global South is that they are serviced by a diversity of 'un-networked' systems of water governance that have emerged as a result of social and spatial fragmentation, differentiated access to networked water and a clear lack of urban planning. Historically, certain sectors of the population (usually the poor or the politically marginalized) have had to rely on a diversity of alternative (and usually informal) systems of water governances for their water provision.

And, indeed, while the aspiration of a 'networked' system of water governance has proven unfeasible, inefficient and difficult to implement especially in developing countries, the diversity of informal or outside-the-state systems of water provision ('un-networked' systems) has succeeded in governing and distributing water even to the most marginalized sectors of society. And, as the case of San Cristobal demonstrates, these alternative systems also offer other advantages that could prove beneficial for tackling San Cristobal's increasing water challenges such as experience with alternative water technologies (i.e., the rainwater harvesting systems and EcoLavaderos in Cinco de Marzo), culturally viable water saving rotation strategies (as evidenced in Chupactic), and valuable strategic alliances (i.e., with water rich rural areas as is the case with La Hormiga).

Thus, perhaps, a more progressive approach to urban water governance would be to foster governance approaches that nourish – rather than abolish – the rich diversity of systems of water provision in cities in the global south.

References

Burgete, A. 2000. *Agua que Nace y Muere: Sistemas Normativos Indigenas y Disputas por el Agua en Chamula y Zinacantan*. Chiapas: Programa de Investigaciones Multidisciplinarias sobre Mesoamerica y el Sureste, UNAM.

Kooy, M., and K. Bakker. 2008. Splintered networks: the colonial and contemporary waters in Jakarta. *Geoforum* 39(6): 1843–1858.

Markman, D. 1984. *Architecture and urbanization in colonial Chiapas, Mexico*. Independence Square, Philadelphia: The American Philosophical Society.

Chapter 2.6
Water Knowledge, Use, and Governance: Tibetan Participatory Development Along the Mekong (Langcangjiang) River, in Yunnan, China

Yin Lun

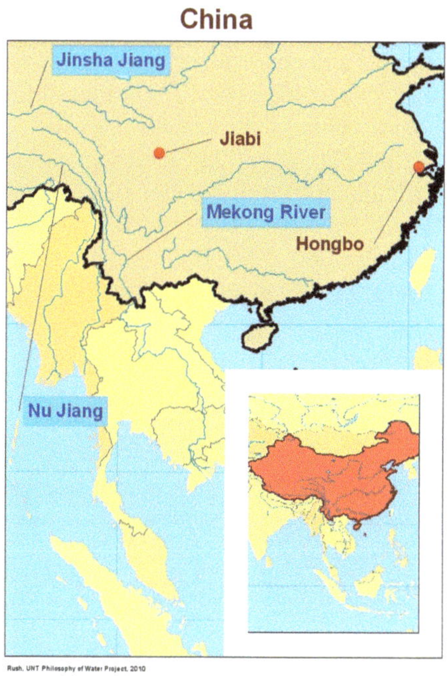

Map 2.6.1 China

2.6.1 Introduction

Water culture traditions reflect the lived experiences of local communities as well as their sense of place. Such traditions evolve over time through ongoing interactions with local environments and enable people to use and manage their water resources effectively. Environmental changes, however, can disrupt previously harmonious relations, requiring additional adaptations from local peoples. The Tibetan villages along the Mekong River, known in China as the Lancangjiang River, have a long history of managing their own water resources, but in recent decades they have experienced more frequent natural disasters such as flooding and mudslides. In order for water resources development and water governance modifications to be successful, they must be designed to meet these new challenges and must be built on foundations of local knowledge and local participation. Participatory development projects

Y. Lun (✉)
Yunnan Academy of Social Science, Kunming, China
e-mail: lun.yin@gmail.com

B.R. Johnston et al. (eds.), *Water, Cultural Diversity, and Global Environmental Change: Emerging Trends, Sustainable Futures?*, DOI 10.1007/978-94-007-1774-9_14,

undertaken in the villages of Hongpo and Jiabi, Deqin County, Diqing Province, are examples of such successful collaborations.

2.6.2 Tibetan River Stories

The Tibetan people describe the nature of their four main rivers in a series of important myths. The Nujiang River, originating in Tibet and flowing south, is the beautiful mother of all rivers, and the Dulongjiang, Jinshajiang, and Lancangjiang Rivers are her three sons. The Dulongjiang is the bad-tempered eldest son who brings severe disasters and diseases to local people. The god of Kawagebo mountain punished such bad temper by changing the Dulongjiang's flow from the south to the west. The Jinshajiang River is the bold and uninhibited second son, bringing fortune and gold to the locals. The huge wave sounds made by this arrogant river enraged the god of Kawagebo, who changed its flow from south to east. The last son is Lancangjiang, the Mekong River. He is both kind and peaceful, bringing abundant fruits and agricultural products to those who live along his banks. The stories of the rivers tell people what to expect from their natural environment and how they should behave towards these primary sources of life-giving water. The Tibetan people of Deqin have always chosen to live and farm on the tablelands and slopelands along both banks of the peaceful Lancangjiang River.

Fig. 2.6.1 Multiple habitats characterize this vast, diverse landscape. A village is located in the far upper left side of this image (Photo courtesy of Yin Lun)

This river's watershed has very rich forest resources, and above 2,800 m it is covered by thick primeval forests. It also has underground water resources, and in the high altitude forests there are many springs. In Tibetan language, they are called as 'ji ni'. When villagers herd animals in the forest, they will build wood houses near the spring water and grow buckwheat and potatoes in the moist soil near the spring. They will also create a natural garden of edible vegetables such as wild brake and mushrooms. Thus, generation after generation of Tibetan people have depended on the river and its tributaries and treated them as an indispensable guarantee of their sustainable subsistence.

However, the Lancangjiang River has recently brought disaster. In 1983 the largest and most terrible mudslide in the history of the People's Republic of China happened in Jiabi Village. Local dirt roads, farmland, and houses were destroyed, and after the disaster the local township government had to relocate the entire village to a new place, Jiunongding. In 1986 Jiabi Village was again enveloped by a mudslide, then by flooding in 1995, then by further mudslides in 2001 and 2003. Each of these disasters destroyed a great deal of productive farmland.

The local villagers addressed the issue in two interrelated ways: they built dykes to prevent further flooding, and they invited the living Buddha to conduct a special river ceremony to quiet and calm the river. During the ceremony, the living Buddha sprinkled highland barley to the river and gave some medicine balls that he made to the villagers. In order to frighten the river, he directed them to put both the medicine balls and some highland barley into a pottery vessel and then put the pottery in a specific place where mudslides will always pass through. Such responses are grounded in both existing cultural traditions and existing systems of water governance.

2.6.3 Water Knowledge in Deqin County

Most Tibetan villages in Deqin lie on the flat land beside the Mekong tributaries, 44 of which flow through the county. The Mekong watershed covers 3,090 km^2 (41%) of the total county and is the heart of agricultural production. The residents of Hongpo and Jiabi villages are primarily farmers and herders, obtaining additional cash income by collecting valuable fungi growing wild in the forests or working outside the village. The villages sit on the banks of rivers of the same name, both rivers originating from different parts of the Baima Snow Mountain. The villages are small, home to fewer than 200 inhabitants, and rely solely on the rivers for their drinking and irrigation water.

Whereas Jiabi River only passes through its eponymous village, Hongpo River passes through six other villages. The two rivers also flow through five different habitat zones: alpine talus, alpine meadow, forests, farmland, and valleys, creating different kinds of habitats and different kinds of water resources. Local people living in these watersheds have developed a rich body of knowledge that incorporates local hydrology, appropriate technologies to capture and store water, and respect for the spirits and powers associated with the water. Such knowledge also underpins a local governance system that regulates water use.

Fig. 2.6.2 Written prayers (Photo courtesy of Yin Lun)

The villagers regard the alpine lakes formed from melting glaciers and lying on the slopes of Baima Snow Mountain as sacred. The gods of each lake, and the god of the mountain itself, play a very important role in people's lives.

Jiabi Village has one sacred lake, Xiru lacuo, while Hongpo Village has two: Jia chuwa and Dawa. The local people believe that if they keep the lake waters totally unpolluted and pure, as well as protecting the animals and plants near the lake, the lake god will maintain favorable farming conditions and protect villagers from natural disasters and man-made calamities. The god of each of the sacred lakes controls the rainfall for the nearby village farms, sending rain when he is angry. Any activity such as cutting trees, destroying the forest, or hunting animals near the sacred lake will enrage the lake god, who will punish people accordingly with rainstorms, flooding, and mudslides. Therefore, the locally held beliefs contribute to environmental protection and good governance of the watershed.

When there is a drought, people organise a ceremony near each lake to pray for rainfall. In Hongpo Village, the villagers choose one representative to go to Gedan Hongpo Temple to ask the one living Buddha for aid. The Buddha then gives them a special medicine ball (*jia sen re bu*) that, along with a piece of wood and stone from the Hongpo river, is carried to the sacred lake. The village representative first burns joss sticks (incense), prays, and asks the god of the lake to forgive the interruption. Then he throws the medicine ball, wood, and stone into the lake, in order to disturb and enrage the god of the lake. Soon the sky above the lake clouds over and heavy rain falls. Tibetan people can also call for rain using the bark and leaves from a special tree or by pouring water from another river into the lake.

Box 2.6a Water culture traditions: Pond myths

—Anwar Islam and Shireen Akhter

Many of the larger ponds or lakes, called *dighi* in Bangladesh and West Bengal, have their own myths. Often one hears of a pond in which some spirit is said to live, a pond that would not hold water unless some type of sacrifice was made, or a pond that gives wealth to people in the vicinity. Three such myths are presented below.

In Delduar Town of Tangail District, there is one huge pond that was dug by a *zamindar*, a wealthy land-holding revenue collector during the British colonial period. His daughter was very ill. All efforts to treat her failed to cure her illness. One night the *zamindar* was told in a dream to dig a big pond and offer it for the use of all the people, not for private use. He was told in the dream that his daughter would be cured soon if he did this. It worked.

One pond in the Ramganj Sub-district of Noakhali District is located at Kanchanpur Dorga Bari. To this place an especially spiritual Muslim man and his sister came from Baghdad to spread Islam. They died in this place. The pond close to the graveyard gave forth all kinds of cooking and eating utensils made of gold and silver every year for use in a festival named Orosh. It is said that once a golden spoon was stolen, and the pond never provided the vessels and utensils again. Many people come to the graveyard to have special wishes granted.

Box Fig. 2.6a.1 Typical large pond or lake, called *dighi* (Madaripur District, Bangladesh, 2009. Photo credit: S. Hanchett)

Some believe that water from a mythic water body can fulfil wishes and cure sickness. In Noakhali District, for example, there is a pond named Fuldan. Once there was a man named Fuldan who dug a pond but found no water. Then he disappeared into the pond, and the hole was instantly filled up with groundwater. People believe this was a 'miracle of God'; and ever since then, the water of that pond has been considered to be holy water. Many people, both male and female, visit the place with 'wishes'. Women from different areas come to visit in hopes of conceiving a child, having a son, finding a missing son, retrieving lost items such as missing gold ornaments, and curing illness. Some women said that whenever they need to, they come for fulfilment of a wish. They bathe in the pond, drink the water, and bring some water to their houses for later use as a healing agent. One woman said she went there after she had a dream in which she was advised to visit. Many agree that their wishes have come true after using that holy pond water.

Almost every village in the region has its own sacred lake, river, or spring, and every year, local residents carry out ceremonies for these bodies of water. Yubeng Village has a number of sacred water sites but also a special sacred waterfall that is shared by people from all villages. Every August 15, Deqin Tibetans go to Yubeng Sacred Waterfall to pray, and during the ceremony they burn a cedar branch and then walk to the waterfall, circling clockwise under the waterfall three times.

Fig. 2.6.3 Yuben Shenpu (Photo credit: Martin Luerssen 2010)

Before they leave, the villagers hang their prayer flags on stones and trees near the waterfall, and they take water from the waterfall back home with them and use it ward off calamities and heal diseases.

At low altitude where the villages are located, there are also mineral water springs. During the dry season local people invite one living Buddha to conduct a river ceremony. After that, the living Buddha gives medicine balls to villagers and asks them to bury them under the mouth of the spring for rainfall. According to local traditions, water from springs located at lower altitudes also has special properties: Qupo Baca spring water can cure stomachache or toothache and improve people's appetite; Yala spring water can cure cholecystitis and rheumatism; Duoqu spring water can cure skin disease; and Xiaqiong spring water can ward off calamities and evil spirits. The waters of Deqin County keep people physically and psychologically healthy.

2.6.4 Management and Use of Water Resources

The villagers of Hongpo and Jiabi have developed their own way to manage and use water resources effectively. First, from a management point of view, villagers divide water resources into two parts: drinking water and irrigation water. In managing drinking water, villagers divide the village into three parts according to land structure and build three ditches drawing water from the nearby river to send to the three parts. A group of three households will rotate the responsibility of cleaning the ditches every month. In managing irrigation water, villagers divide the village into upper and lower parts and during the dry season, villagers from the upper part irrigate farmland in the morning, villagers from the lower part in the afternoon. These divisions form the basis of water management in social terms.

Those villagers who have houses close to the river use a water pipe or wood stove flume to draw water into a water tank in their dwellings. Houses further away from the river use a ditch to draw water for drinking and irrigation. Sometimes ditches are temporarily blocked to help upstream villagers irrigate their farmland, and downstream villagers are not supposed to complain about this. If three households need the water in the ditch to irrigate their farmland simultaneously, usually the household that uses the ditch to draw water first will be allowed to irrigate first. At certain places along the ditch, villagers build traditional water mills to grind wheat and corn, and there are regulations about the use and management of these mills. Social prohibitions also ensure that drinking and irrigation water remain unpolluted and of good quality; an old Tibetan saying is that those drinking dirty water downstream will not fall ill or become degenerate, but those making upstream water dirty will get their punishment from the river god. The local management and utilization of water resources are based on Tibetan traditions that help maintain harmonious relations and meet people's needs.

Box 2.6b Water and culture in the land of the gods

—Tashi Tsering

The Tibetan Plateau holds more fresh water in the form of glacial ice than any other place on earth except the Arctic and Antarctic. In recent years, the increased melting of Tibet's glaciers because of climate warming has attracted significant attention in science and the media. In 2007, the Intergovernmental Panel on Climate Change (IPCC) reported that the Himalayan glaciers 'are receding faster than in any other part of the world' (Parry et al. 2007:493), and in the words of the chairman of IPPC, Dr. Rajendra Pachauri, 'At least 500 million people in Asia and 250 million people in China are at risk from declining glacial flows on the Tibetan Plateau....This is one of the great concerns – a staggering number of people will be affected in the near future' (Schneider and Pope 2008).

There is another, more direct threat confronting Asia's water supply, however, and that is the many large-scale hydro-development projects built or planned on the sources of these rivers on the Tibetan Plateau. China is building hundreds of large dams here, including some of the world's tallest dams and largest water diversion projects. The plateau, in addition to being a critical water and energy resource for nation-states on both sides of the Himalayas, is the sacred ancestral land and 'foodshed' for millions of people. International media and governments, especially those of downstream nations, have paid considerable attention to the environmental changes and human consequences resulting from dam construction, diversions, and subsequent decreased river flows. Yet little to no attention is paid to the social implications of dam development for the local Tibetan people. The perspectives and concerns of these disenfranchised people are overlooked in the interest of science (climate change), economics (hydro-development), and politics (transboundary issues and so on), with the conversation and the agenda framed by powerful interest groups (businesses, states, academia).

In order to understand the impact of climate change and water development on Tibetan society, it is necessary to have at least some appreciation of how Tibetans live with regard to their local water resources. The following example from the Tibetan Buddhist villages of Spiti Valley in the arid western Himalayas shows that the traditional social system and cultural beliefs intimately connect nature and associated water systems with the spirit world and people's well-being.

(continued)

Box 2.6b (continued)

2.6b.1 *Spiti Lha Yi Yul*: 'Spiti, the land of gods'

Spiti is a sparsely populated (10,679 people in a 7,591 km^2 area) high altitude (4,000 m above sea level) desert valley situated on the southwestern tip of the geographical Tibetan Plateau. Local inhabitants are ethnically, linguistically, and culturally Tibetan. Spiti is best known for its ancient Buddhist monasteries, especially Tabo and Dhangkar, considered the oldest in the world. Agriculture and pastoralism are the main sources of livelihood for a majority of the people. Traditionally subsistence farmers, now most agriculturalists grow peas as a cash crop in addition to the traditional crops of barley and wheat. And although the number of tourists visiting this otherwise isolated valley is growing every year, the lifestyle in Spiti is still relatively traditional compared with that of neighbouring regions, thanks to several factors. The vast majority of the local people are Tibetan Buddhist farmers. The Indian state has formally recognised the traditional resource governance customs and regional institutions. India restricted tourist access to this valley between 1962 and 1993 due to international border sensitivity. This high altitude desert valley is surrounded by 6,000 m-tall mountains on all sides, isolating it geographically. The unpaved mountainous roads that are the only way to get into the valley are shut off for 5 months of the year because of snow.

In isolated and arid regions, a major limiting factor for permanent human settlement for farming is availability of water. Water is so important in Spiti that the main households of any village are family groups with legitimate claims to local water sources, groups known as *Khangchen* (khang chen, literally 'large household'). In order not to subdivide resources, the rights to land and water pass to the eldest son alone, rather than shared equally among the members of each generation. The head of the *Khangchen* may give or sell a piece of land to someone, but the land returns to the *Khangchen* after the death of the person who received the land. These practices ensure that the traditional land and water ownership rights continue to belong to the same households. Other classes of household have fewer rights to land and waters; their members often act as labourers for *Khangchen*. However, new *Khangchen* groups called *Sarshugpa* (gsar zhugs pa, 'new members') have also arrived since 1968, when the Indian government began redistributing land to the landless under the Himachal Pradesh *Nautor* Land Rules. Nautor is an ancient right whereby people are allowed to break fresh agricultural land in common land areas. The *Sharshugpa* have plots of land further from the village but participate in village life alongside the *Khangchen*.

(continued)

Box 2.6b (continued)

2.6b.2 The spirits of water

Water plays a significant role in shaping Tibetan religious practices and belief systems. The mythical story of the mountain deity of Kibber Village provides insights into the significance of water in Spitian cultural and religious practices. According to local legend, farmers in Kibber were unsuccessful in building canals (*yura*) to irrigate their fields despite many attempts. To help with the situation, a powerful ghost-spirit (*rtsan*) arrived in their village as a crow. The crow landed high up on the side of a mountain and touched the ground with its beak, from which point a spring miraculously emerged. To this day, the spring (although meagre) is the head source of canal water, and the villagers consider it very sacred. The ghost-spirit came to be known as the Restorer of the Irrigation Source (*Yur mgo log*).

In addition to making the source of irrigation canal a sacred space that must be looked after and respected, the presence of *Yur mgo log*'s spirit on the mountain where it first landed and is believed to reside has rendered the whole mountain sacred. It is customary for local farmers to offer annual prayers to the deity and put up prayer flags at the deity's mountain abode on every seventh day of the fifth month of the Tibetan calendar. The farmers offer the prayers to appease the spirit and request it to grant continued protection and prosperity to the village and its people. Villagers have also built a temple and invited the spirit to live in the temple as the protector deity of Kibber Village. Farmers communicate directly with the spirit through a shamanic medium (*lhapa*) about crop productivity, water flow, weather, health, and other matters. In other words, the deity is the chief consultant, arbitrator, and protector for the villagers. The deity also owns large tracts of the communal land in the village. Farmers have a close and regular relationship with their local mountain deity, which they regularly propitiate through various ceremonies, such as *Sangsol* (bsang gsol) offerings, *khrus* and *dge ba* rites of purification, *chos sil* recitations, and so on.

Although this story of Restorer of the Irrigation Source comes from a single village, the propitiation of mountain deities (*yul lha*) is a common indigenous (pre-Buddhist) religious practice throughout the Tibetan Plateau and is unique to it. Religious practices meant to show respect, serve, or appease the mountain deities constitute a major part of the spiritual and community life of Tibetan farmers (Karmay 1996, 2005).

Other major elements of Tibetan spiritual life include respect for those spirits residing in specific places and water bodies. Thus serpent spirits (*klu*), 'life-force' (*bla*) as well as divine lakes (*lha tsho*) are all sacred. Tibetans believe

(continued)

Box 2.6b (continued)

Box Fig. 2.6b.1 Yur Go Log

Box Fig. 2.6b.2 Farmer praying at the abode of Yur Go Log

Box Fig. 2.6b.3 Spot where *klu* the crow is believed to have landed

water bodies (lakes, ponds, rivers) and sources (headwaters, springs, glaciers) are the dwelling places of serpent spirits (*klu*). Tibetans often prepare *klu khang* ('houses of serpent spirits') decorated with prayer flags and surrounded by painted rocks at certain designated spots, such as headwater springs. If the natural qualities of these dwelling places are disrupted or polluted, the serpent spirits will cause harm to the polluter in the form of disease, accident, or bad luck. If the perpetrator is a farmer, angry serpent spirits may cause heavy hailstones to fall or a drought to destroy the farmer's crop. Such understandings of the powers in the landscape create what would elsewhere be called an ethic of conservation.

Another important aspect of Tibetan belief is related to the concept of *bla*, 'life-force' or 'soul'. Water bodies are often believed to be *bla-gnas* (the dwelling place of *bla*) on which the health and life of a person, a community, or even a whole nation depends. For example, the Turquoise Lake (*Yamdrok Yu Tso*) in south-central Tibet is held to be the *bla-gnas* of the Tibetan nation, and so Chinese plans to build dams on it and other sacred watercourses create much opposition. Destruction or pollution of *bla-gnas* can bring great harm or even death to those whose *bla* resides there.

(continued)

Box 2.6b (continued)

2.6b.3 Concluding concerns

Three-quarters of the indigenous Tibetan people sustain their lives through the practice of agropastoralism. This kind of livelihood is not only extremely vulnerable to climate change but it is also being profoundly transformed through China's Western Development policies, which emphasise urbanization, industrialization, mineral extraction, and infrastructure development. Given such profound change undertaken with minimal or no concern about how local farmers and indigenous peoples relate to their ancestral lands and local water resources, policy-makers and dam proponents are bound to face resistance. When such cultures and voices are overlooked to make way for dams and other such development projects, policy makers not only impose development in harsh and rights abusive ways, they also disrupt spiritual and pro-conservation cultural heritages.

For example, in May 2009 thousands of angry Tibetans protested against the proposed Lianghekou dam project near Tawu County, Karze Tibetan Autonomous Prefecture, part of China's Sichuan Province (Central Tibetan Administration, 2009). Reports describe instances of local people being forced to sign documents for relocation amidst vehement local opposition. Protestors led by a 70-year-old Tibetan woman destroyed a symbolic stone pillar that officials had erected to signify dam construction and relocation plans. Armed police, dispatched to control the protestors, opened fire at the angry protesters, leaving six Tibetan women wounded. Such incidents, sadly, are not isolated cases. Violent clashes between dam protestors and police are increasingly common not only in China but also in other parts of the world, such as India and Thailand.

Whether it is reports of dramatic incidents of local resistance or debates about dam projects in general, what is absent in the public framing of these controversies is an effort to articulate and appreciate local cultural connections to water and land. Instead, reports of local opposition to large hydropower projects from around the world offer images of protestors as poor, primitive, often 'indigenous' villagers, refusing to be displaced by some centrally or internationally funded hydropower project. Development project proponents argue that hydroelectric dams and water diversions will promote economic development in the region and that development and the associated displacement of villagers are a small price to pay for the larger societal good. Sympathisers argue that these communities should be left to themselves, untouched by modern society, which has no 'right' to develop and transform their 'sacred' lands. More progressive sympathisers advocate that affected people should be consulted as 'stakeholders' in development projects and compensated with money and new land for resettlement. However,

(continued)

Box 2.6b (continued)

dam-affected people are often disenfranchised peoples with no right to organise and articulate their demands.

Thus, appreciation of social and cultural significance of ancestral lands and local water resources to the affected people is a necessary, but not sufficient, condition to avoid conflict over dam projects. In the Tibetan Plateau, as elsewhere, hydro-development is framed in the language of national economies and societal progress, yet all too often the driving force behind such development agendas is money and power: How much money will be made from the sale of electricity or saved through flood control? Who benefits and who suffers? How are decisions made?

When development is imposed with disregard for local (and downstream) values, conditions, and livelihoods, protest is inevitable. Because dam proponents are typically more powerful than the local opposition and have the ability to summon the police and military muscle of the state to bulldoze people's homes and lives, the use of brute force to eliminate opposition is increasingly common.

Avoiding such rights-abusive conflict requires rethinking the fundamental questions and priorities that shape and legitimise development. In the hierarchy of priorities, *human* interests, particularly of those directly affected by the projects, ought to be given precedence over those of interest groups such as businesses and bureaucracies (Gurtov 1999; Tsering 2003).

Resources

Central Tibetan Administration. 2009. Six Tibetans seriously wounded in protests against China's hydroelectric dam project. http://www.tibet.net/en/index.php?id=938&articletype=flash&rmenuid=morenews&tab. Accessed 20 Dec 2009.

Gurtov, Mel. 1999. *Global politics in the human interest.* Boulder: Lynne Rienner Publishers, Inc.

Karmay, S. 1996. Mountain Cults and National Identity in Tibet. In R. Barnett and S. Akiner (Eds), *Resistance and Reform in Tibet.* India: Motilal Banarsi Dass Publications..

Karmay, S. 2005. The Arrow and Spindle: Studies in History, Myths, Rituals and Beliefs in Tibet - Volume 2. Kathmandu: Mandala Publications.

Parry, M.L., O.F. Canziani, J.P. Palutikof, J. van der Linden, and C.E. Hanson (eds.). 2007. *Climate change: Impacts, adaptation and vulnerability.* Contribution of Working Group II to the Fourth Assessment Report of the Intergovernmental Panel on Climate Change. Cambridge: Cambridge University Press.

Schneider, K., and C.T. Pope. 2008. China, Tibet, and the strategic power of water. *Circle of blue reports.* http://www.huffingtonpost.com/j-carl-ganter/tibetan-plateau-water-res_b_100818.html. Accessed 15 Sept 2008.

Tsering, T. 2003. China's water politics: In whose interest? MA thesis, Portland State University.

2.6.5 Action Research Projects for Water Resources: Promoting Good Water Governance

Historically, development projects along the Mekong have not incorporated local consultation and involvement in their design, and therefore they have not delivered benefits to local peoples. In many cases such projects have also harmed the environment in the long term, as they were focused on short-term benefits (Lebel and Dore 2007). Recent literature from China has emphasised the problems created by inappropriate development that harms the environment more than it benefits local people (Guo 2008; Zheng 2006; Yin 2002). Proper adaptation to current and future environmental challenges requires attention to local knowledge about water resources and creating the tools to put that knowledge into practice. Such processes lead to an effective model for good governance and co-management, as well as a harmonious relationship between the local community's sustainable development and the natural environment. Until now, too little attention has been paid to the relationship between local knowledge and good governance and how to use local knowledge to establish effective models of water utilization and management. Water development activities undertaken in this region between 2004 and 2008 incorporated these local water management customs as part of the effort to provide sustainable water development at the village level.

2.6.5.1 Jiabi Village

The four main water resources projects in Jiabi Village were the following:

1. Building water tanks and setting up water pipes. Based on their knowledge of the topography of their own village and with project support and technical guidance provided by experts from the Deqin Water Conservancy Bureau, the villagers drew left, middle, and right routes as the three best lines for drawing water from the river. At the same time, they marked places to build water tanks. Further investigation by water conservation experts assisted the villagers in confirming the left route as their main ditch for drawing water, along with supplementary ditches on the middle and right routes. As the three ditches were built in the three directions, every household built a branch ditch connecting its house with one of the three ditches, and three water tanks were also built. The above activities were based on the initial project design. After they were completed, the villagers organised a meeting to discuss and approve the final project design. Once they had approved it, the villagers chose a representative to buy water pipes, cements, bricks, and other materials from the county. These were used to line the ditches to ensure their durability. Today villagers can get water easily and do not have to carry water anymore.

2. Installing solar water heaters. The villagers allocated one-third of the total money for the project to install solar water heaters to provide continuous hot water to the households. Installing the heaters had important additional benefits: women's workload decreased because they did not need to spend as much time in the forest collecting firewood, and forest clearing for firewood also decreased, indirectly improving downstream water quality.
3. Strengthening the village embankment. Using support and advice from project members, villagers rebuilt part of the dyke and strengthened the embankment.
4. Establishing local regulations for additional water system management. At a meeting, the villagers agreed that every month, three of their number in turn would act as a team to clean the ditches and maintain the water tanks.

2.6.5.2 Hongpo Village

The three main activities for managing the water resources in Hongpo Village were as follows:

1. Establishing a water resource management team. Hongpo villagers have their own traditional associations; women have their Sister Association and men have a Sword Association. All men and women in the village are members of the respective associations, and at a meeting the villagers decided to establish the Hongpo Water Resource Management Team (HWRMT), based on the two existing associations. This makes the HWRMT a real grassroots organization growing out of community traditions.
2. Conducting a comparative history of water resources. Following the establishment of HWRMT, villagers evaluated current and past water resources by taking notes, collecting samples, drawing pictures, and taking photos. People recalled the water resource changes of the last 50 years, and some villagers drew pictures of the location of the headwater forest and glacier. They also recorded beliefs and legends about sacred lakes and spring waters, changes in the forest, and natural disasters.
3. Reaffirming good governance of water resources and environment. Project resources and participants assisted local people to build a Tibetan Buddhist pagoda. Invited to conduct a ceremony at the new pagoda, the living Buddha from Gedan Hongpo Temple recited scriptures for the sacred mountain and confirmed the location of an area where villagers could not cut trees. At a village meeting, the villagers made strict regulations about forest protection and the use of timber and firewood. They divided the forest into five zones and agreed than in each year they would cut trees from only one forest area, keeping the other four fully protected. The villagers also planted trees on barren hills along the Jiabi River and cleaned rubbish along the banks of the Hongpo River. Finally, a water environment protection exhibition was held in the village.

Box 2.6c Development, disaster and myth: A case study

—Anwar Islam

In September and October of 1988, a devastating flood overran much of Bangladeshi land. Roads, houses, animals, and human lives were damaged. Some of the damage, however, was the result of human action. The World Bank and other donors had imposed communication infrastructure development programmes that were implemented without any comprehensive national hydrological survey. Local interests were given more emphasis than the national interest. Unethical political and powerful socially reactionary forces supported this situation.

Donors paid for a road through the centre of Shaharatol, a village in Tangail District. A 25 m concrete bridge on this road allowed huge amounts of water to drain from marshy areas surrounding the village. At the time of flooding, some village roads were submerged. Everywhere flood waters threatened. In that situation, one morning as residents awoke to the twittering of birds they were surprised to see a 3-acre pool, a *dou,* under the bridge. A *dou* is a natural body of water with deep, eerily blue water. It is considered frightening, somewhat ghostly. Residents do not allow their children to go near a *dou,* where they might be attacked by ghosts.

The previous night, strong water pressure had eroded the earthen walls supporting the bridge's two ends. As the earth eroded, people heard a murmuring sound as the *dou* was forming. People said it was a 'revenge flood' that created the *dou.* Immediately after the *dou's* creation, some people started saying that God was punishing Shaharatol people for their sins – not praying regularly, taking bribes, and money laundering. Many laid out fruit, lit candles under trees, or set offerings afloat on banana trunk rafts, hoping to regain God's favour.

This account is of Anwar's experience more than 20 years ago.

2.6.6 Conclusion

The fragile Tibetan environment and the heavy dependency of local people on water use create challenges for effective water management and the ongoing subsistence and development of Tibetan people. In recent years, the water situation has become more serious, with more and more watershed forests being destroyed and natural disasters such as droughts, floods, and mudslides increasing in frequency. This situation makes it urgent to explore effective ways to manage water resources, and local knowledge of water must be paid due attention.

Environmental challenges in the Mekong River Basin are many and complex, requiring domestic and international action to achieve efficient water allocation, restore degraded habitat, halt forest conversion, and prevent water pollution. Since the Mekong River provides livelihoods for a significant majority of the basin's

75 million people, it is an important source for economic development. Such development poses both opportunity and threat. Significant development activities along the upper Mekong River in Yunnan, China, include the expansion of agriculture (including rubber plantations), forest clearing and degradation, and large hydropower projects. Such activities have led to an increase in water consumption and a decrease in water availability and quality.

This Deqin County example suggests that integrated watershed management might best be achieved by incorporating local cultural knowledge and values to (1) improve the capacity and capability of indigenous people to utilise local resources for sustainable livelihoods and (2) create effective models of water government in local communities. Good governance is the foundation in establishing a harmonious relationship between human beings and water resources, and good governance flows from consultation, the appropriate sharing of knowledge, and respect for the decision-making power of local communities.

Resources

Jiaji, Guo. 2008. *Development reflections: Anthropological research on ethnic peoples' change along Lancangjiang River*. Yunnan: Yunnan People's Press.

Lebel, Louis, and John Dore. 2007. *Democratizing water governance in the Mekong region*. Bangkok: Mekong Press.

Karmay, S. 1996. Mountain cults and national identity in Tibet. *Resistance and Reform in Tibet*, ed. Barnett, R and Akiner, S. India: Motilal Banarsi Dass Publications.

Karmay, S. 2005. The arrow and spindle: Studies in History, Myths, Rituals and Beliefs in Tibet - Volume 2. Kathmandu: Mandala Publications.

Shaoting, Yin. 2002. *Rain forest or rubber forest*. Yunnan: Yunnan Education Press.

Xiaoyun, Zheng. 2006. *Ethnic peoples' culture and forest protection along Lancangjinag River in Yunnan Province*. Beijing: China Literature Press.

Chapter 2.7
Ecological Change and the Sociocultural Consequences of the Ganges River's Decline

Georgina Drew

2.7.1 Climate Change and Growing Water Scarcity in the Himalayas

Climatic warming poses a threat to much of the freshwater reserves trapped in glaciers worldwide. In the glacier-capped regions of the Himalayas, the shift in water resource availability could be dramatic. Known as the 'abode of snow', the Himalayas are home to thousands of glaciers that form the largest freshwater reserve after the polar ice caps. The runoff generated by these glaciers feeds seven of Asia's greatest rivers, providing water and supporting the production of food for over 1.5 billion people. Waters from the Himalayan glaciers also feed a 'hotspot' region of biodiversity with some 10,000 plant species, an estimated 300 mammals, and almost 1,000 types of birds (Conservation International 2008). Disturbingly in an area of such importance, the Himalayan glaciers are receding. While the timeline for glacial melt

Map 2.7.1 Map of the Ganges River in India

G. Drew (✉)
Department of Anthropology, University of North Carolina, Chapel Hill, NC, USA
e-mail: georginadrew@gmail.com

B.R. Johnston et al. (eds.), *Water, Cultural Diversity, and Global Environmental Change: Emerging Trends, Sustainable Futures?*, DOI 10.1007/978-94-007-1774-9_15,

here is a subject of debate and even some controversy, a wealth of scientific data indicates that we can expect extensive climatic transformations within generations (IPCC 2007; UNEP 2009). Global temperature increases, shifts in precipitation patterns, and increased deposits of dust and black carbon that reduce light deflection from glaciers will drive the anticipated changes (UNEP 2009). If the glaciers deteriorate, on the one hand, and monsoon trends shift, on the other, many Asian countries will likely face a diminished capacity for surface water recharge and a significant shift in freshwater availability. These changes will wreak havoc on agriculture, industries, and domestic livelihoods downstream. If glaciers continue to decline, a United Nations Development Programme (UNDP) report warns that the water and food security of developing nations in Asia will be threatened by the middle of the twenty-first century, signaling the 'reversal of hard won development gains' (Khoday 2007: 8).

The Ganges River and its tributaries are among the expected casualties of glacial retreat. Flowing some 2,500 km and supporting the livelihoods of half a billion people, the Ganges River is one of India's most important freshwater sources. The 18.5-mile-long Gangotri glacier, which in some estimates provides up to 70% of the Ganges' perennial flow, retreated significantly in the twentieth century (IPCC 2007). Although the glacier has been withdrawing since the late 1,700 s, the rate of retreat accelerated as the millennium approached. Between 1962 and 1999, the snout of the glacier withdrew by 1.25 km, about three-fourths of a mile (Naithani et al. 2001: 93). If the glacier continues to lose mass and runoff potential, diminished long-term flows in the Ganges could force millions of people and countless multitudes of non-human lives to make do with significantly less water.

Along the primary tributary of the Ganges, the Bhagirathi (henceforth referred to as the 'Bhagirathi Ganges') flows are reaching extremes between high and low seasons that many Himalayan residents say they have never before witnessed. In addition to irregularities in glacial runoff, the change is related to shifts in snow and rainfall patterns caused by warming global temperatures (Basishta et al. 2009). Residents of Uttarkashi in the northwestern part of the state of Uttarakhand (India), where the author conducted fieldwork from 2008 to 2009, note a paucity of winter snowfall compared to previous decades. The lack of snow combined with reduced glacial mass lessens spring and summertime runoff (Singh et al. 2006). Although precipitation appears to be increasing in the once drier months of March and April, a good thing in terms of household needs, when rain comes so late in the harvest season it can damage mature wheat crops. The weather variability augments a growing sentiment of vulnerability in the region. Hoping to mitigate the deleterious impacts of climate change, a number of activists and environmental groups are working to promote sustainable practices along the river's course.

Fig. 2.7.1 An environmental activist looking down at the scant, polluted waters of the Ganga in Uttarkashi (Photo credit: Georgina Drew, 31 March 2009)

Box 2.7a Mitigating the impacts of Himalayan glacial melt: Trends and policy recommendations

—Georgina Drew

Worldwide, glacial mass could retreat 25% by the year 2050 and up to 50% by the year 2100 because of climatic warming (IPCC 2007). If such predictions hold true, downstream regions will experience a major decrease in the availability of fresh water. Inhabitants along the Ganges River in the Himalayas are already feeling the strain of glacial melt and declining water supplies. In addition to the material implications of shifting water availability in the Ganges, the river's reduced flow poses numerous concerns for the maintenance of cultural traditions and Hindu religious rituals in which the river plays a central role. Because of the high levels of carbon already released into the atmosphere due by unsustainable human consumption and production practices, it will take a concerted international effort to stop the warming and slow the rate of glacial retreat. Although it is not possible to stop glacial melt in the short term, there are prospects for resilience. Key issues entail revising development practices, escalating locally based conservation measures, and providing environmental education activities to promote long-term adaptive measures.

(continued)

Box 2.7a (continued)

2.7a.1 Issues of vulnerability

- Runoff from Himalayan glaciers support livelihoods for two billion people in Asia.
- Water scarcity because of glacial melt jeopardises agricultural productivity and food security.
- Warming trends and erratic snow and rainfall threaten Himalayan biodiversity.
- Hydroelectric development schemes damage fragile Himalayan ecologies for short-term gains.
- Resource management is gendered and women have an important role in adaptation.

2.7a.2 Promising trends

- Reforestation and conservation efforts are increasing near glacial areas.
- Women are active in environmental campaigns to protect the Ganges and regional ecologies.
- Bans on plastic help to create awareness.
- Caps on tourism to glacial areas promote environmental sensitivity.

2.7a.3 Policy recommendations

- Work with downstream populations to provide culturally sensitive environmental education.
- Engage Himalayan residents, especially women, in forming regionally appropriate responses to mitigate short- and long-term vulnerabilities to warming temperatures and glacial melt.
- Revisit and revise development and energy production policies in the Himalayas based on the most up-to-date data on current and projected hydrological flows.
- Promote the production of energy with renewables such as solar and wind.
- Expand existing water storage technologies and encourage innovation in harvesting water.
- Outpace the annual rates of forest loss by planting deep-rooted tree species that store groundwater.
- Revive the cultivation of crops that use minimal water and are regionally suited for resilience.
- Accelerate global efforts to reduce carbon and black soot emissions that contribute to warming trends and impact the fragile balance of Himalayan ecologies.

2.7.2 Hydroelectric Development and Contested Natures

In the context of climatic change and glacial retreat, contemporary approaches to water resource management in the Himalayas are hotly debated. A prominent source of contention centres on hundreds of hydroelectric dams proposed for construction in the Himalayas. Not counting China, the Himalayan regions of India, Pakistan, Nepal, and Bhutan could generate an additional 150,000 MW of hydroelectric energy by 2030. According to some estimates, India may produce as much as two-thirds of the expected total (Dharmadhikary 2008: 6-8). The renewed policy emphasis on dam building is backed by international organizations, many of which are scaling up their support for hydropower after years of cautious lending for the contentious development projects. These dams have government and institutional backing, ostensibly for the 'clean' energy some think that they produce. This expansion of hydroelectric development however concerns many of the people living in project areas. Despite the marginal, unskilled employment opportunities that dam building provides, opponents argue that the ecologically destructive projects are not viable because of climate change and glacial retreat.

Along the Bhagirathi Ganges high up in the Himalayas in India, numerous social movements and non-governmental organizations adamantly oppose four run-of-river hydroelectric projects on a 110-km stretch between the river's glacial source and the district capital of Uttarkashi. Of the contested dams here, two of the four are already under construction. The civil society actors opposing the dams question the environmental damage caused by dynamite blasting in a fragile, zone-five area of seismic activity with frequent landslides (zone five being an area at risk for seismic activity). Challenging claims that run-of-river hydroelectric projects are benign, opponents argue that the dams will threaten the ecological balance of the region by diverting water through several submontane tunnels for energy production. By taking water out of the riverbed, critics charge, downstream sections will be left with minimal flow. Such diversion will hinder groundwater recharge, withhold water needed to support biodiversity, and may possibly alter the river's ability to maintain a self-cleansing rate that is nearly three times as fast as rivers such as the Yamuna (Down to Earth 2008: 31). Another concern is that as the melting glaciers deposit more sediment into the runoff, corrosive, turbine-damaging silt may prevent the power plants from meeting energy production targets (Dharmadhikary 2008: 25). And, the increasingly irregular precipitation patterns in the region contribute additional concerns over energy production and long-term water availability.

In light of these and other issues, civil society initiatives are working to raise public awareness about the proliferation of hydroelectric development projects along the Bhagirathi Ganges. Several social movements and NGOs, for instance, supported two separate 'fast unto death' programmes undertaken by G. D. Agarwal, a retired professor of the renowned Indian Institute of Technology, Kanpur. Agrawal's extended fasts in 2008, 2009, and 2010 were grounded in ecological, cultural, and religious concerns for the well-being of the Ganges and the Bhagirathi Ganges in particular. These fasts – one of which lasted just short of 40 days – helped force the central government to suspend the dam building temporarily. Set in tandem with the agitations of groups such as Ganga Bachao Abhiyan ('Save the Ganges Movement'), Ganga Ahvaan ('Call of the Ganges'), and the Ganga Raksha Manch ('Ganges Defense Forum'), these actions helped to compel the central government to declare the Ganges a 'national river' in November of 2008. As a result of the river's elevated status, a new governing body, the Ganges River Basin Authority (GRBA), was established to monitor projects on the river and promote its sustainable use. Although the measure is heartening for some, the GRBA must learn from earlier fiscally wasteful and only nominally successful schemes such as the Ganges Action Plan (GAP) begun in the 1980s. To be more democratic than past initiatives, the GRBA will also have to engage communities living along the river in planning and monitoring efforts (CSE 2008).

Subsequently, in a major victory for opponents of large hydrodevelopment, the Government of India cancelled two proposed hydrodevelopment projects, the Loharinag Pala and Pala Maneri, in August of 2010. With the announcement, the government declared the 135 km stretch (83 miles) from Gaumukh to Uttarkashi city an "eco-sensitive zone" under the Environment Protection Act of 1986. The decision was taken in deference for the faith that people express for the river's Goddess and the "unique status of the sacred Ganga (Ganges) in our culture and in our daily lives." The historic measure sets a precedent for development practice to honour sociocultural traditions of reverence for water.

2.7.3 Cultural, Religious, and Gendered Dimensions of the River's Decline

The issue of access to water is at the centre of concerns over glacial melt and development along Himalayan rivers such as the Ganges. With a ranking of 133 out of 182 countries for water availability per capita (UNWWAP 2003), India already has significant problems. Ecological change and hydroelectric development projects will likely exacerbate existing global disparities and in-country water inequities. These disparities and inequities are highly gendered, with low-income women bearing the brunt of resource shortages, so the intensification of scarcity threatens to increase women's resource-related burdens.

Along the Bhagirathi Ganges in Uttarakhand, a lack of alternative economic opportunities combined with outward male migration means that women have heavy domestic and agricultural workloads (Mehta 1996). Because daily life depends on the resource base, the projected decline in water availability endangers women's livelihoods in the Himalayas, as well as the welfare of the children and animals under their care. Taking significant roles in the civil society initiatives challenging development and environmental destruction in the Himalayas, many rural and semi-urban women voice concern about the ecological shifts they witness along the river. Their comments show keen observations of environmental change. Beyond material concerns, they also indicate a range of cultural preoccupations and dependencies on the river's flow.

For many rural and semi-urban women, the issue of livelihoods is not merely about the presence of water, it is about the *use* of what they consider a very special source of water to adhere to cultural prescriptions of conduct in life. As numerous Himalayan women readily point out, the river's sociocultural significance extends millennia back into India's history, including key roles in Indian epics such as the *Ramayana* and the *Mahabharata*. Worshiped by Hindus and non-Hindus, the Ganges is a vital element for ritual (Darian 1978). Inability to access the Ganges would impact daily practices connected with its waters as well as the cultural rites of passage in the Hindu life cycle known as *sanskar*. A companion throughout life, water from the Ganges is meant to sustain a person from birth by washing away physical and spiritual impurities. Immersion in the Ganges is especially critical in rituals that mark death. For some followers of the Hindu faith who live along the river, cremation without subsequent immersion of one's ashes into the Ganges River is unthinkable. The fear of dying without merging into the river's flow spurs some to action. Although it is crucial to acknowledge the material implications of glacial decline, therefore, it is also important to remember that cultural and religious traditions of reverence for the Bhagirathi Ganges motivate also human behaviour.

Illustrating the role that culture plays in environmental initiatives, women active in movements to conserve the river consistently cite a need to protect the river's goddess, whom they refer to lovingly as *Ma Ganga* or 'Mother Ganges'. The concept of the Ganges as a mother is more than a term of endearment; women, men, and children across India honour the river and its associated goddess as a loving guardian who protects and provides for her offspring.

The prospect of being without access to the river is deeply disturbing for those who depend upon her support. The leader of a women's committee near Uttarkashi summarised her fears of losing the river with a lament: 'If 1 day She is no more, then what will we do without our mother?'[1] The question was more than rhetorical; as the woman spoke, she motioned to a nearby dam upstream that diverts water out of the riverbed. In lean periods, the water in the Bhagirathi Ganges slows to a trickle, depriving people of water and making ritual observance difficult. Stressing the importance of the river's flow, she continued: 'The [Bhagirathi] Ganges is part of our identity. She is the reason people come here from all over the world. Without her, we will be nothing'. The woman's concerns, grounded in personal as well as practical considerations, demonstrate how intimately the cultural and religious reverence for the river is tied to the region's economy. According to such women, glacial melt and dam building threaten to eliminate the very attraction that lures thousands of tourists and pilgrims every year and provides the bulk of nonagricultural income.

Although climatic shifts and hydroelectric projects have already altered the average flows in the Bhagirathi Ganges, life in the absence of the river is still unthinkable for some. The difficulty of conceptualizing a world without the river's flow is

[1] Comments from a focus group with members of the women's committee, or *mahila mangaldal*, in Mori village on 8 August 2008.

Fig. 2.7.2 Photo taken on Ganga Dussehra, the day that marks the descent of the Ganges from the heavens to the earth (Photo credit: Georgina Drew, 2 June 2009)

particularly pronounced on days marked with cultural and religious rites in which the river plays a central role. On the occasion that commemorates the goddess' descent to earth in liquid form, the *ghats* (steps leading to the river) that dot the landscape of places like Uttarkashi are filled with people engaged in rituals, celebrations, and ablutions. Beginning in the early morning hours and continuing well into the evening devotees make their way to the river's banks to touch the sacred liquid, wash away sin, and offer water to appease various other gods and goddesses who desire the river's blessing. On such days, village women come from all over the district to worship at Manikarnika ghat in the centre of the city. Immersing themselves fully, their shrill cries at contact with the frigid, glacier-fed waters fill the air. These cries are interspersed with occasional screams as regional goddesses enter the bodies of devotees to join in the celebration.

When explaining the importance of these rites, devotees emphasise the relationship between the river, the sacred geography of the Himalayas, and the joint efforts of King Bhagiratha and Lord Shiva to bring the river to earth. As the story goes, a mortal named Bhagiratha did hundreds of years of penance to bring the Ganges down from the heavens and purify the sins of his ancestors. When the river's Goddess Ganga heard the call to descend and bless the world, she replied egotistically that she would break the earth with her fall. It was only when Shiva, a Hindu god, intervened and demanded she flow down into his matted hair that she conceded to the request. For this reason, the Himalayan mountaintops are sometimes associated with the dreaded locks of Shiva, which broke the impact of the river's descent. Retelling the river's history with zeal, devotees marvel at the effort it took to bring the river before them. The blessing marked by the presence of the Bhagirathi Ganges, however, contrasts with the environmental degradation that devotees say epitomises disregard for the river's health in contemporary times.

Although many are ready to talk about the obvious signs of pollution in the river, the issue of water scarcity is harder to broach. When asked what will happen when there is no water to worship, for instance, many devotees reply with furrowed brows. 'The [Bhagirathi] Ganges will never disappear', declared one ethnic Garhwali woman as she sat on Manikarnika ghat watching people celebrate the river's birthday. When pressed to consider the ecological changes rapidly taking place in the region, the woman looked thoughtfully down the steps to the water running below. 'On the day when She is no more', she conceded, 'it will be the end of the world'.[2] The difficulty imagining life without the river demonstrates the monumental value that people place on its existence and the cosmological disruption that a lack of flow could cause. The river's significance underscores the importance of understanding cultural models of nature where 'use-meanings' for the material world reflect diverse cultural values (Escobar 1999).

To sustain relationships with the Ganges, holistic forms of people-centred development are needed to foster economic opportunities in the Himalayas without jeopardizing personal and cultural needs to be in contact with the river. Such efforts should be rooted in the promotion of ecological resilience tempered with respect for cultural identities interlinked with nature. 'Sustainable' initiatives should, likewise, be grounded in conditions of ecological as well as cultural diversity. These cautions are as relevant for government programming as they are for the efforts of social movements and NGOs.

2.7.4 Prospects for Resilience

Despite the challenges, a number of trends offer hope for the resilience of the Himalayan ecologies through which the Bhagirathi Ganges flows. Village-based reforestation efforts, NGO tree nurseries, and forest department programmes operate at multiple scales in numerous locations to rebuild the fragile environment with diverse tree varieties. These reforestation projects strengthen soil and water retention while supporting mountain biodiversity. They also help to combat the extensive pine monocultures that contribute to landslides and seasonal wildfires. In some

[2] Interview with Anu Devi on Manikarnika ghat in Uttarkashi on 2 June 2009.

cases, efforts are made to promote ecological resilience near glaciers. The Gangotri-Gaumukh Conservation Project (2008), for instance, grows and plants tens of thousands of high-altitude trees and medicinal flora around a nursery 6 miles below the glacier that feeds the Bhagirathi Ganges. In addition to repairing a sensitive area harmed by decades of mass tourism, the reforestation project is an important site for environmental education.

Innovative policies also help to protect and conserve the region's resources. Since 2008, for instance, only 150 tourists are allowed entry into the park leading to the glacier each day. Downstream in the district capital of Uttarkashi, a ban on plastics reduces the amount of polythene that clogs sewers and chokes the banks of the river where it flows through the city. Measures such as these will likely expand under the Indian government's climate change action plan.[3]

Although the environmental initiatives underway are encouraging, much more can be done to improve the ecological condition of the Himalayas and the corresponding levels of social welfare. Since contemporary resource use in the region focuses on economic development from resource extraction and tourism, a management shift is needed to support the sustainable use of Himalayan water, forest, soil, flora, and fauna. Efforts to manage the region with the welfare of future generations in mind must also be combined with an aggressive adaptation programme for residents.

To address ecological change in the Himalayas, socially engaged environmental education programmes must be expanded. Misunderstandings about the science of climate change abound, and more education is needed to delineate the issues as well as the behaviours that can mitigate the coming ecological shifts. Although some people living along Ganges know about climate change, warming patterns, and the retreat of the river's glacial source, they are not all aware of the possible reasons. The causes cited for climatic change from some of the less informed range from unchecked pollution to the rise of sin in the epoch of *kali yug*. One of four cycles of time in the Hindu cosmology, this period is characterised by extreme moral, social, and ecological degeneracy. Within this worldview, some believe that the Goddess Ganga – the river deity – is angry with human misconduct in the world and is retreating from the earth back to the heavens to show her displeasure. These responses are poignant denunciations of the human conduct that contributes to the destruction of our ecological balance. Such interpretations of environmental change, however, can have undesirable effect of deflecting responsibility, since blame is placed on a cycle of time about which a person can do nothing. By providing culturally sensitive education on climate change, activists can more easily promote active involvement in environmental conservation and adaptation efforts by showing the relationship between human behaviour and the science of climate change.

[3] For a full copy of India's national plan on climate change, see http://pmindia.nic.in/climate_change.htm

Although environmental education is needed, it is important to keep in mind that many Himalayan residents are keen observers of the signs of change. As citizens with regional expertise, they can be among the most qualified to develop appropriate responses. The daily managers of Himalayan resources, residents who toil in the fields, forests, and waterscapes in which they live have authority to speak to resource-related problems and potential solutions. Forming regional strategies for resilience, therefore, involves increasing access to information about climatic change while inviting participation in planning and decision-making processes. Successful long-term solutions will need to be based in the actions, support, and adaptations of those living downstream from the melting glaciers. In these processes, women have a central role. Women's resource-related knowledge, when incorporated into decision-making processes, can inform adaptation strategies with appropriate regional variations. Given that many Himalayan women are already burdened with the bulk of animal husbandry, domestic chores, and agricultural labour, one way to involve them is to work with the pre-existing *mahila mandaldal* or 'women's committees' that are found in many villages. These venues are a site for dialogue, planning, and decision-making in which women have active involvement.

At the state and national level, an urgent need is to revisit and revise the policy of building large hydroelectric projects near receding Himalayan glaciers. Such revisions should be based on the most accurate hydrological data available for past, current, and future flows to inform cost-benefit analyses. When built, proper care must be taken to reduce the ecological destruction that such dams can cause in fragile Himalayan environments. To meet regional and national energy needs, research should also examine the prospects for building small hydroelectric projects that cause less ecological damage, require less financial investment, and involve local populations in the maintenance and upkeep to promote regional employment. Given the predictions of glacial retreat, it is also important to expand the production of power by tapping into renewable solar and wind sources. If used widely, technologies that harness the sun and wind can reduce the need for costly megaprojects that diminish water quality and availability.

To retain resources, water harvesting projects in rural and mountainous areas of the Himalayas must be promoted. These initiatives should expand upon pre-existing pond and tank storage technologies while encouraging innovation. In the broader context of surface water decline and erratic rainfall patterns, water harvesting is a viable adaptive strategy to mitigate vulnerabilities to water insecurity. To further support the retention of water, reforestation policies should work to outpace annual forest cover losses with regionally appropriate, deep-rooted trees that help store groundwater. In order to maximise water supplies, a major need is to encourage and, wherever not in cultivation, revive the growing of crops that require only minimal water resources. This shift will help cultivators meet the coming challenges with strains of wheat, rice, and other grains that can flourish with nominal rainfall and locally available water supplies.

Box 2.7b Perceptions of chaotic water regimes in northeastern Siberia

—Susan Crate

Box Map 2.7b.1 Russia

Water has a myriad of visceral meanings to Viliui Sakha, Turkic-speaking horse and cattle breeders native to the Viliui regions, Sakha Republic, northeastern Siberia, Russia. Many of these meanings are founded on an historically based belief system, a unique adaptation to the sub-Arctic environment, and an actively evolving local knowledge system. Sakha, like many indigenous, place-based peoples, consider all parts of their natural world sentient or spirit-filled. Sakha cosmology dictates that 'water has spirit' and humans need to pay respect to the water spirit when they interact with it. People refer to lakes, rivers, and all surface water sources as *ebe,* or grandmother. Sakha feed the spirit of the water and say certain words to appease it when taking water for use, crossing it, or using it as a mode of transportation. Similarly, subsistence practices involving water, including Sakha's various seasonal fishing techniques, the draining of lakes or damming of hay areas to control water, and people's arrival in or departure from their summer homes, have specific rituals appeasing the spirit of the water. According to Sakha's creation myth, in the beginning, the world was water. Cows, Sakha's main source of meat and milk, are believed to have come from water. Water takes away what is no longer wanted. For example, it washes away the Bull of Winter to usher in the long awaited spring (Crate 2008). It provides a link to the spirit world. Sakha place a cup of water near a person approaching death so the person's soul can travel to the spirit world via the water.

How are those visceral meanings challenged and called into question by global climate change, and how do they shape local perceptions and responses to climate change? For Viliui Sakha, the most pressing local effect of global climate change is a highly altered water regime, resulting in a growing amount of water on the land that turns hayfields into lakes, ruins transportation networks, and destroys household infrastructure. Additional effects include warmer winters, increased snowfall, too much rain at the wrong times, a prolonged autumn and a delayed spring, and the transformation of their ancestral landscape by degrading permafrost.

Many argue that they will adapt as they always have. We know from history that Viliui Sakha have a highly resilient capacity to adapt to change. Their Turkic ancestors migrated from Central Asia to the Lake Baikal regions

(continued)

Box 2.7b (continued)

Box Fig. 2.7b.1 From hayfields to fishing lakes. Viliui Sakha, agropastoralists of northeastern Siberia, Russia, are increasingly challenged by flooded hayfields turned to lakes, one local effect of global climate change. Most are finding it more and more difficult to harvest the fodder they need to over-winter their animals and therefore questioning the future of their ancestral subsistence ways. There are some, like this man pictured here, who are taking advantage of these newly established lake systems that, due to their high fertility from the underlying hayfield, are proving highly productive for fishing sobo (Carassius carassius), one of Sakha's supplemental foods. Sakha fish for sobo using three techniques: kuyuur (early spring by individuals through the lake ice with a catch net); mungkha (mid-winter by groups using a large sweep net); and ilim (in temperate months using a standing net) (Photo credit: Susan Crate)

of southern Siberia in approximately the tenth century AD. Then during Genghis Kahn's reign, they followed the Lena River north to their present-day sub-Arctic home. Upon moving north, they adapted a southern subsistence livelihood to an extreme climate with annual temperature ranges of up to 100°C, from -60 °C in the winter to +40 °C in the summer. Several centuries following their thirteenth-century move north, they, like the other native inhabitants of Siberia, were subject to Russian colonisers, who demanded a fur tax, annexed native lands, and sought to convert the 'heathen' local populations. Another several centuries later, their plight involved the forced collectivization and sovietization of the twentieth century. By the end of that century, Sakha underwent another transition with the dissolution of the Soviet Union. In many respects they have weathered these changes exceptionally well, thanks to a keen capacity for adaptation to change and a time-tested local knowledge base.

The twenty-first century brings Viliui Sakha a new challenge – namely, the local effects of global climate change, and they are again exercising their resilient adaptive capacity. Inhabitants are adapting to changes in their hay production by pooling hay-land resources with kin in neighbouring villages. Many argue for reviving their ancestral practices of haying on ice (after the water freezes), digging canals, moving their herds and haying to headwater areas, or using other forage, including spruce and pine trees. Some are purchasing hay or slaughtering animals to get through the long winter with insufficient hay stores. The government is attempting to drain the worst affected hayfields. However, many of these fields have been flooded for several years and now are home to diverse aquatic vegetation – it will be years before the haylands come back.

(continued)

Box 2.7b (continued)

Box Fig. 2.7b.2 Water in balance. Long-term ethnographic research with Viliui Sakha, agropastoralists of northeastern Siberia, Russia, reveals how delicate the balance of the 'right' amount of water is for them to continue raising horses and cattle in the extreme sub-Arctic ecosystem. In the last 10 years, this delicate balance is particularly nuanced due to the increasing water on their lands, one local effect of global climate change. Although concerned about the future, their focus remains on doing what they need to do to harvest enough for the immediate winter's needs. This means working twice as hard and long – in order to access hayfields not yet inundated with water, using ancestral ways to lessen water effects, including digging drainage canals, haying on ice, and using other plant materials to feed their herds. This kin group is haying on lands that have decreased by half due to expanding waters (Photo credit: Susan Crate)

At the same time, given the very real and extreme local consequences of global climate change, Viliui Sakha will only be able to successfully adapt to a point. At some point they will have to either move elsewhere to continue cattle and horse breeding or stay where they are and adapt by developing a different livelihood. Either of these changes will be problematic for inhabitants on a cultural level, because both cows and homelands are central to Viliui Sakha lifeways, cosmology, worldview, and identity. It is certain that one or some combination of the two will be necessary in the future. The uncertain part is how to facilitate those changes in a culturally sensitive way, taking into consideration and accommodating the many visceral meanings that these changed water regimes imply.

Viliui Sakha are not alone in their plight. As global climate change proceeds, peoples inhabiting the more climate-sensitive areas of the world face increasing challenges, many of which originate in highly altered water regimes. The irony is that these are the very populations who are least responsible for greenhouse gas emissions. As concerned global citizens, interested in advocacy and social justice, we need to bring to light the cultural implications implicit to the physical transformation of global climate change, not only for Viliui Sakha but for all the earth's place-based peoples (Crate 2008; Crate and Nuttall 2009). In every case, beyond the physical challenges of highly altered water regimes, we need to be diligent in understanding the cultural implications, asking questions such as: What does this increased water on the land signify to inhabitants? How are they perceiving it? What are they seeing these changes *with*?

(continued)

Box 2.7b (continued)

Resources

Crate, S.A. 2008. Gone the bull of winter? Grappling with the cultural implications of and anthropology's role(s) in global climate change. *Current Anthropology* 49(4):569.

Crate, S.A., and M. Nuttall. 2009. *Anthropology and climate change: From encounters to actions*. Walnut Creek: Left Coast Press.

2.7.5 Achieving a Balance: Global Efforts Required

In the long term, international planning and management strategies are needed to reduce humanity's collective carbon footprint and reach the goal of sustainable resource management. In this quest, the leaders and inhabitants of the world – especially the heavy emitters of the so-called developed countries – have a vital role in tackling the emissions they produce within their borders as well as the ones that are produced abroad to feed inequitable consumptions patterns. Local and regional responses to glacial melt and ecological change can help mitigate vulnerabilities downstream from the Himalayan glacier systems, but it is international policy and action to reduce carbon and black soot emissions that is needed to reduce the intensity of climate change and its consequences for future generations.

Resources

Basishta, A., D.S. Arya, and N.K. Goel. 2009. Analysis of historical changes in rainfall in the Indian Himalayas. *International Journal of Climatology* 29: 555–572.

Centre for Science and Environment (CSE). 2008. Will Ganga get its life back? http://www.cseindia.org/misc/ganga/ganga_action.htm. Accessed 15 May 2009.

Conservation International. 2008. Biodiversity hotspots. http://www.biodiversityhotspots.org/xp/hotspots/himalaya/Pages/biodiversity.aspx. Accessed 25 June 2009.

Darian, S.G. 1978. *The Ganges in myth and history*. Honolulu: University Press of Hawaii.

Dharmadhikary, S. 2008. Mountains of concrete: Dam building in the Himalayas. Berkeley: International Rivers. http://www.internationalrivers.org/files/IR_Himalayas_rev.pdf. Accessed 30 July 2010.

Down to Earth. 2008. Myth of power. *Down to Earth: Science and Environment Fortnightly* (1–15 Sept):28–34.

Escobar, A. 1999. After nature: Steps to an anti-essentialist political ecology. *Current Anthropology* 40(1): 1–30.

Gangotri-Gaumukh Conservation Project. 2008. http://gangotri.prayaga.org/. Accessed 14 Oct 2008.

International Panel on Climate Change (IPCC). 2007. *Climate change 2007: Climate change impacts, adaptation, and vulnerability, summary for policy makers*. Contribution of the Working Group II to the Fourth Assessment Report of the Intergovernmental Panel on Climate Change, February 2007. Brussels: IPCC.

Khoday, K. 2007. *Climate change and the right to development: Himalayan glacial melting and the future of development in the Tibetan Plateau*. United Nations Development Programme, Human Development Report 2007/2008: Fighting Climate Change: Human Solidarity in a Divided World. New York: UNDP, Human Development Report Office.

Mehta, M. 1996. Our lives are no different from that of our buffaloes: Agricultural change and gendered spaces in a central Himalayan valley. In *Feminist political ecology: Global issues and local experiences*, ed. D. Rocheleau, B. Thomas-Slayter, and E. Wangari, 180–208. New York: Routledge.

Naithani, A.K., H.C. Nainwal, and K.K. Sati. 2001. Geomorphological evidences of retreat of the Gangotri glacier and its characteristics. *Current Science* 8(1): 87–94.

Singh, P., M. Arora, and N.K. Goel. 2006. Effect of climate change on runoff of a glacerized Himalayan basin. *Hydrological Processes* 20(9): 1979–1992.

United Nations Environment Programme (UNEP). 2009. Recent trends in melting glaciers: Tropospheric temperatures over the Himalayas and summer monsoon rainfall over India. na.unep.net/publications/Himalayas.pdf. Accessed 30 July 2010.

United Nations World Water Development Report. 2003. *Water for people, water for life*. Barcelona: United Nations Educational, Scientific, and Cultural Organization (UNESCO).

Part III
Water Value, Access, Use, and Control: Sociocultural Contexts of Water Scarcity

Cover art by Khoso, Iran, titled "No water"

Chapter 3.0
Introduction: Water Value, Access, Use, and Control: Sociocultural Contexts of Water Scarcity

Daniel Niles

How do human societies decide who gets water? Since all people (and, of course, all plants and animals) must have water to live, it seems cruel to calculate who should have it. Yet all social groups in past and present have had principles and rules – mostly unwritten – guiding access to, and use of, water. Underlying these principles, rules, and practices are culturally specific ideas about people's relationships to one another and to the world that they inhabit.

This section presents several examples of how different human social groups negotiate their needs for, access to, appropriate uses of, and control over water. In total the section does not present a comparison of rules and regulations in different places but instead describes how water permeates culture, so that access to and use of water is a deeply cultural question. In particular, this section describes how a person's access, use, and control of water changes depending on whether the person is female or male, native, immigrant or itinerant, rich or poor, or of a particular ethnic or social group. Perhaps because water is so essential to daily life, water practices in a particular place may appear to be 'natural' to the people concerned – but 'natural' practices may quickly come into question when water availability shifts. This section also describes how changes in water practices ripple throughout a culture or nation, often affecting different groups of people in wholly unanticipated ways or exposing submerged prejudices or discriminations.

If the chapters of this section present very distinct examples of the cultural dynamics at play in water access and use, one important point that can be gathered from the section as a whole is that though water is essential for life, it is also a material good. Water is known as 'liquid gold'; it is as essential to the productivity of nations and industries as any other resource. In fact, water is economically indispensable, and so it is laden with power dynamics. While all people have an interest

D. Niles (✉)
Research Institute for Humanity and Nature, Kyoto, Japan
e-mail: dniles@chikyu.ac.jp

B.R. Johnston et al. (eds.), *Water, Cultural Diversity, and Global Environmental Change: Emerging Trends, Sustainable Futures?*, DOI 10.1007/978-94-007-1774-9_16,

in maintaining safe water sources, in practice the water needs of certain people take precedence over those of others, and if water is used by one person or entity, it is often unavailable for use by others. Water is also a rather delicate material good; it can be spoiled, often with negative effect on its surrounding ecosystems. The ways that water is valued, maintained, accessed, used, and controlled are not simply cultural matters, but also deeply political; again, these water practices extend from the ways people relate (or feel they ought to relate) to one another and the world around them. Water practices therefore reflect ideas of individual need, collective equity, and different cultural perspectives on humanity's place in nature.

Gennie Nguyen describes how water features were critical in shaping the sense of community of a marginalised immigrant population in the United States. She provides an intimate description of how water and culture intersect even in urban contexts. Nguyen also shows how water-based cultural practices created a kind of cultural buffer that countered the Vietnamese population's marginalization in U.S. society and how pollution of water sources – in this case associated with the questionable location of a landfill in the aftermath of Hurricane Katrina – may therefore endanger community identity and continuity.

Judit Bari and Krista Harper provide further illustration of the material and socio-environmental consequences of ethnic discrimination. The Roma people (commonly known as Gypsies) of Hungary do not just suffer from ethnic discrimination, they have less access to clean water and water infrastructure, including indoor plumbing, than non-Roma Hungarians. As a consequence, the Roma are more vulnerable to external shocks (including economic shocks) than surrounding communities, in that their poor access to water is exacerbated by general declines in services following shocks.

Water practices do not just vary between cultures and communities; they also distinguish people within a single community. Farhana Sultana provides a detailed description of how gender affects water access and use. In rural Bangladesh, women and girls are often required to fetch water, but this activity, like all female activities, must also adhere to strict cultural rules regarding appropriate feminine behaviour. Sultana describes how the presence of naturally occurring arsenic in local wells exposes the intricate gender and class dynamics that surround the seemingly mundane task of hauling water. While in some cases families consume poisonous water rather than violate gendered social convention, in other situations a lower caste family with a clean well may find itself with new status, as surrounding members of the community come to depend on it for clean water.

Accompanying Sultana's chapter, the series of text boxes authored and coauthored by Anwar Islam, Suzanne Hanchett, Kazi Rozana Akhter, and Shireen Akhter further illustrate how water and water features are imbued with powerful significance in Bangladesh and how individual and collective behaviour is consistent with these beliefs, for better and worse. The text box by Akhter and Hanchett also provides interesting detail of water practices within a single household and the effect of arsenic on family water use.

If the three preceding chapters seem to indicate the need for some kind of governmental intervention, how should governments act in this highly cultured field of water use and practices? National governments often assert their right to control water resources; as a result they are often assumed to bear the responsibility for managing the resource – and for providing their citizens with basic water services. In recent years, debate about government responsibilities and citizens' rights regarding water has often been focused around privatization – certainly the most contentious issue in water governance today. As water is essential to human life, questions of access, use, and control of water were deliberately framed in terms of 'human rights': if humans have a right to life, they certainly also must have a right to water. However, the rights-based strategy may not avoid the problems of water commoditization because it does not address the ecological processes involved in creating common water resources. Three text boxes describe the limitations in privatization and the potential benefits in collective stewardship of water resources. Between privatization and collective management, however, lie many potential regimes of water management. Sylvia Rodriquez demonstrates the applicability of traditional water management systems in a modern, culturally diverse context with her discussion of the reinvigorated use of acequia systems to manage the water commons in New Mexico, USA. Bill Derman and Emmanuel Manzungu describe the Zimbabwean attempt to make water management more equitable and participatory. Unfortunately, institutional incapacity to address complex competing interests and needs led to a cascading set of negative impacts including declining water quality, outbreaks of disease, and starvation. And, Veronica Strang explores how the emergence of new schemes to manage water transform the social, cultural and economic values assigned to water. In particular, she shows how privatization in Australia converted water into a commodity, a raw material from which profit can, and ought to, be earned. As a consequence, 'efficient use' entails shifting water from low- to high-value enterprises, a condition that tends to privilege industry over food production, capital-intensive over subsistence agriculture, urban uses over rural ones, and wealthy over poor.

The notion that water scarcity is often an end result of anthropogenic activity, and sometimes a manufactured perception, if not experience, is explored in Barbara Rose Johnston's essay on manufactured scarcity and environmental inequity. With examples from mining and the processing of minerals (Tarawera River, New Zealand), the manufacturing and disposal of electronics (California's Silicon Valley and Guiyu, China), nuclear militarism (Marshall Islands), Johnston raises questions of how urban demand or profit-generating industrial use of water should be balanced against the livelihood needs and customary practices of people inhabiting often remote source watershed territories.

Further exploration of these complex issues are contained in textbox vignettes by Marcus Barber and Sue Jackson (mining and hydrodevelopment impacts in the Pilbara region of northwest Australia and its impact on indigenous peoples); by Fenda Akiwumi (mining wastes and its adverse effects on rural livelihood in Sierra Leone); and by Alexander Erwin (mining and energy development in Canada's MacKenzie River drainage). The mining industry requires abundant water supply and has often been associated with large-scale water and ecosystem pollution, with complex consequences for those peoples whose livelihoods are intimately bound to the local environment, as illustrated in the brief discussion of Australian mining and dam development by Barber and Jackson. Valuable minerals and oil – and increasingly valuable water – are often in remote or sparsely populated areas frequently inhabited by indigenous or tribal peoples whose resource uses are only minimally monetarised. Erwin describes how the great MacKenzie River drainage system in north-central Canada became bound up in the powerful narrative of national resource security; the need for weapons-grade radium and uranium and then oil and natural gas trumped the traditional (and legal) land-use claims of Native American peoples, who were offered dangerous wage labour in return. Fenda Akiwumi richly illustrates the complexity of disentangling customary tribal, urban, and national-industrial claims on water and land resources in Sierra Leone. Rural and tribal peoples, who make up the majority of the population, have legal support for their traditional uses but no means of enforcement, especially when their claims come into conflict with state-brokered mining endeavours. The World Bank is attempting to facilitate exploitation of mineral resources but must walk a very fine line in order to do so while also appearing to adhere to its own recommendations of best social and environmental practices.

The overall message of this section? Water is so much a part of culture that to examine a people's relationship to water is to examine its fundamental relationships and beliefs. To debate how a society ought to relate to water is to debate how a society ought to be; similarly, to ask how a society ought to be is to ask how it ought to relate to water.

Chapter 3.1
Culture, Gender, and Vulnerability in a Vietnamese Refugee Community: Aftermath of Hurricane Katrina

Gennie Thi Nguyen

Hurricane Katrina devastated the Gulf Coast of the United States, including the city of New Orleans, on August 29, 2005. Although New Orleans is famous for its French Quarter, Mardi Gras celebrations, riverboats, music, and Creole culture and cuisine, few people are aware of its vibrant Vietnamese community. This paper describes how the Vietnamese community of New Orleans East, an area locally known as Versailles, was affected by Hurricane Katrina and why the task of rebuilding community in Versailles was distinct from rebuilding other areas of the city. In Vietnamese, the word for water, *nuoc*, also means 'country'; this linguistic metaphor indicates the great symbolic importance of 'water' and 'home', especially for a group of displaced people. Environmental changes after the storm, particularly in access to clean water, affected key cultural components of the Versailles Vietnamese community.

Hurricane Katrina was one of the worst environmental disasters in the history of the United States, casting debris across 233,000 km^2 in Louisiana, Mississippi, and Alabama (Luther 2006). An estimated 15.3 million m^3 of debris littered waterways in New Orleans and Mississippi, and 96%, or 13.6 million m^3, of remaining debris littered Orleans, St. Bernard, St. Tammany, Washington, and Plaquemines parishes (Bullard and Wright 2009:26). The daunting prospect of cleanup and recovery raised questions about which neighbourhoods would be cleaned up, in what order, and where debris dumpsites should be located. These were primarily social questions to be addressed in the political sphere, and they had great significance in individual communities of New Orleans. According to Robert Bullard and Beverly Wright, 'What has been cleaned up, what gets left behind, and where the waste is disposed of appear to be linked to more political science and sociology than to toxicology, epidemiology, and hydrology' (2009:26).

G.T. Nguyen (✉)
Department of Anthropology, University of Oregon, Eugene, OR, USA
e-mail: nguyen.gennie@gmail.com

B.R. Johnston et al. (eds.), *Water, Cultural Diversity, and Global Environmental Change: Emerging Trends, Sustainable Futures?*, DOI 10.1007/978-94-007-1774-9_17,

Map 3.1.1 Louisiana

Like everyone else in the aftermath of Katrina, the people of Versailles required dependable access to clean water. In Versailles, however, once the immediate recovery needs were addressed (with trailers and other assistance from the Federal Emergency Management Agency, FEMA), it was critical to re-establish the local water environment, as water is fundamental to the cultural character of the neighbourhood and its place in a reborn New Orleans. Gardening is an important feature of the Versailles community.

Locals use water from its main canal to irrigate their gardens, which produce vegetables that are not easily found in American grocery stores but are essential to Vietnamese food culture. Many gardeners, who tend to be elderly members of the community, sell their vegetables at the early Saturday market, which also plays an important role in maintaining the sense of community. This example of disaster recovery in New Orleans demonstrates how water, food, culture, and community are intimately linked and how disasters may affect specific cultural groups in different ways.

3.1.1 The Vietnamese Community in Versailles, New Orleans

The Vietnamese community of Versailles is part of a broader migration of Vietnamese people who left their homeland in response to specific political-economic realities and politics. Versailles's Vietnamese roots can be traced back to villagers from North Vietnam who migrated south after the country was divided into two states in 1954. After the fall of Saigon in 1975, having lived in South Vietnam for over 30 years, the community migrated to New Orleans. Members of the generation who came of age in Vietnam were generally born between before 1940 and 1965. Members of the second generation (the first generation born outside of Vietnam or who were under the age of ten when they left that country) were generally born between 1965 and 1990. This particular historical experience – of successive displacement in Vietnam and arrival as refugees in the United States – affected the kinds of communities Vietnamese people established in the United States and some of their collective ideas about the role of government in their lives.

Fig. 3.1.1 An elderly Vietnamese woman in her garden outside her FEMA trailer (Photo credit: Gennie Thi Nguyen)

Versailles lies in the most eastern part of New Orleans East. The name of the neighbourhood came from 'Versailles Arms', a subsidized low-income apartment complex in the area where Catholic Charities resettled many Vietnamese soon after they arrived to United States. Although the majority of the population of Versailles is African American, about 7,000 Vietnamese people live within a 1.5 km radius of the Mary Queen of Viet Nam church [Maria Nu Vuong Viet Nam]. The total Vietnamese population here is small relative to other Vietnamese population centres in the United States, such as Houston, Texas, or Westminster, California, but Versailles has the densest population of Vietnamese people outside Vietnam.

Before Katrina, leaders in the community often held libertarian attitudes towards government, meaning 'If you don't bother us, we won't bother you' (Tang 2009:19). Settling in Versailles presented an opportunity for Vietnamese refugees to build a new community without government interference, and they created an environment similar to that of their former villages in Vietnam. Growing up in Versailles, I noticed that many Vietnamese Americans visiting the neighbourhood often compared it to the look, feel, and environment of Vietnam with its lush tropical plants, subtropical hot, humid weather, and most of all, the Vietnamese culture. Versailles' location in New Orleans isolated it from much of the rest of the city (Airriess 1994). The community claimed a space of its own by opening businesses and catering to its member's needs. Currently, the neighbourhood has

Vietnamese-owned businesses that include grocery stores selling Vietnamese-style food and produce, hair salons, tailors, bars and nightclubs, restaurants, family medical practices, cell phone retailers, coffee shops (which also serve as gambling centres and men's spaces), and other businesses oriented specifically to the Vietnamese community.

Before Katrina, 31% of Versailles residents lived in poverty, compared to 18% of the larger New Orleans area and 12% of the U.S. population (Airriess 2008:170). The median household income for Versailles residents was about $31,000, compared to approximately $35,000 earned by New Orleans residents and almost $42,000 earned by Americans nationwide. Only 52% of Versailles residents owned their homes, compared to approximately 62% of resident homeowners in New Orleans and 66% of people in the United States. Only 12% of Versailles residents 25 years and older hold a bachelor's degree, whereas 19% of the New Orleans population and 26% of Americans overall have BA degrees (Airriess 2008:170).

After Hurricane Katrina, members of the Versailles community returned to their homes sooner and in greater numbers than residents of surrounding communities. People of Versailles began to return in October 2005, and by mid-October 500 people had signed a petition asking the local electric company to restore power to the area. By the summer of 2007, nearly 2 years after the storm, almost all of the Versailles residents had returned, whereas less than 50% of pre-Katrina residents of the rest of the city had returned. Currently, community leaders estimate that 95% of pre-Katrina neighbourhood residents have returned.

As the neighbourhood was rebuilding, the mayor established the Chef Menteur Highway Landfill by executive order in order to receive debris from the massive demolition and rebuilding effort in the city. Located less than a mile from the community, debris from the unlined landfill contained toxins and polluted the local groundwater – the same groundwater that flows into the canals that the community elderly use to water their gardens. Since the vegetables grown in these gardens play a significant role in the community's identity, the contaminated water has serious impacts for the entire community of Versailles.

3.1.2 The Chef Menteur Highway Landfill

As the people of Versailles began to rebuild after the storm, community members noticed dump trucks filled with hurricane debris continually entering the community. The Louisiana Department of Environmental Quality and the U.S. Army Corps of Engineers granted permits to Waste Management, Inc., to reopen and operate the Chef Menteur Highway landfill in April 2006. Soon afterwards, Waste Management, Inc. began dumping toxic debris into the unlined landfill, which was designated to hold over one-third of the Hurricane Katrina debris from New Orleans (Tang 2009).

Despite governmental claims that the landfill materials were non-toxic and fit for disposal in the Chef Menteur site, hazardous and unhazardous materials frequently are mixed together in a flood and are very difficult to separate afterwards. Government officials testified on several occasions to the safety of the Chef

Fig. 3.1.2 The main canal, contaminated by the toxins from the Chef Menteur Highway Landfill, is the source of irrigation water for the Vietnamese residents of Versailles (Photo credit: Gennie Thi Nguyen)

Fig. 3.1.3 Dumping of highly toxic Hurricane Katrina debris in the adjacent Chef Menteur Highway Landfill prompted protests from Vietnamese community members. This photo illustrates their call to close the landfill (Photo credit: Gennie Thi Nguyen)

Menteur site. First, they asserted that the risk of toxic materials being dumped at the site was insignificant and that the materials were sorted to keep hazardous waste from entering the landfill. Second, they insisted that protective liners were not needed for construction and demolition (C&D) materials because those materials are cleaner than other rubbish (Eaton 2006). Federal laws do not require C&D landfills to have protective liners, but they do require that municipal landfills have liners because these dumps are expected to contain hazardous household waste (Bullard and Wright 2009:27).

Reports to Congress demonstrated that disaster debris from the storm was indeed a mix of toxic and non-toxic materials and that harmful materials such as oil, pesticides, paints, and cleaning agents were difficult or nearly impossible to separate from other debris before disposal (Luther 2006; Kemp 2006; Pardue 2006). In extensively flooded areas, containers holding household hazardous waste (HHW) may have leaked and contaminated the surrounding debris. Toxins such as arsenic, chromium, and copper can leach from treated lumber, and hydrogen sulphide gas leaches from deteriorating gypsum drywall. Household chemicals, paint, batteries, and other toxic debris were also dumped into the unlined landfill (Lydersen 2009). Indeed, geological and chemical analysis proved that toxins from the landfill were contaminating the groundwater that flowed into the canal (Pardue 2006; Kemp 2006).

Box 3.1a Water in a Hungarian Roma community*

—Judit Bari and Krista Harper

Box Map 3.1a.1 Hungary

In many countries across Europe, the Roma (also known as Gypsies) face environmental inequalities, including lack of access to clean water (Harper et al. 2009; Hajioff and McKee 2000). With the 2004 and 2007 expansions of the European Union, the Roma became one of the largest ethnic minority groups in Europe, with a population of approximately ten million. The Roma are the largest ethnic minority living in Hungary today and number over a half a million, equal to 5% of the country's total population.

Originating in the Indian subcontinent, the Roma migrated west into Eastern Europe during the Middle Ages. Because the majority populations excluded Roma from land ownership and

*See also textbox Box 5.5a "Participatory action research: Water in a Hungarian Roma community" by the same authors.

(continued)

Box 3.1a (continued)

Box Fig. 3.1a.1 Horses are still very a much a source of Roma community pride (Photo credit: Mariann Bari)

membership in guilds, Roma people developed economic niches as itinerant musicians, blacksmiths, horse traders, and craftspeople.

Today, very few Roma still travel to make their living. Hungarian Roma settled in villages over a century ago, choosing places where they had established trading relationships and access to water and other critical resources. Today, the majority of these people speak Hungarian, although some also speak the Lovari Romani or Beash languages.

Hungarian Roma today face widespread discrimination and struggle for recognition as full-fledged citizens with constitutionally guaranteed rights to health, water, and sanitary infrastructure. A recent survey in Hungary, for example, examined environmental indicators, including access to indoor plumbing and distance from waste dumps, and found that Roma neighbourhoods have significantly less infrastructure and more environmental hazards than non-Roma ones (Kosa et al. 2007). The social and environmental inequalities faced by Roma in Hungary are characteristic of the marginalization of their communities throughout Eastern and Western Europe, and access to clean water and sanitation are the most critical issues.

Sajószentpéter is a small city located on the Sajó River, a tributary of the Tisza River, in northeastern Hungary. During the 1970s, the Sajó River was known as one of the most polluted in Eastern Europe, thanks to industrial wastewater pollution from mills located upstream in Czechoslovakia and Hungary (Vári and Kisgyörgy 1998). Following 1989, remediation efforts and the collapse of upstream industries improved the river's water quality. Efforts to further improve water quality now depend on constructing new sewerage lines (Somlyody 1999). Although water quality improved after the collapse of state socialist industries, the region staggers under the highest unemployment rate in the country and lacks funds for infrastructure investments that could make much greater improvements to water quality and availability.

Sajószentpéter's predominantly Roma neighbourhood is located on the Sajó's bank, just across the river from the town square. This community (comprised of over 2,000 residents, or about 15% of the town's total population of 13,343) has been especially affected by the post-1989 economic decline. Prior to the state socialist era, many Roma played music for a livelihood. With the wave of industrialization in the 1960s, the majority of the town's Roma

(continued)

Box 3.1a (continued)

Box Fig. 3.1a.2 Every time you need to wash or cook or clean, someone has to go to the street for water (Photo credit: Mariann Bari)

Box Fig. 3.1a.3 The very same week that this picture was taken, water pipes were being upgraded in neighborhoods across the bridge (Photo credit: Sándor Kelemen)

population found work in local factories and hospitals, but these jobs disappeared in the early 1990s. Although the neighbourhood has been a part of the town for many decades, the municipality has invested fewer resources in building public infrastructure there than it has in neighbourhoods on the other side of the river. As a consequence, residents of the *cigánysór* ('Gypsy row') do not have access to sewerage or indoor water. Instead, they build outhouses in small yards adjacent to their residences, and they carry water for drinking, cooking, household cleaning, washing, and personal hygiene from street taps in buckets.

On the coldest days of the winter, the taps freeze, leaving households without water. When the poorly maintained municipal pipes burst from the cold, runoff creates cesspools in residents' yards, trapping people indoors until the pipes are fixed and the contaminated water subsides.

As in other marginalised neighbourhoods across Hungary, summer brings the threat of hepatitis A outbreaks caused by contamination from sewerage leaks (Hungarian Public Health and Hygiene Service 2007). The neighbourhood was the site of a hepatitis outbreak in summer 2004. Local public health officials contained the spread of the disease through inoculations, but several years later residents complained that the infrastructural problems leading to the outbreak remained unaddressed.

The Sajó River Association for Environment and Community Development, a grassroots organization established by a group of Roma residents, is mobilizing to address the neighbourhood's socioeconomic and environmental problems. The Sajó Association, led by Judit Bari, seeks to build community members' capacity as citizens, to forge multiethnic alliances to improve the Sajó River environment, and to urge policymakers to address community concerns sustainably and equitably.

(continued)

Box 3.1a (continued)

Resources

Hajioff, Steve, and Martin McKee. 2000. The health of the Roma people: a review of the published literature. *Journal of Epidemiology and Community Health* 54: 864–869.

Harper, Krista, Tamara Steger, and Richard Filcak. 2009. Environmental justice and Roma communities in Central and Eastern Europe. *Environmental Policy and Governance* 19(4): 251–268.

Hungarian Public Health and Hygiene Service. 2007. *Hepatitis fertőzés Szabolcsban* ('Hepatitis Outbreak in Szabolcs'). http://www.antsz.hu/portal/portal/hepatitis_20070823.html. Accessed 12 Oct 2009.

Kósa, Zsigmond, György Széles, László Kardos, Karolina Kósa, Renáta Németh, Sándor Országh, Gabriella Fésüs, Martin McKee, Róza Ádány, and Zoltán Vokó. 2007. A comparative health survey of the inhabitants of Roma settlements in Hungary. *American Journal of Public Health* 97(5): 1–7.

Somlyody, L., K. Buzas, A. Clement, and Z. Melicz. 1999. Strategies for approximating EU legislation in Hungary: The Sajó River case. *Water Science and Technology* 40(10): 87–94.

Vári, Anna, and S. Kisgyörgyi. 1998. Public participation in developing water quality legislation and regulation in Hungary. *Water Policy* 1: 223–238.

3.1.3 The Impact of the Landfill: Gardens and the Early Saturday Market in Versailles

Clean water serves more than just a biological need; it is an anchor of the Vietnamese culture of Versailles. Gardens, for example, are a major part of Vietnamese culture in New Orleans East as they provide special foods largely unavailable in standard U.S. produce markets. Community and private house gardens are also important because they are spaces in which the Versailles elders – the generation that fled Vietnam as refugees – find psychological solace in growing the traditional foods of their homeland. Also, the weekly market at which most garden and home-produced products are sold functions as a kind of weekly cultural bazaar in which the Vietnamese heritage of Versailles is on full display.

For many immigrant groups, traditional foods represent a connection to the past and function to maintain ethnic identity (Kalcik 1984). Vietnamese food is not only important for the maintenance of identity for refugees who fled Vietnam in search of asylum, but also for their descendents who were not acculturated in Vietnam. Andrea Nguyen, who was born in Vietnam but left when she was 6 years old, stated: '[Food] was a way my parents made sure we held on to our ethnic heritage' (Haughton 2006). Because many refugees left Vietnam in haste, they were only able to take a few possessions with them (Marino 1998). One important piece of property often saved was a recipe book. Nguyen notes, 'My mother is an excellent cook' who escaped Vietnam with few possessions 'except a small range notebook of recipes that she brought to this country in case she opened a restaurant' (Haughton 2006).

Before Katrina, two types of gardens existed in Versailles: the Versailles garden and backyard gardens. The former was a community garden that lay near the edge of the neighbourhood, close to the Versailles Arms apartments. Flooding and debris from the hurricane destroyed this garden. Backyard gardens were usually simply lawns that property owners transformed into gardens (Airriess and Clawson 1994). Most first-generation members of Versailles created the community as a reflection of their socioeconomic status in Vietnam: agrarian and poor (Airriess and Clawson 1991). According to Christopher Airriess and David Clawson, most gardeners increased the size of their gardens by using the land between their property and the canal, and some gardeners utilised the sloping bank of the canal; the average size of a backyard garden is 167 m^2 (1994:18). These gardens produced 34 leafy green vegetables, tubers, cucurbits, condiments and herbs, legumes, and medicinal plants not common in the Western diet (Airriess and Clawson 1994). Because New Orleans has a climate similar to that of Vietnam (hot, humid, and subtropical), the inhabitants are able to grow the same vegetables they once produced in rural Vietnam.

Most of the vegetables grown in the gardens are sold at the early Saturday market, which is similar to the outdoor village markets found in Vietnam. Vendors begin setting up every Saturday around 4 o'clock in the morning and will typically begin to close up around 10 o'clock. The market in Versailles is full of women wearing the traditional cone-shaped Vietnamese hats, squatting down near the ground and selling everything from baked goods and live ducks and rabbits to homegrown vegetables. The sounds of women haggling fill the early morning air. Vietnamese fishermen or their wives also try to sell fresh-caught shrimp, fish, and other goods.

More than environmental similarity and economic incentives, the residents of Versailles may have other reasons to recreate something of the familiar landscapes of Vietnam (Airriess 2002:241). Elderly members of the community do most of the gardening in Versailles. In many cultural diasporas, the elderly experience psychological adjustment problems in a new environment because they lack local language skills and employment opportunities and become dependent upon their children for the basic daily interactions outside the community, including financial matters (Roberts and Starr 1982; Marino 1998; Airriess 2002). This dependence challenges the traditional social structure. As a result, the elderly can experience a profound sense of helplessness that leads to psychological health problems. Because the Vietnamese community does not formally recognise mental health in the individualised Western style, many of these problems are not addressed formally. Gardening can be seen as a 'hortitherapeutic' activity (Kaplan 1973; Wood 1997; Airriess 2002). Gardening and the Saturday market allow the elderly to reminisce and recreate rural Vietnam and their sense of home. This form of psychological healing eases the trauma of being part of a forced ethnic displacement.

Most Vietnamese Americans of the second generation were born in the United States or were under the age of ten when they left Vietnam. The mass out-migration at the end of the civil war has had a profound effect on Vietnamese culture and identity.

Considering this, how does the second generation of Vietnamese Americans, who have limited or no memories of the 'homeland', create and maintain a Vietnamese culture and identity? Food acts as a medium for the Vietnamese acculturated in Vietnam and later generations born in the United States to share cultural meanings. Corinne Trang noted that 'food is the centre of life'(Trang 1999). This is true both literally and figuratively; food is at the centre of family life and the sense of home and culture. It affects social relations and personal and community health, and is also laden with many ideas, emotions, forms of display, and so forth. For the children of refugees and displaced people, Vietnamese food offers a strong sense of place within the United States and cultural belonging and link to the larger history of Vietnam.

Lily Tsay is a second-generation Vietnamese American who was born in the United States. Her mother was raised and came of age in Vietnam. Born in Chicago but raised in a mostly white area away from a Vietnamese community, Lily expresses the importance of Vietnamese food for her identity. 'Without my mother's food, I would be completely Americanised. The knowledge of Vietnamese food helps connect me with my Vietnamese friends. I wouldn't connect with another Vietnamese person/second generation person without my mother's food embedded into my life as I was growing up. When I first met people who looked like me, I didn't know we shared the same culture until we talked about how our houses smell like fish sauce some nights or how mom uses banana leaves to wrap another dish'.

In Versailles the early Saturday market establishes links between contemporary Vietnamese people in the United States and their cultural homeland. Gardening in the neighbourhood, then, provides more than an availability of Vietnamese food. Older generations pass on culture to younger generations through food. Gardening and food customs are crucial in the continuity of Vietnamese culture outside of Vietnam.

3.1.4 Conclusion

Polluted groundwater from the Chef Menteur Landfill site is not just a threat to public health but also affects the cultural identity of the Vietnamese community in New Orleans East. The story of Versailles demonstrates how many of the political-economic dynamics in place before Katrina, especially the role of race, class, and gender in shaping exposure to risk, were exacerbated after Katrina. In this sense, the creation of the Chef Menteur site can be considered a form of environmental racism, as it involved 'the deliberate targeting of ethnic and minority communities for exposure to toxic and hazardous waste sites and facilities, coupled with the systematic exclusion of minorities in environmental policy making, enforcement, and remediation' (Bullard 1993:15). Environmental racism goes beyond the poisoning of air, water, and soil; it is embedded in discriminatory practices, policies, and procedures. It accumulates over many years, and links pollution, toxic waste, and race (Checker 2005:9). This story thus illustrates the relationships between place, race, and the experience of disaster and 'recovery' after a dramatic environmental disturbance.

References

Airriess, C.A., and D.L. Clawson. 1991. Versailles: A Vietnamese enclave in New Orleans, Louisiana. *Journal of Cultural Geography* 12(1): 1–15.

Airriess, C., and Clawson, D.L. 1994. Vietnamese market gardens in New Orleans, Louisiana. *Geographical Review* 84(1): 16–31.

Airriess, C. 2002. Creating Vietnamese landscapes and place in New Orleans. In *Geographical identities of ethnic America: Race, space, and place*, ed. Kate L. Berry and Martha L. Henderson, 228–254. Reno/Las Vegas: University of Nevada Press.

Airriess, C., Li, W., Leong, K.J., Chen, A.C., and Keith, V.M. 2008. Church-based social capital, networks, and geographical scale: Katrina evacuation, relocation, and recovery in a New Orleans Vietnamese American community. Geoforum, 39(3): 1333–1346.

Bullard, Robert. 1993. *Confronting environmental racism: Voices from the grassroots*. Boston: Southview Press.

Bullard, Robert, and Beverly Wright. 2009. *Race, place, and environmental justice after Hurricane Katrina*. Boulder: Westview Press.

Checker, Melissa. 2005. *Polluted promises: Environmental racism and the search for justice in a southern town*. New York: New York University Press.

Eaton, L. 2006. A new landfill in New Orleans sets off a battle. *New York Times*. http://www.nytimes.com/2006/05/08/us/08landfill.html. Accessed 15 July 2006.

Haughton, Natalie. 2006. The flavors of Vietnam. *Daily News*. http://www.dailynews.com/food/ci_4431504. Accessed 5 Dec 2006.

Kalick, S. 1984. Ethnic foodways in America: Symbol and the performance of identity. In *Ethnic and regional foodways in the United States*, ed. L.K. Brown and K. Mussel, 37–65. Knoxville: University of Tennessee Press.

Kaplan, Rachel. 1973. Some psychological benefits of gardening. *Environment and Behavior* 5: 145–162.

Kemp, G. Paul. 2006. *Geological analysis of Chef Menteur Landfill site, Orleans Parish, Louisiana*. Baton Rouge: Louisiana State University, School of the Coast and Environment.

Luther, Linda. 2006. *Disaster debris removal after Hurricane Katrina. CRS Report for Congress*. New Orleans: The Library of Congress.

Lydersen, Kari. 2009. Landfill worries cloud hope for New Orleans gardens. *The Washington Post*, July 4. http://www.washingtonpost.com/wp-dyn/content/article/2009/07/03/AR2009070302436.html. Accessed 4 July 2009.

Marino, Katherine. 1998. Women Vietnamese refugees in the United States: Maintaining balance between two cultures. *The History Teacher* 32(1): 90–117.

Pardue, John H. 2006. *Anticipating environmental problems facing hurricane debris landfills in New Orleans East*. Baton Rouge: Louisiana State University, Louisiana Water Resources Research Institute.

Roberts, Alden E., and Paul D. Starr. 1982. Community structure and Vietnamese refugee adaptation: The significance of context. *International Migration Review* 13(3): 595–618.

Tang, Eric. 2009. On alternative citizenships: The Vietnamese Americans of Black New Orleans East. Presented at American Crossroads Conference: Migration, Communities, and Race. University of Texas, Austin. 16 Apr.

Trang, Corinne. 1999. *Authentic Vietnamese cooking*. New York: Simon & Schuster.

Wood, Joseph. 1997. Vietnamese place making in North Virginia. *Geographical Review* 87(1): 58–72.

Chapter 3.2
Water, Culture, and Gender: An Analysis from Bangladesh

Farhana Sultana

3.2.1 Introduction

In most of rural Bangladesh, the proliferation of tubewells that pump up groundwater has increased people's access to drinking water over the last couple of decades. Most of the tubewells found in households, markets, schools, mosques, and other locations are privately owned, although the government has also installed some public tubewells. The government and development agencies heavily promoted these devices as 'safe' water sources compared to surface water (e.g., ponds and rivers), which is often chemically and pathogenically contaminated (and frequently led to high morbidity and mortality rates from water-borne diseases). However, the tubewell water that was deemed a public health success story only a few years ago is now poisoning millions of people, as naturally occurring, tasteless, odourless, colourless, carcinogenic arsenic is showing up in drinking water drawn from these wells.[1]

The discovery of arsenic has reduced water security and increased pressure on the tubewells that are still providing safe water. These are often the more expensive deep ones, which access deep aquifers that do not have high concentrations of arsenic. Deep tubewells are generally owned by those who can afford to purchase them and drill that deep. The majority of rural households use shallow tubewells that tap shallow aquifers, where arsenic is present in high concentrations as a naturally-occurring carcinogenic metalloid. Recent government initiatives to alert people to the arsenic in water sources have included painting red the tubewells that are producing water with unsafe levels of arsenic. Those wells deemed safe are

[1] Details of the arsenic situation in Bangladesh can be found in Ahmed and Ahmed (2002), Ahmed (2003) and Sultana (2006, 2007a, b, c, 2009).

F. Sultana (✉)
Department of Geography, Syracuse University, Syracuse, NY, USA
e-mail: sultanaf@syr.edu

B.R. Johnston et al. (eds.), *Water, Cultural Diversity, and Global Environmental Change: Emerging Trends, Sustainable Futures?*, DOI 10.1007/978-94-007-1774-9_18,

painted green.[2] As a result of the considerable heterogeneity in the geologic distribution of arsenic in the aquifer, the rural landscape is dotted with red and green tubewells, sometimes clustered, sometimes scattered, with tubewells in close proximity to each other producing water with different concentrations of arsenic. Households with green tubewells have secure access to safe water, while those with red tubewells have to decide how to obtain their daily water. They have to choose between fetching safe water (calculating the social, personal, and familial costs that may ensue) or consuming contaminated water (risking falling ill from chronic arsenic poisoning, or arsenicosis, which can lead to various health complications over time and eventually to death). In this chapter, I analyze the social and cultural issues that have a direct bearing on people's water consumption habits and the ways that arsenic and water affect men and women in the villages of rural Bangladesh. Analysis of the gender relations of water management sheds light on the multifaceted and profound implications of finding arsenic in drinking water, with people continuing to consume contaminated water amid an escalating public health crisis.

This chapter describes ethnographic research conducted between 2003 and 2005 in 18 villages in four districts of Bangladesh acutely affected by arsenic. The research involved participant observation, 232 semi-structured interviews with men and women of different ages and socio-economic, religious, and educational backgrounds, 15 focus group discussions with men and women, and case studies of individuals and households facing water crises or water poisoning.[3] All the villages in the study were predominantly agricultural, with high percentages of landless subsistence farmers involved in sharecropping arrangements through patron-client relations with a few wealthier farmers. The villages had significant levels of inequality and poverty and did not enjoy infrastructure such as piped water systems. People overwhelmingly obtained water via tubewells that tapped the groundwater in the deltaic landscape, and it was predominantly women and girls who fetched water for their families on a daily basis.

In my study, the villages had clusters of red tubewells at a variety of scales – sometimes a few adjacent households in a neighbourhood had contaminated wells, sometimes entire neighbourhoods, or sometimes an entire village. In all of these instances, households faced the challenge of securing safe water. Some lost access to their own tubewells when the tubewells were identified to be unsafe. Others whose tubewells were labelled safe (with no or low concentrations of arsenic) struggled with the stresses of sharing their water with a greater number of people. The

[2] While the concentration of arsenic in water may vary considerably within short distances, the policy that is being followed by the Bangladesh government is to paint red tubewells that are producing arsenic at concentrations greater than 50 μg/l and paint green those that are at concentrations below 50 μg/l. It is worth mentioning that the WHO (World Health Organization) standard of permissible arsenic in drinking water is stricter at 10 μg/l. A discussion on the politics of such development endeavors is beyond the scope of this article. For more details see Sultana (2006, 2007a, 2009).

[3] For greater detail on the methodologies used and the study sites and research participants, see Sultana (2007a, b).

majority of households reported increases in the time, distance, and energy needed to fetch safe water after arsenic was identified in water sources in their village. It is in such landscapes, where women and girls labour several times every day over various distances to fetch pitchers of water for their families, that access to safe drinking water becomes increasingly contentious (see also Sultana 2006, 2007a, b, 2009).

3.2.2 Gender, Class, and Water in Rural Bangladesh

Although scholars have argued that the role of gender in water management deserves more attention, few have as yet focused on the role that broader societal and ecological factors play in the implication of gender in water management – and the ways by which gendered waterscapes are produced, reproduced, and challenged. In studying gender-water relations, it is important to look at who does what with which type or source of water and why, where, and how these practices relate to gender identities and social relationships in general.

Household structures are quite hierarchical in rural Bangladesh, with a clearly demarcated gendered division of labour and rights. Typically, the patriarch (the oldest brother or father) has greatest say in household decision-making and controls the labour and behaviour of other household members. Men do not participate in fetching domestic water for drinking and cooking as that is deemed a feminine task, one especially suited to younger women and girls (Crow and Sultana 2002). The senior woman (matriarch), who may be the mother, grandmother, or eldest daughter-in-law (*boro bou*), can allocate the arduous task of fetching drinking water to younger daughters-in-law. The weak social power of daughters-in-law often results in their greater subjugation and weakens their bargaining power in the household and community; other people being able to command their labour generally perpetuates their lack of power. Young women, especially new brides, almost never challenge their mothers-in-law's oppressive actions, such as verbal and physical abuse, if the daughter-in-law did not fetch the water on time or in sufficient quantities.[4]

Class and gender relations are intricately intertwined in rural Bangladesh, and one cannot be studied without looking at the other (White 1992). In a hierarchical family structure, differently positioned members command differential access to cash, food, decision-making powers, education, and other resources.[5] Although women within a household generally have less power than men, women are able to

[4] See Kandiyoti (1988) for greater discussion of gender and patriarchy.

[5] I do not have the space in this article to go into detail on the measurements of class or the politics involved in such measurements, but do want to highlight that I recognise it is a contentious, multi-faceted and complex issue. In this chapter, I use three broad categories of class (wealthy, middle, poor) based on overall landholding, income, remunerations and assets. In rural Bangladesh, ownership of land is the largest source of wealth and power and class is closely linked to education and non-agricultural earnings (for further discussion, see Sultana 2007a).

command different powers and resources based on their membership in a particular household depending on the socio-economic bracket it occupies. Although women in wealthier households may be powerless within their own families, they may have access to the family's tubewell and thus easier access to water, which places them at an enormous advantage compared to poorer women of households that do not own tubewells. When women belong to a landowning or powerful family, they are generally able to exercise some control over the women in share-cropping, agricultural labouring, and poorer or dependent kin families, who may be tasked with helping more powerful women to fetch safe water. Thus, class positions are important in the ways that gender relations play out socially with respect to water.

3.2.3 Space, Place, and Gender in Water Management

Generally, public spaces have been historically construed as masculine and private or domestic spaces as feminine. Men or women who intrude in the domain of the other gender are often seen as 'out of place' (Massey 1994; McDowell 1999; Creswell 1996). Females, especially, when found 'out of place' are often thought to be in need of greater control (Domosh and Seager 2001). Notions of *ijjat* (honour) and *lajja or sharam* (shame) are social sanctions used in rural Bangladesh to regulate women's presence in public spaces by limiting their mobility and dictating dress code and behaviour. Similarly, notions of *purdah* (veiling or seclusion) also operate in defining appropriate feminine behaviour (Rozario 2001). Although public-private boundaries may be blurred and often are for various reasons, cultural and material practices with regard to water (e.g., men irrigate farm land; women manage domestic water needs) also help to maintain them. What these concepts and practices mean is that broad sociocultural norms, that are also affected by age, class, education and position in the household, may constrain the mobility of women and girls. Thus, local customs, norms, and endowments of women, as well as class, marital status, and age are all important factors in determining which women will be burdened with the menial and laborious task of water fetching.

Moreover, the mobility involved in fetching water is often circumscribed within specific spaces and places, e.g., within one's own *bari* (a homestead consisting of a kin-based cluster of households around a common courtyard), or a neighbouring *bari*. As a result, it is more difficult for women, especially younger women and unmarried or teenage girls, to fetch water from sources in overtly public and masculine places such as bazaars, mosques, and roadsides. The public-private and home-outside divides become problematic when safe water sources are increasingly located in distinctly public spaces. These gendered constructions of public-masculine and private-feminine space come into conflict with each other when women, in pursuit of a domestic, feminine task, attempt to fetch water from public spaces. To fulfil their domestic duties, they venture out into roads, bazaars, mosques, and schools where the only safe water source may be. The private and public gendered spaces collide as a result of the need for safe water.

Box 3.2a Managing domestic water in Bangladesh after the arsenic crisis: A case study

—Kazi Rozana Akhter and Suzanne Hanchett

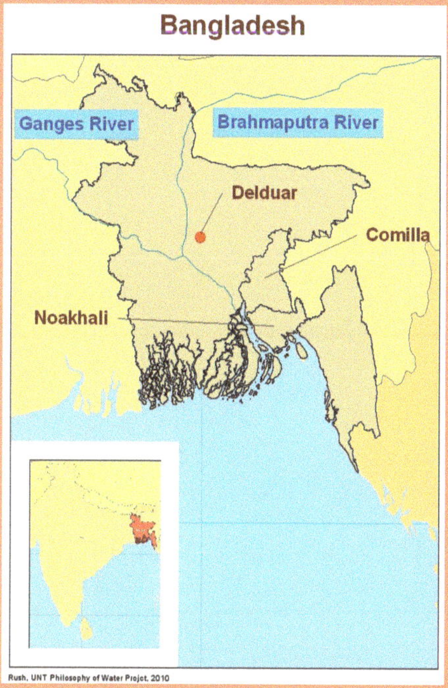

Box Map 3.2a.1 Bangladesh

The day-to-day picture of domestic water use is very complicated in Bengali-speaking rural communities of Bangladesh and West Bengal, India. Especially in well-watered rural areas, several different sources of water are used for various domestic purposes. Although they may not always have access to their ideal sources of water, people have strong preferences. It is typically the responsibility of women to collect and preserve water for domestic purposes, such as cooking, drinking, washing, cleaning, and personal hygiene.

Women tend to be very careful about their in-house methods of preserving water for different purposes. After bringing water to the house, for example, women of southeastern Bangladesh store it in separate pots: one for cooking, one for drinking, and one for washing after latrine use. They never use one pot's water for the other pot's intended purpose. This is a strict observance. Women of many regions generally are careful to maintain this separation. If women do not keep their household waters separated, they will be socially criticised. People say that violation of this norm is 'hateful', making food 'unpleasant to eat' and 'distasteful, causing an unpleasant feeling' (DHV 1998).

Since the 1970s and 1980s, increasingly large numbers of households in the Bengali-speaking region have come to depend on tubewells – easily installed, hand-pumped wells drawing water up from aquifers that are 12–60 m below ground, some even deeper. This technology was widely promoted by virtually all development agencies because the water is free of pathogens; but at the end of the twentieth century, it was found belatedly that in certain regions the water of aquifers less than 60 m deep had dangerously high concentrations of presumably naturally occurring arsenic.

(continued)

Box 3.2a (continued)

Box Fig. 3.2a.1 Girl carrying household water from a neighbor's tube well, Brahmanbana District, 2009 (Photo credit: Kazi Rozana)

Box Fig. 3.2a.2 Tubewell with red painted spout indicating high arsenic content in the water, Chittagong District, Bangladesh, 2000 (Photo credit: S. Hanchett)

The extent of the arsenic problem in Bangladesh was well understood by 2002–2003, by which time the water of many of the nation's tube-wells had been tested. Numerous agencies, governmental, non-profit, and international, set out to educate the public about the arsenic problem and help identify safe water sources. Technical challenges have been daunting. Arsenic dissolves thoroughly in water, and only chemical treatment can remove it. Some promising early arsenic removal methods turned out to be either less effective than hoped or too difficult for most village people to manage properly. A small number of community - and household-level filtration technologies have proved technically viable and have been provisionally approved for experimental distribution in Bangladesh. In West Bengal some organizations, such as the Bengal College of Engineering in Kolkata, have carefully installed community-level arsenic removal plants and monitored them.

Alternative arsenic-free water sources exist, as large and small rivers criss-cross much of the Bengali-speaking region and rain-fed ponds dot it. Several projects have supported a return to obtaining drinking water from ponds or dug wells; but these efforts have faced serious problems with social coordination among multiple users and owners. And these sources tend to have excessive levels of bacterial contamination unless they are carefully managed. The most popular alternative sources are 'deep tubewells', probing old aquifers 90–275 m down, which are mostly (though not all) free of arsenic.

(continued)

Box 3.2a (continued)

Box Fig. 3.2a.3 Narghiz, a housewife, carefully manages multiple domestic water sources and avoids cooking or drinking with arsenic contaminated water, Komilla District, 2009 (Photo credit: Kazi Rozana Akhter)

Box Fig. 3.2a.4 Narghiz's house, 2009 (Photo credit: Kazi Rozana Akhter)

The arsenic situation has increased women's workload as they try to avoid the hazards of arsenic-affected water. Many women, now accustomed to the convenience of their own tubewells, have started making long trips to collect drinking and cooking water from arsenic-free sources. Some have tried out new household water filters, all of which require regular cleaning.

Narghiz, a 38-year-old mother of five living in Comilla District, Bangladesh, is one woman who tries to provide her family with safe water by making long trips each day to collect water from a distant deep tubewell. The trip not only takes time, it also forces her to cross boundaries that are socially uncomfortable for a married Muslim woman who would like to observe some degree of honourable, purdah-type of restriction on her movements around the village.

Narghiz gets up very early every morning, cleans all the rooms of her house, and sweeps her courtyard. She goes out to collect drinking water before daybreak in order to avoid the crowd and also to avoid being seen by men. She says, 'I must strongly maintain purdah as I move around the village. This is my father-in-law's place, and as a wife from a respectable family I have to maintain purdah until I die'. She brings two pitchers of drinking water from a far-away deep tubewell and keeps them in her dining room. Sometimes she also collects drinking water for her neighbour. When her neighbour feels sick, Narghiz helps her, and vice versa.

The household does not have their own pond and their tubewell is contaminated with arsenic, thus every evening Narghiz goes out again to collect water from a distant pond. This she will use for cooking. After returning to her house, she pours the collected pond water into a large clay pot kept in her kitchen. Her family of nine needs much cooking water. She keeps the stored pond

(continued)

Box 3.2a (continued)

water for at least 24 hours without disturbing it, to allow the silt to settle out. The cleared water gives her curries a very nice colour. After it is clear, the water is transferred to a smaller clay pitcher, from which she draws out water as needed. Some of this water is kept in a white plastic bucket in her dining room, where it is used to rinse plates and spoons before each meal. Drinking water from a distant deep well is also stored in the dining room, in aluminium vessels.

She explains, 'I use my own tubewell water to wash my dishes and pans after we eat. But I do not use tubewell water to rinse the dishes before a meal, because the water has a high level of arsenic'.

Box Fig. 3.2a.5 Women enjoying a pond bath on a hot afternoon. Brahmanbaria District, 2009 (Photo credit: Suzanne Hanchett)

Narghiz also keeps some of her own tubewell water stored for toilet use. Before her daily bath she collects some tubewell water and pours it into a large clay pot kept in a small room next to the latrine, which is detached from the rest of the house. When they go to the toilet, household members carry a small, plastic kettle-like water pot called *bodna*, which is filled with water from that clay pot. 'This type of water is never brought inside the living [or] bed rooms, dining room, or kitchen', she says. Water from her own tubewell water is used for cleaning the latrine.

Narghiz takes a bath every day in a distant pond. After returning home from her bath, she washes her arms according to the prescribed Muslim ablution practice (*ozu*) and performs her mid-day prayer. The ablution is done with water from her own tubewell water. She washes her clothes with pond water.

Narghiz confidently declares, 'I like cleanliness, and I maintain it'.

Authors' Note: We called Narghiz and asked her permission to publish photos and information from the interview for this volume. We asked if Narghiz wanted her own name or a pseudonym used. After enquiring about the details and purpose of the volume, she agreed and said she would like her own name used. We explained to her that she was selected to represent the many women who try to manage their household water sources carefully. She was enthusiastic about the idea of having women's story told through her experience.

Most people in my study stated that the problems of collecting water from outside the *bari* or from farther away stemmed from not only to physical distance and time required to travel it, but also the social significance of extended travel. For many of the men interviewed, having a red tubewell in their homestead means that women and girls from their household have to venture out into public spaces to get water, a social transgression of major concern for the men, whose responsibility it is to maintain a proper household. Most women said that the biggest problem caused by having a red tubewell was the necessity of travelling further to get water or needing to use someone else's source. The second most frequently mentioned difficulty was that they must go into public spaces to access water. Both men and women expressed concerns about collecting water in the early morning or evening when it is dark, when the water source is far from the *bari*, and the social insecurity of travelling longer distances.

In some instances, women are restricted by their own family members from venturing too far to get safe water, and the women may be forced to fetch unsafe water for their family from a closer source. As one teenage girl said, 'My father said we'll have to drink this water [from the red tubewell] and that we shouldn't go to the bazaar to get water from the green tubewell. It is not allowed'. Such sensitivities often result in entire families continuing to consume contaminated water in a trade-off between safeguarding family honour and risking family health (which can seem less important because the health impacts of arsenic poisoning are not immediately felt but develop over time). The fear of loss of honour and shame when younger women from a *bari* are seen fetching water in distinctly public spaces or traversing public spaces to access someone else's tubewell, discourage families from accessing safe water. As one older women said, '*Oi barir boura bahir theke pani aney, amader barir bouderke ta korte deina ami*' ('The daughters-in-law of that other household get water from outside, I don't allow our daughters-in-law to do that', implying that it is disgraceful that the womenfolk from the other family go to public places to get water, whereas she does not expose her daughters-in-law to such socially risky practices). It is a sign of family honour to be able to keep daughters-in-law within the *bari* and not subject them to public visibility. This view is more prevalent in wealthier households. Women in poorer households observed that they do not have the luxury for such sentiments: '*Bahir theke pani na anle amaderke ke pani ene dibe?*' ('If we don't get our own water from outside, who will bring water for us?') This woman further noted that her mother-in-law does not have any choice but to let her get on with providing for their livelihood, as the only other option would be for her mother-in-law to do the work herself.

Although opportunities for women to be in public places have dramatically increased in recent years (e.g., women enjoy greater engagement with markets, schools, and jobs), these transgressions into formally male-dominated space are explicitly regulated through proper attire and behaviour. Thus, women in public spaces are required to cover their bodies more carefully than when they are within their homesteads. Usually the custom is for women to put their sari over their heads in public places as a form of proper decorum. Draping the end of the sari over the head is referred to as 'putting on a *ghumta*'. A woman in public without the *ghumta* is often seen in a negative light, as wanton and inappropriately behaving. Whereas *ghumtas* often slip off and people do not give much attention to them when women are working in agricultural fields, people consider the *ghumtas* to be more important when a woman is walking about, going places, or doing less physically demanding work, as fetching water is understood to be, in comparison to field labour. The constant need to pull the *ghumta* back on means that a woman has to keep at least one hand free, which is possible if she is carrying one pitcher of water. If women are carrying more than one, then they will put the pitchers down to fix their *ghumta* before proceeding, especially if men are nearby.[6] Although there is flexibility in such veiling practices, the women of wealthier and middle-class households adhere to the norms of proper attire more readily than poorer women, who often have to engage in physical labour in public places and are less subject to social regulation of their attire.

In short, fetching water is a particularised burden for women, as notions of honour, shame, and decorum affect quite literally their access to water. The decisions that men and women make about where to obtain water reflect a struggle between the purity of women and the purity of water. These daily transgressions represent pollution in both symbolic and material terms.

Box 3.2b The purity problem and access to water in Bangladesh

—Shireen Akhter

Bangladeshi women are restricted from using safe water at critical points in the life cycle. For example, 'Selina', a 19-year-old who had given birth in the previous week, was found secretly cleaning her birth-stained cloths in a very dirty ditch behind her house. She said, 'I am not allowed to use our pond for my unclean cloths. If I do', she explained that according to Muslim birth customs, 'the pond will become as impure as I am. I need 40 days to become pure'. It is also said that a woman who is menstruating or who has just given birth is vulnerable to the threat of evil eye. She is in danger of 'bad wind' (*khaaraap baataash*), and it is assumed that supernatural elements will try to enter into her body, possibly causing her to become mad.

(continued)

[6] Such sentiments are stronger in more remote and conservative areas and less so in areas closer to urban centers, where more women have begun to go about without the *ghumta* and have normalised such attire in line with more urbanite women. A few of the highly educated women or job-holding women in villages may be seen without a *ghumta*, but they are often seen as exceptions to the norm due to their education/earning status. While religion does play a role in this irrespective of social location, as more conservative Muslim families will practice covering than less conservative Muslim or Hindu families, *ghumta* is practiced among Hindus too, but less stringently.

Box 3.2b (continued)

Box Fig. 3.2b.1 Women approaching a small, dirty puddle behind their home to clean menstrual cloths in secret (Noakhali District, 2007) (Photo credit: Shireen Akhter)

Menstruating women are not allowed to use common water bodies such as ponds and tubewells. They cannot even approach the boundaries, platforms, or steps around such water sources. They are even supposed to avoid rivers. According to one study, it is believed that a boat carrying passengers may capsize if a menstruating woman is on it. Women clean their bloody cloths outside of the house or compound in very secret, dirty places with unclean water from sources such as ditches or drains, risking skin and genital diseases. A menstruating woman is called sick (*sarir khaaraap,* literally 'bad body'), or 'impure' (*naapaak*). In every ethnic group we have encountered, women at these times in Bangladesh are restricted from using safe water sources. Because of the shame associated with their polluted condition, women hide their physical condition. Some women said, 'When a woman has menstruation and she goes under a tree with uncovered hair, she will attract illness-producing "bad wind" and djinns.'

Although water is universally agreed to remove pollution, those most in need of its purifying benefits are ironically forced by custom to use hidden and unclean sources.

In this respect, critical resource scarcity challenges social norms and the search for viable solutions affects household gender dynamics. Some women were able to argue successfully that having to fetch water from farther away meant that they must travel through and into public spaces. Such social breaches, which challenge social status and family honour, could be avoided only if their husbands installed a tubewell in their own homestead. Such arguments, of course, rest on the hope that a new tubewell would tap into an arsenic-free part of the aquifer.

Some younger women used the daily necessity to fetch safe water from farther places as a way to get out of the confines of the *bari* and to socialise with others. One development project worker called this '*Pani ante prem korte jay*' (roughly translated, 'Having an affair while fetching water'). Although such a comment may be prejudicial to women's mobility in public spaces and their honour (especially young unmarried women's honour), in this instance, arsenic and water become useful allies in manipulating power relations to increase mobility. However, some families circumvented such situations by continuing to use their contaminated sources or making alternative arrangements (e.g., paying hired labour if they can afford it, sending sons if possible, or sending younger women with other women who will act as chaperones). One mother said 'It is not good to send our unmarried daughters to get water from so far away, people will talk and it is bad for their prospect of marriage'.

In securing access to safe water and its use, women also invoke other identities, depending on the context. They may invoke affiliation with certain powerful or wealthy households when trying to gain privileged access to a safe water source. Similarly, women may invoke their advanced age or status as a widow to claim certain rights to safe water sources. Women also use notions of femininity associated with motherhood to claim safe water, arguing that their children need safe water to survive. Some women also use kin and fictive kin status with other women (*shoi* or 'sisterly friend') to get help in fetching water if they themselves are unable to. Such informal networks and relations help in securing access to water but are increasingly challenged as safe water sources are fewer and further apart.

Beyond these social relations and strategic (albeit limited) manoeuvres by some women and girls, both men and women noted that the prevalent gendered division of labour in water management had a significant bearing on the ways they relate to water. In many instances, irrespective of their social standing within the household, women felt that all family members of their household should fetch water if they were capable of doing it. Fetching water was not as laborious once tubewells were dug inside their own *bari* as it had been in the past when pond water had to be hauled from greater distances. Some women said that with the convenience of tubewells, they would ask whoever was able and available to get a pitcher of water quickly, within reason (i.e., not matriarchs or adult men). Among sisters-in-law, there may be clear-cut delineation of who can fetch water for whom and in exchange for what, but often children and younger men fetched water as needed for whichever hearth within the *bari* needed it. However, with people having to travel greater distances outside of the *bari* to get safe water now, older patterns of gendered divisions of labour are resurfacing, thereby increasing women's burdens in providing water for their families.

Although gender makes most women less powerful than men in households and societies, the differences among women of different households are also noticeable, especially in relation to access to safe water. In a few instances where a safe tubewell was in the homestead of a poor family, they gained an unusual new power through the ownership of a safe water source in a landscape of poisoned tubewells. Although some wealthier women were reluctant to get water from such a well, many were forced to overlook the social status infractions occasioned by depending on the poor. Although some exerted existing power relations in securing access to the tubewell, using a poor household's facility went against the sensitivities of most of the wealthier households. Fetching water from specific places thus holds meaning, especially when from a well in a poorer *bari*. The heterogeneous distribution of arsenic and safe water came to play a role in the overall status that households had, especially for the women of the poorer household. Although a poor family's having the safe water source did not destabilise trenchant patron-client relations and hierarchical class structures, it did provide the poor with some leverage, however small. As such, gender, class, and geographical location intersected in reducing the water insecurity of a few households in unexpected ways.

Box 3.2c A new type of 'social problem' – Comments of some male and female union council members in Ramganj Sub-district, Laksmipur District, Bangladesh

—From Suzanne Hanchett's field notes, 2006

Danida (the Ministry of Foreign Affairs of Denmark) and the Dhaka Ahsania Mission (DAM), an NGO, gave deep tubewells to poor people after doing screening for arsenic content of shallow tubewell water and finding high arsenic levels in the water....

One deep tubewell was given to a group of ten households. 'Social problems' resulted. Some rich people are living in nearby multifamily, residential compounds called *baris*. Their 'egos' absolutely will not allow them to ask for water from the *baris* of poor people. So now the solvent and middle-class people are forced to drink unsafe water. This is not fair. Danida and DAM gave help only to the very poorest people's *baris*, but everyone needs safe water....

More middle-class people are perceiving the need for arsenic-free water nowadays...

3.2.4 Masculinities and Femininities in Water Management

Water and arsenic in rural Bangladesh have come to be key elements in the production of gendered norms, in how people's time and labour are valued, and how different groups of people feel powerless or empowered to change their access to safe water. Certain identities are created vis-à-vis water (safe and unsafe), whether in decision-making about water management, water collection activities, or suffering from water's effects. The constellation of ways that water plays a role in the production of identities and norms can vary by community and context, but overall gendered norms appear to respond to changing water conditions in the following ways: in the ways that gendered labour and gendered spaces in the landscape change with the manifestation of arsenic in water; in the ways that gender is negotiated in terms of water access and use; and in the ways that individuals negotiate a sense of self in relation to the complexities of unsafe waterscapes. Since teenage boys and adult men still resist helping fetch water, entrenched gender ideologies are generally maintained; but for those men who are more open to fetching water and for women who are also supportive of this change, water poisoning is bringing changes in gender roles and norms.

Water and arsenic bring into sharper relief the negotiations of masculinity and femininity in relation to the acquisition of safe water, but they can also blur the boundaries in instances where resistance to such norms is manifest. As one woman argued: 'Even if we are ill, our men will not fetch water for us. It is not a man's job to fetch water. It would be nice if they did sometimes, but we do not ask'. Yet another argued in favour of the gendered control of labour relations in water management: 'Why should men fetch the water? That is a woman's job'. Similarly, a man justified the social norms that regulate embodied relations to water: 'I would die before I fetched water for a woman. If I did, people would think I am mad'. Such socialised norms are common in maintaining the gendered division of labour in relation to water. And yet another man confided: 'Sometimes I help my wife get water, or my son does. This arsenic problem is for all of us'. These sentiments, however, are not commonly expressed.

It is important not only to pay attention to the different gender roles and meanings attached to activities that reinscribe gender in water, but also to the way water struggles themselves come to reconstitute and reinforce different subjectivities (Jackson 1998). Environmental struggles can end up reinforcing gender and power relations and highlighting the inequalities that exist, which are not substantially reconfigured even if some people contest them, as people can both internalise and challenge gender notions. Arsenic has largely tended to worsen patterns of inequality in the division of labour and people's sense of themselves in relation to water. The discovery of arsenic in the water seems to have intensified traditional gender roles, as more women bear the burden of fetching water, which had decreased as tubewells had become available in many homesteads. When tubewells are located in the *bari* or near the kitchen area (i.e., more private spaces), sometimes men get their own water without too much fuss, but now that there is greater dispersal of safe water sources, men are more reluctant to be seen participating in such a gendered task. Thus, notions of 'traditional' femininity are reinforced as a result of tubewell contamination and the spatialised nature of this water crisis.

In responding to a question about whether men should help more because of the arsenic situation, men and women's responses are strikingly similar: 80% of both groups said men should help more, and 20% said that men should not. The reasons given in the affirmative were often qualified by statements such as, 'Men should help only when women are ill, unable, too busy, or it is too difficult for them'. Those opposed argued that fetching water is a woman's job and that society looks down on men for doing a woman's task. In general, older women expressed less eagerness than younger women to have men participate in collecting drinking water; and younger men appeared to be more supportive of helping women than older men. Poorer households were more supportive of gender equality in this respect than the slightly better off; this discrepancy could perhaps be related to perceived social status concerns for the wealthier households if men in their household participated in drinking water collection. In households with people who have fallen ill from arsenic poisoning, men were more open-minded in challenging traditional norms in accessing safe water for their families. Some of the younger, educated men who were more aware of arsenic's impacts were also more willing to help get water once in a while from farther afield if needed, especially if they had bicycles to transport the water (this was not very common, though, as bicycles are less available among poorer households).

Approximately a third of the 232 people interviewed, both male and female, claimed that men do occasionally or sometimes help in getting drinking water for their households in light of increased hardship in procuring water. The majority, however, agreed that men did not help at all. In instances when men got water 'sometimes', it was usually when the water source was within the *bari* and the water was for himself (to drink or to make tea). What was also interesting to note was that although poorer men were more open to the idea that they should help fetch water given the arsenic crisis irrespective of proximity, wealthier men only agreed when the water source was close by and within the *bari*, not in public places. Fetching water in public would threaten their masculinity; within one's own *bari* the threat was less of a problem.

What explains the trend in opinions across classes is that poorer households largely do not have their own tubewell, and so the men are more willing to go outside to fetch water; conversely, richer households tend to have tubewells within their *baris*, and thus more men are willing to get water from such sources, as this does not transgress social norms drastically. The middle-class households that often do not own their own source and worry about social repercussions and gendered identities in fetching domestic water are less willing to have their men get water from other places. Men's visibility and the distances and spaces involved appear to be deterrents, as both the middle-class and wealthier households are generally more concerned about social norms than the poorer households, including the implications of fetching water for masculine identities and thereby family honour and social standing. Thus, the spatial distribution of arsenic and tubewells and the spatialised nature of water collection influence the relationships that men and women have with water, challenging gendered roles and identities and the construction of masculine norms vis-à-vis water. This entrenched gender division of labour and gendered identities in relation to water management may be increasingly challenged in the future as water scarcity forces all household members to participate more actively in procuring safe water, but at the moment only a minority of men are willing to engage in this activity.

3.2.5 Conclusion

Attention to gender relations and norms that are produced through and responsive to complex environmental change demonstrates that struggles over water are not only over access, control, or use, but also about gendered power relations. Such a conceptualization of gender-water relations, in which the spatial distributions of arsenic and contaminated tubewells influence the ecological and spatialised identities negotiated in water management and gendered norms are produced simultaneously socially, spatially, and ecologically, is also useful for practitioners and policymakers in gauging the ways that individuals and households access safe and unsafe water, respond to water contamination, and participate (or not) in water management projects in their locales (see also Sultana 2009). Understanding the social and cultural norms that influence gender relations to water can help explain why so many Bangladeshi households continue to consume poisoned water despite official efforts to increase awareness about arsenic.

Acknowledgments This chapter draws from my publication 'Fluid Lives: subjectivities, gender and water in rural Bangladesh', *Gender, Place, and Culture* 16(4): 427–444, 2009. Thanks to Taylor and Francis Group (http://www.informaworld.com) for permission to reproduce some material from that article.

Resources

Ahmed, Feroze. 2003. *Arsenic contamination: Bangladesh perspective*. Dhaka: ITN-Bangladesh.

Ahmed, Feroze, and Mufad Ahmed. 2002. *Arsenic mitigation in Bangladesh: An outcome of the international workshop on arsenic mitigation in Bangladesh, Jan 14–16*. Dhaka: Local Government Division, Ministry of Local Government, Rural Development and Cooperatives, Government of Bangladesh.

Creswell, Tim. 1996. *In place/out of place: Geography, ideology and transgression*. Minneapolis: University of Minnesota Press.

Crow, Ben, and Farhana Sultana. 2002. Gender, class and access to water: Three cases in a poor and crowded delta. *Society and Natural Resources* 15(8): 709–724.

DHV Consultants Ltd. 1998. Baseline Study, Vol. 1. Five Districts water and sanitation group. Dhaka: DHV Bangladesh and DPHE-Danida.

Domosh, Mona, and Joni Seager. 2001. *Putting women in place: Feminist geographers making sense of the world*. New York: Guilford Press.

Jackson, Cecile. 1998. Gender, irrigation and environment: Arguing for agency. *Agriculture and Human Values* 15(4): 313–324.

Kandiyoti, Deniz. 1988. Bargaining with patriarchy. *Gender and Society* 2(3): 274–290.

Massey, Doreen. 1994. *Space, place and gender*. Minneapolis: University of Minnesota Press.

McDowell, Linda. 1999. *Gender, identity and place: Understanding feminist geographies*. Minneapolis: University of Minnesota Press.

Rozario, Santhi. 2001. *Purity and communal boundaries: Women and social change in a Bangladeshi village*, 2nd ed. Dhaka: University Press Limited.

Sultana, Farhana. 2006. Gendered waters, poisoned wells: Political ecology of the arsenic crisis in Bangladesh. In *Fluid bonds: Views on gender and water*, ed. Lahiri-Dutt Kuntala, 362–386. Kolkata: Stree Publishers.

Sultana, Farhana. 2007a. Suffering for water, suffering from water: Political ecologies of arsenic, water and development in Bangladesh. PhD dissertation, Department of Geography, University of Minnesota.

Sultana, Farhana. 2007b. Reflexivity, positiona lity and participatory ethics: Negotiating fieldwork dilemmas in international research. *ACME: An International E-Journal for Critical Geographies* 6(3): 374–85.

Sultana, Farhana. 2007c. Water, water everywhere but not a drop to drink: Pani politics (water politics) in rural Bangladesh. *International Feminist Journal of Politics* 9(4):494–502.

Sultana, Farhana. 2009. Community and participation in water resources management: Gendering and naturing development debates from Bangladesh. *Transactions of the Institute of British Geographers* 34(3): 346–363.

White, Sarah. 1992. Arguing with the crocodile: Gender and class in Bangladesh. Dhaka: The University Press Limited.

Chapter 3.3
Privatization and Collective Stewardship of Water Resources: Case Studies

Box 3.3a Zimbabwe's water crises: The importance of environmental governance and a right to water

—Bill Derman and Emmanuel Manzungu

Box Fig. 3.3a.1 Testing a faulty borehole during Zimbabwe's dry season in 2000. Boreholes are the main supplier of water in the rural areas (Photo credit: William Derman)

Box Fig. 3.3a.2 Women's Garden, hand irrigated from a shallow well, dry season in 2000, Mhondoro, Zimbabwe (Photo credit: William Derman)

Zimbabwe instituted a series of water reforms in the 1990s on the basis of the four Dublin Principles for Integrated Water Resource Management. A new parastatal institution, the Zimbabwe National Water Authority (ZINWA), was formed to co-manage Zimbabwe's waters with more participatory institutions. The president of Zimbabwe remained the owner of the national waters; ZINWA became their manager. The long-standing distinction between primary or free water (water used by rural household use and in small-scale irrigation) and commercial water (water used for commercial purposes whose use was subject to fees) remained. In urban areas, ZINWA was to supply raw water that would then be purified and delivered for human use, and treated before being returned to rivers.

Two years after the creation of ZINWA, Zimbabwe underwent what has been termed a 'fast track' land

(continued)

B.R. Johnston et al. (eds.), *Water, Cultural Diversity, and Global Environmental Change: Emerging Trends, Sustainable Futures?*, DOI 10.1007/978-94-007-1774-9_19,

Box 3.3a (continued)

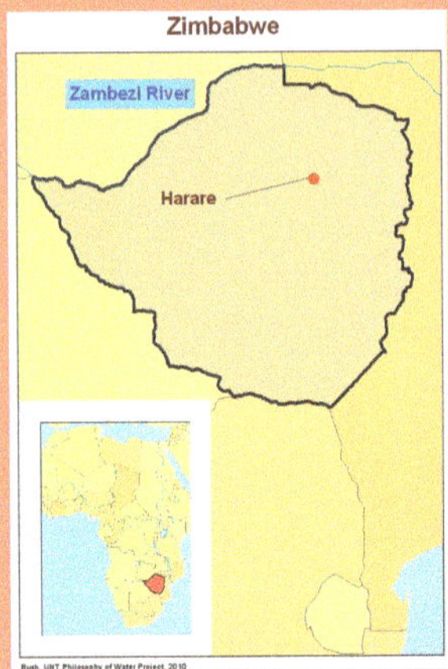

Box Map 3.3a.1 Zimbabwe

reform program, in which large- and small-scale black farmers settled on most of the formerly white-owned farms. The goal of reforming water policy so that users would begin to pay for its use was lost in the process of land reform; the new farmers had far fewer resources than did the owners of the former large estates, and they claimed that they should not pay for water while getting the land for free. In addition, land resettlement took little account of irrigation systems already in place. This omission has led to a rapid decline in the total area of irrigated agriculture and so in production of food for both domestic consumption and export.

The result has been food shortage. Zimbabwe has become increasingly reliant on international food assistance and has become a source of large-scale migration to South Africa and neighbouring countries. Given the national objectives of land reform and dependable agricultural production, declining agricultural production, food shortages, and out-migration highlight the essential need to consider water in agricultural and environmental governance.

The lack of government attention to water issues has been most apparent in the capital city, Harare, and its surrounding area. ZINWA began to have serious financial problems following land reform. This was due to dependence upon commercial farmers who through paying for water were to keep ZINWA solvent. In terms of urban water, the city of Harare received important revenues from the sale of drinking water to city residents. However, ZINWA lacking monies took over water supply to the city of Harare in 2007. It failed to maintain the quality of the water or to the amount of water necessary to meet urban needs. It also continued to neglect the failing sewage system with devastating consequences leading to the cholera epidemic of 2009. The city was controlled by the opposition party, the Movement for Democratic Change (MDC) following the elections of 2000. Water and sanitation had not previously been a significant political issue in urban areas because since 1980 Zimbabwe had made impressive strides in the provision of safe water and improved sanitation.

In the mid-1990s an estimated 99% of the urban population had access to safe water, compared to 64% of people in the rural areas, and sanitation coverage was estimated at 99% and 48%, respectively (Chenje and Johnson 1996:213–214). However, as of 2004 Swatuk (2008) put access to safe water at 81% of the total population and access to adequate sanitation at 53%. These

(continued)

Box 3.3a (continued)

figures have continued to drop since ZINWA assumed control of the Harare water supply in 2005. Because the opposition MDC won the municipal elections in all of Zimbabwe's major cities in 2005, ZANU–PF (Zimbabwe National Action Union Patriotic Front, then Zimbabwe's ruling political party) stripped municipal authorities of control over urban water and sewage services and turned these over to ZINWA. This action deprived cities of a major source of revenue and diverted the funds to ZANU–PF's control. Directed by central government, ZINWA was bankrupt almost from the start because it was to be funded by water revenues, which precipitously declined with land reform, and it had insufficient skilled staff.

One of Harare's two waterworks ceased operating in 2005, when ZINWA took over the management of the water services. ZINWA then began depositing raw sewage into the Mukuvisi and Manyame rivers, from whence it flowed into Harare's main reservoir, Lake Chivero. As a result, Harare's water was heavily polluted. ZINWA then found it difficult to provide clean water to residents, as treating the heavily polluted water was expensive. Eight different chemicals, some imported, were needed to make Harare's water potable again. ZINWA thus became more reliant on foreign exchange for its operations, but this need was not included in national budgets. ZINWA suffered from a lack of funds, and water shortages set in very rapidly. Nearly every part of Harare, a modern city of about two million inhabitants, has suffered continual water cuts, some lasting only a few hours a day, others lasting years.

Box Fig. 3.3a.3 Children collecting surface water in Harare, December 2008. Zimbabwe's cholera outbreak was attributed to a failure to maintain sewage, water treatment, and water delivery systems (Photo credit: Associated Press. Accessed from the Daily Mail online, December 8, 2008, http://www.dailymail.co.uk/news/worldnews/article-1092609/South-Africa-sends-doctors-Zimbabwe-cholera-epidemic-spreads-Africa.html)

Although Zimbabwe has been affected by cholera in the past, the situation worsened during the rainy season of 2008-2009. According to the World Health Organization (WHO), as of 30 May, 2009, there have been a total of 98,424 cases and 4,276 deaths in Zimbabwe related to the outbreak (WHO 2010a). These are remarkably high death rates and

(continued)

Box 3.3a (continued)

reflect the partial collapse of Zimbabwe's health care system. All 10 of the country's provinces have been affected by cholera, according to United Nations (UN) statistics (which remain incomplete because of a lack of local medical personnel and problems in communications) (WHO 2010a). Since there have been few improvements in health care infrastructure or water management, the cholera epidemic is likely to be repeated in the future. In 2010, 18 out of the 62 districts in the country have been affected by a cholera outbreak that started on 4 February 2010 compared to 54 districts at the same time in the previous year (WHO 2010b).

The lack of clean water dramatically affects all health care institutions and especially people living with HIV or AIDS and young children susceptible to diarrhoeal diseases. Poor governance, state-sponsored violence, intimidation, and corruption have prevented Zimbabwe's citizens from exercising their civil and political rights and denied them the right to satisfy the most basic social and economic needs for food, health, and clean water. Government indifference precipitated the health crisis and then exacerbated it by restricting access to food, thereby weakening the population and reducing resistance to disease.

Resources

Chenje, M., and P. Johnson (eds.) 1996. *Water in Southern Africa.* Maseru/Harare: SADC/IUCN/SIRDC.

Derman, B., and F. Gonese. 2003. Water reform: Its multiple interfaces with land reform and resettlement. In *Delivering land and securing livelihood: Post-independence land reform and resettlement in Zimbabwe*, ed. M. Roth and F. Gonese, 287–307. Harare/Madison, WI: Centre for Applied Social Sciences/Land Tenure Center, University of Wisconsin.

Derman, B., A. Hellum, E. Manzungu, P. Sithole, and R. Machiridza. 2007. Intersections of law, human rights and water management in Zimbabwe: Implications for rural livelihood. In *Community-based water law and water resource management reform in developing countries*, ed. B. van Koppen, M. Giordano, and J. Butterworth. Comprehensive assessment of water management in agriculture series, vol. 5, chap. 15. Wallingford: CABI Publications.

Manzungu, E. 2008. Running out of options? The status of service delivery in the Harare–Chitungwiza metropolis. In *Towards a new water creed: water management, governance and livelihood in southern Africa*, ed. E. Manzungu, 11–24. Bonn/Harare: Inwent/Weaver Press.

Mapedza, E., and Kim Geheb. 2010. Power dynamics and water reform in the Zimbabwean context: implications for the poor. *Water Policy* 12: 517–527.

Swatuk, L. 2008. A political economy of water in southern Africa. *Water Alternatives* 1(1): 24–47.

World Health Organization. 2010a. http://www.who.int/csr/don/2009_06_09/en/index.html. Accessed 8 Feb 2010.

World Health Organization. 2010b. http://www.who.int/hac/crises/zwe/sitreps/zimbabwe_7june2010.pdf. Accessed 23 June 2010.

Box 3.3b Extremely private: Water trading and the commoditization of water in Australia

—Veronica Strang

Internationally, there is wide cultural and ideological diversity in views on whether fresh water is a common good or merely an economic asset that can be held privately by individuals or elite groups (Hann 1998). Between these ends of the spectrum lies a process through which water is transformed from a 'natural' common substance into an alienable 'cultural' commodity (Strang 2009). En route it is both literally and metaphorically abstracted from rivers and aquifers, becoming a generic 'resource' in controlled impoundments or being reconstituted as 'virtual' water allocations (Parry 2004).

Australia contains examples of both extremes. Aboriginal Law[1] constitutes a classic limited common property regime (Ostrom 1990). Each indigenous language group regards both water and land as integral parts of its permanent relationship with a particular place. Local water sources contain totemic ancestral beings and the spiritual and social essence of their clans of descendents. These clans own the water sources collectively and cannot commodify or alienate them. Clan members have a moral responsibility to care for their local environment and – as this is an ideologically permanent relationship – to ensure that their use of it is sustainable.

Box Fig. 3.3b.1 Irrigation scheme in South-East Queensland (Photo credit: Veronica Strang)

At the other end of the spectrum is Australia's recent National Water Reform, instigated by John Howard's government in 2004. This program introduced a free-market system of water trading, asserting that this approach would improve the efficiency of water use, producing better economic *and* environmental outcomes. Thousands of water allocations previously defined by related land area were reframed as volumetric 'water access entitlements', separate from the land (COAG 2004). They could be exclusively owned and traded, subdivided, amalgamated, mortgaged, or used as collateral. Thus they became an abstract economic 'asset'.

The introduction of water trading has benefited the small minority of

(continued)

Box 3.3b (continued)

Box Map 3.3b.1 Queensland, Australia

people able to speculate: for example, at Cubbie Station in south west Queensland, one company was able to build a storage dam with a capacity of 500 gigalitres (GL) and buy up allocations comprising approximately a quarter of the water available to the Darling River. Although the National Water Commission continues to present the broader effects of water trading in positive terms, there have been more questionable ecological and social outcomes.

Higher water prices were supposed to make low-value or inefficient crops less viable and decrease demand in general. But instead, the pressure on aquatic systems is now reaching a state of crisis as higher prices have awoken 'sleeper' licenses (unused but still legitimate licenses), encouraging people to trade for previously unviable uses of water, invest in high-value crops, and extend irrigation into new areas (Isaac 2002).

Box Fig. 3.3b.2 Farming in south-east Queensland (Photo credit: Veronica Strang)

The management of water based on free-market principles has also tempted state governments to over-allocate resources at the expense of ecological needs. Few environmental organizations have sufficient funds to 'buy up' allocations, and the trend is primarily in an opposite, production-based direction. In the legislation accompanying the political posturing towards water reform, the focus is similarly on 'consumptive' rather than 'non-consumptive' rights, and there is no provision for strong representation on behalf of the environment. In such a framework, ecological protection inevitably has low priority (Ladson and Finlayson 2004).

Water trading has also introduced social as well as economic volatility. Competing for an increasingly scarce and expensive resource strains relationships in farming communities, and when financial pressures force farmers to trade their allocations to an emerging group of 'water barons', they can be left with dry farms for which no water is available or affordable. Many farmers argue that extreme economic rationalism has a destructive effect on community stability and security, and on the viability of farming as a way of life.

(continued)

Box 3.3b (continued)

They point to the disruption of people's relationships with places and an increased population drain from rural areas into the city. They also question the wisdom of Australia's growing dependence on cheap imported food, which merely externalises the social and ecological costs of production to poorer nations.[2]

By creating a system that narrows water ownership to primarily commercial and private interests, water trading excludes not only many farmers and non-commercial water users, but also rides roughshod over indigenous interests, making little room for alternative and more collective relationships with water and local environments (Altman 2004). Quite apart from the wider moral question as to whether water should be treated as a common good or as a private asset, it is necessary to ask if water trading genuinely serves the interests of the majority of the Australian population or has any real potential to enable more sustainable social and ecological relationships.

Notes

1. There are numerous ethnographic texts describing Aboriginal Law, the worldview or 'cosmology' that underpins indigenous ways of life in Australia (e.g., Morphy 1992; Rose 2000).
2. This material is drawn from extensive interviews with farmers in Queensland, conducted as part of an Australian Research Council-funded project, 2003-2008 (see Strang 2009).

Resources

Altman, J. 2004. Indigenous interests and water property rights. *Dialogue* 23(3): 29–34.

COAG (Council of Australian Governments). 2004. *Council of Australian Governments' Meeting*, 25 June 2004.

Hann, C. (ed.) 1998. *Property relations: Renewing the anthropological tradition.* New York: Cambridge University Press.

Council of Australian Governments. http://www.coag.gov.au/coag_meeting_outcomes/2004-06-25/index.cfm#nwi. Accessed 15 June 2010.

Isaac, M. 2002. *To market, to market: Why dogma hasn't worked with water.* Brisbane: The Brisbane Institute.

Ladson, T., and B. Finlayson. 2004. Specifying the environment's right to water: Lessons from Victoria. *Dialogue* 23(3): 19–28.

Morphy, H. 1992. *Ancestral connections: Art and an aboriginal system of knowledge.* Chicago: Chicago University Press.

Ostrom, E. 1990. *Governing the commons.* Cambridge: Cambridge University Press.

Parry, B. 2004. Bodily transactions: Regulating a new space of flows in 'bio-information'. In *Property in question: Value transformation in the global economy*, ed. K. Verdery and C. Humphrey, 29–48. Oxford: Berg.

(continued)

Box 3.3b (continued)

Rose, D. 2000. *Dingo makes us human: Life and land in an Australian aboriginal culture.* Cambridge/New York: Cambridge University Press.
Strang, V. 1997. *Uncommon ground: Cultural landscapes and environmental values.* Oxford: Berg.
Strang, V. 2009. *Gardening the world: Agency, identity, and the ownership of water.* Oxford: Berghahn.

Box 3.3c Acequia communities and the politics of water

—Sylvia Rodríguez

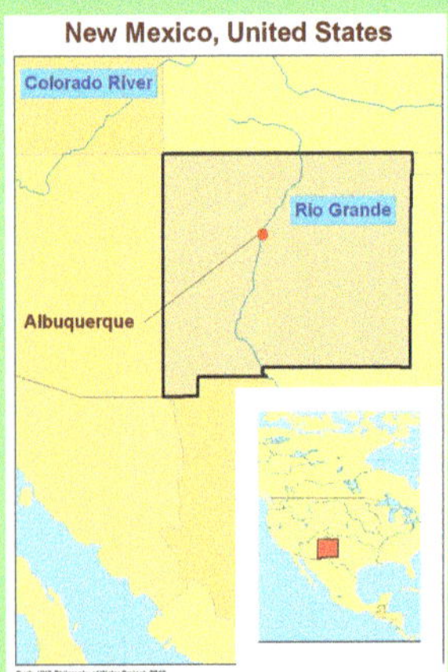

Box Map 3.3c.1 New Mexico, United States

Anthropologist John Wagner argues that a simplistic opposition between the social imaginaries of water-as-commodity and water-as-commons can obscure the complex ways that multileveled systems of governance actually operate. Seen holistically, water governance in any locale will, in Wagner's words, 'encompass a messy clustering of institutions, economic interests, histories, cultures, ethnicities' (2009:6). This observation certainly characterises the complex reality that is water governance in New Mexico (USA). The prospect of capturing the full canvas of New Mexico's divergent, competitive, yet interpenetrating systems of water governance into a single analytic framework is challenging. Significant insight can nevertheless be achieved by paying attention to the human face of water management and exploring the many varied dimensions of water as it is valued, managed, accessed, and controlled (Rodríguez 2006).

The struggle for water has always shaped human interaction in the upper Rio Grande Valley. Securing a safe and reliable water supply in the high desert requires cooperation but generates competition. Competition for this scarce resource and limiting factor grows more intense with every passing year, exacerbated by population growth, urban development, drought, and climate change. Different approaches to the management and allocation of water are historically rooted in different kinds of subsistence technology and social organization. Water management is never a technical matter alone. It is political, social, and ideological. It is famously an exercise of power through governance, in one

(continued)

Box 3.3c (continued)

Box Fig. 3.3c.1 Acequia, Santa Fe, New Mexico (Photo courtesy of Sylvia Rodríguez)

form or another. In the modern world this power is never absolute but infused with tension and conflict between local interests, the state, and market forces.

Centralised and bureaucratic water management policies in New Mexico have been challenged, and to a modest degree, transformed as citizens attempt to reassert the principles of local *acequia* governance and water ownership in a system now maintained and controlled by the state.

In New Mexico, the Arabic-derived word *acequia* refers to both a canal structure and a social institution whereby river water is diverted and distributed via gravity flow among a community of irrigators or water rights owner called *parciantes*.

Historically, acequias made possible the Spanish colonial settlement of the semi-arid upper Rio Grande Valley, starting in the late 16th century. Acequia systems in the northern counties of the state continue to function as autonomous irrigation communities, while at the same time they constitute legal subdivisions of the state. Dating from the 17th, 18th, and early 19th centuries, parciantes' rights depended upon water flows in naturally operating systems fed by snowmelt, healthy watersheds, natural river flow, and replenished aquifers. Such waterscapes are universally endangered. Like much of western U.S., New Mexico has seen massive, federally funded dam and canal construction projects built to support cities and industrial agriculture. Thus, historical rights may exist yet access to water is increasingly threatened or even lost. Many traditional rights-holders turn to – or are forced to – the courts for some semblance of justice, a costly process that can move at a glacial pace. Alternative strategies are, however, emerging.

Consider, for example, what is happening with water management in the city of Albuquerque. New Mexico's most populous city, it is the major beneficiary of two of the state's largest hydraulic engineering achievements: the Middle Rio Grande Conservancy District (MRGCD) and the San Juan-Chama Diversion Project.

(continued)

Box 3.3c (continued)

The MRGCD was established in 1925 as a flood control and agricultural reclamation project that created the massive system of weirs, canals, drains, levees, and jetty jacks that parallel the river that runs through the city from Cochiti Pueblo to Elephant Butte dams. It replaced traditional acequia irrigation systems with a powerful administrative bureaucracy whose inner workings have become so opaque that disgruntled tax-paying constituents, farmers and non-farmers alike, have recently called for greater transparency, accountability, and a selective redrawing of its powers.

The San Juan-Chama Diversion project transports water from the Colorado River through mountain tunnels across the Rocky Mountains, via the San Juan and Chama tributaries into the Rio Grande. This 'new water' is stored in reservoirs on the Chama and the Rio Grande. Because it brought 'new water' into a fully allocated stream system, San Juan-Chama triggered water rights adjudications. Brought by the state engineer, these lawsuits seek to determine the nature and extent of all water right claims on the Rio Grande and its tributaries. These lawsuits require acequia associations to defend in court the water rights claimed by individual parciantes. Because water rights in western U.S. rest on the principle of prior appropriation, in which the first user has priority, these lawsuits also oppose aboriginal (mostly Pueblo Indian) rights to all junior claims, starting with acequia communities. The adjudications are thus designed to produce a ranked priority valuation that constitutes a key objective in the relentless process to 'meter, measure, and market' every drop of water in the state.

All future growth along the Rio Grande corridor depends on leased or purchased San Juan-Chama water or the transfer of water rights away from agricultural to industrial and metropolitan uses. The cumulative impact of these developments, compounded by the steady decline in small-scale agriculture since World War II, certainly seems to portend the demise of New Mexico's acequia communities. Yet many acequia associations are resisting this prospect through an evolving, increasingly sophisticated combination of traditional and innovative organizational tactics.

Box Fig. 3.3c.2 *Sacando la acequia* (digging out the ditch) in Albuquerque's South Valley (Photo credit: Syliva Rodríguez)

Ten years ago for example, no one would have predicted that a grassroots effort to reconstitute acequia associations would emerge in Albuquerque's vast, poor, semi-rural South Valley. In early 2009 organisers of the South

(continued)

Box 3.3c (continued)

Valley Regional Acequia Association began to *sacar* or dig out old lateral irrigation ditches that had not been usable for decades. Such efforts represent the organization's goal of delivering water to small farm plots via ditches dug before the U.S. conquest in 1848, ditches that they claim the MRGCD has failed properly to maintain. Organisers are also helping parciantes declare their water rights to prepare for the impending adjudication of the conservancy district.

Comparable defensive grassroots efforts have emerged elsewhere in the state with the help of the New Mexico Acequia Association (NMAA) and its *El Agua es la Vida* or Water is Life Campaign and its Acequia Governance project. According to the NMAA website, the purpose of the *Agua es la Vida* campaign is

> to defend water as a community resource by challenging the economic and political forces that are resulting in the commodification of water. We view water as a '*don divino*', a divine gift that must be treated with respect and reverence. We are cultivating acequia leaders, allies, and youth leaders around basic principles through our *Agua es Vida Declaration.*

The Acequia Governance project gives workshops for acequia associations to develop or refine existing bylaws and thereby strengthen their legal footing. 'This programme builds on the centuries-old acequia legacy by strengthening self-governance of acequias with regard to water management, water rights defence, and water policy advocacy'.

Acequia proponents explicitly contrast their traditional principles of water-as-commons governance with the water-as-commodity policies of the state engineer (Garcia and Santistevan 2008:117). Such concerns are reflected in the *Bendición de las Aguas* or the Blessing of the Waters that immediately preceded the state engineer's arrival at the 2008 *Congreso* (annual meeting of regional acequia coalitions made up of local associations).

Box Fig. 3.3c.3 *Bendición de las Aguas* (Blessing of the Waters) the 2008 Annual *Congreso de las Acequias* (Photo credit: Sylvia Rodríguez)

This ritualised event (repeated in 2009) involves a ceremonial pouring together of waters brought from streams and acequias around the state. Delegates are asked to bring a small container of water from their acequia, stream, well, or tap. Each stands in line, waiting to state his or her name, community and the ditch or stream the water came from and then pours it into a large *olla* or ceramic pot that sits on a table near

(continued)

Box 3.3c (continued)

the podium. Some delegates invoke a blessing or protection for their community water supply, whereas others identify a threat to its welfare. This symbolic enactment of the water commons reinforces a sense of local, personal participation in a larger whole. It has become a prominent feature of the *Congreso's* 2-day programme, which also includes workshops on practical and policy matters. This merging of the practical and *communitas*-oriented events reflects the emergent, hybrid character of New Mexico's modern acequia movement.

Resources

Garcia, Paula, and Miguel Santistevan. 2008. Acequias: A model for local governance of water. In *Water consciousness*, ed. Tara Lohan, 110–119. San Francisco: Alternet Books.

New Mexico Acequia Association. http://www.lasacequias.org/. Accessed Dec 2009.

Rodríguez, Sylvia. 2006. *Acequia: Water sharing, sanctity and place*. Santa Fe: School of American Research Press.

Wagner, John. 2009. Water as commons and commodity in the twenty-first century: Two competing imaginaries. Paper presented at the American Anthropological Association Annual Meeting, Philadelphia, 5 Dec 2009.

Chapter 3.4
Manufacturing Water Scarcity, Generating Environmental Inequity

Barbara Rose Johnston

3.4.1 Water Scarcity Is Relative…

Water scarcity in popular terms suggests a state of immediate or impending crisis resulting from supply of water inadequate to meet the varied demands of humans and their environment. Scarcity is relative.

Freshwater supply is affected by changes in the hydrologic cycle: the amount of water entering the system; the volume of water captured and stored in surface and subsurface reservoirs; the amount of water that runs off land, enters rivers and streams, and is eventually lost to the oceans; and the amount of water held by vegetation and released into the atmosphere through evapotranspiration. Human activity also affects water availability. Some 69% of the world's freshwater budget is used for irrigated agriculture, which, in turn, is responsible for 70% of the world's water pollution. Fresh water may be in abundant, but safe drinking water may be scarce as a result of biological and chemical contamination from agriculture, extractive and manufacturing industry, and urban life.

Water demand is also a relative construct. Demand may simply reflect numbers – increasing populations require greater amounts of water to meet basic human and household needs. Even in the case of increased population, however, changes in water use behaviour and technology may represent a source of increased demand, but they may also decrease total water consumption. Perception of scarcity of water for irrigation, for example, may dissipate when farmers change the crops they grow and the technology they use, thus reducing their water demands.

B.R. Johnston (✉)
Center for Political Ecology, Santa Cruz, CA, USA
e-mail: bjohnston@igc.org

B.R. Johnston et al. (eds.), *Water, Cultural Diversity, and Global Environmental Change: Emerging Trends, Sustainable Futures?*, DOI 10.1007/978-94-007-1774-9_20,

Perception of water scarcity not only reflects relative supply and demand, but also the cultural and economic values associated with water. Access to water, patterns of water use, and ability to influence water management and distribution all shape an individual or community's perception of water scarcity or abundance. Thus, scarcity might reflect a person's economic ability to pay for water, or the customs, social conditions, and relationships that privilege access for one person or group while withholding access from others. And, while water may be plentiful, human activities can significantly degrade supplies and effectively limit access to safe water.

A case in point is the local consequences of industrial development. Over the past 50 years global industrialization has consumed increasing amounts of water, generating hazardous wastes that pollute rivers and lakes, and introduced toxic chemical and mineral contaminants into groundwater supplies. Often the economic benefits of this industrialization are prioritized over the rights, concerns and well-being of the communities living in affected watersheds. Such priorities often result in degraded water supply and, given historical conditions and social relations, increasing scarcity of safe water, and inequitable exposure to environmental hazards.

3.4.2 Biocultural Effects of Water Pollution – A Case from New Zealand

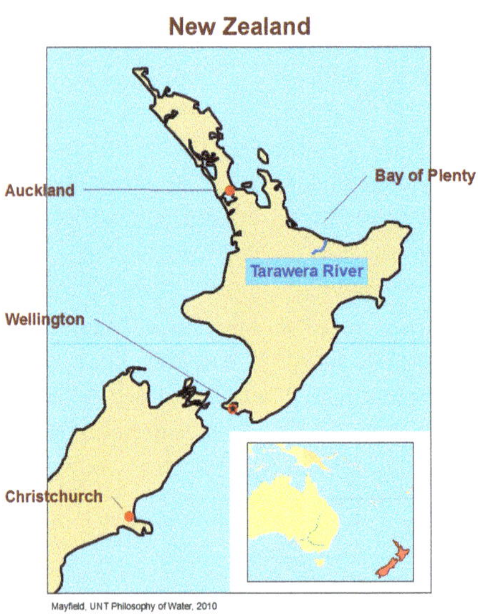

Map 3.4.1 New Zealand

Consider for example the biocultural effects of effluent discharge from pulp and paper mills on the Tarawera River in New Zealand documented by Materoa Dodd (Dodd 2010).

Located in the Bay of Plenty of the North Island of New Zealand, the Tarawera river is home to three iwi (tribes): the Tuwharetoa ki Kawerau who live on its banks and around the greater Kawerau area, the Te Arawa of the subtribe Tuhourangi who inhabit lands near the source of the river at Tarawera and its outlet at the town of Matata, and the Ngāti Awa iwi reside along the tributaries to the east of the river. The river, its tributaries, and its near shore environs are the heart soul of these cultures. The river carries the birthright of chiefly lines, is central to the history and legends of each iwi, is a symbol of mana (prestige) amongst the respective iwi, with each tributary and its landmarks contributing to the social personae of the tribe, and the river's abundance historically fed the community and allowed the gracious hosting of guests. The relationship between iwi and the river was traditionally controlled through customs and practices that conserved the river as a source of social and spiritual life. This meant demonstrating respect for the mauri (life force) of the river; respect for the river's waahi tapu (sacred sites); the gathering of particular foods in appropriate seasons and times; and the avoidance of despoiling or destroying the river and its life. When the iwi cared for the river, it treated them most generously. These points are further illustrated in Box 3.4a below.

Box 3.4a Water, indigenous

—Marcus Barber and Sue Jackson

The Pilbara is a remote arid region of Australia with a significant Indigenous population, rich mineral resources and rapid rates of mineral resource development.

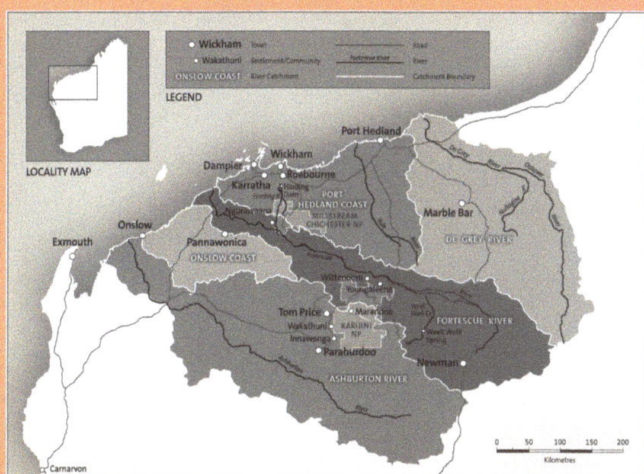

Box Map 3.4a.1 Pilbara region of north-west Australia

Pilbara Indigenous people claim deep ongoing connections to the land and waterscapes of the area and value water sources and features for a range of sociocultural, economic, and environmental reasons. Those water sources have come under increasing pressure from a massive new phase of development in the iron ore and offshore gas mining sectors in the region. Water shortages are already critical, and in places, current supplies are projected to be exhausted in the next few years, even as new mining resource developments continue to be approved to feed growing demand in Asia for raw materials. The Western Australian government recently responded to increased demand by announcing a major new desalination plant, but it is yet to be built, is powered by non-renewable energy and may be sited on the culturally and environmentally

(continued)

Box 3.4a (continued)

Box Fig. 3.4a.1 Mine at Tom Price (Photo credit: Copyright Auscape, reproduced with permission)

sensitive Burrup peninsula. In other parts of the region an excess of water presents problems for miners who require access to mineral ore below the water table. Disposing of excess water at these sites can bring about hydro-ecological changes and consequent social impacts.

At present the mining sector accounts for a substantial 78% of total Pilbara water use and this volume is expected to increase dramatically in coming years (Department of Water 2010). All parties with a stake in the area are searching for solutions to the major challenge this growth presents and in the absence of detailed government research, one company in particular, Rio Tinto Iron Ore, is consulting widely with the Indigenous people of the region about how best to manage the water issues this new phase of development has created.

For Pilbara Indigenous people, water is part of the creative legacy of the ancestral beings, it is an elemental resource for life, and it reflects and constitutes of group and individual identity by relating people across time and space. People directly associated with water places have responsibilities to sustain and protect them; contemporary guardians feel they have obligations to ancestors, subsequent generations and living people with kin and historical connections to those places. Senior elder Slim Parker says:

> Places that are of importance are connected to water, and the water provides for us when we conduct our ceremonies. Important places are known as thalu sites, and Inda is the word for the waterholes made by the snake, where the snake is. These are the most important places for us.

Care of water sites also implies a responsibility to those living downstream – water is not a static resource associated with a particular place alone, but part of an interconnected series of surface and subterranean flows, as another elder Brendan Cook describes:

> Upstream and downstream are responsible to each other. We are all linked together; we are not really independent people. If you are not doing the right thing on your lands, then the effects go downstream.

Pilbara Indigenous people are concerned about the intensifying use of water occurring on top of impacts that they have already experienced from mining and colonisation. These include both impacts in particular sites such as the creation of the Harding dam and the large scale water extractions from the Millstream (Jirndawurrina) aquifer, but also the general drying of the country,

(continued)

Box 3.4a (continued)

Box Fig. 3.4a.2 Drying waterhole on Robe River (Photo credit: Marcus Barber/CSIRO)

as a senior woman from the Thalanyji explains:

> The Pilbara looks different now from when I was younger. Then the river systems were all full, now they are drying up. When you drive down the Fortescue it used to be beautiful, but now it looks like a floodplain. We see the drying and know that the water serpent belongs to the river area. We are wondering what is happening to it. There are now hardly any food sources, fish, native plants, and wild onions anymore. We used to be able to get food all along the river.

Long-term knowledge of the region lies with its Indigenous people, and they feel that they have a responsibility to their descendants to work towards solutions to the current problems:

> And if this happens, what are our future generations going to think of us in terms of allowing this to go ahead? That is something that we have got to carry with us and hope that we are still able, to scientifically, I suppose, work together with people, who are from those professional fields, to enable us to monitor and see what's happening here and to try to avoid a catastrophe (Slim Parker).

The Pilbara is an economically crucial region, not just to Australia but to the wider global economy. The scale of the water use pressures and the depth of feeling amongst its Indigenous residents emphasise the need for greater resource allocations and engagement by those involved in mine water management and regional water planning, as well as by governments at every scale from the local to the national. Appropriate and sustainable new infrastructure is required, but it alone cannot solve the current challenges. Any lasting solutions need to be developed with those with the greatest long term stake in the region; those whose ancestors were residents long before the first mine opened, and who believe their descendants will still be living there long after the last mine has closed.

Reference

Department of Water. 2010. Pilbara Regional Water Plan 2010-2030. Perth. Western Australian Department of Water, Perth.

Further Information

Barber, M. and Jackson, S. in press. Aboriginal water values and resource development pressures in the Pilbara region of north-west Australia. Australian Aboriginal Studies.

Beginning in the 1950s, Tasman Pulp and Paper Mill and the Caxton Mill at Kawerau discharged effluent into the Tarawera River, directly for the first 20 years, and then beginning in the 1970s through slow seepage from sludge settlement bonds. The ecosystemic health of the river changed and the place took on a new name, the Black Drain (sludge seepage blackened the rivers waters). Fish and local plants became contaminated or disappeared. Water became unsafe even for bathing, let alone consumption.

As conditions deteriorated, the relationship between iwi and the river changed. While people still occasionally gather some of the traditional foods, there are hazards in collecting as well as consuming such food. The inability to gather customary food and herbal medicines undermines diet, health and social customs. Similarly, the inability to enjoy the river, to drink it, to swim in it, to play in it, undermines community cohesion as families move away from the river to nearby towns, or further, to avoid living anywhere near the polluted river.

Box 3.4b Legacies and challenges in integrating cultural diversity in the Sierra Leone water sector

—Fenda A. Akiwumi

In Sierra Leone, incorporating cultural diversity into water management entails balancing the needs of industry and of the approximately one million residents of the capital, Freetown, and several thousands of residents in provincial headquarter towns such as Bo and Kenema, with the needs of the many distinct indigenous peoples who have depended on and managed natural water sources throughout their territories for generations. Including such disparate 'stakeholders' in contemporary water management is critical to sustainable oversight of water resources (Akiwumi 2005), but it is a very difficult task in practice.

Of Sierra Leone's population, 67% is rural; as there is no modern water infrastructure in the countryside, rural life and livelihoods depend entirely on surface water sources, springs, and wells. Customary laws typically govern water use in such areas. For example, natural water sources must be freely accessible to all for drinking, animal watering, domestic purposes, fishing, and navigation. Damming rivers is strictly prohibited, as it disrupts fish breeding cycles (Fenton 1948). Natural water bodies such as springs, rivers, and lakes are often sacred sites. Sacred river spaces are central to *Sande,* the traditional religious-administrative organization that governs the affairs of 90% of indigenous women (Akiwumi 2005).

Rural peoples increasingly find that the urban demand for water, state water infrastructure development such as hydroelectric power schemes, and the mining industry, which requires large amounts of fresh water for operation, affect

(continued)

Box 3.4b (continued)

their customary practices of water use and management. Are all uses and users equal? Adjudicating between different water uses and designing a sustainable approach to water management requires disentangling the large number of ad hoc arrangements governing water provision, multiple layers of regulation, contradictions in statutory law, and poorly integrated water resources management.

Historical legacies complicate the contemporary situation. State planning in Sierra Leone under British colonial rule from 1808 to 1963, as in Britain's other colonies, was ad hoc (Crowder 1968), and resource management in the period since independence has been fragmented and poorly conceived. The constitution of 1978 called 'for the respect of customary law and usage as well as for the conservation, development and utilization of natural resources to be based on the principle of community interest' (FAO/UN 1979:163). The current, 1991 Sierra Leone Constitution aims 'to harness all the natural resources of the nation to promote national prosperity… to secure the maximum welfare and freedom of every citizen on the basis of social justice and equality of opportunity' (Sierra Leone Government 1991).

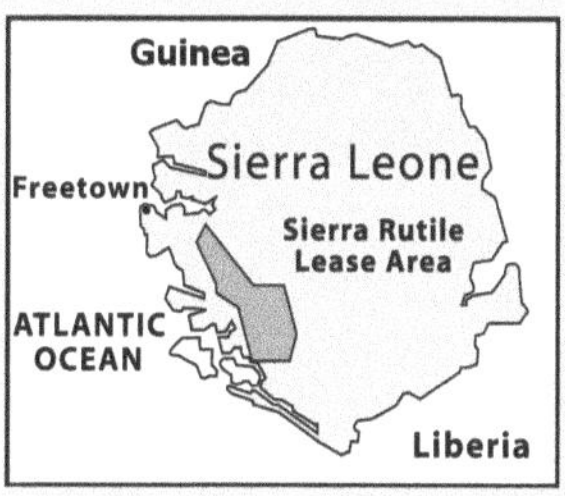

Box Map 3.4b.1 Rutile mining area, Sierra Leone

Multiple layers of legislation, some of it contradictory, apply specifically to water. Under the Water (Control and Supply) Act of 1963, the government claimed all natural water supplies. The Sierra Leone Citizenship (Amendment) Act of 1976 stated that water rights in provincial (and especially indigenous) areas should be regulated by customary law. This law leaves open the question of how conflicts between customary use, 'community interest', 'national prosperity', and 'equality of opportunity' should be balanced.

Sierra Leone is rich in minerals and has long been attractive to international mining interests. Since alluvial mining began in the early 1930s, and water began to be treated as an industrial input, there has been conflict between the indigenous peoples' water practices and attitudes and the institutional view of the colonial or independent state of water as an 'economic good'. The colonial government simply appropriated mineral and water resources, declaring that 'the entire property in all minerals and of all rivers and watercourses resides in the Crown' (Sierra Leone Government 1932). The Republic of Sierra Leone continues to hold such powers through current mining laws, such as the Mines and Mineral Act of 1994, in spite of some legislative advocacy in favour of respecting customary water and other resource rights and practices. Nevertheless, state legislation often contradicts and overrides customary use.

(continued)

Box 3.4b (continued)

Box Fig. 3.4b.1 This dredge-pond, contaminated with mine tailings, is used as the village water supply in Kpetema, located 100 m from Mogbwemo (Photo credit: Fenda Akiwumi)

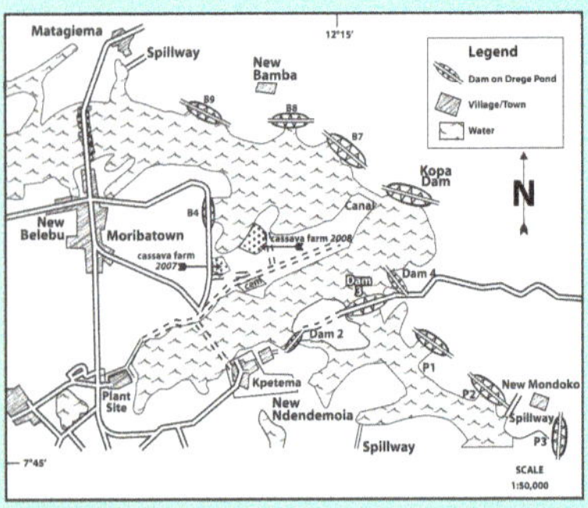

Box Fig. 3.4b.2 A cassava garden is growing on the mine tailings that extend to the edge of the village

Today, indigenous people and their livelihoods are at the heart of conflict over the desecration of sacred water sites, the sources of domestic water supply, the loss of swampland for rice cultivation, and the loss of riparian resources and livelihoods. The source of this conflict involves rutile mining in southwestern Sierra Leone. Rutile is a mineral that can be processed into titanium. Reservoirs critical to the mining effort submerge riparian resources and hamper livelihoods like the shallow water scoop-net fishing practised by some women.

Legislation is often rife with contradiction and loopholes. Dredge-mining laws, for example, allow the damming of rivers and water diversion (Sierra Leone Government 2002), practices that are strictly prohibited under customary law in order to prevent disruption of fish breeding cycles (Fenton 1948). In other cases, state laws restrict access to water bodies in alluvial mines. Such laws simply override customary regulations that require natural water sources to be freely accessible (Wohlwend 1978). Again, there is conflicting legislation: the Chiefdoms Councils (Farming Areas) Order of 1955 gives a holder of mining rights the freedom to manipulate water sources ‘provided he does not interfere with existing rights’ (FAO/UN 1979:172–173).

Currently, no single administrative or legislative body coordinates water issues; responsibility for water management remains divided among various ministries and departments. The unequal power relations between traditional systems of political and social authority (chiefs) and the centralised state bureaucracy make it difficult to merge statutory and customary law. Though the bureaucracy often acknowledges customary rights, the state officials simply overlook them when customary claims conflict with goals of industrial

(continued)

Box 3.4b (continued)

development. Customary authority rests on consent and consensus within and between social groups; it has little capacity for rule enforcement. National authorities, on the other hand, have extensive infrastructures for rule enforcement and can overrule customary law; chiefs mostly function as facilitators rather than partners in decision making. Even when specific statutory laws drawn up to support customary practices with regard to fishing rights, the watering of livestock, or water use, other national government legislation may supersede such laws (FAO/UN 1979). A comprehensive review and harmonization of water resources legislation is necessary (Ndomahina and Kabia 2004).

Continuing international interest in extracting Sierra Leone's minerals, which would require large amounts of water, highlights the challenge of integrating different cultural perspectives in water management. Industrial demand for water is not clearly integrated into overall water planning and management, and the Ministry of Mineral Resources does not have clear jurisdiction in water policy or use (Ndomahina and Kabia 2004). Nevertheless, much of the country's water-related legislation revolves around the alluvial ores of interest to the mineral industry (Akiwumi 2003).

In 2008, the World Bank sponsored the Mineral Sector Technical Assistant Project (MTAP) to strengthen the Ministry of Mineral Resources in Sierra Leone and to create a blueprint for future mining ventures (World Bank 2008). The Environmental and Social Impact Assessment (ESIA) recommended that mining endeavours in Sierra Leone comply with relevant World Bank safeguard policies, especially those regarding management of cultural property and projects in disputed areas, though not indigenous peoples. The cultural recommendations in the report were limited to educating expatriates and other non-indigenes about local cultural practices. Exactly *how* conflicting resource claims should be resolved remains ambiguous in the report: 'any cultural or sacred resources when impacted shall be mitigated in line with traditional rites and national laws' (Ndomahina 2008:17). ESIA made no linkages between traditional culture and water.

The UNESCO-International Hydrological Programme's expert advisory group for the project on Water and Cultural Diversity advised that 'governments…should create an enabling environment through appropriate reforms of water policy and legislation in such a way that customary law, informal water use practices and cultural diversity are formally recognised and accommodated' (UNESCO-IHP 2008). Including cultural diversity in contemporary water planning – that is, giving equal legal status to customary water use – remains a challenging task. Clearly, the Sierra Leone government must address the amalgam of customary and statutory law in order to reform the water sector and successfully integrate cultural diversity into water management.

(continued)

Box 3.4b (continued)

Resources

Akiwumi, Fenda A. 2003. Roots of conflict: Water, worldview and sustainability in a West African environment. Paper presented at the UCOWR, NIWR, and EWRI 2003 Joint Conference: Water Security in the 21st Century. Washington, DC, 30 July–1 Aug.

Akiwumi, Fenda A. 2005. Indigenous people, women and water: The importance of local knowledge in water project planning in an African context. In *The business of water and sustainable development*, ed. Jonathan Chenoweth and Juliet Bird, 188–197. Sheffield: Greenleaf Publishing.

Crowder, Michael. 1968. *West Africa under colonial rule*. London: Hutchinson and Co., Ltd.

FAO/UN (Food and Agricultural Organisation/United Nations). 1979. *Water law in selected African countries (Benin, Burundi, Ethiopia, Gabon, Kenya, Mauritius, Sierra Leone, Swaziland, Upper Volta, Zambia)*. Legislative study no. 17, ed. Dante A. Caponera, 162–205. Rome: Food and Agricultural Organisation of the United Nations.

Fenton, J.S. 1948. *Outline of native law in Sierra Leone*. Freetown: Government Printer.

Ndomahina, E.T. 2008. *Environment and social impact assessment (ESIA) study: Republic of Sierra Leone Mineral Sector Technical Assistant Project (MTAP)*. Government of the Republic of Sierra Leone and World Bank. http://www.slmineralresources.org/int_perspective/ESIA_mineralfinal.pdf. Accessed June 2010.

Ndomahina, E.T., and I.S. Kabia. 2004. *Review of the water supply and sanitation sector in Sierra Leone*. Pasco: Government of Sierra Leone, Poverty Alleviation Strategy Coordinating Office.

Sierra Leone Government. 1934. *Sierra Leone: Reports of the geological and mines department for the year 1932*. Freetown: Government Printer.

Sierra Leone Government. 1955. *The chiefdoms councils (Farming areas) order*, 1955 (Cap. 61) Section 2. Freetown: Government Printer.

Sierra Leone Government. 1960. *Laws of Sierra Leone*, vol. VII. Freetown: Government Printer.

Sierra Leone Government. 1978. *The constitution of Sierra Leone*. Freetown: Government Printer.

Sierra Leone Government. 1991. *The constitution of Sierra Leone*. Freetown: Government Printer.

Sierra Leone Government. 2002. *The Sierra Leone Rutile agreement, 2002*. Freetown: Government Printer.

UNESCO-IHP. 2008. Report of the second meeting of the expert advisory group of UNESCO-IHP project on Water and Cultural Diversity, held at Cranfield University, Cranfield (3–4 July 2008). Available online http://www.unesco.org/water/ihp/pdf/0708WCD_EAGmtgdraft2_report_final_with%20annexes.pdf. Accessed June 2010.

Wohlwend, B. 1978. *Water legislation in Sierra Leone*. Consultancy report to FAO-UNDP SIL/73/002/LRSP-MAF.

World Bank. 2008. *Sierra Leone mining sector reform: A strategic environmental and social assessment*. Report no. 46655-SL. Washington, DC: World Bank, Sustainable Development Unit, West Africa and Environment Department.

The chemicals required to process pulp and paper and the nature of this industrial process meant that high rates of toxic substances – organochlorines, dioxin, and petroleumwere discharged into the air, soils and the river. Exposure through contact, respiration, ingestion and immersion has undermined the health of surrounding Maori communities. Complaints include increased rates of asthma and other respiratory diseases, skin diseases, particularly skin cancers, early death and foetal and birth problems.

While there are provisions in New Zealand's Resource Management Act that recognize Maori rights and responsibilities as stewards and guardians of the land and thus their rights to participate in the management and use of that land, the right to discharge toxics was granted by the State as recognition of the economic benefits of the industry and the local jobs that such industry brings. Local and national complaints over industrial practice and its obvious impact on the ecosystem were, until very recently, ignored.

After decades of protest, in March 2010, the Ngati Awa, Ngati Rangitihi and Ngati Tuwharetoa side-stepped the Resource Management Act process and signed a memorandum of agreement with the owners of the Tasman Pulp and Paper Mill in Kawerau, one of the key sources of toxic effluent discharged into the Tarawera River. The agreement establishes a formal consultation process between the industry and the iwi acknowledging their status as rights holders (tangata whenua, people of the land). In establishing a formal consultation process, the company is demonstrating its compliance with the terms of its new 25-year pollution permit signed in October 2009. Through consultation it is hoped that the industry take direct remedial action, such as discontinuing dumping of chemicals directly into the river and installing new technologies to clean effluent before it is discharged (Davison 2010). Whether consultation truly respects iwi rights and leads to ecosystemic restoration remains to be seen.

3.4.3 Transnational Inequities and Water Pollution – Electronics and E-Waste

A second example of the sociocultural consequences of manufactured scarcity is demonstrated in the emergence of the high tech industry (Johnston 2006). In Silicon Valley, California, USA, electronics companies consumed 24% of the city of Santa Clara's water (1994-1995) and produced 65% of the wastewater discharged. By the late 1990s, the production of each 15 centimetres silicon wafer chip required 8611 litres of de-ionized water, 9 kilogrammes of chemicals and 623 litres of various gases. Arsenic, trichloroethylene, and 1,000 or more chemicals were used at that time to manufacture silicon chips (Trichloroethylene, for example, is a known carcinogen, and suspected cardiovascular or blood toxicant, developmental toxicant, gastrointestinal or liver toxicant, kidney toxicant, neurotoxicant, reproductive toxicant, respiratory toxicant and skin or sensory organ toxicant). This chemical waste was stored in underground tanks and, when treated with solvents, leached, leaked, or slipped through the molecules of storage containers, percolating into the valley's water table (Schoenberger 2002).

The resulting contamination of Silicon Valley groundwater and its related health consequences were first publicly reported in 1982, when the Fairchild Semiconductor facility in San Jose was charged with leaking underground tanks and resulting trichloroethane contamination of groundwater supplies. Residents sued Fairchild Corporation, and the lawsuit stimulated epidemiological, environmental geology and toxicology studies that demonstrated health consequences, including a documented pattern of miscarriages, birth defects, increased cancers, and a host of debilitating disorders. The California State Department found three times the expected number of birth defects in the neighbourhood near the plant. The Regional Water Quality Control Board found 85% of the underground tanks in Silicon Valley to be leaking. In 1983, the County of Santa Clara developed the first Hazardous Materials Storage Ordinance in the country, regulating underground storage tanks and enacting public-right-to-know legislation. A statewide initiative was passed in 1984 based on the county ordinance, and similar federal legislation was adopted in 1986 (Johnston 2006; SVTC 2001).

Lawsuits, the eventual plant closure of the responsible party, and increased regulations involving the use of solvents and other hazardous chemicals and underground facilities resulted in profound changes in the way high tech businesses are run. When Fairchild closed its plant other high tech companies took notice: increased environmental protection brought increased costs. Companies began to move the dirtier aspects of manufacturing to more hospitable political settings. The groundwater problems that were originally concentrated in Silicon Valley are now emerging in the many manufacturing sites of a globalised industry. By 2003, almost all components of the desktop personal computer, including microprocessor and software, were manufactured in places like China and shipped back to the United States for final processing and sale.

The problems of high tech-generated groundwater pollution and related environmental health risks are not restricted to the manufacturing process. Rapid improvements in technology produce a limited life for computers and products containing computer components, and many functional products are discarded as "out of date" electronic waste. Landfills and recycling centres are typically unable to accept such material, however, as state and federal regulations prohibit disposal of e-waste given its hazardous content and threat to subsurface aquifers. Thus, an estimated 80% of electronic waste collected in the United States is shipped to scrap brokers in developing countries, mostly located in Africa and Asia, with the largest portion going back to China and India.

Recycling electronic waste is a messy and dangerous business, with by-products including heavy metals, radioactive elements, and a wide variety of toxic chemicals. Often this recycling occurs in primitive conditions, among the living space of the poorest populations. The experience of the Chinese town of Guiyu is illustrative. According to a study by Huo et al. (2007), some 60-80% of the town is involved in stripping down parts for their valuable elements, most of which is done in small family-run workshops. Tons of e-waste materials and process residues have been dumped in workshops, yards, roadsides, irrigation canals, riverbanks, ponds and rivers. The resulting extensive contamination of soil and water has produced serious health implications. Samples taken in 2002 from the Lianjiang River – Guiyu's public water supply – showed levels of lead to be 190 times higher than World Health Organization's drinking water thresholds, and river sediment samples demonstrated high rates of lead, zinc and chromium. Environmental health studies have found high levels of skin damage, headaches, nausea, chronic gastritis, ulcers and, most notably, significantly elevated levels of lead in children's blood. Children are especially vulnerable to permanent brain damage and other injury as a result of continued exposure to lead (Qiu et al. 2004; Huo et al. 2007; Puckett et al. 2002).

In the above example, the United States, while developing high domestic standards for disposal of hazardous waste and minimizing contamination of water sources, has yet to join the international community in banning the export of hazardous wastes. The Basel Convention on Control of Transboundary Movement of Hazardous Wastes and their Disposal opened for signature in 1989 and entered into force in 1992. As of 2010, Afghanistan, Haiti, and the United States are the only signatories of the Basel Convention that have not deposed instruments of ratification (UNEP 2010). Without ratification, there are no implementing laws, and the U.S. can continue to legally export hazardous waste. The result is transnational environmental inequity, where first world consumers dispose of e-waste toxins in the impoverished recipient communities of export nations. Such practice reflects the socioeconomic inequities between nations as well as the contradictions of a cultural mindset that deploys a "NIMBY" (not in my backyard) rationale. The notion that 'what is unacceptable in my backyard is perfectly acceptable when disposed of in a less-developed nation' suggests that the fundamental human right to water, as currently experienced, is a highly privileged right. These points are further illustrated in the following example from the Marshall Islands.

Box 3.4c The vulnerabilities of native peoples in the Mackenzie and Athabasca drainage systems: Tar sands, gas, and uranium

—Alexander M. Ervin

3.4c.1 The drainage system and uranium

The Mackenzie River drainage system, after those of the Amazon and the Mississippi rivers, is the third largest in the western hemisphere. Flowing south to north, the Mackenzie River is 1,738 km long; its drainage encompasses 1,805,200 km^2, or one-fifth of the entire area of Canada, the world's second largest country. It contains the Firth, Liard, Peace, and Athabasca rivers and

(continued)

Box 3.4c (continued)

some of the largest freshwater lakes in the world, including the Great Bear, Slave, and Athabasca lakes, as well as hundreds of thousands of smaller lakes, muskegs, and tributaries. Boreal forest dominates the area, except for a few places in the Rocky Mountains, a small amount of farmland, and a tiny amount of tundra. The great forested area is an important carbon sink. It is a world of bountiful water and a habitat and breeding ground for many migratory water birds and fish. Many of the mammal species found there, such as beaver, moose, and muskrat, are especially well adapted to water environments.

Regarding humans, this vast area contains few – about 500,000 people live here, mostly upstream in the province of Alberta. There are a small number of large settlements (Fort McMurray at 80,000 people is the biggest by far), but only a few hundreds or thousands of people reside in most settlements. Much higher proportions of Native peoples live in the Mackenzie area than in most regions of North America; Natives are the majority in numerous northern communities. These include Inuvialuit in the Mackenzie Delta region, Dene along the Mackenzie Valley and into northern Alberta and Saskatchewan provinces, and Woodland Cree in northern Saskatchewan and Alberta. Métis peoples of mixed Native and European ancestry are everywhere. These indigenous peoples still depend on the waterways of the Mackenzie for hunting, fishing, trapping, collecting, and travelling.

Box Fig. 3.4c.1 Water is life (Artwork by Beryl Rink, Canada, grade 9)

Until recently this wilderness was relatively unspoiled, but this condition is changing more rapidly than can be managed or monitored. Development threatens on a massive scale: the zone contains or is adjacent to billions of cubic metres of natural gas, oil reserves second in size only to those of Saudi Arabia, and much of the world's high-grade uranium, as well as significant quantities of diamonds, copper, zinc, and gold. Ominously for many of its inhabitants, this region is now the centre of the 'Western Energy Corridor' that is supposed to make the United States and Canada 'energy independent'. When large-scale projects are realised in wilderness zones, especially fragile subarctic ones, they can be extremely disruptive to local and regional environments, as well as the communities of people living there.

(continued)

Box 3.4c (continued)

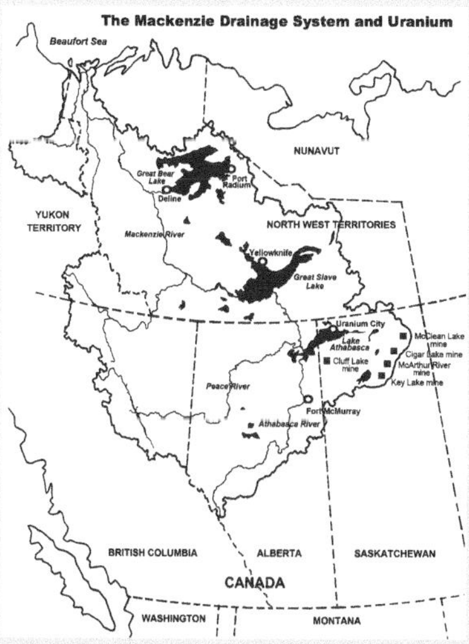

Box Map 3.4c.1 The Mackenzie drainage system and uranium

When these projects are located in water environments, they should attract special scrutiny, as these landscapes are particularly vulnerable.

Radium and uranium first attracted developers to the Mackenzie area. At one time, radium was the most valuable mineral on earth. In 1930 a lucrative deposit was discovered along the western edge of Great Bear Lake and mined at Port Radium. Though the mine was closed by 1940, the government-owned company Eldorado secretly reopened it in 1942. From this company the U.S. government ordered 850 tons of uranium (often associated with radium) for the Manhattan Project; 220 tons of this material were mixed with uranium from Colorado (USA) and the Belgian Congo (today the Democratic Republic of Congo) to supply the atomic weapons programme that culminated with the bombs dropped on Hiroshima and Nagasaki.

Eldorado employed Sahtugot'ine Dene men to carry uranium concentrate in cloth bags, load it onto boats, transport it across the lake, and then transfer the bags onto barges going southward. Though the men were in direct physical contact with the porous sacks for hours at a time, the company did not inform them in any way of the health risks posed by the material, did not issue them protective clothing of any kind, or tell them to shower carefully after work.

Later, the settlement of Deline on the eastern shore became known as the 'village of widows' because of the unusually high rates of cancer among men who had worked for Eldorado. Out of 30 former employees, 14 died young, their deaths depriving the community of able hunters and future elders to teach oral traditions and knowledge to younger men. The Canadian government has denied any responsibility for the deaths. Nonetheless, in 2002 the federal government agreed to clean up the abandoned mines and their toxic waste, or tailings, which were still leaking into the lake.

By the early 1950s, intensive uranium mining moved to Lake Athabasca in northern Saskatchewan. This province is now the world's largest uranium producer. Abandoned now, Uranium City was a major source through the late 1970s. Questions linger about uranium from tailings leaching into bays and about who is responsible for cleaning up the abandoned mines bordering Lake Athabasca.

(continued)

Box 3.4c (continued)

Mines still operate in the Athabasca region in the early 21st century, but they are located relatively far from settlements. Nonetheless water contamination disputes have arisen because of leaky tailing pits at McLean Lake and the 2003 mine cave-in and flooding of radioactive water at McArthur River. The Cigar Lake mine, which was considered one of the most advanced uranium mines in the world, has had to be shut down indefinitely because of massive flooding of the mine by groundwater.

The Saskatchewan provincial government, a major supporter of the industry, minimises the health and environmental hazards associated with mining, and a baseline health study of the affected Indigenous population has never been done. Nor has there ever been a solid study of the sociocultural impacts of over fifty years of uranium mining in northern Saskatchewan. These developments need to be much more rigorously monitored because of the long-term implications of the release of radionuclides and heavy metals into the complex hydrology of the groundwater, aquifer, river, lake, and muskeg systems in northern Canada.

3.4c.2 Gas and oil

Proposed in the 1970s, the Mackenzie Valley Pipeline was designed to deliver billions of cubic metres of natural gas southward from the Beaufort Sea, on Canada's northwestern border. Running for 1,800 km, it was to be the world's largest private construction project. A federal commission put a moratorium on the project, however, when it determined that the pipeline would threaten the fragile environment of the Mackenzie Valley and have a negative impact on Native peoples. The commission also noted the need to consolidate Indigenous treaty rights and protect wildlife.

The project is alive again and even this time with Dene First Nations governments as potential partners. Licensing and hearings have caused delays, but the pipeline should be operable by 2020. It is certain to have significant impact on the valley and its inhabitants and will also open up other gas and oil developments where there are known reserves along the route.

The recognised ecological impacts of the pipeline, though, will pale in comparison to the effects of the world's current largest private industrial project: the Athabasca heavy oil and tar sand extractions that stretch over an area of 94,000 km^2, an area roughly the size of the state of Florida, USA, and produce 1,000,000 barrels of oil a day. One big problem with oil and tar sand extraction operations is that it takes the equivalent of one barrel of oil to produce four to five barrels of oil. Some propose that natural gas from the Beaufort

(continued)

Box 3.4c (continued)

Sea would be a cheaper and cleaner fuel for processing these heavy oils and bitumen. Ominously, however, it has already been proposed that nuclear power be used, and communities willing to accept reactors along the Peace and Athabasca Rivers are already being sought. Such proposals raise further environmental questions about safety, radioactivity, and nuclear waste.

Box Map 3.4c.2 The Mackenzie drainage system and oil & gas

The environmental and social impacts of extracting oil from sand are already profound and likely to get worse, especially as they affect indigenous peoples. These facilities pollute: oil sands facilities are the leading producer of acid rain and carbon dioxide in all of Canada. Well-pads, pipelines, seismic cut-lines, industrial roads, as well as gargantuan clear-cuts, strip mines, and tailing ponds disrupt and blight thousands of square kilometres of the boreal landscape. Nevertheless, future oil and tar sands projects have already been approved that increase the size of operations five-fold.

Water environments are highly vulnerable to pollution and disruption, yet the long-term consequences of this production are still hardly known. Oil sands operations require huge amounts of water to remove bitumen. The existing projects draw up to 2.3 billion litres from the Athabasca River each year, an amount equivalent to that used by a city of 2,000,000. Once used, little of this water is recyclable and most may have to be kept in massive tailing ponds for an unknown number of years. When there is too much 'overgrowth' – when the deposits are located so far underground that clear cutting and strip-mining are not feasible – companies send pressurised steam through underground pipes to loosen and melt the bitumen, which is then brought to the surface using more conventional drilling methods. This process also requires extremely great volumes of water. Can the Athabasca River sustain this kind of intensive use, given reduced glacier melting because of climate change, low winter flows, and a five-fold increase in production?

For all life forms, the tailing ponds are lethally toxic. To an extent not yet fully determined, these tailings have leaked from ponds adjacent to the Athabasca River, with serious effects on wildlife and people downstream.

(continued)

Box 3.4c (continued)

One indicator of the downstream impacts of oil sands operations comes from Fort Chipewyan, a community of 1,200 Dene, Cree, and Métis peoples on Lake Athabasca. Highly dependent upon lake water and wildlife, they have noted deformed, soft-fleshed, and blemished fish that do not taste as they should, and the town residents have expressed deep unease with their drinking water. The local doctor discovered high rates of renal failure, lupus, leukaemia, hyperthyroidism, and cancer, including five incidences of a rare bile duct cancer not usually found in populations of even 100,000. The provincial government denied that any problem existed and attempted to discredit the doctor. In response, the town's community health board hired an independent ecologist (Timoney 2007), who established the presence of many contaminants in the river delta located just above Lake Athabasca. Contaminants of particular concern were arsenic, mercury, and polycyclic aromatic hydrocarbons, whose incidence in the delta was high and increasing in comparison to upstream locations. The people most at risk of poisoning were those who ate an abundance of wild, rather than store-bought, foods and consumed surface water. According to Timoney (2007), if the standards of the U.S. Environmental Protection Agency (U.S. EPA) were applied with regard to these contaminants, all of the abundant whitefish and walleye would be considered inedible.

As an example of the severe impact of this kind of development on socio-cultural viability, consider the case of the Lubicon Lake Cree First Nation. During the surveying and treaty-making period of the late 19th century, the government overlooked this group and did not clarify their land rights. Though the Lubicon Lake Cree claim an ancestral trapping and hunting territory of 10,000 km^2, since 1979 thousands of oil industry employees, almost 2,000 oil wells, and thousands of kilometres of pipelines and cut lines have invaded this land. In spite of entreaties, negotiations, roadblocks, boycotts, and censures by the World Council of Churches, the United Nations, and Amnesty International, the provincial and federal governments have failed to negotiate an agreement with the Lubicon people. Meanwhile, oil development has polluted fisheries, game has become scarce, and the future of Lubicon people's bases of subsistence and culture is at risk. The number of people suffering from tuberculosis, a bellwether of poverty and stress, has risen.

It is far too early to understand the full impact of ongoing oil, gas, and uranium development on the ecology of the Mackenzie and Athabasca drainage systems or the peoples who have long inhabited them. Past experience, however, leads one to suspect that the impacts and vulnerabilities are profound and need close monitoring. There are legitimate reasons for imposing moratoriums on any new oil sands projects until their full impact on local and regional environments is better understood.

(continued)

Box 3.4c (continued)

Resources

Edwards, Gordon. 2010. How uranium from Great Bear Lake ended up in A-bombs: A chronology. Canadian Coalition for Nuclear Responsibility. www.ccnr.org/uranium_events.html. Accessed 17 June 2010.

Martin-Hill, Dawn. 2008. *The Lubicon Indian nation: Indigenous knowledge and power*. Toronto: University of Toronto Press.

MiningWatch, Canada. http://www.miningwatch.ca. Accessed 20 June 2010.

Nikiforuk, Andrew. 1998. Echoes of the atomic age: Uranium haunts a Northern Aboriginal Village. *Calgary Herald*, March 14:1.

Nikiforuk, Andrew. 2008. *Tar sands: Dirty oil and the future of a continent*. Vancouver: Greystone Books.

Outaouais Lubicon Solidarity. http://www.lubiconsolidarity.ca/. Accessed 10 June 2010.

Timoney, Kevin. 2007. *A study of water and sediment quality as related to public health issues, Fort Chipewyan, Alberta*. Fort Chipewyan: Nunee Health Board Society.

3.4.4 Water/Culture/Power – Manufactured Scarcity in the Marshall Islands

The Republic of the Marshall Islands is made up of 29 coral atolls, spread across a watery nation the size of Mexico. Land is scarce – dry land on 1,225 islands and reefs totals 200 square kilometres. There are no rivers, streams, or lakes. In past times, the residents of these islands met their water needs by using natural or handcrafted catchments and storing water in giant clamshells. During the Japanese occupation (1914-1944), the Marshallese sometimes used ground coral for catchments and cisterns. On some islands, people maintained springs or dug wells to tap into the brackish waters that lie beneath the soil.

Although water was scarce, the Marshallese were not water-starved. Most of their daily water needs were met by consuming coconut water, pandanus, and succulent plants, and eating seafood – shellfish, small reef fish, tuna and other large fish, and turtles. People lived in small communities, moving several times a year to access critical resources. The low density and high mobility of the Marshallese minimised surface water contamination. All people had the right to access water, and exercising this right also meant respecting traditions – asking permission, honouring taboos, and accepting stewardship responsibilities. Everyone knew where the water sources were, the stories of the gods who watched over these sources and the potential consequences if taboos were broken. In customary law, the penalty for destroying a freshwater spring was death.

Conditions changed during the Cold War following the introduction of new hazards and the imposition of new taboos. During the Cold War, the United States conducted 67 atmospheric tests of nuclear weapons in the Marshall Islands. The fallout included high levels of radioiodine, strontium, cesium, radioiron, and other isotopes. Declassified data from the 1954 Bravo test reported dangerous levels of radioactive deposition on each of the 22 populated atolls. Extensive environmental monitoring by U.S. scientists occurred before and after the nuclear weapons tests. Afterward, radiation permeated the soil, plants, and the entire food chain. Many traditional foods – reef fish, coconut crabs, arrowroot, and coconut – became dangerously radioactive. Following the 1954 Bravo test of the hydrogen bomb, U.S. officials evacuated the exposed people of Rongelap and, after studying the effects of the test on them for 3 years, returned them to their homeland. Up until the 1990s U.S. scientists continued to visit and document the health effects of the Rongelap residents' acute radiation exposure and the long-term consequences of living in a radioactive environment (Johnston and Barker 2008).

Nuclear testing introduced new taboos: foreign scientists declared some islands and foods off-limits; marriage to certain people involved new social stigmas; birthing presented new fears and health risks; family life often involved the psychological, social, and economic burden of caring for the chronically ill and disabled. The failure of the U.S. government to provide accurate information to the Marshallese about environmental hazards and risks, coupled with contradictory pronouncements on what was and was not safe to eat, created taboos that were incomprehensible yet dominated every facet of life. This transformation in the loci of control over taboos from a Marshallese cultural realm to a U.S. scientific realm undermined rules and the customary power structures that shaped, interpreted, and reproduced strategies for living.

This history plays a complicated role in current effort to manage resources and provide a safe supply of water for the nation. Rural Marshallese rely on rain and groundwater. The urban population also has access to desalination and importation. Most households harvest rain from rooftops and store it in found or built containers (discarded jet fuel tanks from Cold War-era flights-over the islands or ground coral cement cisterns).

People on Majuro also harvest rainwater using the airport runway as an 80-acre catchment and storing this runoff in a shallow, plastic-lined reservoir just off the airfield. Other than small springs and seeps, this reservoir is the only body of fresh water in the nation.

In an effort to craft a sustainable plan for the nation, an Integrated Water Resource Management (IWRM) assessment was undertaken in 2007 (RMI and US EPA 2009). This assessment identified a number of key constraints. Pollution sources at the household level include unsanitary roofs, guttering, and catchments. For homes using wells, pollution sources include human and solid waste. At the airport catchment, fuel spills, residue from asphalt, salt spray, and inundation all compromise water quality. The lack of a sewerage system and a viable waste disposal system is a nationwide problem. Water resource assessment, monitoring, and management is largely limited to external aid-driven projects. And, the assessment recognized that the Marshall Islands are extremely vulnerable to natural disasters, from typhoons, storm surges, flooding, sealevel rise and other climate change effects as well as droughts, fires, marine oil spills, water supply pollution, hazardous chemicals, and disease outbreaks.

Fig. 3.4.1 Freshwater reservoir created by runoff from Majuro airport, Marshall Islands (Photo credit: Barbara Rose Johnston)

While all these are all important factors, and serious issues that affect Marshallese water supply, IWRM assessment was not tailored to the cultural context and historical circumstances of the Republic of the Marshall Islands. Thus, scientists examined water samples for a standardised set of biological or chemical contaminants, not the place-specific pollutants associated with U.S. militarism, including radiogenic materials, heavy metals, and complex chemical residues from nuclear weapons testing and the missile base activities at Kwajalein. Similarly, the vulnerability assessment focused on natural disasters and their potential impact on water quality (including saltwater intrusion associated with rising sea levels, violent storms and other issues associated with climate change), largely ignoring contamination the question of non-biological hazards in the water supply. Thus, consideration of the pollution and resulting health problems stemming from militarism, especially the persistent, long-lived contaminants from nuclear weapons testing, are not part of the national water resource management plan. Nevertheless, given the bioaccumulative nature of nuclear contaminants, population-wide low-level exposure continues via the consumption of local shell and reef fish, fruits, vegetables, coconut crab, pigs, chickens, and other animals; inhalation of dust and cooking fire smoke; and drinking contaminated water. Moreover, radioactive materials were not the only poisons deposited on the islands during the U.S. use of the islands as a nuclear weapons testing ground. Fallout from the Kwajalein military testing facility includes the release of unburned solid rocket propellant and ammonium perchlorate, which is now detected in the soil, reef fishery, and freshwater lenses of islands in Kwajalein Atoll. Perchlorate blocks the accumulation of iodide in human thyroid glands, and iodide insufficiencies in pregnant women are associated with permanent mental deficits in children.

IWRM recommendations include strengthening of government capacity to monitor, regulate, and remediate water supply and quality problems and improving public knowledge and behaviour. Addressing these challenges requires an understanding of the cultural norms and structure of power in the Marshall Islands. However, because the IWRM framework defines and addresses water management problems as fundamentally local problems, it cannot comprehend the most serious problems confronting the Marshallese – the legacy of militarism and the looming threats from climate change. Marshallese water culture traditions are based upon notions of stewardship, embedded in a cultural framework of taboo, within which people accept and respect the rules that govern behaviour because they enable for community survival. As mentioned, violation of taboos can result in harsh consequences, including death. In today's world, in which forces far behind the Marshallese shores drive environmental change, access to safe water and a healthy environment requires full knowledge of environmental conditions, hazards, and risks; the means and ability to remediate; and the assurance that stewardship principles – and taboos – guide the actions of both local and global actors.

3.4.5 Conclusion

Water scarcity often has less to do with the total quantity of available water in a given time and place than it does with a range of human actions, cultural norms, historical conditions, societal inequities, and the loci of control over water and other critical resources. The struggles over water in Botswana's Kalahari Desert and the 2011 overturning of the High Court decision by the Botswana Court of Appeals illustrate a case in point. In 2010, a High Court ruling confirmed the legality of the government's decision to withhold water in San communities, who in 2002 were evicted from the Central Kalahari Game Reserve by their government whose development policies favoured the expansion of ecotourism and diamond mining in the Kalahari. In this case, the water needed to sustain the San living in the Kalahari, drawn from subsurface aquifers via boreholes, has relatively minimal impact on overall supply, given the small number of people and their subsistence-oriented way of life. At issue is a contested notion of development rights to a region legally recognized to be the indigenous homeland of the San. Water quality conditions, awareness, remedial actions, and preventive measures vary from community to community and nation to nation, reflecting varied technical capabilities as well as broader factors of government priority and assessment of acceptable levels of societal exposure. Where such conditions and experiences once reflected localized norms related to local water resources, globalization, militarism, and other transnational factors now play a determining role in manufacturing water scarcity and environmental inequities. Recognition of the complex interplay between water, culture, and power at local and global scales is increasingly necessary to fully identify the driving forces and biocultural consequences of water scarcity.

References

Davison, Isaac. 2010. Accord settles decades long feud over river management New Zealand Herald, 5 Mar 2010. http://www.nzherald.co.nz/maori/news/article.cfm?c_id=252&objectid=10629900. Accessed 25 Sept 2010.

Dodd, Materoa. 2010. Effects of industry on Maori cultural values: The case of the Tarawera river. In *Indigenous voices, indigenous research*, ed. Veronica Arbon. World Indigenous Nations Higher Education Consortium, Journal 2010. Sámi University College, Guovdageaidnu. http://www.winhec.org/files/World_Indigenous_Nations_Higher_Education_Consortium_formatted2.pdf. Accessed 25 Sept 2010.

Huo, Xia, Lin Peng, Xijin Xu, Liangkai Zheng, Bo Qiu, Zongli Qi, Bao Zhang, Dai Han, and Zhongxian Piao. 2007. Elevated blood lead levels of children in Guiyu, an electronic waste recycling town in China. *Environmental Health Perspectives* 115(7): 1113–1117.

Johnston, Barbara Rose. 2006. The commodification of water and the human dimensions of manufactured scarcity. In *Globalization, water and health: Resource management in times of scarcity*, eds. Linda Whiteford and Scott Whiteford, 2006: 133–152. Santa Fe: School for American Research.

Johnston, Barbara Rose, and Holly M. Barker. 2008. *Consequential damages of nuclear war: The Rongelap report*. Walnut Creek: Left Coast.

Puckett, J., L. Byster., S. Westervelt, R. Gutierrez, S. Davis, A. Hussain, et al. 2002. Exporting harm: The high-tech trashing of Asia. Basel action network, Silicon valley toxics coalition. Available: http://ban.org/E-waste/technotrashfinalcomp.pdf. Accessed 24 Sept 2010.

Qiu, B., L. Peng, X. Xu, X. Lin, J. Hong, and X. Huo. 2004. Medical investigation of e-waste demanufacturing industry in Guiyu town. In *Proceedings of the international conference on electronic waste and extended producer responsibility, 21–22 April 2004*, 79–83. Beijing: Greenpeace and Chinese Society for Environmental Sciences.

Republic of the Marshall Islands (RMI) Environmental Protection Agency and the United States Environmental Protection Agency. 2009. Integrated Water Resource Management (IWRM) Republic of the Marshall Islands Diagnostic Report, 2007. http://www.epa.gov/safewater/contaminants/unregulated/perchlorate.html, 15 Aug 2009.

Schoenberger, Karl. 2002. Silicon valley's dark side: A mercury news special report. *San Jose Mercury News*. 24 Nov 2002: 1A, 16–18A.

Silicon Valley Toxics Coalition (SVTC). 2001. Poison PCs and toxic TVs. Silicon Valley Toxics Coalition Report, San Jose, 2001. http://www.svtc.org/site/DocServer/ppc-ttv1.pdf?docID=124. Accessed 26 Sept 2010.

United Nations Environment Programme. 2010. Parties to the Basel Convention. Secretariat of the Basel Convention, United Nations Environment Programme. 2010. http://www.basel.int/ratif/convention.htm#113. Accessed 27 Sept 2010.

Part IV
Hydrodevelopment, Cultural Diversity and Sustainability

Cover art by Javier Sterling Ruiz, Cuba

Chapter 4.0
Introduction: Hydrodevelopment, Cultural Diversity, and Sustainability

Barbara Rose Johnston

In this section we examine how large-scale hydrodevelopment involves synergistic processes and produce cumulative effects that result in the degradation of rivers and the complex human environmental systems they support. This critical focus is not an outright rejection of hydrodevelopment. On the contrary, contributing authors argue that sustainability is achievable when river basins are developed and managed in ways that sustain diverse human and ecological needs. This coupled bio/social systems approach forces consideration of complex concerns: WHOSE livelihoods are sustained? WHO defines sustainability? On WHAT terms? Towards WHAT prioritized goals?

Such concerns are the focus of the overview essay by Barbara Rose Johnston. Reviewing the history and consequential damages of hydrodevelopment, this essay argues that such development has been a driving force in global displacement, loss of the means to sustain a cultural way of life, and related impoverishment. Emerging trends in hydrodevelopment are also briefly considered, including the resurgence of large dam-development and the related threats to vast numbers of people, many of whom are indigenous or ethnic minorities. What this means in terms of human misery and social response is suggested in a brief vignette on Chixoy Dam development in Guatemala. The political fallout and resulting social protests following the announcement of unpopular and destructive development decisions is described by anthropologist Terrence Turner in his analysis of the events that led to Kayapo protests over Belo Monte Dam construction on Brazil's Xingu River. Strategies that may, in the long run, transform power/equity dynamics in decision making by empowering and engaging affected peoples in the planning and decision making-process are suggested with the emergence of consultations in Latin America, described by Monti Aguirre. This review

B.R. Johnston (✉)
Center for Political Ecology, Santa Cruz, CA, USA
e-mail: bjohnston@igc.org

B.R. Johnston et al. (eds.), *Water, Cultural Diversity, and Global Environmental Change: Emerging Trends, Sustainable Futures?*, DOI 10.1007/978-94-007-1774-9_21,

ends with a series of provocative questions that are taken up by the authors of case-specific chapters and vignettes in this section.

The complicated issues surrounding hydrodevelopment-induced displacement are the focus of Robert Hitchcock's case study treatment of the Lesotho's Highlands Water Project. This chapter suggests some of the many problems associated with assessing potential social impact, working with affected communities to shape appropriate compensation and resettlement agreements, and the huge distance between contractual agreements and on-the-ground reality. Hitchcock offers some of the detail that explains a core maxim in hydrodevelopment (WCD 2000), the experience of project-affected peoples is one where stated goals and social program terms articulated in development plans and financing agreements are rarely achieved. The devastation to communities and their environments leaves people further impoverished (Cernea and McDowell 2000). A rare exception to the rule is suggested in the brief description of Swaziland's Mugaga Dam resettlement.

This question of relative costs and benefits in hydrodevelopment is explored from the perspective of the dam-affected peoples movement in a series of vignettes. Robert Kugonza offers a description of the NGO African Rivers Network, its praxis and its advocacy concerns. Ikal Angelei offers a brief summation of the threats posed to Lake Turkana and its peoples if the Gibe III project is completed. These advocacy voices from Africa echo concerns raised earlier in this section.

A more focused look at the synergistic links between hydrodevelopment, riverine ecosystems and the health and well being of human systems is presented in Nathanial Matthew's case study of the Mekong Basin. There, a series of dams on the Mekong and its tributaries threaten freshwater fisheries, food security and ways of life for tens of millions of people. Further north, in China, Bryan Tilt examines similar hydrodevelopment and its inequitable impact on culturally diverse communities in Yunnan, China. Still further north, Tashi Tsering considers efforts to protest hydrodevelopment in Tibet, offering an example where collaboration between international and national Tibetan activists and Chinese scientists and environmentalists resulted in the decision to halt development plans that threatened Megoe Tso, a sacred lake on the Tibetan Plateau.

And finally, in chapters by Colleen Boyd and John Boyd, and the associated vignette by Benedict Colombi, we consider the long term biocultural impacts of hosting dam development in the American northwest, and recent ecopolitics that seek restoration not only of the river and life within, but the ways of life that a free-flowing river supports. These cases illustrate efforts to acknowledge rights-holder status and document, negotiate, and manage water resources in ways that restore and sustain environmental and cultural life.

Collectively this sampling of global experience with hydrodevelopment illustrates these points: If development is to be sustainable, local people must have access to the resources, information, and technical assistance that makes it possible for them to engage in the planning process at the earliest stages of development. They must be able to meaningfully participate in decisions that affect them with the status of rightsholders. Their engagement must be in processes that prioritize stewardship principles, where water development and management prioritizes environmental and cultural flows and, thus, sustains both the environment and human communities. When hydrodevelopment poses a demonstrable unsustainable threat to biological and cultural diversity, such evidence must be considered, the power to say no must be respected, and their obligation as stewards to participate in remediation and restoration must be supported.

References

World Commission on Dams. 2000. *Dams and development: A new framework for decision-making. The report of the World Commission on Dams*. London: Earthscan Publications.

Cernea, M.M. and C. McDowell, eds. 2000. *Risks and Reconstruction: Experiences of Resettlers and Refugees*. Washington DC: World Bank.

Chapter 4.1
Water, Culture, Power: Hydrodevelopment Dynamics

Barbara Rose Johnston

Dams have been built for thousands of years, capturing water for crop production in the dry season, reducing the risks of flooding by controlling the water level in the rainy season, and storing water that can be transported to irrigate distant crops and sustain large urban populations. Dams can provide a barrier between the land and the sea. And dams allow the generation of power, amplifying the force of water to allow the generation of electrical energy.

While hydroelectric power can be generated from a run-of-river project (using part of the stream flow), a regular flow is required. The construction of large dams creates a reservoir to allow management of flow, and thus a more reliable generation of power. According to the World Commission on Dams, 630 large dams were commissioned over the first decade of the last century, many of which were hydroelectric dams that served a small community or industrial plant. During the 1950s a post-World War II reconstruction boom helped provide sources of power to revitalize and sustain war-torn nations, and some 2,735 dams were commissioned. By the 1960s a 'green revolution' was well underway, and large dam construction not only provided the means to transform newly independent nations with electrical energy, but also allowed the storage and diversion of water to irrigate high-yield crops such as rice, maize and wheat. In that decade, some 4,788 large dams were built (WCD 2000).

How did this relatively rapid expansion in hydrodevelopment occur, and to what effect?

B.R. Johnston (✉)
Center for Political Ecology, Santa Cruz, CA, USA
e-mail: bjohnston@igc.org

B.R. Johnston et al. (eds.), *Water, Cultural Diversity, and Global Environmental Change: Emerging Trends, Sustainable Futures?*, DOI 10.1007/978-94-007-1774-9_22,

4.1.1 Internationally Financed Development

In the early years of World War II, leaders of the United States and the United Kingdom met to strategize on the key economic questions of the day. How to structure and rebuild a free market economy that was in depression before the war and was in worse straits because of the war? How can the international community finance an economic stimulus when most of the world's nations lacked even the minimal resources to repair the bombed-out infrastructure that supported agriculture and industry? How to encourage post-war governance based upon capitalism in the face of widespread popular appeal for the alternative model: communism? Clearly some sort of international financial system was needed to allow nations to collaborate, lend money, and support the exchange and consumption of goods.

Over the next 3 years the architecture for an international financial system was drafted by Special Assistant to the U.S. Secretary of the Treasury Harry Dexter White and by John Maynard Keynes, advisor to the British Treasury. In July 1944, these plans were offered to diplomats from some 44 nations attending the United Nations Monetary and Financial Conference in Bretton Woods, New Hampshire. Three weeks of negotiations refined proposals for an international financial system that, with signing of the final conference act, created a system of fixed exchange rates tied to gold and the US dollar, the International Monetary Fund (IMF), and the International Bank for Reconstruction and Development (IBRD). The IMF would manage the system of fixed exchange rates and help expand world trade by providing short-term loans or modifying exchange rates to countries struggling with temporary deficits in their balance of payments. The IBRD, first in what would eventually grow to be five separate banks forming the World Bank Group, would provide the financial assistance for the reconstruction of war-ravaged nations and the economic development of less developed countries (George and Sabelli 1994).

The stated goals of this system were laudatory. The IBRD would finance projects to build roads, bridges, buildings, dams, public water supplies and sewage treatment systems, railroads, electricity generation and transmission grids, and telephone and telegraph networks. This infrastructure development, in turn, would allow access and the means to commercially exploit timber, minerals, oil, gas, hydroelectric energy, and agricultural land. The profit made from such enterprises would allow governments to provide basic social services for the civilian population, open up local economies to global capital, and allow the repayment of loans with interest. The coffers of the IDRB could expand and further finance modernity, and the economies around the world would stabilize and thrive.

In the decades since World War II, the IBRD, IMF and other members of the World Bank group have functioned as a core pillar in the international economy and, for many nations, a factor in determining national economic stability. The World Bank, as an independent specialized agency of the UN, is forbidden by its charter to take a political stance in carrying out its responsibilities to promote economic development and reduce poverty. This notion of the Bank as an apolitical actor is a key element of its organizational culture.

Nevertheless, World Bank actors and actions often have profound political impact. The distance between organizational mandate and societal impact is, in part, a consequence of the internal structure of power and the historical dominance of United States actors and interests in the exercising of power. Development projects are identified and shaped by Bank staff, with the Executive Board power limited to a yes or no vote. The World Bank President, who shapes the Executive Board agenda, is nominated by the United States, which also has a permanent seat on the Executive Board. Changes to the Bank's charter require an 85% vote, and at present the United States holds veto power with its 16.4% ownership share. While member shares have varied over the years, the United States has always been the only country to hold the power of veto.

Loans from the World Bank and other multilateral development banks allow bankrupt nations to build basic infrastructure and stimulate the local economy. The level of support and the nature of supported activities reflect the needs of recipient nations, though often in ways that stimulate the growth of the global economy and serve the economic and political interests of donor nations. Thus, economic assistance often comes with conditions, a set of policies, known collectively as structural adjustment policies. This kind of approach, termed conditionality, attempts transformative change in national economies through, for example, the privatization of industries, reduction of domestic spending, raising of interest rates, devaluation of the currency, slashing of expenditures on social programs such as health and education, lowering or doing away with food and energy subsidies, and cutting the size of the civil service. Such efforts can have profound societal impact in terms of increased human misery and political strife (Bodley 2008).

Societal impact can be further exacerbated when development involves corruption. Development requires immense sums of money, provided through loans granted within a broader context of fungibility. Fungibility means that money, once allocated to the donor party, can be used for any purpose. While oversight mechanisms have emerged over the years in an effort to ensure that financed projects achieve stated goals, the fungible nature of foreign aid means that oversight has a tendency to serve more as a means to reduce donor liability, than to demonstrably protect the health and well being of vulnerable groups and the environment in which they live (Clark et al. 2003).

These factors and this structure of power have allowed the World Bank and other international financial institutions to play a significant role in controlling the direction and shape of economic development with far reaching positive and negative consequences. The development that has occurred under this international financial system, especially the construction of large dam and water diversion projects that disrupted riverine systems and displaced large numbers of people, has had profound societal consequences (Cernea and McDowell 2000; Scudder 2005; WCD 2000).

4.1.2 Hydrodevelopment and Displacement

Hydrodevelopment, especially in the early decades of World Bank work, reflected the rule of law, or lack thereof, in recipient nations. Thus, in 1957, when construction of the Kariba Dam on the Zambezi River in what was then Rhodesia caused the forced eviction of some 57,000 Tongans from their homelands, no adequate land or compensation was provided to rebuild their lives (Scudder 2005). The dam-displaced Tongan nation now numbers some 250,000 people. The post-colonial creation of two states, Zambia and Zimbabwe, has split the Tongan nation into two separate refugee communities and their isolation and extreme poverty has had a devastating impact on language, kinship ties, and cultural solidarity.

The Kariba Dam was the first large infrastructure project built with World Bank financing, and it was built in partnership with the United Kingdom. In 2000, following the World Commission on Dams review of their case, Tongan refugees in Zambia began to document their conditions and consequential damages of dam displacement with hopes of negotiating a meaningful remedy. While governments and financiers have drafted remedial social and economic development plans, implementation has been timid, haphazard, and ineffective (Scudder 2009). According to the UN Office for the Coordination of Humanitarian Affairs, the highest levels of unemployment and poverty in Zimbabwe can now be found in the Tonga nation. And in Zambia, where most of the population survives as subsistence farmers, the Tongan refugees are among the poorest of the poor.

Elsewhere in Africa, Thayer Scudder has documented instances where hydrodevelopment has been the vehicle by which governing elites and their supporters forced local households and communities off their land, resulting in societal degeneration, loss of a way of life, and in several well-documented cases, ethnic cleansing. For example, the 1975 Agricultural Land Law adopted by Somalia as a means to develop agricultural lands irrigated by water from the Baardheere Dam downstream from the Ethiopian border, resulted in the widespread loss of customary land rights, factors that led to 1990s-era inter-clan and inter-ethnic warfare, conflict that still has repercussions today (Scudder 2005:227–228).

Such cases, while the not the norm, are not isolated instances, as illustrated by hydrodevelopment and forced displacement in Guatemala. There, the government of Guatemala, with World Bank and Inter-American Development Bank financing, built the Chixoy hydroelectric dam (1976-1982) with forced displacement, targeted assassination, and widespread massacre of Mayan residents living in dam-affected villages, including half the residents of the Río Negro village (Johnston 2005, 2010).

Box 4.1a Consequential damages of dam development: Maya A'chi experiences in Guatemala

—Barbara Rose Johnston

The Chixoy Dam legacy study, with its focus on the displacement of indigenous peoples in Guatemala, offers clear evidence of degenerative change in the quality of life and health of the community before and after hydroelectric dam development.

(continued)

Box 4.1a (continued)

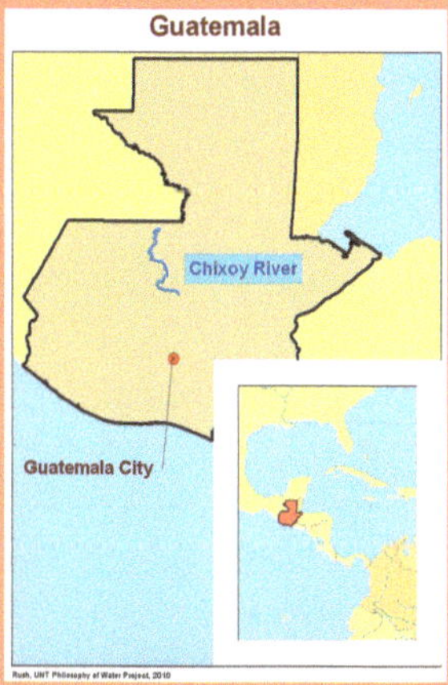

Box Map 4.1a.1 Guatemala

The Chixoy Dam was built in the late 1970s and early 1980s. For the people living in the Chixoy River Basin, land rights were secure, with legal title and communal rights in many cases dating back to the 1800s. These people lived in the same region where their Maya A'chi ancestors lived, working the fields and conducting prayers, dances and other rituals in the crumbling ruins of pyramids, part of a complex of some 25 pyramids built at the crossroads of ancient trade routes connecting the Guatemalan highlands and the coastal valleys. Fertile river basin lands provided a biannual harvest, fish was plentiful and available year round, and common property lands supported livestock and harvesting of palms and other resources as saleable goods. The sociocultural fabric of life was tightly woven across a landscape maintained by trade, familial ties, cultural beliefs, and historical relationships.

Development occurred without a comprehensive assessment of affected peoples, without a compensation or resettlement plan, and with forced displacement that included violence, assassinations, and massacre at the village of Río Negro and in surrounding villages in the river basin. Years later, a United Nations sponsored truth commission determined that the massacres in this case occurred as a result of the Chixoy dam construction, were an example of state-sponsored violence against a civilian population, and constituted evidence of genocide. The Río Negro massacre was cited by the United Nations-sponsored Commission on Historical Clarification (CEH) as a key exemplary case of state-sponsored policy of violence against the civilian population—violence that constituted genocide. In the CEH summation of the Río Negro case, Chixoy dam construction and related forced displacement were specifically recognized as causal factors. Other cited evidence of the Guatemalan Army's intent to destroy the Río Negro community through a genocidal campaign includes four massacres, arbitrary executions of other community members before and after massacres, and harsh living conditions due to flight from massacres and forced resettlement. Hydrodevelopment financing served as the key source of foreign aid to the Guatemalan Government during this period (CEH 1999: Vol. 1, Annex 1, Chap. VI: Exemplary Case No. 10).

In subsequent years massacre survivors from Río Negro and surrounding communities organized and collaborated with national and international advocacy groups to document human rights violations associated with the Chixoy dam development.

Household surveys conducted in 1977 (as part of the preliminary efforts to develop a resettlement agreement) found that in communities upstream and

(continued)

Box 4.1a (continued)

Box Fig. 4.1a.1 Young Maya A'chi boy fishing, upstream from the Chixoy dam. Fishing provided river basin residents with a major source of protein and, through local market sales, generated significant household income (Photo credit: Bert Janssens)

downstream from the dam, prior to construction, household production provided for all food needs for some 79% of the surveyed population, and river fish was the most significant source of dietary protein. Household surveys conducted in 2004 found that only 28% of surveyed households still living adjacent to the dam and its reservoir were able to produce, through farming, fishing, hunting and other means, all their household food needs. For those communities whose homes were now underwater and were now living in resettlement villages, 93% reported the ability to provide all household food needs before the dam, and only 26% reported this ability in 2004. The declining ability to produce food related directly to the loss of productive agricultural land, pastureland, and access to viable river and forest resources.

Not surprisingly, these changes have produced a measurable decline in dietary protein. Households reporting regular consumption of fish several times each week dropped from 74% before the dam to 23%. The consumption of meat several times each week dropped from 30% to 21%. Some 82% of the population raised pigs before the dam; in 2004, only 26% have the household space to raise pigs. Before the dam, 96% of the surveyed population reported keeping an average of 34 poultry per household. In 2004, only 69% report the ability to keep poultry, with the average dropping to 14 per household. Access to milk and other dairy products has also significantly declined, as evidenced by the reduced ability to keep and feed dairy cows. Before the dam, 70% of the households reported ownership of 1,115 cows; in 2004, 21% report ownership of 121 cows. These changes are a contributing factor to the region's extraordinarily high rates of malnutrition and infant mortality.

Years after the completion of the Chixoy Dam, life's essentials can only be acquired with money: money is needed for water, power, firewood, commercial fertilizer, household food, clothing, school fees and supplies, land taxes, roofing and other materials to repair crumbling homes and community halls. Money is needed to travel to distant farmlands. Money is needed to pay for the time and assistance of lawyers and others who help prepare claims to secure long-promised compensation and other entitlements. And, people now lack access to the critical resources that once supported household and community income generation.

(continued)

Box 4.1a (continued)

Yet, the loss of access to fertile lands, pasture, and river and forest resources severely constrains the household ability to generate monetary income. Surplus production and sale of garden products, eggs, chickens, and livestock dropped from 44% to 12%. The number of households catching and selling fish dropped from 49% to 3%. The number of households producing surplus crops to sell in the marketplace diminished from 37% to 7%. Also, the number of households involved in the harvest and sale of forest products changed: collection and sale of palm leaves dropped from 81% to 32%; ocote torches, from 56% to 2%; firewood, from 29% to 11%; and construction timber, from 25% to 1%.

For the communities that hosted the nations' main source of electricity, development has been a disaster. When I first visited the area some 20 years after the dam was completed, villages at the dam site and in the surrounding region still lacked electricity. People in the resettlement villages live in extreme poverty, in crumbling homes, with few economic opportunities. For those who remain in the Chixoy river basin, periodic flooding has seriously affected the length of the agricultural season and the number of harvests per year. Downstream, land is periodically flooded, so erosion is severe. Varied river flow in the summer and winter has created stagnant pools and breeding grounds for mosquitoes, which transmit malaria. Up and down the river fishing villages have seen the complete loss of local fisheries. When dam repairs were made in the mid-1990s, no social impact assessment occurred, and no program was created to protect downstream communities from the hazards of dam operations. Those communities have experienced water shortages, crop failures, and loss of fisheries as a result of inadequate or interrupted stream flow in dry seasons. Flash floods from unannounced dam releases have caused at least three documented fatalities in downstream communities.

Box Fig. 4.1a.2 Farmland upstream and downstream is periodically flooded as reservoir waters rise, and as dam releases occur. Legal title to the reservoir lands and to a portion of the land beneath the dam and hydroelectric facility was never secured. To maintain communal rights to adjacent lands, one community has been forced to pay annual property taxes for their communal allocation, including the submerged lands beneath the reservoir (Photo credit: Bert Janssens)

These findings demonstrate how a close relationship to land and river basin resources through secure

(continued)

Box 4.1a (continued)

tenure contributes to a community's well-being. When that relationship is severed, and the river basin can no longer sustain household and community ways of life, when the market economy dictates livelihood strategies and there is restricted access to critical resources, well-being is severely reduced. Yet, despite such obvious indicators of impoverishment and misery, because the Chixoy Dam facility generates so much of the nation's electricity, this project is viewed by many as a successful example of development.

Misery, and the failure of prior protests to result in a meaningful remedy, eventually led to more public protests. In September 2004, after 29 hours of peaceful protest at the hydroelectric facility involving some 1,000 dam-displaced Mayans, the Guatemalan Government agreed to establish a reparations negotiation process. In July 2005, the independent study of the consequential damages associated with Chixoy Dam development, inadequate compensation, involuntary displacement, violence and massacre was completed and presented by dam-affected community leaders to their Government, project financers, the InterAmerican Human Rights Court, and the UN Special Rapporteur on the Rights of Indigenous Peoples. In a 2008 decision, the Government of Guatemala announced their acceptance of the independent study as a legitimate statement of damage, and their intent to proceed with reparation negotiations (Dill 2009; Johnston 2005). In 2010, following years of negotiation with dam affected communities facilitated by the Organization of American States' InterAmerican Human Rights Commission, the Government of Guatemala agreed to a plan to provide reparation in the form of development assistance to the communities adversely affected by the Chixoy Dam, social and cultural support, and environmental restoration of the river basin. This Agreement, signed by authorized representatives of the Guatemalan Government and all affected parties, has the force of law. However, the political will to implement this Agreement has not been demonstrated. In March 2011 President Colom announced his decision to reverse his position on the negotiated Reparation Agreement, and withdrew his support.

References

CEH (Comisión para el Esclarecimiento Histórico). 1999. Caso ilustrativo número 10: Masacre y eliminación de la comunidad de Río Negro, Capítulo VI, Casos ilustrativos – Anexo 1, Volumen 1, p. 48. In *Informe de la Comisión para el Esclarecimiento Histórico, Guatemala: Memoria del silencio*. Guatemala: Oficina de Servicios para Proyectos de las Naciones Unidas.

Dill, K. 2009. Reparations and the illusive meaning of justice in Guatemala. In *Waging war, making peace – reparations and human rights*, ed. B.R. Johnston and S. Slyomovics, 182–204. Walnut Creek: Left Coast Press.

Johnston, Barbara Rose. 2005. Chixoy dam legacy issues study, vol. 3. *Consequential damage assessment of Chixoy river basin communities*. Center for Political Ecology, March 17. http://www.centerforpoliticalecology.org/chixoy.html. Accessed July 2010.

Like other large infrastructure development, hydrodevelopment is the end result of a culture of governance and modernization mindset, one that typically views the installation of dams and diversions systems as a societal good and the associated environmental and human harms necessary trade-offs. Thus, in the period following World War II through 1997, more than 54,000 large dams were built in the push towards modernity, generating an estimated 20% of the world's electricity and providing irrigation to fields that produce some 10% of the world's food. These achievements have occurred at considerable cost as dams have flooded some of the most productive agricultural lands in the world. Siltation and sedimentation has reduced their operating life and many large dams failed to meet projected energy and economic goals. Changes in downstream water quality have decimated the fisheries, waterfowl and mammals of the world's deltas. They have caused the endangerment or extinction for some 30% of the world's freshwater fish. Damming and flood control have also resulted in increases in the frequency and severity of floods and increases in diseases such as malaria and schistosomiasis (WCD 2000).

Box 4.1b Schistosomiasis

—Barbara Rose Johnston

Some 779 million people are at risk of schistosomiasis, a parasitic disease common in tropical and subtropical areas. This disease involves infection with a type of schistosoma parasite that is transmitted through contact with contaminated water. On contact with humans, the parasite burrows into the skin, matures into another larval stage then migrates to the lungs and liver, where it matures into the adult form. The adult worm then migrates to its preferred body part, depending on its species (bladder, rectum, intestines, liver, spleen, and lungs). This is a debilitating disease that causes fever, chills, anemia and other conditions that inhibit productivity and, if untreated, results in progressive damage to critical organs.

A 2006 review of African studies examining the relationship between water resources development projects and urinary and intestinal schistosomiasis, showed significant risk among some 106 million people living adjacent to dam reservoirs, concluding that the development and management of water resources is an important risk factor for schistosomiasis. Their recommendation: strategies to mitigate negative effects should become integral parts in the planning, implementation, and operation of future water projects (Steinmann et al. 2006).

Reference

Steinmann, P., J. Keiser, R. Bos, M. Tanner, and J. Utzinger. 2006. Schistosomiasis and water resources development: Systematic review, meta-analysis, and estimates of people at risk. *The Lancet Infectious Diseases* 6(7): 411–425.

For the people who live in the river valleys, mountains, and deltas that are degraded and destroyed by water diversions and hydroelectric dam developments, dam development has literally destroyed their health, economy, and culture and thus has had a profoundly negative impact on the world's cultural diversity. For example, in India, while dams have forcibly displaced 2% of the entire population, at least 40% of those displaced since 1947 are categorized as tribals and other ethnic minorities. These, like other displaced peoples, are fishers, farmers, pastoralists: people who live off the land and have done so for generations in largely self-sustaining communities.

Worldwide, it is these members of an ethnic minority or indigenous group that have disproportionately suffered displacement, alienation, and loss of a way of life in the name of economic development. Lacking fertile lands and the means to reproduce a way life, development refugees—those people forced out of areas where development projects are undertaken—typically live in impoverished resettlement compounds and camps or migrate from the countryside to the city to seek employment within their country and abroad (Johnston 1997; Scudder 2005; WCD 2000).

In their assessment of the performance of large dams worldwide, the World Commission on Dams (WCD) calculated that between 1945 and 2000, some 40-80 million people worldwide were forcibly evicted to make way for large dam development, the majority of which are indigenous peoples and ethnic minorities (WCD 2000). These figures were admittedly conservative and are still being revised (Scudder 2005:22). In India, for example, scholars reexamined hydrodevelopment documents and state compensation files, finding that the reported figure of 14 million people displaced between 1947 and 1999 failed to include people who lived in communal fashion on common lands, typically tribal and ethnic minorities who lived in forests and other lands claimed by the state. This reassessment concluded that at least 60 million people had been actually evicted from their rural homes to make way for water development projects (Fernandes 2007).

Similar reassessment is occurring in China. For the WCD study, the Chinese government reported some 13 million people as dam-displaced (Reservoir Development Bureau 1999). More recent government-sponsored reassessment supports a larger figure (Shi et al. 2008). By 2006 China had built some 85,160 dams, and the 2007 budget announced by Chinese Prime Minister Wen included an annual stipend for some 23 million rural migrants officially recognized as displaced without adequate compensation to make way for water development. The displaced numbers will continue to grow. In September 2009, Chinese state media reported that 1.3 million people had been officially relocated by the Three Gorges Dam, and another four million will be encouraged to move for this and related hydrodevelopment in the near future.

On a global scale, tens of millions of people have suffered the loss of homes, community, and traditional livelihood in the name of hydrodevelopment, and in far too many instances resettlement programs have failed to deliver the development promise (Oliver-Smith 2009). This global tally is further expanded by the findings of a 2010 study of dam-induced alterations of river flow and their consequence for downstream communities. Richter et al. (2010) found that hydrodevelopment has devastated natural riverine-production systems and adversely affected the livelihoods of a conservatively-estimated 472 million river-dependent people living downstream.

Given these numbers, it appears that hydrodevelopment–in the enclosure and destruction of the world's riverine ecosystems commons–may be one of the most significant factors driving global poverty rates. And, given the location of dam projects in remote regions and the documented demographics of development-refugees, these figures suggest that water development and associated displacement of culturally diverse communities are one of the major factors in the global decline of cultural diversity.

4.1.3 Social Response

We are just beginning to understand the many and profound intergenerational consequences of this massive uprooting of place-based peoples and ways of life. Despite recent and emerging efforts by international financers to revise and implement development and resettlement guidelines in ways that protect the fundamental rights of project-affected peoples and the environments in which they live, significant barriers exist. Impact assessment processes often prioritize the economic benefits, with short-term gains often outweighing the long-term, synergistic, or cumulative costs that development poses to local communities and the environment. Operational policy, guidelines and other mechanisms employed by financers ideally allow an orderly and legally defensible approach to development, and can be achieved with the assumption that the rule of law in host countries exists. Yet, for indigenous peoples and ethnic minorities who live in the wild margins of society the rule of law that protects customary rights and the health of the environment is easily ignored in favour of opportunities to exploit lucrative reserves of oil, gas, minerals, timber and hydroelectric potential.

Dam-affected peoples around the world are responding to these conditions in a number of different ways. They are participating in social movements aimed at promoting human rights and sustainable development. They were successful in arguing for the right to free and prior informed consent, an element of the Declaration on the Rights of Indigenous Peoples adopted by the United Nations General Assembly in September 2007. And they engage in direct action, taking part in demonstrations at the local, national, regional, and international levels, and taking governments and transnational corporations to court at the national and international levels seeking remedy: ensure adequate environmental and social impact assessments to take place, halt projects, modify water management to sustain the environment, decommission dams, and other remedial measures (Blaser et al. 2004).

Increased communication between project-affected communities and engagement with more distant worlds (downstream communities, national and international civil society) has transformed dam-affected peoples and, to some degree, created rights-protective spaces. Increased awareness of development impacts and the rights of dam-affected peoples has historically led to a rise in protests, calls for remediation, and strengthening of local and regional movements, such as the Movement of Dam Affected Peoples in Brazil (MAB); African Rivers Network; Latin American Network Against Dams and for Rivers, Communities and Water (REDLAR); Himalayan and Peninsular Hydro-Ecological Network (HYPHEN); and, the 1997, 2003 and 2010 "Rivers for Life" international meetings of dam-affected peoples. These movements over the past few years have strengthened in reaction to increasingly oppressive actions by the state.

In some cases, public protests over corruption and the ulcerating social impacts of development-induced displacement have led to changes in governance, resulting in a transformation of the compensation principle from simple asset replacement to restoring a sustainable way of life. In China, for example, public protests over failed resettlement programs and evidence of corruption encouraged the formation of a national policy to address the immediate and long-term societal costs for communities displaced by hydrodevelopment. For example, relocation assistance for the about 330,000 people being moved to make way for the Danjiangkou reservoir, part of the long-planned North-South project, not only includes compensation for physical property (unmovable property with old homes, an allotment of 0.1 ha dry farmland or 0.07 ha irrigation farmland per person, which is similar to the amount of farmland of host farmers of new arable land, and new infrastructures such as road, school, clinics, water supply and electricity etc. in a newly built village) but also compensation and entitlements to address the way in which loss of land affects the ability to support a way of life. Thus, the state will provide an annual subsidy of 600 Yuan (about 88 U.S. dollars) per person for 20 years, and local governments will provide education, job training, employment opportunities and other forms of assistance to dam-displaced households. This project-specific entitlement is part of a broader national post-relocation supporting policy for all households, with a total number of 22.88 million affected persons (including growth population) whose lands were taken for dam projects from October 1949 to June 2006. In establishing an annual pension and entitlements for dam-displaced people in the countryside, the government is acknowledging and providing redress for the abuses accompanying past development projects.

In other cases, efforts to implement rights in new development project proposals, especially the right to free and prior informed consent, are also making some headway, as evidenced by the recent rise of *consultas* in Latin America. Originally an element of financing agreements that required public participation, the *consultas* are increasingly taking on the status of plebiscites, meaning a legally binding vote by communities in support of or opposition to development proposals as a result of the incorporation of consultation and informed consent language in the national legislation. And when the results are ignored by the state, the InterAmerican Human Rights Commission has increasingly taken a stand (McGee 2010).

Box 4.1c One person, one vote – The voice of communities in development decisions

—Monti Aguirre

Across Latin America a slew of large and small dams are planned for the rivers of the region, lined up for construction almost as if dams were being manufactured in a factory. These projects are planned and implemented by governments and companies with little or no meaningful participation by the communities that are most likely to be affected. Consultation with communities is usually a requirement for project approval, yet the process is often deeply flawed. People often lack timely access to project information. Some affected people may be bribed or deceived into signing agreements endorsing the project. Those who speak out against dams and mines are subject to violence, repression or death.

Despite these political conditions, rather than participating in flawed development processes, communities across Latin America are organizing local referenda to record their votes on dams, mines and other large development projects. These referenda, often called consultations, also involve detailed discussions so community members can understand the likely impacts of these projects on their lives.

For example, after the Guatemalan government began promoting the Xalalá Dam as a national priority, community members began organizing meetings to learn about the project, talk about its impacts and ultimately record their vote on whether they approved or disapproved of the project. Women and men, young and old, met in local plazas and communal meeting rooms. They discussed what was at stake if the proposed dam was built—how they would lose their rivers, lands, ancient burial grounds, crops, hunting grounds and sense of community. They told the sad story of how development of the Chixoy dam had drastically changed the lives of their Maya-Achi relatives. And, in April of 2007 they held a "Consulta Comunitaria de Buena Fe" (Community Consultation in Good Faith).

Box Fig. 4.1c.1 Community members, young and old, recorded their votes on whether Guatemala's Xalala Dam should proceed, 2007 (Photo credit: Commission on Community Consultation, Photo courtesy of International Rivers)

Victor Caal, a local teacher who served as a facilitator during the popular consultation around the Xalala Dam in the Ixcán region of Guatemala, describes the event: "We got together early on the day of the

(continued)

Box 4.1c (continued)

popular consultation. The elderly, women, men and children all came. Several national and international observers were present." The consultation considered the construction of the Xalala Dam and whether to grant permission for oil exploration and extraction in the region. Of the 19,911 people who participated, 91% rejected both developments. While the Guatemalan government has yet to officially recognize the vote, the demonstrated local opposition discouraged companies from bidding on the Xalalá Dam in November 2009.

The resurgence of popular consultations in Latin America reflects a convergence of customary and emergent human rights law. Mayan communities, like other indigenous groups in Latin America, have traditionally held formal consultations as a means to ensure that decisions are legitimized through sharing of information, dialogue and consensus. This customary law reflects the same goals and principles of emergent international human rights law, specifically those instruments which call for free, prior and informed consent.

For example, Convention 169 of the International Labour Organisation states both that "governments should respect and protect the property rights and possession of lands that are traditionally occupied by indigenous people" and that "governments must consult the affected communities before authorising any program that plans to exploit the natural resources in their territories." The right to free and prior informed consent was more specifically stated in the UN Declaration on the Rights of Indigenous Peoples adopted by the General Assembly in 2007. Such rights are increasingly recognized as a crucial element of the planning and decision-making process shaping the development of resource extraction projects such as dams and mines, many of which are situated on indigenous lands.

Popular consultations not only represent an opportunity to secure consent, when plans are pushed through without free and prior informed consent consultations demonstrate widespread condemnation of the development process.

At this writing, in Guatemala alone, more than 500,000 people have participated in 35 community consultations on mining, oil and dam projects. In 2005, the Municipality of Río Hondo, Guatemala, held a popular consultation on three dams proposed on the Colorado River near the headwaters of the Sierras de las Minas mountain range. The vote, proposed by the Mayor and Municipal Council and conducted by the Supreme Electoral Tribunal, overwhelmingly rejected the dams due to their potential environmental impacts, and irregularities in the environmental impact study. This vote was recognized and respected by the Guatemalan government.

(continued)

Box 4.1c (continued)

One of the most successful efforts resulting from a plebiscite campaign in Latin America took place in 1996 around construction of the Paraná Medio Dam in the Entre Ríos Province in Argentina. The project spurred heated debates in the Provincial Senate between environmentalists and dam proponents, which resulted in a request by two radical senators for a plebiscite. The request was voted on, passed and sent to the executive branch to authorize it, where it was blocked. River defenders did not get a plebiscite, but after defeating many obstacles obtained full victory in September 1997, when their efforts resulted in Argentina's first ever anti-dam law. The law, which is still standing, forbids new dam projects from being constructed on the Paraná and Uruguay rivers in the entire province of Entre Ríos. The middle Paraná River supports the livelihood of more than 30,000 people and indirectly benefits more than 100,000 people. Popular consultations on dams and mines are now taking place in many countries.

We have yet to see the final impact community consultations and referenda will have in the defence of rivers and the livelihoods of local people. It is, however, clear that consultations challenge current development practices and offer a viable mechanism for the direct participation of communities in the development process. Using local referenda to record the voices of local communities is a powerful democratic tool. Consultations represent a consent mechanism that allows for the direct participation of communities in the development process, and thus empowers local communities to determine their own path of development.

In recent years, efforts to transform the status of dam-affected peoples in development projects from stakeholder to rights-holder—with the right to benefit materially from development income—have gained traction. For example, some host communities in the James Bay region of Canada have negotiated a share of hydroelectric energy profits as a means to mitigate the social and environmental damages of large infrastructure development. This notion of resident peoples as rights-holders with an equity stake in the development project has been explored by the World Commission in Dams, the World Bank (2004), endorsed by development and resettlement experts (Oliver-Smith 2009), and operationalized by the Australian state (as described in the Murray River case by Monica Morgan in Chap. 5.3).

4.1.4 Backlash

The human rights/social justice initiatives mentioned above involve confrontation with the State and other powerful actors whose economic interests may be curtailed, and backlash is inevitable. Increasingly, governments have imposed a security framework over water to insure the priority of sovereign rights to manage and use strategic resources. Defining water as a national security resource militarizes the water commons, and people who protest the consequences of dams can be charged with crimes against the state under legislation modelled after the US "Patriot Act," which the US pressured other nations adopt following the September 11 2001 attacks in the US. In Chile, for example, at least seven Mapuche indigenous activists protesting state hydrodevelopment and the related loss of traditional lands and resources have been arrested and charged under Pinochet-era anti-terrorism laws since 2006, with most of these arrests taking place in 2009. Similar protests and arrests with charges of violating the national security act were reported in 2009 for indigenous and other environmental activists in Turkey, El Salvador, China, India, Bangladesh, Burma/Myanmar, Thailand, Borneo, Sudan, the United States, Canada, Australia, Greece, and Brazil.

Many critics see the Southeast Anatolia project in Turkey as a prime example of this militarization of the water commons. The Southeast Anatolia Project consists of 22 dams in the Tigris and Euphrates river basins, in the heart of ancient Kurdistan. Turkey's water development projects will create large reservoirs on their national borders, a watery fence that is easily patrolled by boat. The rising waters will also drown ancient cities, create a watery barrier between Kurdish populations in Iraq and Syria, and forcibly displace many of the remaining Kurdish communities within Turkey. The construction of the Ataturk Dam alone displaced some 11,000 Kurds. The Ilisu Dam project proposes the dispossession of about another 36,000 people. These reservoirs will also capture and store a critical resource of potential marketable value for thirsty downstream and distant countries that have historically relied on the Tigris and Euphrates for water, energy, and food (Ronayne 2006; Warner 2008).

4.1.5 Emerging Trends and Cautionary Concerns

Following the release of the World Commission on Dams report in 2000 and the resurgence of national security and militarism in the United States and other nations in 2001, the number of projects financed by the World Bank and other development banks declined in Africa, Latin America, and Southeast Asia (as did bilateral aid to these regions). Development, however, did not cease, as project financing, engineering, and construction increasingly involved bilateral deals with China, especially in Africa, Latin America, and Southeast Asia. By 2010 China was financing, or Chinese companies were building, 257 large dam projects in some 59 countries (data compiled by International Rivers 2010). As a major player international development, this involvement has largely been on economic, rather than political, terms. China is putting its engineering and dam-construction companies to work on the international stage building the means to extract and process minerals and other critical resources, the rights to which are often granted to China as a part of the bilateral agreement.

In the past few years, thousands of new large hydroelectric dam and river basin projects around the world have been announced or are in the planning and early construction phases. Hydroelectric energy has been adopted by the world's nations as an appropriate strategy to combat global climate change and offset predicted water shortages. As the World Bank notes in a 2009 report justifying this resurgence of large dam building:

> Hydropower currently accounts for about 20 percent of the world's electricity supply and over 80 percent of the supply from (nonbiomass) renewable resources. Scaling up hydropower is not limited by physical or engineering potential: OECD countries have exploited over 70 percent of economically feasible potential. Yet only 23 percent of hydropower potential in developing countries has been exploited. Indeed, 91 percent of unexploited economically feasible potential worldwide is located in developing countries, with one quarter in China (2009:9).

The nations of the world, their bankers, and their builders are engaged in what may be the final act of containing and controlling the world's freshwater systems. Large dams and water diversions are being built, planned, or proposed for every major river and tributary on both sides of the Himalayas, throughout Africa, the Americas, Australia, Southeast Asia, and truly, across the world. For example, at this writing, more than 140 large dams are in various stages of planning in the Amazon basin, in an area that contains 60% of the world's remaining tropical rainforest and hundreds of thousands of indigenous peoples. And, projects that were formerly proposed and rejected because of their adverse social and environmental impact are back again—like the Belo Monte Dam on the Xingu River in the Brazilian Amazon. The World Bank financing for this dam was cancelled in 1989 following Amazonian Indian protests. Today, financing is available, and some 15,000 Kayapo and other Amazonian groups are again threatened and they are protesting over the possibility of forced eviction and the inevitable loss of a way of life.

Box 4.1d Power struggles – dams on Brazil's Xingú River

—Terence Turner

As I write this in the spring of 2010, once again, the indigenous peoples of the Xingú valley in the Brazilian Amazon are planning to make the long journey to the town of Altamira, where the Trans-Amazonica highway crosses the Xingú. Their ultimate destination will be the island of Pimental, a short distance downriver from the town, where the Brazilian government plans to build a huge hydroelectric dam they call Belo Monte after the nearest Brazilian village. The Indians' bold plan is to prevent the construction of the dam by building a new village directly on top of the proposed dam site and maintaining

(continued)

Box 4.1d (continued)

their occupation until the government abandons its plans for the dam. The planning for the encampment is being led by the Kayapo, the largest and most politically organized of the indigenous nations of the region, but other indigenous groups are also participating.

The Kayapo, however, are not waiting for the discussion of the plan for the encampment among the 23 indigenous groups of the Xingú Valley to reach consensus. They have already seized the ferry that carries Brazil Route 80, an important link in the Trans-Amazonica highway system, across the Xingú River at the Kayapo village of Piaraçú. The ferry and the river crossing are now under guard by armed Kayapo warriors, who have announced that they will continue their blockade until the government negotiates with them about their plans for the Belo Monte dam.

This will not be the first indigenous encampment organized by the Kayapo in their effort to stop the building of dams on the Xingú. In 1989, when the government first set out to implement its plan for a giant hydroelectric complex on the Xingú, with financial support from the World Bank, the Kayapo led a great rally of 40 indigenous nations at Altamira against the scheme, setting up an encampment of several hundred Indians at a Catholic retreat center just outside the town. The five-day rally was extensively covered by national and international media, and succeeded in persuading the World Bank to withdraw its planned loan for the construction of the dams.

After the 1989 Altamira meeting, the Xingú dam scheme remained dormant, but not dead, for two decades. Two years ago it was revived as the centerpiece of the Lula government's Project for Accelerated Development. The Xingú River is one of the major tributaries of the Amazon. With its numerous affluents it has created a valley larger than Texas in the US that remains perhaps the least disturbed and most diverse ecosystem in Brazilian Amazonia. It is unquestionably the most culturally diverse. Twenty-three indigenous peoples of distinct cultures and languages make their homes there, most of them among the headwaters of the Upper Xingú, which has been made a national park by the Brazilian state. In the Middle Xingú region just to the north (downriver) of the National Park, the large and politically dynamic Kayapo people have their territory, consisting of seven mostly contiguous reserves with a combined area of 150,000 km^2 (roughly the size of Austria).

Further downriver, between the Kayapo reserves and the mouth of the Xingú where it empties into the Amazon, several other indigenous peoples live in varying degrees of proximity with Brazilian settlers, some of them "river people" who subsist on a technology little different from that of the Indians, but others dwelling in towns they have established along the river and the Trans-Amazonica highway, which crosses the Xingú near the largest town, the regional capital of Altamira.

(continued)

Box 4.1d (continued)

Over the years, this variegated system of social and cultural groups has evolved a relatively sustainable pattern of coexistence with one another and the even more varied riverine and forest ecosystems of the Xingú valley. All of these systems, however, have now been imperiled by the Federal government's plan to build a series of six giant hydroelectric dams along the Xingu and its largest tributary, the Irirí. The largest of these dams, Belo Monte, is to be the first built. Construction is scheduled to start in January 2011.

The master plan for damming the Amazon river system, which includes Belo Monte and the Xingú dams, was originally created in the 1970s by the military dictatorship then in power. It essentially treats the Amazon as a reservoir of natural resources to be extracted without regard for the destruction of its riverine and forest environment or the displacement and pauperization of its indigenous and local Brazilian inhabitants. Recognized by many scientists and engineers as technologically problematic and inefficient, this mega-dam proposal is now the centerpiece of Brazil's "Accelerated Development Project."

If built, Belo Monte will be the third largest hydroelectric dam complex in the world, comprising one huge dam and two smaller dams, and requiring the diversion of the water from a 100-km stretch of the river's channel through canals and underground tunnels to two massive arrays of turbines. The whole system will have a peak generating capacity of 11,200 kilowatt hour. Critics of the project, however, have pointed out that this level of output would be attainable only for four months out of the year at the height of the rainy season. For the remaining eight months, during the dry season, the level of the river falls by nine metres or more, so that much less water would be available to flow through the turbines, and the average output would fall to an annual rate of only 4,000 kilowatt hour. This means that the electricity that the dam would generate, measured against the enormous cost of the dam, would be considerably more expensive than that potentially produced by alternative means.

Taking into additional consideration the relatively short life-expectancy of dams in the Amazon because of silting and acidic erosion of turbine blades, the Belo Monte does not appear to be economically viable as a stand-alone dam, without another big dam upriver with a large enough reservoir to release a sufficient volume of water during the dry season to keep Belo Monte producing at close to its peak capacity all year. There are plans for such a dam, called Altamira, which would have an enormous reservoir that would flood a vast area of forest. Upriver from that, four other sites have been selected for a whole series of dams that could feed into the reservoir of the Altamira dam. Despite these plans, the government insists that Belo Monte as viable by itself, that it is currently planning only to build one dam on the Xingú. Such assurances mean that environmental and social assessments do not need to consider, and development does to need to address, the cumulative effects of a series of dams.

(continued)

Box 4.1d (continued)

These objections, however, are not the only serious problems of the Belo Monte project. The 100-km section of river that would be diverted to pass through the turbines and thus drained of its water now passes through two indigenous reservations (Arara and Paquiçamba-Juruna), whose people depend on the river for fish and transportation. The villages they currently occupy would thus become unviable. The Brazilian constitution mandates that indigenous communities must be consulted in advance before development projects are carried out within their reserved territories, and that all local peoples must be given a chance to discuss with responsible officials any government projects that will affect their livelihoods. The government agencies charged with building the dams have refused to comply with this legal requirement in the cases of the two indigenous communities affected, as they have in those of the other indigenous peoples of the Xingú. They have also failed to produce a satisfactory environmental impact evaluation, which is legally required as the prerequisite for the issue of a license to build the dam. Instead, the license was released, under intense political pressure, in the absence of a completed Environmental Impact Report.

The Brazilian state is far from monolithically behind the Xingú dam Project. The Public Ministry, an autonomous governmental agency empowered to decide on the constitutionality and legality of government projects and actions, openly denounced the Belo Monte dam project as illegal and in violation of the constitution, and moreover as likely to produce an environmental catastrophe in the Xingú. On April 7, 2010, the Public Ministry announced its decisions that the government's plan to hold the auction at Altamira unconstitutional and in violation of several existing laws, and that the Belo Monte Project would violate the constitutional and legal rights of indigenous peoples whose territories and communities it would either flood or cut off from access to the river. As a consequence of these decisions, the Public Ministry called for annulling the government's decision to hold an auction on April 20 for Belo Monte construction bids.

In reaction to the Public Ministry findings, the Attorney General of Brazil threatened to have Public Ministry attorneys arrested and imprisoned for interfering with the project (As of late 2010, they have not been arrested, though the threat of this illegal attempt at repression of political opposition to state policies remains open). In the week before the April 20 auction, a Federal judge in Altamira handed down a judgment based on one of the Public Ministry's two briefs annulling the government's decision to hold the auction. This decision was then reversed by the Regional Appeals Court in Brasília. The Altamira court judge then handed down a second order to cancel the auction on April 19. His decision, a 50-page document with extensive legal arguments, precedents and references, was also based squarely on the

(continued)

Box 4.1d (continued)

Box Fig. 4.1d.1 Dito Kayapo, of the indigenous Kayapo tribe, works with his laptop during a public hearing at the Commission of Human Rights of the Federal Senate in Brasilia, Wednesday, December 2, 2009. Native communities of the Amazon rain forest are protesting the Brazilian government's decision to build the massive hydroelectric Belo Monte dam in the Xingu River (Photo credit: AP Photo/Eraldo Peres)

Public Ministry's documents. This decision was also reversed by the Appeals Court and the auction was held. The cited rationale for the Court of Appeals reversals was Brazil's need for energy and the demands of the Project for Accelerated Development.

These ecopolitics further aroused the opposition of the broad and growing array of elements of Brazilian civil society who have been organizing against Belo Monte and the other planned Xingú dams. Tens of thousands took to the streets in Brasilia on April 12, 2010 to call for the cancellation of the project.

The protest gained international attention, in part because of the involvement of James Cameron, the writer and director of the movie *Avatar*, and members of the cast of the film. In March 2010, James Cameron visited the site of the planned Belo Monte dam, and some of the indigenous villages that it would affect. He was so struck by the similarities in their situation with that of the Navi, the indigenous natives in his film *Avatar*, that he committed himself to support their movement against the dams. His return to Brazil with members of the cast to take part in the march in Brasilia, was a public affirmation of his support for their cause. Sigourney Weaver, of the *Avatar* cast, later led a similar march in New York against the Xingú dams.

There are clear parallels between the battle of the fictional indigenous people of Avatar against the attempt by a giant corporation to extract precious minerals from their planet. The film is modeled on the Amazon rainforest and the decades-long struggle of the inhabitants of the Xingú valley against the damming of their rivers to generate power, much of which is intended for the production of minerals such as aluminium for export. In both cases—film and reality—the collateral damage of extractive industry threatens to destroy the ecosystem and way of life of the native people. And, in both cases, they resist.

As this is written, a Kayapo delegation led by Chief "Raoni" (or as he pronounces it, Rop-ni) of the Xingu Kayapo, is travelling through France, Belgium, and Luxembourg, visiting government ministers and heads of state and appealing for support of the indigenous campaign against the Xingú dams. Other campaigns, some involving other tours by indigenous leaders, are getting under

(continued)

Box 4.1d (continued)

way in other European countries and in North America. The Brazilian government's attempt to push ahead the Xingú dam scheme in the face of the mounting storm of opposition from local settlers, indigenous peoples, environmentalist and human rights NGOs, other sectors of Brazilian civil society and important elements of the state itself (such as significant parts of the judicial system and political opposition) is thus becoming a problem for Brazil's foreign relations. Within Brazil, it has already moved from its original status as a localized problem involving indigenous rights and ecological impacts of a dam in a remote part of the Amazon to a major legal, political and constitutional crisis involving Brazil's political conduct as a democratic state.

At stake in this crisis is Brazil's political ability to reconcile and accommodate the demands of its capital-intensive policy of economic growth, epitomized by its Accelerated Development project, with the principles of constitutional legality and democracy supported by its rapidly growing middle class, in alliance with the indigenous and settler groups of its vast Amazonian interior. An irony of the Xingú dam project is that it has done much to bring this historically unique alliance into political being, and in so doing has inadvertently made a profoundly hopeful contribution to the development of Brazilian democratic civil society. This contribution, however, has only been realized thanks to the courage, leadership and political resourcefulness of the Kayapo, other indigenous groups who have supported them, and the Brazilian social movements of the Xingú Valley. Whatever the immediate outcome of the struggle over the Belo Monte project, the broad alliance of indigenous peoples, Brazilian settlers and social movements, environmentalists, human rights organizations and elements of the Brazilian state committed to democratic legality and constitutionality in common opposition to the dam scheme the movement has built, will continue the fight against the other dams the government hopes to build in the Xingú, with catastrophic effects on the flora, fauna and human inhabitants of the Xingú valley.

This new boom in large-scale development project development is largely promoted as measures that will improve local and national economies, strengthen national security, and combat global warming. Global warming, coupled with deforestation, mining, agriculture, and urban sprawl in watersheds, threaten water supplies for more than half the world's people, especially the poor. One of the major ways in which governments are responding to threats posed by global environmental change is to propose or authorize massive efforts to capture, hold, and re-route rainfall in the form of a series of dams and river-linking projects. Such endeavours, if completed, threaten the viability a huge portion of the world's remaining biological and cultural diversity.

Closer examination finds that the driving force behind many plans is not green power or food security. Hydropower generation instead will enable energy and extractive resource development, allow export of electricity and water to sustain the high-energy and high-water consumer lifestyle of some first-world nations, and permit the development of a market in *water futures*, meaning the capturing of water to sell in a future world water market.

In Canada, for example, hydrodevelopment is closely linked with mining and energy development, and some 63% of all surface water is used for energy production. Some of the proposals floating around include water diversions and containment to support nuclear power plant operations, which in turn produce the electricity and the steam to support oil extraction from the Alberta Tar Sands deposits.

In Guatemala, existing and planned hydrodevelopment is promoted as poverty alleviation—providing electricity to rural households—yet in fact it supports the urban and regional grids (with power flowing north through Mexico to the United States), while local power is often used to support the mining and processing of nickel, gold, and other minerals. The Guatemalan initiatives are part of the multi-billion-dollar Plan Puebla-Panama, formally initiated in 2001 and retitled the Mesoamerican Integration and Development Project in 2008, which drew on the decades-old templates of modernization:

- Build highways, 381 hydroelectric dams, and new utility grids in the rural hinterlands from Guatemala to Panama to create the infrastructure that attracts foreign investment and taps the mineral and energy wealth of Central America.
- Put the rural peasantry to work in manufacturing, export agriculture, and biofuel production.
- Move people off communal lands and individually title them, and with privatization allow the financing of agro-loans and the production of "boutique crops" destined for the wealthy consumers in the United States and Canada.
- Bring the region (Southern Mexico, Central America, and Colombia) into modernity.

Other projects, like the aforementioned Southeast Anatolia Project in Turkey, deploy a similar neoliberal development template—a product of 1970s project planning. Dams in Turkey will generate electricity for the European Union (cf. Warner 2008).

How will affected people respond to the threats posed by this new boom in hydrodevelopment? In the event that protests are futile, where will these people go? How will they survive? Will the world see a continued escalation in global rates of poverty, health, misery, and violence, as a result of the collective experiences with this new round of development and displacement? Or, in the effort to incorporate broader participatory processes and manage rivershed resources more sustainably, will we see large-scale hydrodevelopment as the increasingly rare alternative?

References Cited

Blaser, M., H.A. Feit, and G. McRae (eds.). 2004. *The way of development: Indigenous peoples, life projects, and globalization*. London/New York: Zed Books.

Bodley, J.H. 2008. *Victims of progress*, 5th ed. Lanham: AltaMira Press.

Cernea, M.M., and C. McDowell (eds.). 2000. *Risks and reconstruction: Experiences of resettlers and refugees*. Washington, DC: World Bank.

Clark, D.L., J. Fox, and K. Treakle (eds.). 2003. *Demanding accountability: Civil society claims and the World Bank Inspection Panel*. Lanham: Rowman & Littlefield.

Fernandes, W. 2007. Singur and the displacement scenario. *Economic and Political Weekly* 42(3 Jan 20–26): 203–206.

George, S., and F. Sabelli. 1994. *Faith & credit: The World Bank's secular empire*. Boulder/San Francisco: Westview Press.

Johnston, B.R. (ed.). 1997. *Life and death matters: Human rights and the environment at the end of the millennium*. Walnut Creek: AltaMira Press.

Johnston, B.R. 2005. *Chixoy Dam Legacy Issues Study*. Vol. 1: *Executive summary: Consequential damages and reparation – recommendations for remedy*. Vol. 2: *Document review and chronology of relevant actions and events*. Vol. 3: *Consequential damage assessment of Chixoy river basin communities*. Santa Cruz: Center for Political Ecology. http://shr.aaas.org/guatemala and www.centerforpoliticalecology.org

Johnston, B.R. 2010. Chixoy dam legacies – The struggle to secure reparation and the right to remedy in Guatemala. *Water Alternatives* 3(2): 341–361.

McGee, B. 2010. Participation with a punch: Community referenda on dam projects and the right to free, prior, and informed consent to development. *Water Alternatives* 3(2): 162–184.

Oliver-Smith, A. (ed.). 2009. *Development and dispossession: The crisis of forced displacement and resettlement*. Santa Fe: School for Advanced Research.

Reservoir Development Bureau of the Ministry of Water Resources of China. 1999. Probe Chinese characterized reservoir resettlement ways. *Water Resources of China* 2(23).

Richter, B.D., S. Postel, C. Revenga, T. Scudder, B. Lehner, A. Churchill, and M. Chow. 2010. Lost in development's shadow: The downstream human consequences of dams. *Water Alternatives* 3(2): 14–42.

Ronayne, M. 2006. Archaeology against cultural destruction: The case of the Ilisu dam in the Kurdish region of Turkey. *Public Archaeology* 5: 223–236.

Scudder, T. 2005. *The future of large dams: Dealing with social, environmental, institutional, and political costs*. London: James and James Science Publishers.

Scudder, T. 2009. Resettlement theory and the Kariba case: An anthropology of resettlement. In *Development and dispossession: The crisis of forced displacement and resettlement*, ed. A. Oliver-Smith, 25–48. Santa Fe: School for Advanced Research.

Shi, Guoqing, Xiaonan Chen, and Qingian Yu. 2008. Post-resettlement support: Policy aimed at poverty alleviation and sustainable development. In *Proceedings of the 16th IAHR-APD Congress and 3rd Symposium of IAHR-ISHS*, 1456–1462. Beijing: Tsinghua University Press.

Warner, J. 2008. Contested hydrohegemony: Hydraulic control and security in Turkey. *Water Alternatives* 1(2): 271–288.

World Bank. 2004. *Involuntary resettlement sourcebook: Planning and implementation in development projects*. Washington, DC: World Bank.

World Bank. 2009. *Directions in hydropower: Scaling up development*. Washington, DC: Water Sector Board of the Sustainable Development Network of the World Bank Group.

World Commission on Dams. 2000. *Dams and development: A new framework for decision-making. The report of the World Commission on Dams*. London: Earthscan Publications.

Chapter 4.2
The Lesotho Highlands Water Project: Water, Culture, and Environmental Change

Robert K. Hitchcock

Southern Africa is facing tremendous challenges when it comes to water. In 2009, half of the 15 states of the Southern African Development Community were considered to be facing serious water-related constraints, and some have been classified as water-scarce. As the South African Minister for Water Affairs and Forestry, Kadar Asmal, said in 1996,

> We have a potential for long-term water crisis in the region which could cripple the sustainability of our development if not handled with great foresight. Conflict, both internally and between countries, could arise unless the challenge is dealt with in a progressive and transparent manner (Asmal, quoted in Chenje and Johnson 1996:177).

There are some reasons to think that the challenge may be met without conflict. One of the states facing severe water shortages, South Africa, worked with Lesotho, which had water to spare. Now some of that abundant water is being channelled to South Africa through a major water transfer scheme, the Lesotho Highlands Water Project (LHWP). This project may also serve as a model for how to handle involuntary relocation.

The impacts of global and local environmental change, growth in demand for energy, and shifts in land use patterns have caused major concerns in southern Africa. As Simms and Reid (2005:2) point out, the continent of Africa is particularly vulnerable to the impacts of climate change. This vulnerability results in part to its geographic location on the planet – the majority of the continent falling in the tropics and sub-tropics – as well as from the fact that a significant percentage of Africa's people depend on natural resources to provide for their household needs. Work on the impacts of climate change in southern Africa predicts that the eastern part of the region will receive greater rainfall and potentially have more floods,

R.K. Hitchcock (✉)
Center for Global Change and Earth Observations (CGCEO), Michigan State University, East Lansing, MI, USA
e-mail: hitchc16@msu.edu

B.R. Johnston et al. (eds.), *Water, Cultural Diversity, and Global Environmental Change: Emerging Trends, Sustainable Futures?*, DOI 10.1007/978-94-007-1774-9_23,

Map 4.2.1 South Africa and Lesotho

whereas the opposite was predicted for the western half, where droughts, rising temperatures, and crop failures are said to be likely in the near and longer term (Low 2005; Simms and Reid 2005; Dinar et al. 2008).

In order to meet water and energy demands, southern African governments have initiated both large-scale and small-scale water development projects. The largest water user in southern Africa is agriculture, and irrigation is on the increase in some areas, with some significant environmental and health effects. Many of the larger rivers in southern Africa have been dammed. There are nearly 30 dams on the Gariep (Orange) River and nearly the same number on the Limpopo. The construction of the Kariba Dam on the Zambezi River in the mid-1950s led to a massive resettlement effort, with some 57,000 Tongans and others being moved out of the basin, many of them into the plateau region of Zambia (Clark et al. 1995).

As the Kariba and other large dam projects in Africa showed, people who are relocated by dams experience serious problems (Scudder 2005). Project-affected peoples often are moved to upland areas that are less productive ecologically. As a result, their incomes and agricultural yields decline, and they have to cope with various kinds of social, psychological, and physiological stresses. In many cases, the people who are relocated are forced to become at least partially dependent on food relief and other livelihood supports provided by international relief agencies, governments, and civil society.

One of the areas in which southern African nations have made significant progress is in the rules and procedures governing involuntary relocation or resettlement resulting from the construction of large-scale water infrastructure projects. This progress can be seen in the case of the LHWP, currently one of Africa's largest hydroelectric projects. It provides water to South Africa and electricity and revenues to Lesotho. Detailed social and environmental impact assessments and socioeconomic and ecological monitoring characterized this project from its inception in the 1980s. Particularly useful were the efforts of the project authorities, consultants, non-government organizations (NGOs), researchers, and a panel of environmental experts to collect both quantitative data on project-affected people and qualitative data on individuals' perspectives. Local people characterized Lesotho's water as "white gold" because of the many benefits it brings.

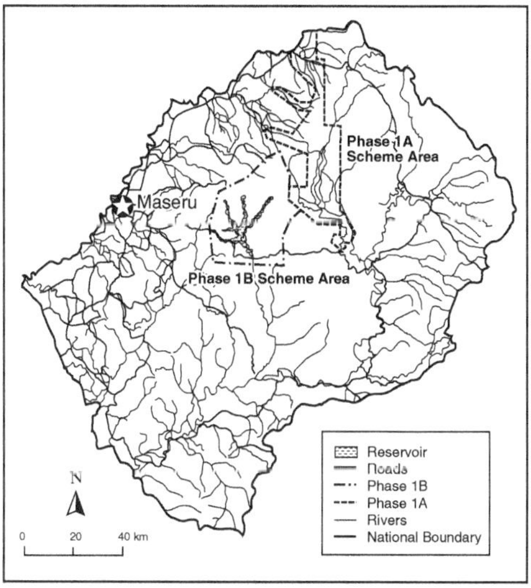

Map 4.2.2 Lesotho Highlands Water Project

4.2.1 The Lesotho Highlands Water Project

Lesotho is a land-locked country in southern Africa; it is slightly smaller than the U.S. state of Maryland (30,355 km^2). Surrounded completely by South Africa, Lesotho today is a constitutional parliamentary monarchy that has had complex relations with its giant neighbour. One of the front-line states that opposed apartheid in South Africa, it experienced raids, assassinations, and political and economic pressures. In 1986, after a military junta took over the Lesotho government, the country signed a treaty with South Africa setting up the LHWP. The project's purpose is to transfer water from the headwaters of the Gariep River (called the Senqu in Lesotho) to the Vaal River catchment in South Africa and on to the commercial and industrial heartland in Gauteng province. Besides water transfer, the LHWP also aims to provide hydroelectric power to Lesotho. Phases 1A and 1B of the project, implemented between 1987 and 2008, saw the development of infrastructure, including roads, power lines, administrative and engineering facilities, dams, and tunnels.

Phase 1A, constructed at a cost of approximately 20 billion South African Rands, included the erection of a large dam at Katse on the Malibamatso River, the highest dam in Africa at 180 m; a 45 km transfer tunnel to the 'Muela hydropower station and tail-pond; and a further 37 km delivery tunnel to the Ash River in South Africa.

Phase 1B includes the Mohale Dam, the highest rock-filled dam in Africa at 145 m, and the Mohale Reservoir, located on the Senqunyane River. A 32 km tunnel connects the Mohale Reservoir to the Katse Reservoir, and a weir was constructed at Matsoku near 'Muela on the Matsoku River, from which a 5.6 km transfer tunnel runs to the Katse Reservoir. The 72 milliwatts (MW) hydropower station at 'Muela connects with the southern African power pool and supplies Lesotho's electricity needs without producing any greenhouse gases.

However, all this development had its costs for the people living in the region. Lesotho is largely mountainous. The Maluti Mountains, the so-called highlands, make up some 61% of the country. The Malutis can be viewed as 'water towers' out of which rivers and streams flow to the foothills and lowlands and on to South Africa and beyond. The highlands of Lesotho in the past were remote, with the resident Basotho people mainly engaging in livestock production and subsistence agriculture, and during the past century, migrant labour, especially in the mines of South Africa (Murray 1981; Ferguson 1994).

Fig. 4.2.1 Basotho sitting outside a traditional Basotho hut in the Lesotho Highlands (Photo credit: Robert Hitchcock)

Only about 11% of Lesotho comprises arable soils, primarily in the foothills and lowlands as well as in the deep mountain valleys. People raise cattle, horses, donkeys, and small stock (sheep and goats) in various parts of the country, with the mountains serving as summer pastures for herds from the lowlands and foothills. The LHWP required some of the people of the highlands in the Katse and Mohale areas to relocate. The project would deprive them of substantial amounts of arable land, grazing, trees, and other natural resources crucial to their economy.

The highlands population consists of socially integrated Basotho communities who hold fast to their customs and traditions, and 99% of directly-affected population were Basotho. Social and environmental impact assessments conducted for this project included baseline studies of people and natural resources likely to be affected by the project, proposed a valuation scheme and the payment of compensation to project-affected people, and called for the relocation of 71 households in the Katse Basin in Phase 1A (1988-1996) and the resettlement and relocation of 321 households in the Mohale Basin and in the lowlands and foothills of Lesotho in Phase 1B (1996-2006). Once the preferred options of the various project-affected households were determined, the planning for the actual movement of the households was carried out. People who were relocated were those whose homes would be inundated by dams or whose homesteads were directly affected by roads, tunnels, gravel pits or other project-related infrastructure.

The LHWP provides for compensation, resettlement, and development initiatives aimed at ensuring that project-affected people will be "enabled to maintain a standard of living not inferior to that obtaining at the time of first disturbance" (Government of Lesotho and Government of South Africa 1986: art. 7, par. 18). Compensation is done both in kind (in the form of grains and pulses) and in cash. The project replaced houses, schools, and churches and made payments for losses of other assets, including brush wood, trees, thatching grass, and stones used for construction. In addition, the LHWP provided communal compensation in the form of cash payments to community cooperatives or local legal entities (LLEs). Both upstream and downstream populations received compensation.

In addition, the LHWP provided for environmental rehabilitation, natural resources management, and conservation. Efforts were undertaken to conserve endangered species such as the Maloti minnow (Pseudobarbus quathlambae), the bearded vulture (lammergeier, Gypaetus barbarus), and the spiral aloe (Aloe polyphylla). Nevertheless, the project had negative impacts on biodiversity and habitats. To cope with some of these impacts, the project authorities set aside some protected areas, including a new national park, Ts'ehlenyane in the 'Muela area, and the Bokong Nature Reserve, located in the highlands in an alpine wetland with rare vegetation.

The project also made an effort to protect culturally important sites and to relocate the graves of the deceased relatives of people affected by the construction. The cultural heritage management program included in the LHWP included the establishment of a small cultural park at a site of historic importance to the Basotho people in Liphofung, consisting of a small museum, shop, and walkway that winds down to a rock shelter used long ago by San (Bushmen), who painted on the walls. Subsequently, it was used to shelter members of the Basotho nation by its founder, Moshoeshoe. More recently, young herders kept cattle and sheep in the shelter to protect them from the elements. This cultural park is now run in cooperation with a Community Consultation Forum from the 'Muela area.

4.2.2 Resettlement and Compensation

Examples from around the world demonstrate that compensation programs for involuntary resettlement, damage to the land and assets of individuals and communities, and the re-establishment or improvement of the lives of people affected by development projects have varied tremendously over the past 50 years. The systems for handling people's losses have ranged from providing no compensation whatsoever to offering relatively substantial compensation packages aimed at improving the standards of living of project-affected people. The main conclusion about these compensation programs is that, in general, they have failed to restore the livelihoods of people affected by projects, and in a number of cases they have made people worse off.

The projects that typically have had the most comprehensive compensation and resettlement programs are large-scale water projects, primarily dams, and big development efforts involving relocation, such as the Transmigration Program in Indonesia (which moved millions of landless Indonesians from the densely populated islands of Java, Bali and Madura, to less-populated outer islands such as Kalimantan, Sumatra, Sulawesi, Maluku and West Papua). Post-project analysis of these large-scale development programs has revealed that in most cases, safeguard measures failed, in large part because environmental impact assessments and especially social impact assessments did not get accurate counts of the people and land adversely affected by those projects. Exceptions exist in those cases in which projects affect a relatively easily defined area, one example being the James Bay Hydroelectric Project, which affected 11,000 Cree and Inuit in northern Quebec, Canada (Scudder 2005).

Resettlement is a complicated process, one that is often extremely hard on the people who are relocated. A major problem with development-related resettlement projects is that planners generally have tended to focus attention on loss of residence (i.e., homes) rather than on loss of access to the means of production (especially land, grazing resources, and wild resources on which people depend for subsistence and income). A second major problem is that in nearly all cases, planners have seriously underestimated the degree of impact on populations. The World Bank arguably has had the most comprehensive guidelines on resettlement and social impact assessment. However, these guidelines call for the *restoration* of livelihoods of people affected by projects, but they do not argue for *improvement* of the standards of living of people who have been affected by projects. The World Commission on Dams (2000) argued for the need to improve the livelihoods of project-affected people as well as those downstream of the project.

Many of the programs involving compensation have tended to use cash as the primary compensation for lost assets. It should be stressed that there are some drawbacks to cash compensation. Based on global experience in resettlement and compensation programs, cash payments have generally not served to restore the incomes of the people who were resettled. One of the reasons for the difficulties with cash compensation is that recipients sometimes expend their money very quickly. Frequently, few opportunities to invest locally exist. A third problem relates to the control of the cash. In some cases, adult males in the household appropriate the cash for their own use while women and children end up being disadvantaged.

An annuity system is an attractive alternative because it allows for investment and can be managed with relative ease. An annuity system has the advantage of being able to accommodate the various sources of individual and community income. It allows people the flexibility to save their money, divide it among designated kin or other people, spend it as they choose, or pool their funds for use in community projects. A 'nest egg' plan can allow for the banking of funds indefinitely. Individuals could choose to invest a portion of their annuities in a special-purpose activity such as a revolving credit fund. These schemes have worked reasonably well in a number of developing countries, and they have served as a means of providing people with the capital necessary to initiate businesses and both on-farm and off-farm income-generating activities. If development is to be sustainable, local people must have access to the resources, information, and technical assistance, and they must be able to participate in decisions that affect them. This is a position taken by many Basotho, both rural and urban.

4.2.3 Economic and Social Rehabilitation

Large-scale water projects like the LHWP almost inevitably have not only economic and environmental impacts but also significant social and cultural effects. The LHWP led to the loss of religious sites, the inundation or removal of graves, and the separation of families that had long lived together as a result of resettlement. For many families a difficult period of adaptation and uncertainty followed physical resettlement. The resettled households had simultaneously to reconstruct their sources of household income and their relations with their neighbours, settle their children into new schools, and adapt to new climatic and environmental conditions. They also had to learn to manage their new financial situations, which for many had been modified substantially by the substitution of their familiar land and natural resources for cash payments.

The development-displaced households that moved to the Maseru, the capital of Lesotho, found new opportunities as well as difficulties. On the positive side, some people found new opportunity in the urban economy. Some relocated individuals took their lump sum compensation payments and invested in taxis, a highly competitive business. Others invested in *malaene*, rental property located close to the industrial area where there were garment factories. The local economy in these neighbourhoods were flourishing at the time, but by 2005 the massive influx of Chinese-made garments into world markets, including Lesotho's, and China's admission to the World Trade Organization, led to a crisis in Lesotho's garment industry. Most factories closed down, leaving large numbers of workers unemployed. A decline in the demand for rental accommodation resulted.

The people who resettled in Maseru found that living costs in the city were far higher than in the mountains. The transition from an agricultural and pastoral life to urban life was by no means easy, judging from comments of resettlers. One group who had moved to Makhoakhoeng, a village on the edge of the city of Maseru, found that their hosts refused to allow them to bury one of their members who had died there. Although this problem was eventually resolved, before that happened the hosts called on the Lesotho Highlands Development Authority to relocate the settlers to another area.

For the most part, the resettlers who moved to the foothills of Lesotho integrated reasonably well with their hosts, entering into sharecropping and land rental agreements with them and intermarrying. Host villagers sometimes saw the resettlers as resources since they arrived with cash from their compensation payments. A concern of some in the host villages was that having additional people in their midst would place greater strains upon their grazing, firewood, and water resources. People were particularly concerned about the addition of substantial numbers of livestock on the range.

Initially, resettlers moving to the foothills had difficulties getting access to land for crops, in part because host populations asked high prices for the land or were reluctant to allow the incomers to rent or sharecrop. As familiarity and mutual trust developed, the situation began to change and within a few years virtually all of the resettled households had land of their own or rented land or sharecropped. Also, settlers who started off renting or sharecropping tended to buy land as they settled in to their new surroundings.

Interviews with people who had resettled in the foothills and lowlands or who had relocated within the highlands revealed a variety of views on whether they felt themselves to be better or worse off as a result of the LHWP. Half of the households who had moved to the foothills said that their agricultural yields and commercial sales of agricultural products were now higher. At the same time, they said that their cash expenditures for agricultural inputs such as seeds and fertilizers were far greater. In interviews conducted annually over a 10-year period, I found that approximately one fifth of the resettlers living in the foothills experienced problems in maintaining their livestock, as most of their flock had been left to pasture the highlands, and the great distance between households and their livestock left many vulnerable to theft. Furthermore, nearly all of the people resettled in both the foothills and lowlands reported that while they now had greater access to social services and markets their costs of living had also significantly increased.

Of the 321 households that had to move as a result of the Mohale Reservoir, dam, and other infrastructure development, 61 opted to relocate locally, staying in the Mohale Basin. Some of them said that they then had to contend with greater competition for pasture, arable land, and natural resources such as firewood. People from several communities in the Mohale Reservoir inundation area relocated to a nearby village, Ha Koporale, above the water line, in part because of the good grazing there. On these steep slopes there was hardly any land suitable for farming and gardening, and as a result some people opted to move elsewhere in the basin.

Households headed by women or elderly people were usually short of labour and found themselves worse off after the LHWP than they were before, in spite of the Minimum Threshold Payments that were provided to people who fell below a specific income level set by the Lesotho Highlands Development Authority (LHDA). The relatives, friends, and neighbours who had previously helped these households out in the fields and with herding had moved away. Many households split up, with younger members moving to town or starting new homesteads, leaving behind the elderly people.

The surveys and reports on the affected families (e.g., Panos Institute 2001; Transformation Resource Centre 2004, 2005; Thamae and Pottinger 2006) show that the LHWP had widely disparate effects. Some of the resettled people expressed strong views about the effects of the LHWP, including complaints that the project had led to local-level climatic changes, such as a reduction of rainfall and a drying up of springs. One man, Sebili, said that life in the past was better, with good rains. 'This land was very good for crops but now it is…changing and is causing us hunger during these times of drought' (Panos Institute 2001:3). He went on to say that with resettlement, "the wisdom of living in this place will be lost" (Panos Institute 2001:4). Another man, Lipolo, from the Molika-Liko area, said that "there was drought, but not this bad" (Panos Institute 2001:7).

Some households increased their incomes temporarily by having members work on the dam, tunnels, feeder roads, or the water and sanitation projects carried out in the project area. Over time, some households were able to capitalize on development opportunities provided by the project or to invest their compensation funds in small business enterprises. Many other households saw their income levels fall substantially, especially immediately after resettlement. Households reported food shortages more frequently than before relocation.

A critical constraint for many of these households was that they lacked members with the skills to cope successfully in the complex and diversified economy of Maseru and its environs. In a post-relocation survey carried out in 2004, respondents assessed their own situations as follows: 8% said that they felt very stressed and that they missed what they termed "the good old days" in the highlands. In contrast, 21% said that they liked their new situations and that they were settling in well and making strides toward raising their incomes and living standards. This positive outcome was particularly true for those households in the foothills that owned land and those living in the lowlands who owned taxis. Of the households in the foothills, 60% had the use of arable land. Even more, 66%, had diversified their sources of income. Eight households at Thetsane, an industrial area in urban Maseru, for example, were selling water, growing vegetables for sale, selling telephone cards, and renting telephone facilities.

People who had moved to the foothills now had to rent or purchase land, whereas in the mountains they had access to land that had been granted to them by traditional authorities. Even when they did get land from local land committees in the foothills, they had difficulties in getting what are known as Form Cs, which are proof of land rights in Lesotho. In some cases, people who had made sharecropping or rental agreements learned that the landowners had denied making those agreements and had refused them the use the land. Nevertheless, 95% of the resettled households at one foothills village (Ha Koporale) had been able to plant crops by 2009.

The fact that the displaced people ended up in different places, some of which were in new communities in the Mohale Basin and others in the foothills or lowlands, meant that their old communities were thoroughly dismantled. One result was it became much more difficult for the affected families to articulate their shared difficulties and grievances to the authorities. They attempted to do so by going to local organizations known as Community Area Liaison Committees, traditional authorities, the Lesotho Highlands Development Authority, and the Lesotho Ombudsman. They also spoke frequently to various NGOs working in the area, such as the Transformation Resource Center, which publicized their complaints.

A number of Basotho in the project area and in the resettlement locations mentioned greater vulnerability to crop failure and reduced crop yields. Fishing, they said, was harder now in the reservoir, and even the large-scale fisheries pilot projects, one of them led by a company known as Katse Fresh Fish, were having problems with fish production. These occurred, for example, when high winds on the reservoir surface broke the trout cages, releasing the caged fish into the reservoir, or when fish died off because of high temperatures and lower water levels. Local fishermen and women said told researchers and Panel of Environmental Experts members that it was harder to get fish in some of the rivers than it used to be. Local people also said that wild medicinal and food plants were less accessible than they used to be. Some of the scarcity, they said, was because there were more people on the landscape using these resources. They also said that local-level habitat changes were responsible for shifts in the availability of economically and ritually important plants.

Box 4.2a The fight over fish in the Lesotho Highlands

—John A. Ledger and Robert K. Hitchcock

Box Fig. 4.2a.1 Mosotho fishing group member with her catch, 2009 (Photo credit: Robert K. Hitchcock)

The implementation of the Lesotho Highlands Water Project (LHWP) has led to a struggle over biodiversity conservation versus development in the highlands of Lesotho. Preserving the Maloti minnow (Pseudobarbus quathlambae), a critically endangered fish, was identified early on in the implementation of the LHWP as the single most important component of the natural heritage and environment programme. The population of minnows in the Mohale catchment is genetically distinct and has been protected for perhaps 250,000 years by a waterfall on the Senqunyane River, now situated 3 km below the Mohale Dam wall.

The completion of the tunnel between Mohale and Katse has enabled movement of fish into the Senqunyane catchment for the first time in 250,000 years. The construction of two large dams, Katse on the Molibamatso River and Mohale on the Senqunyane River, has resulted in an increase in yellowfish and trout populations in the two reservoirs and the rivers that lead into them. Smallmouth yellowfish have already invaded the dam; trout will certainly follow, and when this occurs, the extinction of the Maloti minnow population will simply be a matter of time, unless appropriate interventions are made.

At the same time that these events have occurred, the Lesotho Highlands Development Authority (LHDA) allowed a private South Africa company, Katse Fresh Fish (KFF), to establish a commercial caged trout aquaculture operation in the Katse Reservoir. There have already been instances when the cages were breached by winds and large numbers of young trout escaped into the dam. Local artisanal fishers have benefitted from the presence of the KFF operation, since some of them have been able to catch yellowfish and trout in the reservoir. But the presence of this commercial fishery operation and LHDA plans to expand it to other parts of the Katse Reservoir and possibly to the Mohale Reservoir pose grave risks for the well-being of the Maluti minnow. It is possible that for the first time in decades, a large dam project might lead to the extinction of a species. The extinction of the Maloti minnow would be a significant failure in a project that has otherwise achieved mainly successes.

To emphasize the importance of the issue, the LHDA published a Maloti Minnow Policy in 2002. The policy document noted that the government of

(continued)

Box 4.2a (continued)

Lesotho is a signatory to the Convention on Biological Diversity (CBD) and said that the country is mindful of the concerns of the international biodiversity community about the extinction of species. It went on to state that as an agency of the Government of Lesotho, the LHDA is expected to take reasonable steps to conserve the relict Maloti minnow population in the Mohale project area. One possible intervention is to construct a physical barrier that would prevent trout and yellowfish from moving upriver to the areas where Maloti minnows live.

It has long been recognised that the construction of physical barriers to prevent alien fish from entering Maloti minnow habitat in the Senqunyane and Bokong rivers is the only lasting measure that would permit the establishment of minnow sanctuaries upstream and allow commercial and subsistence fisheries development to proceed downstream in the Mohale Reservoir. The only structure deemed able to withstand a 100-year flood while remaining effective as a barrier to trout is an artificial waterfall created by making a 'meander cut', emulating what nature has done in other parts of the catchment. The current cost of such an engineering project is estimated to be between 20 and 23 million maloti (the maloti, which is tied to the South African rand, was trading at M1 = US$7.65 in October 2009). The cost of the project could be brought down with the use of local labour in the construction, something that would bring much-needed employment and income into the Mohale area.

Because of the cost and the desire for commercial fishing to proceed, the two governments overseeing the LHWP have been reluctant to build the barrier. Currently, there are debates within the Lesotho Highlands Water Commission, the LHWP board, and the LHDA about the 'economic value' of the Maloti minnow compared with that of the commercial fisheries. What the project authorities do not realize is that the damage to the international reputations of Lesotho and South Africa for failing to act on this issue could make it harder for them to find international funding for Phase 2 of the LHWP. Both private banks and international financial institutions such as the World Bank will be reluctant to support a project that has failed to protect biodiversity. Both local and international environmental NGOs will flag the project as one that is environmentally damaging.

The question remains: should Lesotho and South Africa invest the funds to protect the critically endangered Maloti minnow? The answer, we believe, is an unqualified yes.

A particularly problematic aspect of development projects involving resettlement is that the relocated people leave behind places with which they have long been familiar. Some of the people bemoaned the fact that they had to leave the graves of their ancestors and places where they and their predecessors had practiced culturally significant ceremonies such as initiations and rain-making. Discussions with various categories of people, such as traditional healers and herdboys, revealed different kinds of information. It was found, for example, that healers were concerned about leaving places where they had obtained ritually important plants, minerals, and other items. Herdboys knew paths through mountain passes on which they could move cattle, goats and sheep safely from one point to another. A wide range of people in the Lesotho highlands believed that they would suffer individual and cultural loss as a result of resettlement. It would not have been possible to assess some of the perceived social and psychological impacts of resettlement had field workers not engaged in detailed discussions with local people, combining interviews with participant observation.

4.2.4 Project Impacts

The displacement experiences summarized above suggest that life, on all fronts, has been made much more difficult. Some post-project evaluations suggest contrary findings, especially studies that explore a limited set of economic variables or limited time frame. For example, a survey of resettled families undertaken by a contractor in 2000-2001 (see Tshabalala and Kisubi 2003) found that the average cash income had increased almost by a factor of three. And, in some instances holistic measures demonstrate relative success in the resettlement, as illustrated in the case of Swaziland's Maguga Dam.

Box 4.2b Dams, diversions and transboundary river management

—Barbara Rose Johnston and Robert K. Hitchcock

Hydrodevelopment of the Komati River basin was initiated with the signing of two treaties in 1992 recognizing water resources as matters of common interest between the Kingdom of Swaziland and the Republic of South Africa. These treaties established mechanisms that allow common development and use, and were developed in consultation with Mozambique which also shares this international river basin. The binational Komati River Basin Management Authority (KOBWA) was created as a means to implement the river basin development plan.

(continued)

Box 4.2b (continued)

Construction for the first dam, Driekoppies Dam, was begun by South Africa in 1993 as a post-Apartheid project with the stated goal of reducing floods and providing water for small-scale irrigated agriculture (7 ha plots for sugar cane, providing an income for black communities who had previously been excluded from water resources). Driekoppies Dam was completed in 1998, though the relocation plan took many more years to complete. Construction caused the displacement of 239 homesteads (2,000 people), loss of arable land to 700 farmers, loss of grazing land to 800 cattle owners, and the loss of critical natural resources to some 55,000 people. The initial environmental impact assessment focused on short-term relocation needs, but not on compensation for loss of arable land (an issue that was eventually addressed with a new environmental impact assessment and compensation program after the Maguga Dam Project included this element). Compensation and resettlement took placed in direct consultation with the affected communities, though the institutions created to address affected peoples concerns failed to include traditional authorities and respect cultural norms. The resulting complaints, community disputes and conflicting land claims delayed the project's completion. In 2004, the responsibility for completing the resettlement and sustainable development programmes was handed over to KOBWA by South Africa. Subsequent post-project evaluation of the Driekoppies Dam Project found that "the loss of ancestral land by the communities and the non-recognition of women as important players in the provision, management and safeguarding of water are still topical issues. Other challenges include: maintenance of the water quality standards, obtaining the actual water use and meeting the water allocation demand" (Okonkwo 2008, see also Woodhouse 2008).

The Maguga dam, the second of two dams built in the phase 1 plan for the Komati River Basin Development Project, was built in Swaziland in 2001. As with the Driekoppies Dam, the impact assessment and consultation process reflected the Resettlement and Compensation Policy Document signed by the two Governments in 1996. The primary objective of this policy was to ensure that the affected people are made "better off" than they were prior to Project implementation. This priority was further refined, as a result of the pragmatic lessons learned from the problems encountered with the Driekoppies Relocation Action Plan and the added input from the World Commission Dams (WCD 2000) best practice recommendations. Meaningful remediation included social and economic development in addition to compensation and relocation assistance.

(continued)

Box 4.2b (continued)

In its original conception, the Maguga Dam project was meant to store and provide water for commercial agroforestry (citrus groves and sugar cane in Swaziland and sugarcane in South Africa) and produce hydroelectricity (a 15 megawatt (MW) installed capacity). Such development was meant to stimulate large-scale commercial agriculture for an export market and transform an area that then supported subsistence-oriented small-farmer agropastoral production.

The approach eventually taken emphasized the water needs of small farmer production, a labour-intensive approach that reduced the number of households displaced by downstream improvements from an initial proposed 3,300 to 3. Resettlement policy not only attempted to replace lost assets, it attempted to improve the livelihoods of affected people through social and economic development. Compensation included a commitment to replace comparable land for land lost. And, affected people were recognized as project beneficiaries, right-holders whose status was protected by a Resettlement and Compensation Policy. The importance of affected people in relation to the project is illustrated by the fact that at one point Maguga Dam construction was suspended as there was no agreement on where displaced people would be relocated. Construction only continued once this issue had been settled and households relocated from the reservoir area received comparable arable lands.

To ensure the resulting resettlement and development process was both equitable and meaningful, a participatory approach that embraced traditional norms was adopted. The consultation process led to the establishment of the Ekuvinjelweni Resettlement Committee, an organization that both represented the affected people and was made up of affected people, and served as a partner in shaping an implementing the relocation action plan. This involvement of local people and their leaders was a major driving force in the evolution of the Maguga project. (Other factors include the tremendous decline in the global citrus market, which led to a rethinking of how diverted water was to be used, and by whom).

After much struggle and effort, affected communities were able to receive water, electricity, and jobs from the project, assistance with setting up farming cooperatives, and health and sports facilities. Households displaced by the reservoir, instead of moving into contractor-built resettlement homes, were able to build their own shelters with local labour, and in some instances, used left over funds to develop businesses or purchase communal equipment. An independent dispute resolution process was established, and resulting judgments were binding on the Komati Basin Water Authority. Years later, post-project evaluation found that most people say they are better off than they

(continued)

Box 4.2b (continued)

were before the dam – an extremely rare situation in the history of dam-induced resettlement. Notably, the equitable and decentralized water management principles employed in this project were adopted as national policy in the Kingdom of Swaziland's 2003 Water Resources Act.

The success of this hydrodevelopment reflects:

- Flexibility in project design, coupled with the growing commitment of project authorities in Swaziland and South Africa to avoid repeating the mistakes of other water projects in the region;
- The strong leadership, organization and involvement of affected peoples and their advocates, including resident peoples, members of the Swaziland royal family and government, engineers and other members of the technical support and water management teams;
- A development philosophy that evolved over time to embrace the notion of displaced and adversely affected communities as project beneficiaries and partners who are entitled to play a role in and receive the benefits from hydrodevelopment and water management.

The Maguga Dam resettlement experience is cited by many as an illustrative case of doing development right (Yan 2010). Ensuring that adversely affected people share in project benefits is a key sustainable development tenet, and one that goes beyond merely compensating people for their losses. Such goals are difficult, but not impossible to achieve. The Maguga project in the Kingdom of Swaziland represents an example to apply this principle in practice. The resulting outcome is a hydrodevelopment that, first and foremost, addresses the development needs of project-affected peoples and respects their rights to participate in and profit from development and water resource management decisions.

References

Application of integrated water resources management approach to the management of shared Komati River Basin (Case Study-Good Practice), Okonkwo, Adaora. 2008. IV international symposium on transboundary waters management, 15–18 Oct 2008, Thessaloniki, Greece. http://www.inweb.gr/twm4/abs/OKONKWO%20Adaora.pdf. Accessed 10 Sept 2010.

Swaziland Government, 2003. The Water Act, 2003. Mbabane, Swaziland.

Woodhouse, Philip. 2008. "Water rights in South Africa: Insights from Legislative Reforms" Brooks World Poverty Institute Working Paper 36. Institute for Development Policy and

(continued)

Box 4.2b (continued)

Management, University of Manchester, Manchester. April 2008. http://www.bwpi.manchester.ac.uk/resources/Working-Papers/bwpi-wp-3608.pdf. Accessed 10 Sept 2010.
World Commission on Dams. 2000. Dams and development: A new framework for decision-making. The report of the World Commission on Dams. London: Earthscan Publications
Yan, Katy. 2010. "Doing dams right: The WCD in practice" in *World Rivers Review*, June 2010: 8–9.

My 20 years of research with the displaced communities – conducted in my capacity as a member of the Panel of Environmental Experts for the Lesotho Highlands Water Project – found that 80% of this increase in income was derived from compensation payments (not a sustaining source of income). Furthermore, despite moving to the centres of employment and trade, the income-generating capacity of most households had declined. In part this decline reflected the massive reduction in employment opportunities for Basotho in the South African mines and cutbacks in employment in the garment industry. And, in part this decline reflected the fact that jobs created by development were largely temporary in nature. Once construction of dams, roads, and other associated infrastructure supporting the water-transfer project was completed, only a few people were able to find employment in the area, cleaning, for example, roads into the project areas or working on the construction of water and sanitation systems (Hitchcock et al. 2006).

As is often the case with reporting on the outcome of a resettlement program, the people who have made a conspicuous success of the opportunities offered by the LHWP were readily identifiable, easily accessible, and quite legitimately held up by the LHDA as examples of how well the project has worked. Those who have fared less well, however, are easily lost from sight unless researchers make determined efforts to locate them. Interviews with members of what were defined as vulnerable households (those who were extremely poor, female-headed and child-headed households, households made up of people who were elderly or physically incapacitated, or who otherwise were at risk) indicated that many of them were fearful about the future and uncertain about how they were going to make a living in the face of all the changes in their lives.

The post-resettlement economic indicators of household welfare have proved very difficult to assess. Furthermore, reliable evaluations of the cultural, social, spiritual and personal losses are almost entirely missing, and had these been investigated and reported, few, if any, remedial actions, had been recommended. Some losses are irreparable and cannot be made good by the normal means of compensation. Numerous people said, "How can our lives, which were dependent on the soil and on rainfall, be the same if we have no land?" Some also said, "Our lives depended in part on our cattle and sheep; now we do not live close to our animals, and some of them have been stolen."

One of the unintended consequences of the LHWP was an earthquake that occurred in November 1995 and caused damage in seven villages around the Katse Reservoir. It destroyed 11 homes and, according to local people, caused a crack in the earth at Mapaleng. This earthquake may have been a consequence of reservoir-induced seismicity (RIS) induced by the weight of the water in the Katse Reservoir, which had begun to fill in October 1995. Some local people blamed the project for the earthquake and subsequent seismic activity, while others placed the blame on a large 'river snake' that they said had been disturbed by the construction activity in the Katse area.

Fig. 4.2.2 Katse Dam and Katse reservoir in the Phase 1A area of the Lesotho Highlands Water project (Photo credit: Robert Hitchcock)

A second unintended consequence of the LHWP was the drying up of springs in the catchment areas of the project. Such an event occurred, for example, at the village of Ha Mensel near Katse, close to the Katse township and the administrative offices that were built to oversee the project. It was ironic, villagers said, that there was a large water tank built by LHDA in the village to provide water to the engineers

Fig. 4.2.3 Many springs and streams in the catchment area have dried up. Young Bosotho playing in a tributary to the Malibamotso River which leads into the Mohale reservoir, Lesotho (Photo credit: Robert Hitchcock)

and their families and to workers in Katse, but the local people themselves had less access to water now than they did before the project began.

A third unintended consequence was an increase in HIV/AIDS in the population in the highlands of Lesotho. Estimates of the HIV/AIDS prevalence rate in the project area in the late 1980s was 0.9% of the population. Recent figures suggest that the HIV/AIDS prevalence rate is now some 22%, which is approximately the rate found in urban areas in Lesotho, such as Maseru (Human Sciences Research Council 2007). A question that needs to be addressed is whether the LHWP itself is responsible for the increase in HIV/AIDS, and, if so, what kinds of mitigation measures can be implemented to ensure that the project-affected population is at least as well off as it was prior to the inception of the project, in keeping with the treaty obligations between the two governments of Lesotho and South Africa.

A fourth unintended consequence of the project is an increase in social stratification along class, socioeconomic, gender, and age lines resulting in an expansion in the numbers of people at risk. In short, development has induced impoverishment. For many of the households physical resettlement was followed by a difficult period of adaptation and uncertainty. They had to reconstruct their sources of household income and subsistence and to work out relationships with new neighbours. There was greater ethnic and class diversity in the resettlement areas, with cultural and economic differences introducing new tensions and conflict. In addition to Basotho from a number of different clans (e.g., Taung and Phuthi), there were also Xhosa, Zulu, Tswana, and Chinese. Some of the people who were resettled felt bereft. Although they were able to get cash compensation and have their dead relatives moved to their new locations, they felt that their new ways of life were seriously lacking compared to what they had experienced prior to the project. According to some of the resettlers, there were fewer traditional ceremonies being conducted in the new locations, and people had to go long distances to take part in cultural activities. The Lesotho Highlands Water Project, they maintained, represented a serious threat to Basotho culture and community cohesion. Household cohesion was also undermined. Many households split up, with younger members moving to town or starting new homesteads, sometimes leaving behind the elderly and the infirm. Households headed by women or by elderly people were typically short of labour and as a result worse off after the LHWP than they were before, in spite of the livelihood supports provided by the Lesotho Highlands Development Authority. Traditional sources of support – relatives, friends and neighbours who had previously helped out in the fields and with herding – had moved away.

The LHWP offers a number of lessons that are applicable to projects in other areas of the world.

- Lesson no. 1. The success of water management activities and institutions depends on transparency, openness, accountability, and flexibility.
- Lesson no. 2. To obtain a well-rounded picture of how a project impacts people, it is necessary to employ a variety of types of data collection, both qualitative and quantitative, and to get detailed baseline data against which changes can be measured.
- Lesson no. 3. Social impact assessments should ensure that diverse categories of people are interviewed and monitored, breaking the population down along gender, age, class, occupational, and other lines.
- Lesson no. 4. It is necessary to have a policy environment that is appropriate and positive for all concerned, one that takes into careful consideration international, regional, national, and local policies and practices and places significant emphasis on local culture and traditions.
- Lesson no. 5. Parties involved in water resource management must pay close attention to the social, political, economic, and environmental situations in the various project areas.
- Lesson no. 6. No matter how good a development policy is, it is likely to fail if it goes against the interests of the local populations and if local people are not involved in decision-making and planning. Public participation, therefore, is crucial to the success of large-scale water projects.
- Lesson no. 7. Compensation, resettlement, and development programs of large-scale water projects should be integrated effectively with existing government programs if project-affected people, including those that are especially vulnerable, are to have their livelihoods and well-being sustained over the long term.

References

Chenje, Munyaradzi, and Phyllis Johnson (eds.). 1996. *Water in southern Africa*. Maseru/Harare: Southern African Development Community, the International Union for the Conservation of Nature/the Southern African Research and Documentation Center.

Clark, S.J., E. Colson, J. Lee, and T. Scudder. 1995. Ten thousand Tonga: A longitudinal anthropological study from Southern Zambia, 1956–1991. *Population Studies* 49(1): 91–109.

Dinar, Ariel, Rashid Hassan, Robert Mendelsohn, and James Benhin (eds.). 2008. *Climate change and agriculture in Africa: Impact assessment and adaptation strategies*. London: Earthscan Publishers.

Ferguson, James. 1994. *The anti-politics machine: 'Development', depoliticization, and bureaucratic power in Lesotho*. Minneapolis: University of Minnesota Press.

Government of Lesotho and Government of South Africa. 1986. *Treaty on the Lesotho Highlands Water Project between the Government of the Kingdom of Lesotho and the Government of the Republic of South Africa*. Maseru/Pretoria/Capedown: Government of Lesotho/Government of the Republic of South Africa.

Hitchcock, Robert, A. Inambao, J. Ledger, and M. Mentis. 2006. *Lesotho Highlands Water Project*. Panel of Experts Report No. 45.to the Lesotho Highlands Development Authority, Maseru, Dec 2006.

Human Sciences Research Council. 2007. *Socio-economic and related findings of the 2005/2006 monitoring and evaluation survey. Vol. II: Phase 1B, Mohale Project Area*. Contract 1204, Lesotho Highlands Development Authority. Pretoria/Lesotho: Human Sciences Research Council/Lesotho Highlands Development Authority.

Low, Pak Sum. 2005. *Climate change and Africa*. Cambridge: Cambridge University Press.

Murray, C. 1981. *Families divided: The impact of migrant labour in Lesotho*. Cambridge: Cambridge University Press.

Panos Institute. 2001. *Voices from the mountain: Oral testimonies from the Lesotho Highlands*. London: Panos Institute.

Scudder, Thayer. 2005. *The future of large dams: Dealing with social, environmental, institutional and political costs*. London: Earthscan.

Simms, Andrew, and Hannah Reid. 2005. *Africa – Up in smoke? The second report from the Working Group on Climate Change and Development*. London: New Economics Foundation and International Institute for Environment and Development.

Thamae, M.L., and L. Pottinger (eds.). 2006. *On the wrong side of development: Lessons learned from the Lesotho Highlands Water Project*. Maseru: Transformation Resource Centre.

Transformation Resource Centre. 2004. *The irony of the 'white gold*. Maseru: Transformation Resource Centre.

Transformation Resource Centre. 2005. *Lives of resettled communities after resettlement: A survey on the lives of communities affected by the Lesotho Highlands Water Project*. Maseru: Transformation Resource Centre.

Tshabala, Mavuso, and Wilson Kisubi. 2003. *Socioeconomic monitoring of households affected by phase 1 of Lesotho Highlands Water Project: A synthesis of 2000/01 results*. Maseru: Lesotho Highlands Development Authority.

World Commission on Dams. 2000. *Dams and development: A new framework for decision-making. The report of the World Commission on Dams*. London: Earthscan Publications.

Chapter 4.3
Not All Dams in Africa Are Developmental: Advocacy Perspectives from the African Rivers Network

Robert Kugonza Akiiki

4.3.1 Introduction

In the continent of Africa, the word *development* attracts a lot attention. Why? Because the African nations are in a hurry to develop, looking to be where other developed nations are now. On the road to development are a lot of choices – whether to engage in land reform, privatization, globalization, and regional cooperation, among others. However, one element that has been defined as crucial is energy infrastructure, often achieved through hydroelectric development. Generating reliable and large sources of power often provides the means to secure other goals.

According to the World Bank, only 25% of people living in Africa have household access to energy. To better address these needs, African States and their international partners are pursuing an energy development vision that largely reflects a first-world model of centralized grids and large energy infrastructure. In this vision of the future, hydroelectricity has been deemed by many as the best way to achieve the social and economic needs of the continent.

What are the costs and benefits of implementing this vision? A review of the human and environmental costs and benefits from past dam development offers important and timely lessons. Dam-building experience in the African continent too often demonstrates the destructive rather than constructive nature of dams, especially the social upheaval, environmental devastation, and economic damages generated: forced resettlement, lost livelihoods, flooded farmlands, damaged downstream ecosystems, and a huge debt load. In nearly all cases, affected people have had very little voice in these developments. And in too many cases, these dams were not even planned to supply the citizens of the nation in which they were built. Most of Africa's

R.K. Akiiki (✉)
African Rivers Network (ARN), Kampala, Uganda
e-mail: robertk@nape.or.ug

B.R. Johnston et al. (eds.), *Water, Cultural Diversity, and Global Environmental Change: Emerging Trends, Sustainable Futures?*, DOI 10.1007/978-94-007-1774-9_24,

Map 4.3.1 Hydrodevelopment in Africa: Dams and water diversions that are proposed, under construction, under rehabilitation, or being expanded (Credit: International Rivers 2010). Large dams play a critical role in providing water and energy to sustain social, economic, and industrial needs. Every large dam, however, also carries its own distinct set of economic, social, and environmental costs. Dams increase a country's debt burden, displace whole communities and cultural nations, destroy livelihoods, alter ecosystems, and increase disease. Dams also can all to often fall far short of achieving their purpose, especially in a warming world. Climate change and increasingly erratic rainfall can reduce energy and water benefits from dams and increase risks of deadly floods. Today, billions of development dollars are earmarked for large dams and associated project infrastructure in Africa. This map illustrates the location of large dam projects that are proposed, under construction, under rehabilitation, or being expanded as of June 2010. Whether such projects are actually completed and whether they fulfill their societal goals remains to be seen (Source: International Rivers, "African Dams Briefing 2010" June 2010)

dams were built to supply mines, aluminium smelters, and other big businesses, not the basic needs of the people. The high costs involved in large dam construction also often end up making the power produced unaffordable for the local people, especially for the majority of those who are not already connected to the grid. This destructiveness has left many wondering whether focusing on large dams is the right approach to Africa's development.

To help encourage public understanding and debate over the role of hydroelectric dams in Africa's future, leaders from dam-affected communities across the continent met in 2003 in South Africa, forming the African Rivers Network (ARN). The resulting organization is collaboration between African civil society organizations and dam-affected communities. ARN's primary mission is to advocate for people-centred development.

ARN works to influence development decision-making, amplify community voices and rights, and build capacity in civil society and with community partners. Doing this work means promoting the rights of dam-affected communities by helping dam-affected communities in Africa to understand their rights and engage in the decision making process of dam building.

In Africa, it is common that communities affected by dams lack any mechanism to effectively address their grievances regarding resettlement, compensation and/or other local impacts, and to legally protect their rights. Thus, ARN helps to organize dam-affected communities into larger coalitions to better achieve access to information, protection of rights, and meaningful engagement in decision-making processes with regard to the construction and operation of dams. ARN works in collaboration with

Box 4.3a People-centred development

—Robert Kugonza Akiiki

People-centred development means:

Water and energy development that sustains the livelihoods of local people first and includes alternatives to large dams,
Respect for fundamental human rights, including respect for community decisions,
Reparation of outstanding social injustices associated with historical development,
Respect for the rights of future generations through ecologically-sustainable development.

other organizations in the affected areas giving workshops, enhancing the research capacity of members, conducting research, documenting and disseminating data on the impact of existing dams, and sharing success stories where alternatives to dams have been taken. Where projects are proposed, ARN seeks to ensure considerable attention is given to project alternatives, especially the consideration of renewable energy sources and water options on an equal playing field to dams and against equal criteria in addressing the needs for the people of Africa.

4.3.2 The Bujagali Dam, Uganda

At ARN, we believe that before forging on with the development of new large dams, we should take the time to consider and learn from the mistakes of the past. Some of this development has been grossly inept, as suggested in the case of Uganda's Bujagali Dam, a project under construction near Bujagali Falls on the Nile River. Funded by the African Development Bank (AfDB) with the promise of generating electricity for Uganda and other East African nations, the dam was located some 5 km downstream from two parallel dams on the Owen Falls, the Nalubaale Dam, built in the 1950s, and the Kiira Extension Dam, built in 2002. Construction of the Kiira Extension Dam has caused water levels in Lake Victoria to significantly fall, seriously threatening the lake's ecosystem and livelihoods of the fishing communities and activities on its banks. This problem led to tensions between Uganda and the members of the East African Community, especially Kenya and Tanzania.

After completing the Kiira extension Dam, Uganda approved an economic impact assessment and the World Bank announced its decision to fund the Bujagali Dam. Site preparation began immediately, including the forced displacement of affected communities. Traditionally, the river divided two major tribes, the Baganda and the Basoga, each group with their own traditions, beliefs, language, and territory.

Fig. 4.3.1 Bujagali Falls (Credit: Photo courtesy of International Rivers)

Displacement occurred with only one resettlement camp established on the Buganda side, thus forcing the two groups to merge or, as was the case, excluding many Basoga from the camp. Once the dam is completed, the reservoir will submerge several significant cultural sites. One of the areas to be submerged by the water behind the dam will be the shrine for the Basoga, headed by Jaja Bdaghali, the spiritual leader for the Basoga. Efforts to provide redress included an agreement to move a primary shrine and protect the Kalagala falls where cultural rituals take place. This strategy has also generated a new conflict, as the Kalagala site is located on Buganda soil and the Bujagali cultural/ spiritual site was on Busoga land. Basoga cannot put their spiritual house in another land, but also the Baganda do not want the spirits of Basoga to be hosted on their land. The issue of culture in Bujagali dam project remains an unsolved one.

Civil society advocates issued protests and complaints, arguing that the Bujugali Dam would further have a devastating effect on environmental flows, displace even more people without meaningful redress, cause irreparably harm to the cultural beliefs and way of life of affected communities, and – due to the close proximity of the two pre-existing dams – have little chance of generating the promised 250 megawatts (MW) of energy. The National Association of Professional Environmentalists (NAPE), a member of the ARN, argued that project funders were in violation of their own policies and filed a complaint. Investigation by the World Bank Inspection Panel and the Independent Review Mechanism Unit (RMU) of the AfDB found evidence to support more than 80% of the civil society complaints. Violations

Fig. 4.3.2 Jaja Budhagali – the Basoga cultural and spiritual leader – raising concerns over development plans (Credit: Robert Kugonza and African Rivers Network)

included the failure to consider climate change and its potential impact on the dam's economic viability, recognition that the need for constant flows to generate hydroelectric power will further stress an already depleted Lake Victoria, recognition of the failure to consider other energy alternatives, costs for resettlement and biodiversity losses were not fully accounted for, nor are plans in place to ensure that mitigation of these impacts will be achieved. Despite the formal recognition by its own Inspection Panel, the Bank continued to disperse funding and development of the dam proceeds.

4.3.3 Kainji Dam in Nigeria

It is not uncommon that development projects implemented in the African continent involves the displacement of people with no meaningful assistance. Increasingly, this experience is prompting the organization of civil society and meaningful protest. In Nigeria, communities affected by the Kainji Dam are renewing their struggle for compensation from the government, a battle they have been fighting since the dam's commissioning in 1968. Although Kainji produces 760 MW, up to 90% of the affected communities have no access to electricity – a direct failure of government to fulfil the promises made to the communities before construction. The government has also failed to fulfil other promises for housing, roads, schools, water supply and others. The communities are calling for a percentage of the profits from the sale of electricity generated from hydropower dams to be given back to the communities for their development.

Often, civil society demands to address the legacy issues of past dams are intertwined with protests against new upstream or downstream projects. When new dams are proposed without addressing the failures of previous work in the watershed, a new level of synergistic and cumulative assault into an already ulcerating human and environmental context. Such is the case with the Gibe Dam Development in Ethiopia.

Box 4.3b Ethiopia's Gibe 3 dam endangers Kenya's lake Turkana: An activist's view

—Ikal Angelei*

Lake Turkana is a miraculous anomaly of life-giving water in a parched and unforgiving land. Formed millions of years ago in the tectonic upheavals that created East Africa's Great Rift Valley, Turkana is the largest permanent desert lake in the world. Extinct volcanoes enclose the horizon, and the heat is so intense that when the blustery wind from Mount Kulal on the eastern shore temporarily ceases and clouds gather overhead, raindrops sometimes evaporate before they even reach the lake. It is called "ghost rain." The lake contains the world's largest population of Nile crocodiles which survive on the giant perch – the largest freshwater fish in the word – which, in turn, feed on a profusion of blue-green algae. This prehistoric lake has also long been a focus for humans: its shores have revealed the oldest-known fossil remains of *Homo habilis*. Today, some 300,000 residents from at least ten tribes have become masters of wresting sustenance from the harsh landscape. Without the lake, life here would be virtually impossible.

But Lake Turkana and its inhabitants, already stressed by long-term drought, now face a greater environmental catastrophe in the making. Lake levels are projected to rapidly decline when its main source, the Omo River, is depleted by a huge hydrodevelopment project across the border in Ethiopia. This project, built by the state-owned Ethiopian Electric Power Corporation with financing from the Industrial and Commercial Bank of China, involves a series of four dams, including the Gibe III, a 243-m high dam and hydropower plant that at a projected 2,870 MW will, if completed, be the largest hydropower plant in Africa. To power this project, some 80% of the flow from the Omo River will be diverted.

There is no question that Ethiopia needs power. But the irony of the Gibe III dam is that while it threatens the economy of the Turkana region, a large share of its electricity will be sold to consumers in other parts of Kenya. For

*Abstracted with permission from the author and World Rivers Review, March 2009, with revisions.

(continued)

Box 4.3b (continued)

Ethiopia, the project is a commercial one: they want to make money selling the power elsewhere, not provide power to their own people. For Kenya, it's a matter of allowing one part of the country to be devastated so that others may get a little more power.

The assault on the Omo River and Lake Turkana by the Gibe dam is projected to result in a drop of 7-10 m in the lake's depth in the first 5 years alone (the lake has already receded by about 5-8 m in recent years, a fact many attribute to impacts of climate change). Resulting changes in the chemical balance of the water threatens the region's tremendous biodiversity, including large populations of Nile crocodiles, hippopotamus, and over 40 different species of fish and snakes. The riparian forest, one of the last pristine dryland forests in Africa, would also be in grave danger.

The saga of Gibe III Dam is just the latest episode of human pressures contributing to the dying of Africa's biggest lakes. Lake Chad has nearly disappeared from diversions that stopped its flow, and Lake Victoria has seen major drops in its water levels from dams that let too much water out. Climate change could add several more to the list of dead or dying lakes and depleted rivers across the continent. Losing our precious water resources will make us less able to adapt to climate change.

4.3b.1 Endangered people

The lower valley of the Omo was inscribed on UNESCO's World Heritage list in 1980 in recognition of the many fossils of ancient life found in this region, including early hominid remains and artifacts that date back some 2.4 million years. Some 200,000 agro-pastoralists from eight distinct indigenous nations live in this area and depend upon the annual floods of the Omo to support river-bank cultivation and livestock grazing.

Further downstream, the river feeds into the world's largest desert lake, Kenya's Lake Turkana. There, indigenous communities are highly dependent on the lake for their food crops, livestock grazing and watering, and fishing. Any impacts to the lake's ecosystem would disrupt the economy, leading to an increase in conflicts in the area. Considering the unstable state of peace in northern Kenya, such damage to the local economies would invoke a threat to regional stability.

While a power purchase agreement outlining the terms of electricity sales was reportedly signed between Ethiopia and Kenya in 2006, no bilateral agreements on the use of the Omo-Turkana waterway and the dam's downstream effects to Kenya are publicly known. The 300,000 people who live

(continued)

Box 4.3b (continued)

around Lake Turkana in Kenya were neither informed of the project's impacts nor consulted on their priorities. Their situation mirrors that in Ethiopia, where the traditional economy of the Lower Omo Valley supports the above mentioned agro-pastoralists. The Ethiopian dam-affected people – who, like those on the Kenyan side, are largely indigenous peoples leading traditional lifestyles – have heard little or nothing about the project and their options, even though the changes to the Omo will upset the fragile balance of river bank cultivation and herding they maintain, unraveling the valley's best strategy against food insecurity.

The Gibe III threatens the biodiversity, livelihoods, and development of Northern Kenya, yet these potential risks have not been taken into account in the project planning. It is our fear that the consequences of this hydrodevelopment will include immense scarcity, suffering and resulting rise in violent conflict in one of the most culturally diverse areas of Africa.

It is our hope that nations in this region and the agencies and entities that support the development of this region will proceed with notions of sustainable development in mind. The Friends of Lake Turkana celebrate the Turkana basin ecosystem, its unique floral and faunal systems, and cultural biodiversity. We believe that both environmental and personal security are human rights, and are fundamental to the achievement of any form of development in our society.

4.3.4 Conclusion

As Africa braces itself for a boom in large infrastructure development, such projects should not come at the expense of the poor. Those who are directly affected by a project should be the ones who benefit the most and be better off after than before construction (ICOLD 1997). There is increasing realization of the fact that, in assessing the development effectiveness of a major infrastructure project, social and environmental aspects have the same significance as economic and financial factors. Historically, social and environmental issues have been overlooked, undermined, or couched in corporate and political symbolism. This seems to have been the case in the Bujagali and most other similar dam processes around the world. Holistic and integrated thinking and practice, which does not exclude alternative approaches to development, will be necessary to identify and capture all the problems, issues and challenges of the victims of development and deal with them effectively in order to enhance the development effectiveness and human content of large infrastructure development.

The African Rivers Network recognizes that across the continent, there is a great need for clean energy development that will improve livelihoods and economic productivity in both rural and urban settings. Many parts of Africa have immense untapped clean energy potential, as pointed out in a recent UN-sponsored survey. For example, in Uganda, Ethiopia, Kenya and Tanzania, there is geothermal potential of the East African Rift Valley. Similarly in Ghana, there is enough wind potential to generate significant wind energy along the coastal belt, Volta, and Central regions. Southern Africa, too, has big wind reserves, and the potential for tidal and wave power along its coasts. And almost everywhere in the continent, there is huge solar-power potential.

Even with large dams already distributed all over, too many people living in African have been left in the dark. Today, we need solutions that reach these people. Renewable energy can just do that and save rivers as well.

Resources

BBC News. 2002. Work starts on giant Ethiopian dam, 12 Aug.

BBC News. 2008. Africa plans biggest dam project, 21 Apr..

Bank Information Centre. 2007. *Investors hold roundtable to discuss Inga Dam*, 12 Apr. Washington, DC: Bank Information Centre.

British Museum. 2006. *Merowe dam archaeological salvage project: Amri to Kurbekan survey*. London: British Museum

EAWAG. 2006. *Independent review of the environmental impact assessment for the Merowe dam project (Nile River, Sudan)*. Kastanienbaum: EAWAG.

Hildyard, N., and P. Bosshard. 2005. *A critical juncture for peace, democracy and the environment: Sudan and the Merowe/Hamadab project*. Berkeley: International Rivers Network.

Hoering, Uwe. 2006. *Ethiopia's water dilemma*. Berkeley: International Rivers Network.

ICOLD. 1997. Dams and the Environment. ICOLD Position Paper. Paris: International Commission on Large Dams.

International Rivers Network. 2006a. *Damning the Zambezi: risks outweigh benefits of proposed Mphanda Nkuwa dam*. Berkeley: International Rivers Network.

International Rivers Network. 2006b. *Ethiopia's water dilemma*. Berkeley: IRN.

International Rivers Network. 2008. *Sudanese government forcibly displaces more than 6000 families affected by dam*. Berkeley: IRN.

Nile Basin Initiative. 2008. Nile Basin development forum final communique, Nov 17–19, 2008.

World Bank. 2008. *New financiers are narrowing Africa's infrastructure deficit*. From www.worldbank.org

World Energy Council. 2008. *Report on the high level WEC workshop on financing the Inga hydropower projects, held in London*, 20–21 Apr 2008.

Chapter 4.4
Drowning Under Progress: Water, Culture, and Development in the Greater Mekong Subregion

Nathanial Matthews

The Mekong River is an iconic river, the twelfth longest in the world, stretching over 4,300 km and draining an area just under 800,000 km². From its headwaters in the plateaus of Tibet, the river travels through China, Myanmar, Thailand, Lao PDR, Cambodia, and the southeast coast of Vietnam, where it spills into the South China Sea. The Mekong is a flood pulse river affected strongly by seasonal flows; during the flood season a journey through its system takes approximately 3 weeks, whereas that same trip during the dry season could take 3 months (Hori 2000). The region surrounding the river is the breadbasket of Southeast Asia, providing rice and other agricultural products for the inhabitants and the rest of the world. Through agriculture, fisheries, and forestry, the Mekong sustains the livelihoods of millions of people and supports a wealth of biodiversity, including many endemic species of flora and fauna. To the economies of Southeast Asia the Mekong is a source of wealth and power that provides hydropower, transport, and irrigation. The region is also home to hundreds of ethnic groups, making it one of the most culturally diverse regions in the world.

The Mekong is also a river under threat. There are currently four mainstream dams on the upper Mekong (also know as the Lancang in China), the Manwan dam that was completed in 1994, the Dachaoshan dam which was completed in 2002, the Jinghong dam that was completed in 2006 and the giant Xiaowan dam (4,200 megawatts (MW) of electricity) completed in 2010. These dams are part of the Lancang Cascade, a series of eight dams planned or under construction in southern China. When completed these dams will produce nearly as much electricity as the Three Gorges project.

China has come under increased criticism due to the record low flows in the Mekong in 2010 and for completing all the existing dams on the Lancang without consulting the downstream countries. By controlling the flow of the river in the dry season, these dams could have devastating consequences for the ecosystems that rely on the

N. Matthews (✉)
Kings College London, London, UK
e-mail: nathanial.matthews@kcl.ac.uk

B.R. Johnston et al. (eds.), *Water, Cultural Diversity, and Global Environmental Change: Emerging Trends, Sustainable Futures?*, DOI 10.1007/978-94-007-1774-9_25,

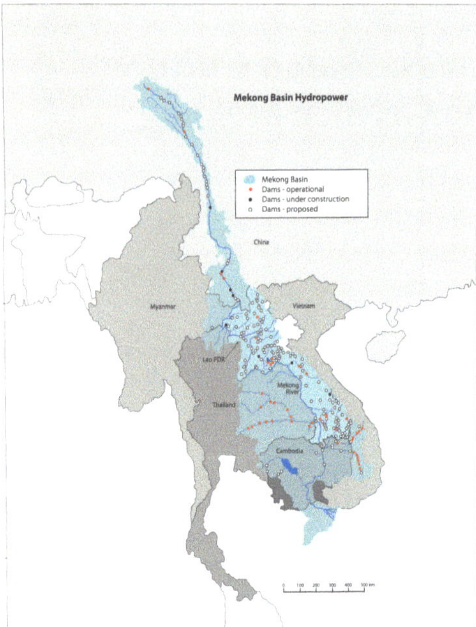

Map 4.4.1 Mekong mainstream dams (CGIAR Challenge Program on Water and Food 2011)

flood-pulse nature of the river and the livelihoods and cultures that rely on these ecosystems. Although China denies these dams are causing major disruptions to the Mekong's hydrology there is hope that it will respond to criticism with a more open and consultative policy (Asia Times online 2010). Of equal or greater concern for the river and the people who depend on it, are the plans for 11 mainstream dams below China in Thailand, Lao PDR and Cambodia. Currently, there are no mainstream dams on the Mekong below the Chinese border. Previous proposals for lower mainstream dams, dating as early as 1960 and as recently as 1994, were dismissed because of the predicted impact on people and the environment. However, a number of factors have changed since 2005. Increased energy demands in Thailand, China and Vietnam, rising oil prices coupled with a growing tendency to view hydropower as a green, modern energy source and an influx of private and state backed investors, have all contributed to the strong likelihood that a number of mainstream dams will be constructed in the near future. The planning for these 11 mainstream dams, funded by Malaysian, Thai, Vietnamese and Chinese firms backed by the state or private capital, is progressing with an alarming disregard for local livelihoods, cultures, and the environment.

Box 4.4a Rasi Salai, Thailand

—Nathanial Matthews

The case of Thailand's Rasi Salai Dam provides an example of the human and environmental damage resulting from poorly planned water infrastructure projects, and the role that affected citizens might play in documenting these damages and using this local knowledge to advocate for meaningful change. The Rasi Salai Dam was approved in 1989 and built without the completion of an environmental impact assessment. The project was built by the Electricity Generating Authority of Thailand (EGAT) with assistance from the World

(continued)

Box 4.4a (continued)

Bank and completed in 1994. It was one of 13 dams to be built as part of the Kong-Chi-Mun Water Diversion Project to irrigate land and reduce flooding in northeastern Thailand. The dam inundated between 5,000 and 8,000 ha of land that the government classified as degraded forest or wasteland. The inundation caused more than 15,000 people to lose farmland, blocked fish migration routes, and destroyed a large freshwater swamp. Many of the people affected by the dam construction received little or no compensation.

After the dam was completed, the water in the inundated land turned out to be highly saline, making it useless for irrigation. To make matters worse, the land that the government deemed wasteland was an important source of livelihoods for the local villages. People used this so-called wasteland for agriculture and saltpan farming, and they gathered non-timber forest products and traditional medicines on it. The river was also an important source of fish and animal protein. EGAT's disregard for the local people's relationship with the land and their local knowledge resulted in the destruction of the area's culture, livelihoods, and environment.

Box Fig. 4.4a.1 With reservoir waters rising and their refusal to move to higher ground, the plight of Rasi Salai villagers captured the eye of the nation (Photo courtesy of International Rivers)

In response to this disaster, local villagers organized, worked with others to document the adverse environmental impacts to migrating fisheries, changes in water quality, and impact on livelihoods, and began to demand the decommissioning of the dam. In August 1999 more than 1,800 people created a village on an island in the dam reservoir, daring the government to drown them or open the sluice gates. In March 2000 another 700 villagers occupied the dam site.

Local and international non-governmental organizations (NGOs and INGOs), including International Rivers, supported these local protests. Having focused national attention and global pressure on their cause, in July 2000 the Science Minister agreed to open all seven sluice gates of the Rasi Salai Dam on July 6, 2000. In the years since the gates have remained open while the Government completes a Social Impact Assessment, a process still under negotiation with the affected peoples, with contentious issues including appropriate compensation for flooded lands.

This case has a number of important lessons. First, it demonstrates that local people can influence large water infrastructure decisions even after the projects are completed. The decommissioning of the dam has become an

(continued)

Box 4.4a (continued)

important source of pride for local people and international and local NGOs. Local people's accomplishments in Rasi Salai have helped to inspire other groups and NGOs across the region and the world to stand up against unintegrated water management practices. The Rasi Salai case also shows the importance of local knowledge and proper Environmental Impact Assessments (EIAs). If local people had been properly consulted before dam construction began and a comprehensive EIA had been conducted, then millions of dollars would have been saved and livelihoods, culture, and the environment would have been preserved.

The protests that led to the decommissioning of Rasi Salai were possible because of the relative freedom that NGOs and INGOs have in Thailand. In other nations in the Greater Mekong Subregion, the situation is very different.

The prioritisation of economics and energy supply in dam construction will have far-reaching consequences for the region's multibillion dollar fishing industry, which is an important source of income and protein for local people, as the dams offer no proven measures to allow fish to move upstream to spawn. Other profound consequences will include the forced relocation of tens of thousands of local people, the loss of livelihoods and cultures, the flooding of valuable agricultural and forest land, drops in water levels, losses of biodiversity, and saltwater intrusion affecting farmlands in the lower delta. This case study explores the interconnections between culture, livelihoods, and biodiversity in the Greater Mekong Region, exposing its vulnerabilities and demonstrating how an unsustainable approach to water management leaves no long-term winners.

The Greater Mekong Subregion comprises the People's Republic of China, Myanmar, Thailand, Lao PDR, Cambodia, and Vietnam. Of the subregion's estimated 316 million people, approximately two-thirds live in rural areas and 30% remain below the poverty line, many without access to fresh water or sanitation (Asian Development Bank 2007). Much of the subregion's population relies on fishing and subsistence or semi-subsistence agriculture that is irrigated by the Mekong and its tributaries (World Bank and Asian Development Bank 2006). Over the past decade the Mekong area has experienced high economic and population growth rates and large cross-border migrations, much of which has come at a cost to the region's traditional livelihoods and environment (Cornford and Matthews 2007).

The keystone program of the Asian Development Bank (ADB) in the Mekong is the Greater Mekong Subregion (GMS) Program, an economic cooperation program with a long-term goal 'to promote development through closer economic linkages' (ADB 2009). The GMS Program has helped to showcase the region to the global

Box 4.4b Myanmar

—Nathanial Matthews

Box Fig. 4.4b.1 Buddhist ceremony and protest against the Hut Gyi Dam on the banks of the Salween River, Burma, 2008. Some 250 villagers from 18 villages in the area affected by the proposed dam took part in the ceremony, and were joined by Burmese opposition politicians, environmentalists and student activists (Photo courtesy of International Rivers)

The political realities in Myanmar are very different from those in Thailand and other Mekong countries. Dam construction in Myanmar has increased significantly in recent years, with China, Thailand, and India investing in hydropower projects designed to supply electricity to neighbouring countries, not the deprived population in Myanmar. China alone has invested in over 60 hydropower projects there.

Despite the strong efforts by local NGOs like the Burmese Rivers Network to organise local protests and to call on the governments of Thailand, China, and India to stop dam construction, local people in Burma have limited options. The country's military dictatorship allows no public participation in decision-making. The military also plays a strong role in dam preparation, with forced labour, forced relocation with little or no notice, and torture, rape, and executions occurring in some cases. Many dams are located in remote regions where ethnic minority groups reside. The military often targets these groups, which are being systematically displaced.

The demand for electricity in Thailand, India, and China and the profit that Burma's military stands to gain from hydropower have devalued culture, livelihoods, and biodiversity. The future of rivers and the people who depend on them in Myanmar relies on neighbouring countries' recognising that they should not meet their energy demands at the cost of any country's biological and cultural diversity.

economy and to develop infrastructure, encouraging unimpeded flows of people and goods across borders. A key aspect of the Program, the Mekong Power Grid, is designed to develop a regional power market by promoting cooperation in hydropower and cross-border high voltage transmission lines. As with many of the ADB's programs, this one is fuelled by a market-driven culture in which billions of dollars of loans and co-financing fund large infrastructure projects designed to stimulate economic growth, reduce poverty, and increase productivity through improved resource access. Although the GMS Program mandate prioritises environmental protection and holistic approaches to development, economic development has always been the primary focus.

This program is part of a history of foreign aid and development projects in the region that dates back to the late 1950s, beginning with the formation of the Committee for the Coordination of Investigations in the Lower Mekong Basin (Mekong Committee) under the auspices of the United Nations Development Programme (UNDP). From the outset the Mekong Committee's task was to "promote, coordinate, supervise, and control water resource development projects in the lower Mekong basin" (Nguyen 1999: 126). In 1995 the Mekong Committee was replaced by the Mekong River Commission (MRC), in an agreement between the governments of Thailand, Cambodia, Lao People's Democratic Republic (PDR), and Vietnam, with China and Myanmar as dialogue members. The MRC's current mandate is development focused as seen in their mandate 'to cooperate in all fields of sustainable development, utilisation, management and conservation of the water and related resources of the Mekong River Basin' (MRC 2009). One of the MRC's main initiatives is the Mekong Program, the objectives of which are to 'achieve more effective use of water and related resources to alleviate poverty while protecting the environment' using the concept of integrated water resources management (IWRM) (MRC 2009). However, the MRC's power is limited because China and Myanmar are not formal members. As a result, China constructed the Lancang Cascade with no downstream consultation and weak environmental impact assessment.

Despite this history of development in the region, political tension, war, and difficult geography impeded large scale engineering projects, so much of the region did not suffer from the construction of the large-scale dams and engineering projects that were part of the 20th century's engineering paradigm (Economist 2004). In the last 10 years, the situation has begun to change. Expanding cross-border trade promoted by the ADB and the World Bank; increasing energy demands in China, Thailand and Vietnam; advances in engineering technology; and increased private investment have helped to stimulate many developments. World Bank estimates from 2006 state that only only 10% of the GMS's hydropower potential is in use (Asian Development Bank and World Bank 2006). These estimates also imply that the basin has the 'flexibility and tolerance' (World Bank 2006: 4) to handle an increase in hydropower development including mainstream dams. However, the competency of this information and other reports that promote the current intensification of hydropower in the region has been strongly criticised (Käkönen and Hirsch 2009). Included in these new developments are the dams mentioned above – the four of eight completed mainstream dams on the upper Mekong in China and the 11 proposed mainstream dams on the lower Mekong in Thailand, Lao PDR, and Cambodia. Although the governments of the region and the ADB, the World Bank, and the MRC cite economic growth, poverty reduction, and improved livelihoods as outcomes of these large-scale developments, the increased regional cooperation and stimulus has produced many negative effects.

The negative consequences of this regional source-management power structure can be seen in such seemingly innocuous things as place-names. The term 'GMS' has entered the mainstream vocabulary throughout the region. This regional place-name suggests a homogenous, fixed space, erasing from the public mind a sense of the unique ecosystems and diverse cultures and livelihoods that exist within the Mekong River Basin. As a name, 'GMS' reflects and facilitates a focus on the development agendas and empowers actors working on behalf of those agendas who are increasingly distant from state citizens and local environmental concerns. Thus, prioritising broad regional concerns can occur with minimal public attention to the cost endured by the most vulnerable groups in society, such as ethnic minorities (Lebel et al. 2005; Cornford and Matthews 2007). In the Mekong, as in other regions, we see that enhancing relations, development and fostering cooperation among countries can lead to degenerative human and environmental consequences (Sneddon and Fox 2006).

In the Mekong subregion, cooperation amongst decision-makers has continually emphasised economic gain over sociocultural and environmental values. This economic focus has created a culture of water resource management that views water solely as a resource and other factors, such as environmental health and social and cultural well-being, receive scant attention. Recently, cooperation between nation-states and the private sector has driven the development of hydropower dams in the region (Molle et al. 2009). For many decision-makers and corporations the Mekong is a working river. Under this paradigm, any water from the river that discharges into the South China Sea untouched by humans is considered a waste.

Asymmetrical power relations and economic growth in the region have also influenced where, how, and what type of development is occurring. In the upper Mekong Basin, China, and Thailand's fast-growing economies, with their high energy demands, have encouraged less developed countries like Burma, Lao PDR, and Cambodia to secure loans from private investors, China, and development banks to construct dams to supply hydropower energy to these countries, with little regard for the environment and people of the region. China and Thailand's influence in the region and the potential profits to be made from these developments has caused Lao PDR and Cambodia's governments to disregard biocultural factors.

To understand the need to dedicate proper investment in time, people, and money to manage water in the region in an holistic way that includes the environment and culture, it is important to examine the growing pool of evidence that identifies the significance of culture, livelihoods, and environmental stewardship for the success of development projects and for the sustainability of our earth's freshwater supplies.

4.4.1 Prioritising Culture

The word 'culture' can take on many meanings and forms. In the widest understanding, it can be defined as 'the whole complex of distinctive spiritual, intellectual and emotional features that characterise a society or social group. It includes not only the arts and letters, but also modes of life, the fundamental rights of the human being, value systems, traditions and beliefs' (Mexico City Declaration on Cultural Policies 1982: art. 10).

The protection and inclusion of culture in development is essential to its success. At the 2007 Commonwealth People's forum, 1,500 representatives from 600 civil society organisations (CSOs) from 59 countries identified culture as a vital component of successful development. The representatives concluded that development is about humans and that culture is an essential part of being human. Some of their key findings emphasised that culture helps to resolve conflict, allowing people to deal with crises; culture can provide identity and encourage the generation of wealth through creative economies and sustainable livelihoods; culture can be both a tool and a process for effective development; culture is important to people's lives and livelihoods; and when culture is protected, people will be happier (Putting Culture First 2008).

Water is an important part of culture in the Mekong subregion. State technocrats, private investors, and multilateral development agencies in the region follow a culture of water resources in which water is a natural resource to be exploited for modernisation and profit. This water resources culture silences and devalues other cultural meanings, such as local knowledge and local livelihoods.

Knowledge plays an important part in culture. In the subregion's management, a culture of scientific and technical knowledge is the dominant paradigm. Decision-makers rely heavily on the hydrological models and scientific and technical expertise of the MRC and the World Bank to inform much of the development in the region (Kakonen and Hirsch 2009). These models, however, do not take into account the complex social, cultural, or ecological aspects of the region and do not acknowledge or represent the majority of people living in the region who depend on natural resources for their livelihoods. Yet the knowledge that belongs to rural people and ethnic minorities, the local and traditional knowledge that decision-makers deem unscientific and therefore ignore, often offers a holistic vision of the complexities of the basin and a much more integrated understanding of water infrastructure challenges.

For local people, water and culture take on many meanings. Local livelihoods are intimately bound up with water and culture. Material culture in the form of fish traps and nets, traditional fishing practices, and traditional knowledge of fish ecology are all aspects of culture. Religious beliefs and practices are another important example of culture. The Naga spirit that protects the rivers and fish across Southeast Asia exemplifies the religious beliefs and practices linked with water, especially among ethnic minorities. For such people, culture is an important source of identity. Not only do ethnic minority groups differentiate themselves though their customs, music, clothing, and language, but they also have strong cultural ties to the places, such as rivers and watersheds, on which their livelihoods have depended for many generations. When large water infrastructure projects like dams alter rivers, flood traditional lands, or force relocation, they often destroy local livelihoods, traditional knowledge, religion, and other important aspects of culture.

Water is an essential part of life, and so it is intrinsically linked to culture. Water permeates culture through religion, livelihoods, and people's relationship with the environment. How we manage water and the environment has direct consequences for how we protect culture.

4.4.2 Prioritising the Environment

The earth's environment is under more pressure now than at any time in human history. Given growing populations, land use changes, increased pollution, and climate change, the need to prioritise the environment and protect biodiversity is stronger than ever. According to the Intergovernmental Panel on Climate Change (IPCC 2007), tropical and subtropical areas, along with river basins, freshwater systems, and flood plains have little capacity to adapt to climate change and therefore will all be severely affected. The Greater Mekong Subregion encompasses all of the above features. Many people living near the river and the coast are at risk of flooding, inundation of coastal areas due to rising sea levels, salinity intrusion, and unpredictable monsoon events. In the Tonle Sap riparian wetlands in Cambodia, climate change coupled with unsustainable mainstream dams could dramatically change the lake's hydrology, leading to devastating effects on agriculture, biodiversity, and the fish resources that make up 80% of the animal protein in the Cambodian diet.

Increased rainfall in the region is expected to raise runoff in the Mekong by 25-40% by the end of the century, potentially causing catastrophic floods (Falloon and Betts 2006). Governments in the region have justified dam building as a way to reduce floods by regulating the river's flow in the wet and dry seasons. This technocratic solution to flooding does not consider the many benefits that floods

Fig. 4.4.1 Fishing near Thonglom Village on the Hinboun River, Laos (Photo credit: David J.H. Blake, photo courtesy of International Rivers)

bring to the region. Although large floods can be devastating for people living close to the river, numerous studies have shown that the flood pulse nature of the Mekong is essential for its biodiversity and therefore the livelihoods and culture of millions of people who rely on this biodiversity (Junk et al. 1989; Arthington et al. 2004; Baran 2006).

Commercial logging and fishing, intensive agricultural practices including high pesticide and fertilizer use, hydropower dams, and monoculture plantations such as palm oil have combined to push the natural resources in the Greater Mekong Subregion to the breaking point. These problems, coupled with rising temperatures and longer dry seasons, could spell disaster for biodiversity, hydrology, and the livelihoods of the subregion's 316 million people.

4.4.3 Integrated Water Resources Management (IWRM)

To address the environmental degradation and loss of livelihoods and culture in the region, policymakers, academics, and non-governmental organisations (NGOs) increasingly encourage the use of IWRM. This approach to water management involves 'a process which promotes the coordinated development and management of water, land and related resources in order to maximise the resultant economic and social welfare in an equitable manner without compromising the sustainability of vital ecosystems' (Global Water Partnership-Technical Advisory Committee 2000).

The growing popularity of IWRM has sprung from the recognition that economists and engineers alone can no longer manage water problems. Many see IWRM as the best strategy to handle the current and predicted effects of climate change on water resources, the negative effects that past water management has had on the environment and people because of economically-focused management styles, and the growing water security problems associated with population growth and development.

Despite the logic of IWRM, true integration of multiple approaches is time-consuming, complex and political. The need to integrate a broad range of social, cultural, economic, and environmental issues and a lack of consensus on how to integrate means minimal guidance and no boundaries for policymakers and planners. This lack of direction and boundaries coupled with a poor understanding of the political support required for implementation has caused many IWRM plans to fail, and the Mekong is no exception. IWRM plans are limited by the realities of time, budgets, staff, political will, and the extent of integration is restricted by the backgrounds of the decision-makers and planners. Unfortunately, those involved in developing many of the Mekong IWRM plans are frequently the same engineers, scientists, and policymakers who participated in the failed single-approach management styles of the past. This continuity of leadership means that the term IWRM is often mentioned but not really implemented.

The term IWRM can be problematic because it assumes that water is a resource to be managed and does not prioritise other aspects, such as biodiversity, livelihoods, and culture. To move away from the focus on water as a resource and to include issues like climate change, revised approaches have been suggested, including Adaptive Water Management (AWM) and Integrated Water Management (IWM). Regardless of whether we write it as AWM, IWM, or IWRM, the value in the term is that it creates a space to engage on key issues like the integration of social, cultural, economic, and environmental values despite the inherent complexities.

If development is about improving people's lives and culture and water are an essential part of people's lives, then development must prioritise culture and the environment, especially water management, to be effective. The following examples from the Greater Mekong Subregion provide a glimpse of how the current unintegrated water resource management practices have affected the local culture and environment.

4.4.4 Examples from the Region: Lao PDR, Cambodia, and Vietnam

4.4.4.1 Lao PDR

The Mekong River defines much of Lao PDR's geography, meandering through 1,860 km of the country and gaining 35% of its flow in this nation. Here, as in many other countries of the subregion, wild fish stocks are important for the livelihoods of the predominantly rural population. A study in the Xe Bang Fai Basin in central Lao PDR found that people's dependence on wild fish was increasing because of the escalating scarcity of forest meat and an increase in people's desire to earn money for new consumer goods (Shoemaker et al. 2001). Recent declines in fisheries, partly attributed to water infrastructure projects, have adversely affected those poor households that are more dependent on fisheries for their livelihoods.

Barney's (2007) study of power, progress and impoverishment in Hinboun District, Lao PDR, details the negative consequences of that nation's policy of rapid rural modernisation through hydropower and land reform. In Ban Pak Veng village, common property rights, loss of livelihood and culture, environmental degradation, and cross-border migration into illegal labour markets in Thailand are some of the results of government concession-based plantation forestry and the large-scale Theun-Hinboun Power Company Hydropower Project, which was designed to provide power for Thailand's growing economy. The Ban Pak Veng village case is one of dozens of examples in which culture, livelihoods, and downstream issues counted for little next to economic gain.

The ADB has recognised these negative effects of development in Laos. The bank's participatory poverty assessment (PPA) states that 'from the outset of the PPA...it was evident that poverty in the Lao PDR is inextricably related to culture and ethnicity'; it goes on to state that 'poverty in the Lao PDR...cannot be studied without reference to culture' (Asian Development Bank 2001: xiii). A study of relocated groups in Lao PDR demonstrated that the cultural disturbance associated with relocation causes increased mortality rates, poor health, and a deep sense of hopelessness and despair (Romagny 2004).

4.4.4.2 Cambodia

Cambodia is home to many inland lakes and important fisheries. Tonle Sap Lake is one of the most important features of the country. Growing from approximately 2,500 km^2 in the dry season to up to 13,000 km^2 in the wet season, the lake provides vital waterflows and habitat throughout the entire southern Mekong basin. Tonle Sap is also home to many Cambodians; the lake hosts a number of floating villages that rely on it for food, transportation, water supply, and sewerage. Millions of people's lives depend on Tonle Sap, and their cultures have developed on its banks and floodplains over thousands of years. The cultures found on and around Tonle Sap are varied and distinctive; many ethnic minority groups base their identities on this place. Cambodians who have lived by the lake for generations have specific names for the fish and other resources that are harvested each season.

Moreover, 'virtually all Cambodians benefit directly in some way from inland fisheries' (Van Acker 2003:3) in this lake and the Mekong River. However, Van Acker explains that fisheries are only part of the aquatic resources that Cambodians rely on. Shrimp, frogs, vegetables, and 400 other different products all feed and earn income for an estimated four million rural Cambodians who depend on aquatic resources. However, these livelihoods, which are essential for the survival of Cambodia's poor, are at risk from other cultures, such as cultures of unsustainable consumption and profit-oriented economies. Destructive fishing practices, such as the use of gill nets; the privatization of fisheries previously held as common property; the blasting of rapids to improve navigation; and the construction of large dams have all led to a decline in fish stocks and a subsequent loss of livelihood and culture for local people. The growth of a for-profit fishing sector in Cambodia and the hundreds of millions of dollars of income it brings annually to the businesses and government have created a culture of 'total harvest mentality' in which sustainability is not considered (Van Acker 2003:18).

4.4.4.3 Vietnam

In Vietnam, upland ethnic minorities have lost subsistence fisheries and forestry-based livelihoods because of deforestation by commercial logging companies and an increase in coffee growing. Vietnam's coffee exports grew rapidly in the last two decades because of government subsidies and increased global demand. By 1995, in the Cu M'Gar district all forest cover had been removed for coffee plantations. This

Fig. 4.4.2 Villagers from Chiang Yen village move their houses piece by piece to make way for the Son La Dam (Photo credit: Hoãi Thanh, photo courtesy of International Rivers)

new commercial crop exacerbated droughts and caused river and reservoir waters and groundwater to drop. In some areas the water levels in rivers and reservoirs fell 2-10 m below average. Locals also believed that the loss of forest cover resulted in increased flooding, causing losses to rice crops (Lidskog et al. 2005). As in Cambodia and Lao PDR, the negative effects of market forces and development were most strongly felt by the poor.

The country-specific literature from the region demonstrates that many ethnic minority groups experienced a loss of or rapid change in culture because of changed environments and livelihoods due to development. Development that is unintegrated or profit focused causes the loss of livelihoods, especially those livelihoods linked to water management. Cultural upheaval in the region often results in social dislocation, psychological trauma, and increased health risks. Despite the recognition of the links between culture, ethnicity, and poverty, the ADB and other decision-makers continue to consider these factors secondary to development and economic growth.

4.4.5 Standing Up for Culture and Biodiversity

Although the outlook for the subregion and the Mekong River may appear bleak, many people and organisations are challenging the unintegrated water management practices that have dominated the large-scale development in the region. The

Fig. 4.4.3 Pak Mun dam protests, 1999 (Photo Courtesy of International Rivers)

advocacy, publicity, and environmental protection work being done by thousands of volunteers and international and local NGOs have provided essential checks on the development projects of international financial institutions and intergovernmental organisations such as the MRC.

In 1997 more than 25,000 rural villagers and urban slum-dwellers known as the Assembly of the Poor carried out a 99-day mass demonstration in front of Government House in Bangkok, Thailand, against large-scale development projects that had devastated the local environment and their culture and livelihoods. The protest resulted in government commitments to address the people's grievances and a significant compensation package for those who lost their livelihoods and culture as a result of these projects (Missingham 2003). Riding the success of the Assembly of the Poor, many protests and much international attention have been directed at unintegrated water resource management and development projects narrowly focused on economic benefits.

There are now estimated to be 800 international NGOs and over 11,000 local NGOs working in the subregion, with the majority of these having formed in the 1990s. The bulk of these organisations work at grassroots and community levels. They combined with international and local academics have been increasingly successful in focusing international attention on the environmental and social problems caused by large-scale development projects. Their success is to be applauded, but there is still a great need for advocacy in the region. In some of the subregion's member countries, NGOs and minority groups have little room or power to protest, and as a consequence the environment, local people, ethnic minorities, and vulnerable groups have no voice. In other instances the NGOs and local people's concerns are simply ignored or dismissed.

The transboundary nature of the Mekong presents other challenges for local communities and environmental protection groups. When the policies of one nation, as with China's Lancang Cascades project, affect the downstream conditions and livelihoods of neighbouring countries, local groups can do little, especially since China's policy in the region to date has been non-consultative. This situation means that local protests and independent reports from the region must reach a wide international audience to place pressure on states and investors to change.

4.4.6 Lessons Learned and Conclusion

The work of local and international NGOs and ordinary citizens is starting to focus attention on the importance of culture and sustainable water management. International financial institutions like the ADB and the World Bank, along with the Mekong River Commission, are making efforts to prioritise the environment and social and cultural well-being. However, there is still much to be done. A review of the literature and case studies suggests the following lessons.

1. We must develop realistic economic assumptions and IWM plans with political support, that take into account the effects on livelihoods and culture and the true costs of relocation and disruption of living conditions. Early planning that incorporates and leaves space for IWM plans of a truly integrated nature with multidisciplinary representation, despite their complexities, will help to ensure holistic approaches to basin management.
2. All stakeholders need a legitimate voice in development projects and must be involved from the earliest stages. Including vulnerable groups and local knowledge in the decision-making process may provide insights into environmental impacts and feasibility. Currently, IWRM plans are not implemented in a way that enables participation. Too often vulnerable and disadvantaged stakeholders are invited to participate only after decisions have been finalised, or the consultation process does not give them a meaningful say in development options. Often, if the option that disadvantaged groups prefer does not align with the economic goals of policymakers, it is simply ignored.
3. It is essential to prioritise environmental health, social and cultural well-being, and human happiness over economic growth, in alignment with the UN Millennium Development Goals. Water permeates all culture, and planners and policymakers need to recognise that economic development does not necessarily confer human happiness. In fact, rising incomes can mean decreased quality of life and declining living standards.

4. Thinking globally and long term are keys to a sustainable future. Incorporating long-term outlooks into development projects will help to ensure that short-term economic gains do not come at unsustainable human and environmental costs. Planners and policymakers need to examine alternatives to hydropower and rethink energy needs. Recognition of the Greater Mekong Subregion's biodiversity and freshwater supply for their environmental value and native forests as important stores of sequestered carbon should be prioritised and economic models developed to show the long-term costs of unintegrated development.
5. Promoting cooperation that benefits all stakeholders will lead to a truly integrated water management strategy for the subregion. Cooperation between countries here has increased economic growth, but this improvement has often come at a cost to the environment and local cultures of the region. Fostering cooperation that not only protects and respects all cultures and the environment but also encourages a common understanding of the importance of these factors and how they link to successful development will go a long way toward ensuring a bright future for the region.

This chapter has attempted to show that protecting and respecting all cultures and the environment in the Greater Mekong Subregion is essential for its sustainable future. For many people protecting the environment and culture are intuitive. IWM plans that emphasise the significance of social and cultural well-being and environmental health are in common use. Yet case studies from the region demonstrate that too often cost benefit analysis, budgets, and timelines reduce cultural well-being and environmental health to tokens. It is time to step back and ask pointed questions, such as 'What is the true price of rapid development?' 'Who are the long-term winners?' 'What are the long-term effects of current development decisions?' 'What are the real costs of unintegrated water resource management?'

The challenges ahead for the subregion's people and environment are great. With much of the population reliant on natural resources for their livelihoods, current development strategies spell disaster for many cultures, priceless biodiversity, and one of the world's great rivers. In a 2006 working paper, the ADB and World Bank strongly supported an aggressive increase of hydropower development in the region (Asian Development Bank and World Bank 2006). Although hydropower dams are often perceived as green and modern energy, when they are constructed without holistic and integrated management plans, they cause extensive environmental and social damage. It is critical for policymakers and planners to recognise that their technocratic culture, which leads to unintegrated development, extinguishes other cultures, especially local ones, thereby making the developments unsustainable and long-term failures.

There needs to be a paradigm shift away from the belief that large-scale infrastructure is panacea for development and towards a philosophy that recognises the value of culture, livelihoods, and the environment. It is also vitally important for local people and the international community to continue to stand up for the environment and vulnerable groups, to give a voice to those who have trouble being heard or can't speak, and to remind everyone of the importance of culture, water, and biodiversity. The economic paradigm that drove the water development agenda of yesteryear is one we can no longer afford to sustain. Prioritising environmental health, sociocultural well-being, and human happiness will result, in the long run, in development that supports a truly sustainable economy.

Resources

Arthington, A.H., K. Lorenzen, B.J. Pusey, R. Abell, A.S. Halls, K.O. Winemiller, D.A. Arrington, and E. Baran. 2004. River fisheries: Ecological basis for management and conservation. In *Proceedings of the second international symposium on the management of large rivers for fisheries*, vol. 1, ed. R.L. Welcomme and T. Petr, 21–60. Bangkok: FAO.

Asia Times. 2010. Marwaan Macan-Markar, China, US angle for Mekong influence. http://www.atimes.com/atimes/China/LF15Ad01.html. 30 June 2010.

Asian Development Bank. 2001. *Participatory poverty assessment*. Vientiane: State Planning Committee, National Statistic Center and Asian Development Bank.

Asian Development Bank. 2007. *Mid-term review of the Greater Mekong Subregion strategic framework, 2002–2012*, Manila: Asian Development Bank, June.

Asian Development Bank. 2009. The GMS Program. http://www.adb.org/GMS/Program/default.asp. Accessed 21 Aug 2009.

Asian Development Bank and World Bank. 2006. *Future directions for water resources management in the Mekong River Basin (Mekong water resources assistance strategy)*. Manila: Asian Development Bank. www.adb.org/water/operations/partnerships/mwras-June2006.pdf. Accessed 30 June 2010.

Baran, E. 2006. *Fish migration triggers in the Lower Mekong Basin and other freshwater tropical systems*. MRC Technical Paper no. 14. Vientiane: Mekong River Commission.

Barney, Keith. 2007. *Power, progress and impoverishment: Plantations, hydropower, ecological change and community transformation in Hinboun District, Lao PDR*. York Centre for Asian Research papers, no.1. Toronto: York University, June.

Commonwealth Foundation. 2008. *Putting culture first*. London: Commonwealth Foundation. 68 pp.

Cornford, J., and N. Matthews. 2007. *Hidden costs: The underside of economic transformation in the greater Mekong subregion*. Melbourne: Oxfam-Australia.

Dieu, Nguyen Thi. 1999. *The Mekong River and the struggle for Indochina: Water, war and peace*. Westport: Praeger.

Economist. 2004. The sweet serpent of South-east Asia: How much longer will the Mekong remain the world's last great unspoilt river? *The Economist* 370(8356 (January)): 28–30.

Falloon, P.D., and R.A. Betts. 2006. The impact of climate change on global river flow in HadGEMI simulations. *Atmospheric Science Letters* 7: 62–68.

Global Water Partnership. 2000. *Integrated water resources management*. Paper no. 4. Stockholm: Global Water Partnership: Technical Advisory Committee.

Hori, H. 2000. *The Mekong: Environment and development*. New York: United Nations University Press.

Intergovernmental Panel on Climate Change. 2007. *Climate change 2007: Impacts, adaptation and vulnerability*. Contribution of Working Group II to the Fourth Assessment Report of the IPCC, ed. M.L. Parry, O.F. Canziani, J.P. Palutikof, P.J. van der Linden, and C.E. Hanson. Cambridge: Cambridge University Press.

Junk, W.J., P.B. Bayley, and R.E. Sparks. 1989. The flood pulse concept in river-floodplain systems. In *Proceedings of the International Large River Symposium*, Canadian Special Publication of Fisheries and Aquatic Science 106, ed. D.P. Dodge, 110–127.

Käkönen, M., and P. Hirsch. 2009. The anti-politics of Mekong knowledge production. In *Contested waterscapes in the Mekong region: Hydropower, livelihoods and governance*, ed. F. Molle, T. Foran, and M. Käkönen, 1–13. London: Earthscan.

Lebel, L., P. Garden, and M. Imamura. 2005. The politics of scale, position, and place in the governance of water resources in the Mekong region. *Ecology and Society* 10(2): 18.

Lindskog, E., K. Dow, G. Nilsson Axberg, F. Miller, and A. Hancock. 2005. *When rapid changes in environmental, social and economic conditions converge: Challenges to sustainable livelihoods in Dak Lak, Vietnam*. Stockholm: Stockholm Environment Institute.

Mekong River Commission. 2009. The Mekong program. http://www.mrcmekong.org/mekong_program_ceo.htm. Accessed 28 Aug 2009.

Missingham, Bruce. 2003. *The assembly of the poor: From local struggles to national protest movement*. Chiang Mai: Silkworm Books.

Molle, F., T. Foran, and P. Floch. 2009. Changing waterscapes in the Mekong region: Historical background and context. In *Contested waterscapes in the Mekong region: Hydropower, livelihoods and governance*, ed. F. Molle, T. Foran, and M. Käkönen, 1–13. London: Earthscan.

Romagny, L. 2004. Resettlement: An alternative for upland development? *Action Contre la Faim Laos* (Jan): 1–16.

Shoemaker, B., I. Baird, and M. Baird. 2001. *The people and their river: A survey of river-based livelihoods in the Xe Bang Fai River Basin in central Lao PDR*. Vientiane: Lao PDR Canada Fund for Local Initiatives, 15 Nov.

Sneddon, C., and C. Fox. 2006. Rethinking transboundary waters: A critical hydropolitics of the Mekong basin. *Political Geography* 25(2): 181–202.

UNESCO. 1982. *Mexico City declaration on cultural policies*. Mexico City: UNESCO.

Van Acker, F. 2003. *Cambodia's commons: changing governance, shifting entitlements?* Centre for ASEAN Studies Discussion paper No. 42. Antwerp: Centre for International Management and Development Antwerp. University of Antwerp.

Chapter 4.5
Damming China's Angry River: Vulnerability in a Culturally and Biologically Diverse Watershed

Bryan Tilt

Of the 50,000 large dams that exist in the world today, nearly half of them are in China. They provide flood protection, supply water for irrigation, and produce hydroelectric power in a nation with a seemingly insatiable appetite for energy. The southwest region, on the edge of the Qinghai-Tibet Plateau, is home to the major share of China's vast hydropower potential. Currently, a 13-dam hydropower development plan is underway on the region's Nu River, in a remote corner of Yunnan Province that is renowned for its cultural and biological diversity. The dam project has a total hydropower potential of 21,000 megawatts (MW), which is slightly more than the mammoth Three Gorges Dam. Should all 13 dams in the cascade be built, it is estimated that more than 50,000 people will be displaced.

The Nu River originates at more than 5,500 m high on the Qinghai-Tibet Plateau, before flowing through the province of Yunnan and continuing through Myanmar (Burma), where it forms part of the Thai-Burmese border and is known as the Salween (see Map 4.5.1). With a total length of 2,018 km, the Nu winds through the Gaoligong mountain range, a series of deep gorges and glaciated peaks created by the collision of the Indian tectonic plate with the Eurasian plate approximately 50 million years ago. It is often referred to as 'Asia's longest free-flowing river'.

The name 'Nu', which translates literally as 'angry', is actually a phonetic rendering of 'Nong', the name given to the river by the local Lisu ethnic group, one of China's many 'minority nationalities' (Mertha 2008). But it is also an apt metaphor for the controversy surrounding the proposed dam development project in this region, home to 7,000 plant species, 80 species of rare or endangered animals, and 22 of China's 55 officially recognised minority nationalities. One of the most biodiverse temperate ecosystems in the world, northwest Yunnan is a focal point for global conservation efforts. Fifteen protected areas in eight clusters, totalling nearly

B. Tilt (✉)
Anthropology Department, Oregon State University, Corvallis, OR, USA
e-mail: Bryan.Tilt@oregonstate.edu

B.R. Johnston et al. (eds.), *Water, Cultural Diversity, and Global Environmental Change: Emerging Trends, Sustainable Futures?*, DOI 10.1007/978-94-007-1774-9_26,

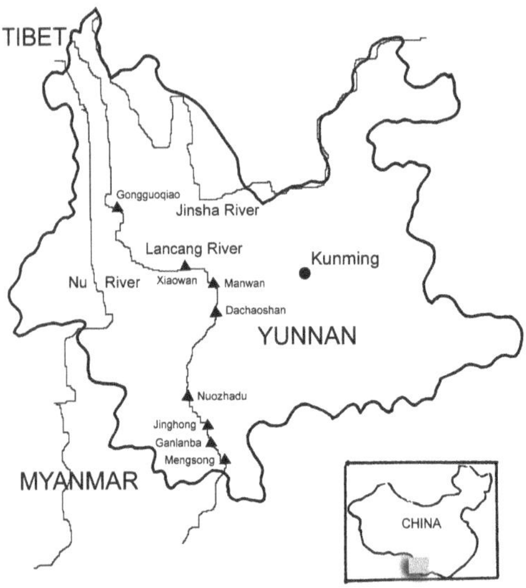

Map 4.5.1 Dams along the Lancang (Mekong) River, China (Credit: Bryan Tilt)

one million ha, were enlisted on UNESCO's World Heritage List in 2003 (UNESCO 2009). The Nature Conservancy, working in close association with the Yunnan provincial government, is also active in the area, working to make Xianggelila County (formerly Zhongdian, the county changed its name to the mythical 'Shangrila' in 2003) into a national park.

In this chapter, I provide information on hydropower development issues in China, review the current status of the Nu River Project, and illustrate some of the potential impacts for local ethnic minority communities in this region of immense cultural and biological diversity. My analysis focuses on three areas of vulnerability: economy, governance and public participation, and cultural autonomy.

4.5.1 Ethnicity and Culture in the Nu River Valley

Northwest Yunnan's Nujiang Prefecture, where eight of the 13 dams are slated to be built, was established in 1954 as an autonomous prefecture for the Lisu people. The Lisu language is Tibeto-Burman in origin and has about 600,000 speakers in southwest China, with thousands more in Burma and Thailand (Ethologue 2009). A traditional Lisu folktale provides a colourful origin story for the region's indigenous inhabitants. According to the story, in the dim mist of an indeterminate past, a mythical brother and sister survived 99 days and nights of torrential flooding by hiding in a gourd, floating about the landscape, and witnessing the destruction of the rest of humankind in the rising floodwaters. Lacking other alternatives and persuaded by a pair of dazzling, golden-coloured birds who possess the ability to talk, they reluctantly got married. In time, the woman gave birth to 12 children, 6 sons and 6 daughters, each of whom married a sibling and went off in search of their livelihood and their destiny. The couple who travelled north became the Tibetans; the couple who travelled south became the Bai; the couple who went east are the Han; the couple who went west are the Keqin; the couple who settled on the banks of the river are the Nu; and the couple who chose to stay with their parents are the Lisu (Miller 1994: 74–84).

Such a story provides a basic framework for making sense of the complex mélange of ethnic identity in this corner of southwest China. But ethnicity in the Nu River valley is often not so clear-cut. Individuals from different ethnic backgrounds have intermarried for a long time; during fieldwork, when I inquire about the ethnic identity of a given individual, I often hear a long pause, followed by an intricate recitation of genealogy: 'My father is Lisu, but my mother is Tibetan, and many members of my extended clan are Nu'. As a result of this prolonged, intimate contact between ethnic groups, facility in two or three languages is common among local residents.

Further complicating the story of ethnic identity is the fact that Tibetan Lamaistic Buddhism and animism, long the dominant religious beliefs of the area, were supplanted by Roman Catholicism about a century ago, thanks to a group of hardy Swiss missionaries. Dozens of rustic churches dot the Nu River valley. Mass is held in several languages at least once each month, and important life-cycle rituals such as weddings and funerals are by and large Catholic affairs (Clark 2009). During the summer of 2008, I interviewed a Lisu couple with five children who lived on the right bank of the Nu River in Fugong County. The government's planned births programme stipulates a maximum of two children for residents of the county, and the couple had paid a fine of 1,450 yuan for each additional child, borrowing money from their extended family.[1] Their house consisted of two rooms: a kitchen with a dirt floor and a hearth where meals were cooked over a wood fire, and a larger bedroom with a cement floor and several wooden beds covered in mosquito netting. The father, a strong, compact man in his 40s, showed me his most prized possession: the family Bible, written in the Lisu orthography that had been invented by Catholic missionaries a century ago.

Like many of the highland ethnic groups in southwest China, the Lisu historically practiced swidden agriculture. The rise in population over the past several hundred years, coupled with the market-based agricultural reforms of the past several decades, induced them to practice sedentary agriculture. They grow a variety of grain crops, including rice, corn, barley, and buckwheat, and a wide array of vegetable crops. Agricultural fields are carved out of incredibly steep hillsides or situated on alluvial fans where tributaries to the Nu River have deposited rich, fertile sediment. Maintaining this cultivation pattern requires a tremendous amount of effort, which the anthropologist Marco Clark has referred to as a 'delicate balance between gravity and human determination' (Clark 2009: 23).

China considers itself a 'unified, multiethnic state (*tongyi duo minzu guojia*)'. In addition to the dominant Han majority, which constitutes about 92% of the nation's

[1] China's planned-births programme (*jihua shengyu*) strictly limits urban households to one child but provides less-stringent regulations for rural areas with large minority populations and areas where a significant percentage of residents live below the poverty level (These two conditions, in fact, often coincide.) Families in Fugong County, most of whom relied on labour-intensive agriculture to make their living, are presently allowed two children.

Fig. 4.5.1 Ethnic minority woman, in her garden near the banks of the Nu River (Photo courtesy of International Rivers)

population, there are 45 'minority nationalities (*shaoshu minzu*)' that received formal recognition by the central government following an ethnic-identification project conducted between 1950 and 1956. The minority nationalities, which numbered 105 million as of the 2000 census, represent a special development problem for the government bureaucracy. On the one hand, their perceived backwardness provides normative justification for targeted development, economic assistance, educational subsidies, and national welfare policies. On the other hand, high concentrations of minority nationalities are perceived by many as a barrier to actually achieving development at a level consistent with the national average.

The anthropologist Fei Xiaotong, who studied in Britain under Branislaw Malinowski, played a key role in the ethnic-identification project of the 1950s. The general aim of the project was to classify each minority group according to Marx's schema as primitive, slave, feudal, bourgeois-capitalist, socialist, or communist (Harrell 1995). The government also borrowed classificatory criteria from the Soviet Union's system, developed by Stalin, which included a common territory, language, mode of subsistence, and psychological makeup. In practice, the process of affording recognised minority status to some groups while denying it to others was highly politically charged, since it was tied to the establishment of ethnic regional autonomy for governance in minority areas (Wang and Young 2006).

Various places with high concentrations of ethnic minorities have received the designation of 'autonomous region (*zizhi qu*)', 'autonomous prefecture (*zizhi zhou*)', or 'autonomous county (*zizhi xian*)', but the ability of officials within these entities to practice self-governance or to influence central policy remains extremely limited (see Rossabi 2004). Tibet, for example is recognised as a provincial-level autonomous region (*zizhi qu*), but the potential for a secessionist movement causes the central government to keep a tight rein on the region. Moreover, China has been reluctant to accord its minority nationalities 'indigenous' status under such frameworks as the UN Convention on Biological Diversity (CBD), which recognises the rights of indigenous people for autonomy and self-determination, as will be discussed below.

4.5.2 Damming the Angry River

The World Commission on Dams (WCD) published a landmark study in 2000 that noted that, although dams had contributed significantly to human development over the years, their deleterious impacts – social, cultural, environmental, and economic – had long eluded meaningful scrutiny (WCD 2000). The WCD report estimates that about 50,000 large dams, which the International Commission on Large Dams defines as those greater than 15 m in height or having a storage capacity greater than three million cubic metres, exist worldwide (Scudder 2005: 2–3). Dams deliver hydropower, provide sources of irrigation water, enhance navigability, and protect against flooding. Proponents of large dams also point out that, amid uncertainty about the future climate regime, dams provide an opportunity to store water as rainfall and snow pack levels change (see Tullos et al. 2009).

Since 1978, Chinese leaders have followed a path of economic liberalization and privatization known as Reform and Opening (*gaige kaifang*). During the reform period, the nation's annual rate of GDP growth has been at or near double-digit figures, and this push for economic and industrial development has been largely powered by fossil fuels. Coal alone provides more than two-thirds of the nation's energy supply, with industries, power plants and households gobbling up 1.5 billion metric tons per year (Smil 2004). The emissions from such activities contribute to widespread environmental degradation and public health risks at home, and to global climate change. Hydropower, for all of its ills, is thus seen by many officials in China as a potentially 'clean' source of renewable energy.

Hydropower development is also a key part of China's 'Develop the West' strategy (*Xibu Dakaifa*) enshrined in the tenth and eleventh national 5-Year plans for economic development (2001-2010), the goal of which is to narrow the economic and social disparities between the prosperous east coast and the relatively impoverished western regions, including Yunnan. The hydropower resources of this region are vast; most of the nation's great rivers – the Yellow, the Yangtze, and the Mekong, among others – have their headwaters in this expansive, arid interior region of southwestern China. It is precisely these regions, moreover, that have been largely left behind in China's rush to develop its east coast, leading political leaders and hydropower officials to argue that dams can contribute both to the nation's energy supply and to jobs, revenue and improved infrastructure in Yunnan.

As with many sectors of the Chinese economy during the reform era, the political economy of hydropower development has followed a neoliberal path. In 2002, the State Power Company of China was disbanded and its assets distributed among 11 limited-liability stock corporations, five of which were responsible for power generation. One of these 'big five', China Huadian Corporation, holds a state-granted monopoly on the right to develop dam projects on the Nu River. Huadian enjoys close ties to the provincial government in Yunnan (Magee and McDonald 2009).

This complex network of public-private relationships is illustrative of the culture of economic liberalism that predominates, somewhat ironically, within the Chinese Communist Party. The rights to develop water management infrastructure have been turned over to private interests (albeit those headed by economic elites with close government ties), while the overall logic and priorities of water management remain under the control of the state. The National Development and Reform Commission and other government agencies have the ability to approve particular projects and block others, or to provide tax incentives to projects that further the economic goals of the state. In the case of the Nu River, this has resulted in a legitimization of hydroelectricity generation in a remote and sensitive region, not for local consumption but for supplying the eastern cities that serve as the world's manufacturing hub.

Information on each of the 13 dams in the Nu River Project is given in Table 4.5.1, from Songta Dam in the north, which is located in Tibet, to Guangpo Dam in the south. Design and operation specifications differ considerably from one dam to the next. Some, such as the Songta and Maji dams, are very large (more than 300 m high) and will include reservoirs with enormous storage capacities, displacing thousands of people; others, such as Bingzhongluo and Liuku, will be 'run of the river' dams with minimal storage capacity and will thus displace far fewer people. The hydropower potential of the project totals 21,000 MW, larger than the total capacity

Table 4.5.1 Design and operation specifications for the 13 Dams in the Nu River Project (Sources: Magee and McDonald 2009; He et al. 2009: 147–148)

Dam name	Height (m)	Installed electrical capacity (MW)	Reservoir storage capacity (million m^3)	Estimated population displaced
Songta	307	4,200	6,312	3,633
Bingzhongluo	55	1,600	13.7	0
Maji	300	4,200	4,696	19,830
Lumadeng	165	2,000	663.6	6,092
Fugong	60	400	18.4	682
Bijiang	118	1,500	280	5,186
Yabiluo	133	1,800	344	3,982
Lushui	175	2,400	1,288	6,190
Liuku	35.5	180	8.1	411
Shitouzai	59	440	700	687
Saige	79	1,000	270	1,882
Yansangshu	84	1,000	391	2,470
Guangpo	58	600	124	34
Total for 13 Dams		**21,320**	**15,108.8**	**51,079**

of the Three Gorges Dam (Magee and McDonald 2009). Current estimates suggest that 51,079 people would be displaced, should all 13 dams be built.

From the beginning, plans to dam the Nu River have been highly controversial. After considerable opposition from Chinese and international conservation organizations, Premier Wen Jiabao ordered a temporary halt to the project on February 18, 2004. This was considered the first serious test of China's new Environmental Impact Assessment Law, which was promulgated in 2003. Under Premier Wen Jiabao's advisement, the National Development and Reform Commission, along with the State Environmental Protection Administration, conducted a review of hydropower development on the Nu. Scholars, activists, and local communities waited and watched.

Early in 2006, officials decided to allow construction to commence on a scaled-down version of the project, beginning with four of the 13 dams: Maji, Yabiluo, Liuku, and Saige. Two of the four, Liuku and Saige, were set to begin construction during the eleventh 5-Year Plan period (2006–2010), according to a document released by the National Development and Reform Commission entitled 'The Plan on the Development of Renewable Energy during the 'Eleventh 5-Year Plan Period' (see Brown and Xu 2009). Based on recent field observations, preparatory work including site surveying and road construction were underway in 2007 and 2008.

4.5.3 Cultural Diversity and Vulnerability

Recent studies on the social impacts of dam construction around the world suggest that displacement and resettlement, particularly by force or coercion, result in a cascade of subsequent negative impacts on employment and income, social networks, and health and well-being (see, for example, Tilt et al. 2009; Scudder 2005). The World Bank and other multilateral agencies involved in hydropower development have established clear guidelines for assessing the social impacts of dam construction. The landmark report published by the WCD in 2000 emphasises the application of a 'rights and risks' approach to evaluating dam projects, which entails, among other things:

> Self-determination and the right to consultation in matters that affect people's lives, the right to democratic representation of people's views on such matters, the right to an adequate standard of living, freedom from arbitrary deprivation of property (WCD 2000: 200).

But such a framework is not easily implemented in China. Different social groups experience the impacts of development in disparate ways, in part based on their vulnerability, which Wisner et al. (2004) have defined as 'the characteristics of a person or group that influence their capacity to anticipate, cope with, resist and recover from the impact of a hazard'. I would like to examine three key areas of vulnerability experienced by local ethnic minority communities as they relate to the Nu River Project: economic vulnerability, vulnerability in terms of governance and decision-making, and vulnerability as it relates to cultural autonomy.

First, from an economic perspective, the Western regions are comparatively poor. None of the provinces in which more than 10% of the population consists of ethnic minorities is listed among the ranks of high-income or middle-income provinces (Wang and Hu 1999). The United Nations Development Programme, for example, uses a Human Development Index (HDI), which includes a measure of economic productivity, life expectancy and education, to identify development needs. In the most recent HDI calculation, Yunnan ranks 29th out of 31 provinces and administrative regions (UNDP 2008: 22). In Nujiang Prefecture specifically, minority nationality people account for 92.4% of the 520,600 residents, and all four counties in the prefecture are designated national-level impoverished counties. This is a common trend in reform-era China; measured a variety of ways, income and wealth inequality has continued to grow throughout the economic reform period, widening gaps between rural and urban communities, and between individuals within communities.

Not surprisingly, many of the official government documents on the Nu River Project describe the dams as part of a poverty alleviation strategy in a region that is otherwise culturally and economically backward. Even the local government in Nujiang Prefecture, which is accustomed to relying on central government subsidies as a main revenue source, is strongly in favour of the projects, which would generate an estimated 36 billion Chinese yuan per year in revenue, one billion of which would go to Nujiang Prefecture, effectively increasing the prefecture's revenue flow tenfold (Magee and McDonald 2009: 50). But careful inspections of the planning documents reveal an interesting fact: the vast majority of electrical power will be sent eastward to coastal cities such as Shanghai and Guangzhou, where industrial, commercial and residential demands are high. This plan, officially termed 'Send Western Electricity East' (*xi dian dong song*) will be facilitated by high-voltage direct-current (DC) transmission lines, cutting-edge technology which Chinese engineers have helped to develop. Thus, existing economic disparities between regions are likely to be exacerbated by the Nu River projects (Magee 2006).

Second, in terms of governance, major questions about public participation in decision-making, and compensation for lost assets, remain unanswered. Recent studies have investigated the compensation structure for hydropower development in China and its application thus far in the Nu River basin (Brown and Xu 2009). On September 1, 2006, the State Council, China's chief legislative body, adopted the 'Regulations on Land Acquisition Compensation and Resettlement of Migrants for Construction of Large and Medium Scale Water Conservancy and Hydropower Projects'. These regulations represent a dramatic step forward from China's historically inadequate compensation structure. For example, the regulations stipulate that resettlers are to receive 16 times the value of average annual income, in addition to compensation for housing of the 'same scale, same standard, and same function' (Brown and Xu 2009).

However, as is often the case, the trouble lies in the implementation of regulations. Just upstream from the Liuku Dam site, where preparatory work is underway, a total of 144 households from Xiaoshaba Village (literally 'small sand bar') were relocated to New Xiaoshaba Village in 2007. Although public hearings were held in 2006, most villagers reported feeling intimidated and effectively shut out of the decision-making process. In addition, when resettlement took place, residents were required to purchase their new houses at an exorbitant price. More than a year after resettlement, no steps had been taken to allocate new farmland to the resettlers, and residents reported a lack of long-term support programmes such as job training (Brown and Xu 2009). Much of the land that will be inundated is low-lying, fertile farmland adjacent to the river. As the anthropologist Heather Lazrus (2009: 248) has observed, 'As much as it is social, political, and physical, vulnerability is also a matter of representation to which questions of agency are central: Who is doing the representing, under what conditions, and for what purposes?'

What little systematic data has been collected on residents' attitudes toward the dam projects seems to be mixed. Recent interviews found that about one-third of respondents supported the projects, one-third opposed them, and the remaining third was either neutral of did not wish to respond (Magee and McDonald 2009: 53). This is likely due to a combination of factors, including a general lack of detailed information about the projects, and hesitation to be seen opposing the government or allying with anti-dam NGOs, which could be politically dangerous.

Box 4.5a Lessons from Tibet's only successful anti-dam campaign —Tashi Tsering

Box Map 4.5a.1 Nyagchu/Yalong River: dams completed, under construction, and planned by Ertan, 2009 (Credit: Tashi Tsering, tibetanplateau.blogspot.com)

Representatives of international environmental and Tibetan rights groups often approach me with ideas for campaigns against dam projects in Tibet. In a series of maps published in 2010 on the Tibetan Plateau blog (www.tibetanplateau.blogspot.com), I identified 198 hydropower projects built, planned, or proposed on the Tibetan Plateau, which is the source of a constellation of international rivers, including the Indus, Karnali, Arun, Brahmaputra, Mekong, Indus, and Salween.

Understandably, international environmental groups are concerned about the environmental and downstream flow implications of these projects, and Tibetan rights groups are concerned about the rights of

(continued)

Box 4.5a (continued)

the affected Tibetans. A key problem with these activist groups is that their campaign approaches typically involve press releases and reports picked up by the international media with the assumption that such efforts will generate political and economic pressure. A media-based strategy, however, is rarely effective in achieving the end goal of stopping the projects as in China, international (mainly Western) media are perceived as being biased, so critical reports are often dismissed. Moreover, such international coverage has little impact on project financing, as most dams in China are financed with state funding through domestic companies.

China has built half of the world's large dams, and its leaders, most of whom are engineers by training and who must deal with serious energy and water crises, continue to depend on large structural solutions like building more dams and water diversion projects. As the world's highest and largest plateau with the third-largest store of glacial fresh water after the north and south poles, Tibet is regarded as 'China's Water Tower' and has some of the world's best sites for hydropower development. Hydropower and large-scale infrastructure development are key components of China's plan to develop Tibet and integrate the restive Tibetan Plateau with the rest of China. With growing international confidence in China's surging economy and the government's intolerance for political opposition, especially from Western and exile Tibetan sources, political campaigns by NGOs based outside of China to stop these projects are not only unlikely to succeed but also often end up causing problems for domestic campaigns. If people are genuinely interested in making a difference to conditions on the ground, I believe they must first support the work of local activists.

4.5a.1 Learning from successful local campaigns

Although protests and campaigns against dam projects in China, such as those agitating against the Lianghekou,[1] Pubugou,[2] and Three Gorges[3] dams are often met with the iron hand of the government, in recent years there have been some remarkable developments in the extent to which Chinese civil society leaders have been able to work with the State Environmental Protection Agency and the media to publicise the social and environmental costs of dam projects. Scientists, journalists, environmental activists, and other citizens have worked together to provide information on the adverse costs of dam development and to advocate for a reversal of government-approved projects. Premier Wen Jiabao's direct intervention on 1 April 2004, to suspend construction on a series of 13 dams on the Salween River was the most notable result of these campaigns. One of these dams, the Songta, was planned in Tibetan-inhabited areas.

(continued)

Box 4.5a (continued)

The premier made his decision in response to efforts by civil society leaders and media to educate local peoples, media, and the government about the social and environmental costs of these dams. One of the activists' most important demands was to subject these projects to a proper Environmental Impact Assessment (EIA), as required by the 2003 IEA Law. After the premier's intervention, all 13 projects had to go through EIA, and the review committee decided that nine of them, including the Songta Dam, should not be built as planned to avoid disastrous environmental and social consequences.[4]

Following this success, Chinese environmentalists and reporters turned their attention towards another dam project in Yunnan Province, the Tiger Leaping Gorge.[5] In addition to threatening the natural integrity of the Three Parallel Rivers World Heritage Site, near which the project is located, the proposed dam would also displace about 100,000 people. Some Chinese experts on water resource politics, such as Ma Jun and Dr. Yu Xiaogang, saw this activist campaign as a significant test of China's budding environmental movement. The outcome would both demonstrate the relative effectiveness of an environmental advocacy that promotes healthy rivers and illustrate the degree to which the government is serious in its commitment to consider their concerns.[6] The project was scrapped in 2007 following widespread Chinese opposition. What is remarkable about the campaigns against the Salween and the Tiger Leaping Gorge dams is not only the unprecedented level of participation and support from Chinese civil society leaders but also the fact that the government tolerated the protests.

During the same time, a smaller and more subtle campaign was being brewed to stop a dam project on a lake on the eastern fringes of the Tibetan Plateau called Megoe Tso. I was a key member of a coalition of Chinese, Tibetan, and international activists who decided to work together in a low-profile manner and use subtle campaign strategies to stop this dam project. The Megoe Tso campaign is significant because it is the first and only one to successfully stop a dam project on the Tibetan Plateau.

4.5a.2 Harnessing Tibet's sacred lake, Megoe Tso, for hydroelectricity[7]

Megoe Tso, known as Mugecuo or Mugetso in Chinese and Savage or Yeti Lake in English, is a high mountain lake situated on the eastern fringes of the Tibetan Plateau, in Kartse Tibetan Autonomous Prefecture of Sichuan Province.

(continued)

Box 4.5a (continued)

Box Fig. 4.5a.1 A local saying describes this sacred lake as "four seasons in one day and different weather in the morning and evening" (Photo credit: Tashi Tsering)

The lake is connected to the Yala River, a tributary of the Wasi River, which drains into a major tributary of the Yangtze River known to Tibetans as Gyarong Gyalmo Ngulchu (Chinese: Dadu He). *Renmin Zhengxie Bao,* a government daily newspaper, reported the status of hydropower development on the Dadu River in September 2004: 'Forty-eight dams have been built along the main stream and branches of the river and a total of 356 dams are in the development pipeline'.[8]

From an environmental perspective, Megoe Tso is situated in a 'biodiversity hotspot' – one of the richest and most threatened reservoirs of plants and animals on earth, according to Conservation International.[9] *China Youth Daily* reports that the lake 'nurtures over 1,000 sub-tropical mountainous plants, over 2,000 vertebrata within the humid river valley, over 100 bird species and many other animals, fish and insects, and over 30 species of azalea'. In addition to its ecological significance, the local Tibetans consider Megoe Tso sacred.[10] The Central Tibetan Administration of the Dalai Lama describes the lake as the most sacred lake in the eastern Tibetan region of Kham.[11] Nearby, Minyak Ghangkar (Gongga) and Sharchok Shara mountains are the two main sacred mountains of the region. These mountains and Megoe Tso Lake have traditionally served as pilgrimage sites for Tibetan spiritual practitioners.[12]

Although local government claims that plans to develop Megoe Tso for hydroelectricity were drafted as early as 1956, the Sichuan Communist Party office and provincial government only listed the Wasi River Basin hydropower development, including the Megoe Tso dam project, as the State Council's main project to support the development of Sichuan's Tibetan area in 1995. According to project designs, a dam 50.5 m high and 260.5 m wide will be built near where the Megoe Tso pours into the Yala River. The dam will be connected to another pumped storage power plant and the Jin'gai hydropower plant through tunnels and diversion channels. The project was to be funded to the tune of US$300 million jointly by the local government and Huaneng Power International, China's largest independent power producer. Huaneng and local government had an agreement to share the revenue generated from the station 60:40.[13]

(continued)

Box 4.5a (continued)

4.5a.3 Opposition to the project

Two key groups, local Tibetans and Chinese civil society leaders, vehemently opposed the project. The former opposed the project because the lake is sacred to them, and any kind of development of the lake for economic gain was simply unacceptable. Their concerns were submitted in a letter to Premier Wen Jiabao through a senior Tibetan Party official, Bapa Phuntsok Wangyal, in May 2003. A month later, the premier's office sent a special 'interagency task force' to investigate the project. The response from the premier's office was partly influenced by initial campaigns by Chinese civil society leaders and media, with support from groups such as Conservation International Beijing and World Wildlife Fund China. These groups mainly argued that the project would be disastrous for biodiversity and the livelihood of minority groups. When the interagency task force met with local Tibetans, the latter wasted no time in expressing their opposition to the project.

On 28 May 2003, *China Youth Daily* published an article titled 'Mugecuo Lake calling for help', followed by another article in June 2003.[14] This coverage attracted the interest of international Tibet rights groups and Radio Free Asia, both of which ran news stories and commentaries on the project. The government of Kartse (Ganzi) Tibetan Autonomous Prefecture reproached *China Youth Daily* for publishing an 'exaggerated' and 'biased' article that created negative political effects and 'internationalization' of the dam project. Weeks later, the Tibetan government in exile issued a press release urging China's central government to 'step in to save this scenic jewel of world importance [Megoe Tso], so it can be protected for all humanity'.

Almost a decade ago, another sacred Tibetan lake, this one in south-central Tibet, Yamdrok Tso, was proposed for hydroelectric generation. A major international campaign formed to stop the project, including an International Union for Conservation of Nature (IUCN) resolution, but to no avail. Considering the fate of Yamdrok Tso, believed to be more important to Tibetans than Megoe Tso, the chances of saving the latter seemed bleak. Not only was Huaneng actively lobbying for the project, authorities were also pressuring experts responsible for project feasibility study to give their approval. After the few Chinese civil society leaders who were initially involved in raising concerns about Megoe Tso project decided to focus on the Salween River and Tiger Leaping Gorge dams, analysts began to consider this a 'failed'[15] campaign (Birnbaum and Xiubo 2006:190).

(continued)

Box 4.5a (continued)

4.5a.4 Networking, partnerships, and strategic decisions

The Chinese environmental activists involved in the struggle to halt dams on the Salween and Yangtze Rivers collaborated with the US-based International Rivers Network (IRN), which helped encourage the attention of international media and environmental advocacy groups. At the time, in 2005, I was working as the Environment Project Director of a Tibetan rights group called Tibet Justice Center, based in Berkeley, California (USA), and I was also a board member of the International Tibet Support Network (ITSN), a coordinating and facilitating body for the global network of more than 135 Tibet-related NGOs.[16] I met with several of these activists along with staff from the IRN to develop a campaign strategy specifically for the Megoe Tso project.

Given the Chinese government's animosity against exile Tibetans, we took great care to ensure that my involvement was not counterproductive for the campaign. I also helped secure an agreement with ITSN leaders – that they would take action *only* if their involvement was needed to support local activists. In this case, the strategic aim of advocacy was to demonstrate the many other values that this sacred lake had, values that may, in fact, eclipse the projected gain from hydroelectric development. This effort to identify environmental values relevant to a national Chinese perspective was not what local Tibetans and Chinese environmentalists had been saying.

Box Fig. 4.5a.2 This lake is a popular destination for pilgrims and tourists. According to information provided to local tourists, Megoe Tso "is said to origin from the tears of a beautiful Tibetan girl Baimanazin and her true love who are forced to part from each other. Later, Master Padmadambhava (Guru Rinpoche, who introduced Buddhist teachings in Tibet) carried his meditation here and pronounced his blessing that people who plight their troth at this place, as true lovers, will get married. The white stones and prayer offerings on the far side of the lake commemorate the site of Padmadambhava's blessing." (Photo credit: Tashi Tsering)

Local Tibetans opposed the Megoe Tso dam project for cultural and spiritual reasons. The idea of a hydroelectric dam on their sacred lake was simply unacceptable to them. Chinese environmentalists were primarily concerned that construction of the dam would seriously degrade or destroy the lake's rich biodiversity and pristine scenery. Such concerns, while important, were not, we felt, sufficient to reverse government decisions about the proj-

(continued)

Box 4.5a (continued)

ect. Thus, we focused on what this lake meant or could mean for the local tourist economy. This approach reflected the pragmatic reality that decision-makers (high-level provincial government authorities) largely view the lake as a means to generate economic development, thus other values – culture, spirituality, or biodiversity – are less important.

We conducted outreach and educational activities to introduce and support the notion that Megoe Tso is a destination of great interest and promise for the local tourism economy. A letter-writing campaign was initiated to approach the Sichuan Tourism Bureau and other relevant state and local government offices that have an economic and bureaucratic interest in protecting the natural integrity of Megoe Tso. We also worked with international celebrities, politicians, and the IUCN to educate them about the project, the lake, and its many potentials so that they might express their appreciation for Megoe Tso's pristine ecology or concern about the dam project and the need to conserve the lake during their exchanges with Chinese officials and the media. And we sent a team of Chinese academics and reporters on a field trip to the dam site and other projects in western Sichuan Province. The resulting news reports informed the government and the nation about the environmental and social costs of the dam development in this corner of China. Their stories made it onto CCTV (China Central Television) and national radio, as well as the daily newspaper of the Chinese parliament.[17]

Finally, in November of 2007, the government of Kartse Tibetan Autonomous Prefecture announced that the Megoe Tso dam project has been cancelled. According to *Sichuan Daily*, the authorities who made the decision believe that it is in the long-term interest of the local economy to use the lake for the promotion of tourism and conservation rather than hydroelectric development.[18] This decision is extremely significant in many respects. It shows a change in local government priorities in water development policy. It is also the first and only time that an initially approved hydropower project on the Tibetan Plateau has been cancelled.

Box Fig. 4.5a.3 Megoe Tso Lake tourist destination, 2010 (Photo credit: Tashi Tsering)

(continued)

Box 4.5a (continued)

4.5a.5 Conclusion: Lessons from a unique and successful campaign

Documenting the consequences of past dam development and developing data to suggest the full array of human and environmental costs in new hydrodevelopment are important and, at times, effective strategies in contesting development decisions. Of equal importance is the identification and assertion of viable economic alternatives to valuing water resources. The ability to contest, protest, and offer viable alternatives to government-approved construction demonstrates that in China civil society is increasingly a significant source of political reform.

The success of the Megoe Tso campaign also provides an important lesson for international environmental and human rights activists. The context and ways in which Western activist groups work is not only different from those in politically sensitive regions like Tibet, but they can also be counterproductive to local efforts. The Megoe Tso example shows that in order to be successful in places like Tibet, activist groups must work in unconventional ways, such as being discreet about their involvement (activist groups in the West generally need a certain level of public credit for their work, and media attention is often seen as an indicator of successful effort) and limiting their important role to supporting informed local activists.

Notes

1. "New Map Throws Light on Tawu Protest and the Future of Eastern Tibet." The Tibetan Plateau Blog. Retrieved May 22, 2010, http://tibetanplateau.blogspot.com/2009/06/new-map-throws-light-on-tawu-protest.html
2. "Controversial SW China dam sparks new showdown." April 28, 2010. Reuters. Retrieved 22 May 2010, http://in.reuters.com/article/worldNews/idINIndia-48065520100428
3. D. Qing, ed., *The River Dragon Has Come!* (New York: M. E. Sharpe, 1998).
4. "News according to Hong Kong newspaper," *Wen Wei Po*, January 11, 2006.
5. Antoaneta Bezlova, "Tiger Leaping Gorge draws strength from Nu River activists," *Inter Press Service*, October 12, 2004.
6. Rachel Beitarie, "Media and the future of Tiger Leaping Gorge." Danwei.org. (accessed March 5, 2008). http://www.danwei.org/media_and_advertising/media_and_the_future_of_tiger.php; Ma Jun, Promoting River Protection in China, Woodrow Wilson International Center for Scholars, 24 January 2006. See http://www.wilsoncenter.org/index.cfm?event_id=168755&fuseaction=events.event_summary
7. For a more detailed discussion of the project, see Tashi Tsering, *Megoe Tso: The Damming Tibet's Sacred Lake* (Berkeley: Tibet Justice Center, 2005).
8. See Zhang Xiaoping, "Dadu River Valley Hydropower Development Craze Out of Control," *Renmin Zhengxie Bao*, September 8, 2004.

(continued)

Box 4.5a (continued)

9. http://web.biodiversityhotspots.org/xp/Hotspots/china/
10. See 14 July 2003 press release of a Tibetan white paper response, Tibet 2003: State of the Environment, online at http://tibet.net/flash/2003/0703/140703.html
11. See 14 July 2003 press release of a Tibetan white paper response, Tibet 2003: State of the Environment, online at http://tibet.net/flash/2003/0703/140703.html
12. For further readings on the cultural significance of mountains and lakes, see: Ngawann Zangpo, *Sacred Ground: Jamgon Kongtrul on Pilgrimage and Sacred Geography* (Ithaca: Snow Lion Publications, 2001); John Bellezza, *Divine Dyads* (Dharamsala: Library of Tibetan Works & Archives, 1997); A.W. McDonald, ed., *Mandala and Landscape* (New Delhi: D.K., 1997); Toni Huber, *The Cult of Pure Crystal Mountain: Popular pilgrimage and visionary landscape in Southeastern Tibet* (Oxford: Oxford University Press, 1999), Toni Huber, ed., *Sacred Spaces and Powerful Places in Tibetan Culture* (Dharamsala: Library of Tibetan Works & Archives, 1999).
13. Wen Huang, "Destroying a Natural Treasure in the Name of Progress," *South China Morning Post*, August 16, 2003.
14. "Experts Involved in Reviewing the Hydroelectric Dam Construction Project Sincerely Hope to Preserve the Rare Resources in Mugecuo and Cancel the Project," *China Youth Daily*, June 11, 2003.
15. S.E. Birnbaum and Y. Xiubo, *NGO Strategies to Promote River Protection and Restoration. China Environment Series* (Washington, DC: Woodrow Wilson International Center for Scholars), 185–190.
16. See www.tibetnetwork.org
17. Liexie, "Do Not Forget to Use Scientific Practice and National Ethnic Policy to Guide Development." in *Renmin Zhengxie Bao,* English trans. Dolker Tenzin, September 8, 2004. http://www.tibetjustice.org/tringyiphonya/num8.html#ge (accessed May 10, 2010)
18. *Sichuan Daily* online, November 8, 2006, http://sichuan.scol.com.cn/gzxw/20061108/200611882013.htm

The final area of vulnerability I would like to consider relates to the issue of cultural autonomy. China is a party to the CBD, one of the key outcome documents from the 1992 UN Conference on Environment and Development, article 8(j) of which states that each contracting party shall 'respect, preserve and maintain knowledge, innovations and practices of indigenous and local communities embodying traditional lifestyles relevant for the conservation and sustainable use of biological diversity'. Furthermore, the Declaration on the Rights of Indigenous Peoples, under consideration since 1985, was finally adopted by the UN General Assembly in 2007, and enjoyed the support of the Chinese delegation. None of these agreements, of course, has the status of international law or treaty. More troublingly, the Chinese representative to the UN Commission on Human Rights, Mr. Long Xuequn, speaking at the 53rd session of that body in 1997, is quoted as saying that, 'In China, there are no indigenous people and therefore no indigenous issues' (Erni 2008: 358).

This has been regarded by the international community as a declaration by fiat that China does not have to provide its minority nationalities with the rights commonly expected by indigenous peoples. China has had at least one delegate serving on the UN Permanent Forum on Indigenous Issues, a key advisory body to the UN's

Economic and Social Council. However, the delegate was nominated by the Chinese government, in contrast to most of the delegates who are elected or appointed by the indigenous groups from the nations they represent (UNPFII 2009).

4.5.4 An Uncertain Future for the Nu River

The Liuku hydropower station was the first dam slated to be completed in the Nu River. During fieldwork in 2008, our research team observed personnel as they went about their work surveying and conducting other preparatory work at the site. The village of Xiaoshaba has been permanently resettled, and other villages will likely follow.

It is difficult to follow the progress of the Nu River Project, since much of the information on the Nu is tightly guarded as a 'state secret' because of the Nu's status as transboundary river. Downstream in Myanmar (Burma), a series of five controversial dams are already underway on the Salween with investment from Thai, Burmese and Chinese companies, including Sinohydro Corporation, the largest of China's 'big five' hydropower corporations to emerge from the privatization process (International Rivers Network 2009).

In Nujiang Prefecture, there are already at least 27 small- and medium-scale hydropower stations from Liuku in the south to Gongshan in the north, a distance of

Fig. 4.5.2 On March 14, 2010 some 600 villagers gathered on the banks of the Salween River at the Thailand and Burma border to perform a traditional ritual to bless the river's fertility and to announce their opposition to the large dams being built up and downstream (Photo courtesy of Painporn Deetes and International Rivers)

about 250 river kilometres. These small projects are nowhere near the scale of the larger dams planned, but their impact on the ecosystem is significant nonetheless. Water is diverted from the mainstream via a cement channel or galvanized pipe, sent along the mountainside for several kilometers, then dropped tens of metres through turbines at a power station on the river before rejoining the mainstream. These hydropower stations are owned and managed by small, private hydropower interests and largely independent from the five large hydropower conglomerates. Most of these projects, because of their small scale, undergo scrutiny only by county-level officials. Owing to the mountainous terrain and the monsoon climate, landslides are common on slopes that have been disturbed by such projects, and their scars can be seen all along the banks of the river.

As for the large-scale Nu River Project, its future remains uncertain. Plans were stalled yet again by Premier Wen Jiabao in April of 2009, who declared that authorities should 'widely heed opinions, expound on [the plan] thoroughly and make prudent decisions' (Shi 2009). Activists and conservation organizations undoubtedly see this as a move in their favour, and as a positive sign that public voices are being considered in the development process. For now, the Nu River Project remains a point of contention where the ideals and goals of international conservation organizations, governmental agencies, hydropower corporations and local communities collide.

Resources

Brown, Philip H., and Yilin Xu. 2009. *Hydropower development and resettlement policy on China's Nu River*. Colby College Working Paper.

Clark, Marco. 2009. Climbing the mountain within: Understanding development impacts and understanding change in Southwest China. MA thesis, Oregon State University, Corvallis.

Erni, Christian. 2008. Country profile: China. In *The concept of indigenous peoples in Asia: A resource book*, ed. Christian Erni, 357–364. Copenhagen: International Working Group for Indigenous Affairs.

Ethologue. 2009. Lisu: A language of China. http://www.ethnologue.com. Accessed 1 July 2009.

Harrell, Stevan. 1995. Introduction: Civilizing projects and the reaction to them. In *Cultural encounters on China's ethnic frontiers*, ed. Stevan Harrell, 3–36. Seattle: University of Washington Press.

He, Daming, Feng Yan, and Hu Jinming. 2009. *Zhongguo Xinan Guoji Heliu Shui Ziyuan Liyong yu Shengtai Baohu (Utilization of water resources and environmental conservation in the international rivers, southwest China)*. Beijing: Science Press.

International Rivers Network. 2009. Salween Dams. http://internationalrivers.org/en/southeast-asia/burma/salween-dams. Accessed 5 June 2009.

Jiangtao, Shi. 2009. Wen calls halt to Yunnan Dam plan. *South China Morning Post* (May 21).

Jianmin, Wang, and John A. Young. 2006. Applied anthropology in China. *NAPA Bulletin* 25: 70–81.

Lazrus, Heather. 2009. The governance of vulnerability: Climate change and agency in Tuvalu, South Pacific. In *Anthropology and climate change: From encounters to actions*, ed. Susan A. Crate and Nuttall Mark, 240–249. Walnut Creek: Left Coast Press.

Magee, D. 2006. Powershed politics: Hydropower and interprovincial relations under great western development. *The China Quarterly* 185: 23–41.

Magee, Darrin, and Kristen McDonald. 2009. Beyond Three Gorges: Nu River hydropower and energy decision politics in China. *Asian Geographer* 25(1–2): 39–60.

Mertha, Andrew C. 2008. *China's water warriors: Citizen action and policy change*. Ithaca: Cornell University Press.
Miller, Lucien (ed.). 1994. *South of the clouds: Tales from Yunnan*. Seattle: University of Washington Press.
Rossabi, Morris. 2004. Introduction. In *Governing China's multiethnic frontiers*, ed. Morris Rossabi, 3–18. Seattle: University of Washington Press.
Scudder, Thayer. 2005. *The future of large dams: Dealing with social, environmental, institutional and political costs*. London: Earthscan.
Smil, Vaclav. 2004. *China's past, China's future: Energy, food, environment*. New York: Routledge.
Tilt, Bryan, Yvonne A. Braun, and Daming He. 2009. Social impact assessment of large dams: A comparison of international case studies and implications for best practice. *Journal of Environmental Management* 90(Suppl 3): S249–S257.
Tullos, Desiree, Bryan Tilt, and Katherine Reidy-Lierman. 2009. Introduction to the special issue: Understanding and linking the biophysical, socioeconomic and geopolitical effects of dams. *Journal of Environmental Management* 90(Suppl 3): S203–S207.
United Nations Development Programme. 2008. *Human development report: China*. Beijing: UNDP.
United Nations Education, Scientific, and Cultural Organization. 2009. Three Parallel Rivers Area. http://whc.unesco.org/en/list/1083/. Accessed 5 June 2009.
United Nations Permanent Forum on Indigenous Issues. 2009. http://www.un.org/esa/socdev/unpfii/. Accessed 24 June 2009.
Wang, S., and A. Hu. 1999. *The political economy of uneven development: The case of China*. Armonk: M.E. Sharpe.
Wisner, Ben, Piers Blaikie, Terry Cannon, and Ian Davis. 2004. *At risk: Natural hazards, people's vulnerability and disasters*. London: Routledge.
World Commission on Dams. 2000. *Dams and development: A new framework for decision-making*. London: Earthscan.

Chapter 4.6
Cultural Survival, Tribal Sovereignty and River Restoration on the Central Northwest Coast, North America

Colleen E. Boyd and John B. Boyd

4.6.1 Landscapes and Stories

The Elwha River system on the Olympic Peninsula in Washington state (USA) is a storied land. For the Klallam (Coast Salish) people who claim it as their homeland, it is a place filled with narratives about culture, place, and the past. Even so, they have not been able to access many of their sacred sites for several generations because of the development of two hydroelectric dams on the Elwha River. In 1992 the U.S. Congress passed the Elwha River Ecosystem and Fisheries Restoration Act. This legislation brings together tribal, federal, and regional partners in an effort to restore the Elwha River through dam removal, which will allow the river's salmon and steelhead populations to access pristine spawning ground in the upper reaches of the river, rehabilitate salmon habitat, and replenish beaches starved by the loss of the sediment now trapped behind the dams. For the last two decades, the Elwha Klallam and the U.S. National Park Service have been intergovernmental partners in the effort to implement this act.

The ability to put stories to work through interactions with specific places is a cornerstone for effective cultural survival and environmental restoration, and it fashions a link between these related endeavours. Restoring the Elwha River also reveals the Klallam community's cultural heritage. The many Klallam stories about their relationships with place, space, and time present a perspective that allows collaboration among the stakeholders in restoring the natural environment and reviving cultural values. This educational process provides access to both Natives and non-natives.

C.E. Boyd (✉)
Department of Anthropology, Ball State University, Muncie, IN, USA
e-mail: ceboyd@bsu.edu

J.B. Boyd
Department of English, Ball State University, Muncie, IN, USA

B.R. Johnston et al. (eds.), *Water, Cultural Diversity, and Global Environmental Change: Emerging Trends, Sustainable Futures?*, DOI 10.1007/978-94-007-1774-9_27,

Box 4.6a Declining salmon, large dams, and the capitalist world-economy: A case study of biocultural diversity in the Nez Perce homeland

—Benedict J. Colombi

Box Fig. 4.6a.1 Salute (Artist: Michael Blackstock)

The Nez Perce [*niimiipuu*][1] homeland is a place of extreme power and beauty. It is a place where migrating salmon swim nearly 800 miles from the Pacific Ocean to spawn in rivers and lakes born from mountainous snow and ice, ancient cedars and rich pine forests meet fertile grasslands, and North America's deepest river gorge, Hell's Canyon, plummets into a hot inland desert.

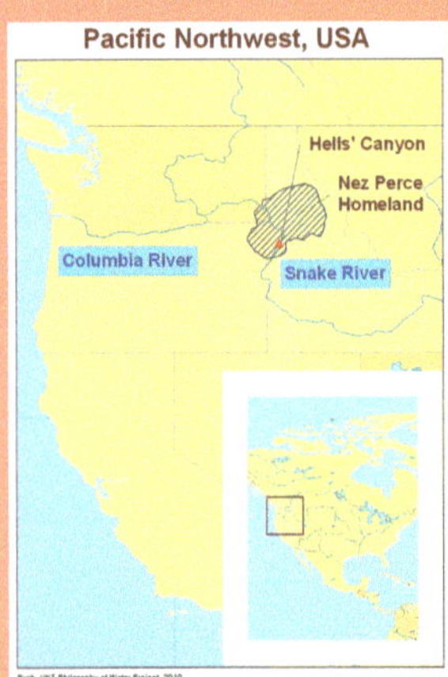

Box Map 4.6a.1 Pacific Northwest

The Nez Perce have sustained a rich tribal culture in these places since the last Ice Age. Salmon [*léwliks*] and water [*kúus*] constitute the ideological and material foundations of both the world and humanity, and these two things express a particular history of place and environment. Nez Perce explain interactions in terms of salmon and water, and without them the Nez Perce world is non-existent. Thus, without the annual return of the salmon and sustainable water resources, the Nez Perce say they will cease to be Indian people.

Salmon are biologically a keystone species. They link biodiversity and productivity because they are "transport vector[s]" for the movement of "materials, and energy and nutrients between

(continued)

Box 4.6a (continued)

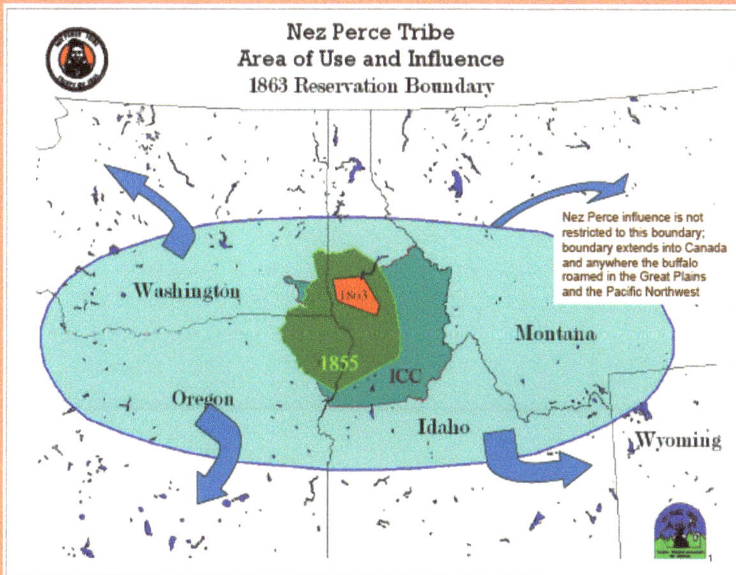

Box Map 4.6a.2 The Columbia River Basin and Lower Snake River Basin, Northwest North America (Source: Nez Perce Tribe Land Services Program, Jeff Cronce 2005)

Box Fig. 4.6a.2 Salmon is the primary source of protein for indigenous peoples throughout the Northern Pacific Rim, as illustrated by this Koryak family eating their dinner outside their home in Russia's Kamchatka Peninsula (Photo credit: A. Shemaev; Source: Pacific Environment http://pacificenvironment.org/img/original/kam_subsistence.jpg)

marine, aquatic, and terrestrial ecosystems" (Cederholm et al. 2001:652). In the Columbia and Snake rivers of the Nez Perce homeland, salmon annually transported over 100 million kg of energy and materials from marine to terrestrial ecosystems, making these river systems the world's richest inland fishery (Lichatowich 1999). Before the arrival of Europeans in the U.S. Pacific Northwest region, these rich ecosystems supported some 700,000 indigenous peoples diversified into 47 cultural sub-areas and representing 11 language families (Kroeber 1939). Allowing for the impact of European disease, the Native population before the arrival and influence of settler Americans and Europeans may have been twice this size (Boyd 1985). In the Columbia River Basin, Native peoples consumed nearly 36 billion kcal in returning salmon per year during the

(continued)

Box 4.6a (continued)

pre-colonial period (Bodley 2006). The five species of Pacific Northwest salmon (*Oncorhynchus sp*) have presumably existed in their present form for six million years and have been a food source for indigenous peoples in the Pacific Basin for millennia.

At the beginning of the 21st century, salmon runs are 2% of their historic levels. Prior to European contact, nearly 10-16 million salmon returned each year to the Columbia Basin. Today, as few as 200,000 salmon return annually (Augerot 2005). The current decline in Columbia Basin salmon can be attributed to the impact of hydroelectric dams, irrigation projects, and overall habitat loss (Blumm 2002; Lichatowich 1999).

Box Fig. 4.6a.3 This Koryak boy is smoking salmon in his family's traditional smokehouse (Photo credit: A. Shemaev; Source: Pacific Environment http://pacificenvironment.org/img/original/kam_koryak%20 boy.jpg)

Box Fig. 4.6a.4 The wheat terminal Almota, located at the northernmost reach of the Snake River just below the Little Goose Dam. The name Almota is derived from the Nez Perce Indians meaning torchlight or moon light fishing (Photo credit: Dan McShane)

The development of water resources in turn fuels the expansion of the capitalist world-economy. After World War II, U.S. federal agencies and private companies constructed eight large dams near the Nez Perce reservation, just a few of the more than 400 dams erected in the Columbia River Basin. The control of water enabled a non-Indian society to support urban populations, increase industrial output, and raise the production and consumption of agricultural commodities.

For example, the dams make the Columbia and Snake rivers into an important transportation corridor. Wheat growers ship commodities downstream to global markets via river barges on the Columbia and Snake rivers, while agricultural inputs such as petroleum fuel and chemical fertilizers move upstream to production centres.

China consumes over 90% of the wheat produced on the Nez Perce reservation. None of the wheat producers on the Nez Perce reservation are Indian. Deep inequalities of this magnitude are a social creation with a history (Colombi 2005). Thus, the transboundary movement of global commodities places negative costs

(continued)

Box 4.6a (continued)

to the environment and Nez Perce salmon-centric culture with its reliance on dams and river barges.

However, the Nez Perce never expressly surrendered the right to fish at all their usual and accustomed places, which was retained by the Nez Perce in signing a treaty with the U.S. in 1855. Measurable declines in salmon and water are immediate dangers to Nez Perce rights and their way of life. Effective adaptation to such changes requires long-term action by a community of citizens that includes Native communities and non-Native peoples, allying with each other to work together to strengthen and improve a common place and a common watershed.

Realizing social and environmental justice in this context may require placing limits on commercial growth, developing alternative sources of renewable energy, and maintaining lower levels of consumption. Problems of this magnitude are culturally mediated and therefore can only be solved using cultural solutions. Such solutions necessarily entail management by highly autonomous local communities and effective global institutions to safeguard critical water resources and enable a more sustainable future.

Note

1. All Nez Perce language [*nimiipuutimt*] words are in brackets and italicized. The spellings are derived from, Haruo Aoki (1994) Nez Perce Dictionary.

Resources

Aoki, Haruo. 1994. *Nez Perce dictionary*. Berkeley: University of California Press.

Augerot, Xanthippe. 2005. *Atlas of Pacific salmon: The first map-based status assessment of salmon in the North Pacific*. Berkeley: University of California Press.

Blumm, Michael. 2002. *Sacrificing the salmon: A legal and policy history of the decline of Columbia Basin salmon*. Den Bosch: BookWorld Publications.

Bodley, John H. 2006. Scale, power, and sustainability in the Pacific Northwest. Paper presented at the annual meeting of the Society for Applied Anthropology. Vancouver, 27 Mar.

Boyd, Robert T. 1985. The introduction of infectious diseases among the Indians of the Pacific Northwest, 1774–1874. PhD dissertation, Department of Anthropology, University of Washington, Seattle.

Cederholm, C. Jeff, David H. Johnson, Robert E. Bilby, Lawrence G. Dominguez, Ann M. Garrett, William H. Graeber, Eva L. Greda, Matt D. Kunze, Bruce G. Marcot, John F. Palmisano, Rob W. Plotnikoff, William G. Pearcy, Charles A. Simenstad, and Patrick C. Trotter. 2001. Pacific salmon and wildlife-ecological contexts, relationships, and implications for management. In *Wildlife-habitat relationships in Oregon and Washington*, ed. David H. Johnson and Thomas A. O'Neil, 628–684. Corvallis: Oregon State University.

Colombi, Benedict J. 2005. Dammed in region six: The Nez Perce Tribe, agricultural development, and the inequality of scale. *American Indian Quarterly* 29(3/4):560–589.

Kroeber, Alfred L. 1939. *Cultural and natural areas of native North America*. Berkeley: University of California Press.

Lichatowich, Jim. 1999. *Salmon without rivers: A history of the Pacific salmon crises*. Washington, DC: Island Press.

Coast Salish stories of the Elwha Valley and the mountains that sustain the watershed define these relationships. Stories about them invite dialogue between community members and provide their governmental partners with the information necessary to address cultural survival and the multiple values associated with conservation areas such as the Olympic National Park and Olympic Coast Marine Sanctuary. Indigenous stories underscore the unique relations of the First Peoples with the natural environment before the development and implementation of resource management plans by federal and state bureaucracies and private industries. Many Klallam people view environmental restoration as a part of a network founded on ancestral rights that links humans to non-humans in relationships of kinship and obligation.

Understanding the Elwha River watershed requires an interdisciplinary and cross-cultural approach, including scholarly and legal research (cf. Boyd 2001, 2006, 2009; Busch 2008; Wray et al. 2002), journalistic inquiry (Mapes 2009), and creative endeavours (cf. Boyd 1995, 1998, 2008; Lundahl 2002, 2004). The story of the river's transformation from the "beautiful stream" chronicled by 18th-century Spanish explorers (see Quimper 1791) into nature harnessed for industrial development has been documented through primary sources now scattered throughout North American archives and libraries and also through the oral testimony and stories of tribal citizens. This study integrates a substantial body of facts, narratives, and opinions acquired over the years about the ever-changing relationship of humans to their environment. At its core, it is about Lower Elwha Klallam people and their relationship to the Elwha River's "unnatural history." It is about how the watershed was transformed by 19th-century newcomers with radically different ideas about "nature," "wilderness" and "progress," and in the 21st century, how stakeholders from diverse walks of life are now uniting to address the excesses of a previous era. At the forefront are tribal citizens who are leading collaborative efforts to restore the Olympic Peninsula's polluted and damaged streams, beaches, and forests.

4.6.2 Knowing the Elwha River Through Time and Space

Entwined with the story of the Elwha River and its fisheries is the story of the Elwha Klallam tribe, the indigenous community whose ancestors fished the river and its tributaries for thousands of years. The Elwha Klallam tribe is one of four Klallam bands that occupy the U.S.-Canadian borderlands along the shores of the Strait of Juan de Fuca, the narrow saltwater bottleneck connecting the Pacific Ocean to the Puget Sound. This strait separates southern Vancouver Island (British Columbia, Canada) from Washington state's Olympic Peninsula (USA). The four bands are culturally related but politically distinct, with their own governments and membership.

The Lower Elwha Klallam possess the oldest and strongest claim to the river. Not only was it once their most prolific salmon stream, it is also the landscape in which important events have occurred throughout Klallam history. Stories told in Klallam and English evoke ancient ties to the land and water and speak to the unique and varied ways that people experience place. The river's headwaters are the home of Thunderbird, who sent thunder and lightening to chase the salmon upstream and

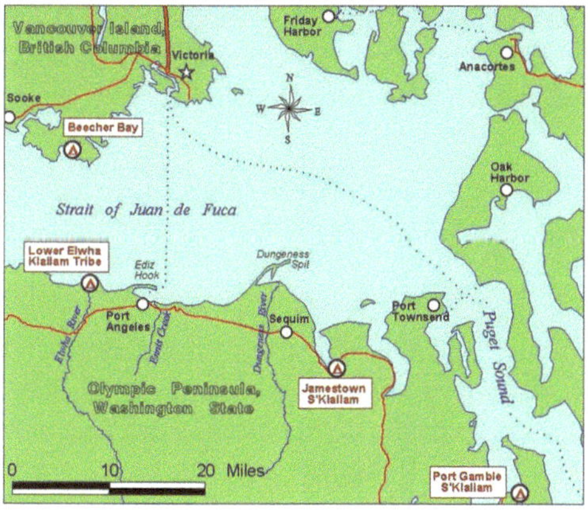

Map 4.6.1 "Contemporary Klallam Communities" by Randall McCoy, LEKT

alert the people when it was time to fish. The river leads up to the Olympic Mountains, where the people once tied their canoes to the tallest peaks to escape rising flood waters (Valadez 2002:21-24). Grandmothers on the Klallam people's reservation sang of how cedar bark ropes saved their ancestors from drowning in the flood. Recent stories dwell alongside ancient teachings about specific places where rocks, creeks, and clay hold the key to individual, family, and village histories. Stories such as how the Klallam received their name, nəxʷsƛ̓áy̓əm̓ ("Strong People"), or the creation of the first Klallam people coexist with narratives about the destruction of the Deep Creek village by a logging company in the early 20th century, the 1912 "blow out" of the first Elwha Dam which flooded out downstream villages, and the arrests of Klallam men for exercising their treaty-protected right to fish. These stories provide a foundation for cultural and historic identity by revealing what it has meant to be Klallam and, more recently, to be from the Lower Elwha Klallam Reservation.

Elwha Klallam families retain a keen and subtle awareness of their historic ties to village sites at places like pə́sct (Pysht), ʔiʔínəs (Ennis Creek), č̓ixʷícən (Port Angeles), and x̣ŋínt (Clallam Bay). Even if Klallam people no longer inhabit them, the sites remain significant to cultural and historical identities, and tribal citizens still know and visit them. Each was a traditional use area, and people still carry memories of the seasonal resources that were available. It is an important aspect of storytelling and resource preservation to identify those places that provided cultural and subsistence resources. Access to and knowledge of various resources, like prime fishing sites, was also a function of cultural wealth and family prestige. Such knowledge and access was the foundation for Klallam governance, economy, and social life and found its ultimate expression in the potlatch – the feasting and giving away of wealth associated with ceremonial rites of passage.

Therefore, as a place, the Elwha River watershed is rich in meaning, as it nourishes the bodies, hearts, and minds of the people who claim it as a portion of their ancestral legacy. For Elwha Klallam people, the physical environment, which includes bodies of salt water and fresh water, mediates their relationships to animals, plants, culture, and history. It constitutes a sense of kinship and continuity even as boundaries between people, cultures, and countries shift and merge. When Elwha Klallam discuss both contemporary and ancestral relationships, they make subtle and obvious references to the primacy of water. As canoe people, the Elwha Klallam navigate the salt waters bounding their lands and the freshwater streams feeding the Strait of Juan de Fuca. The river cuts a path through lands leading up to old mountain trails where people peel cedar bark to make baskets and ceremonial attire. Submerged beneath the waters between the two hydroelectric dams is the precise location where the Creator scooped mud from a basket-shaped depression called spčùl. The anthropologist T. T. Waterman, in 1920, described this site as follows:

> The pits or hollows are the place from which dirt was scooped, out of which the human race was formed. Sometimes people go to these pits to get information about their future life. If a man thrusts his hand into this water, and brings out deer hair for example, he knows he will be a good hunter (Waterman 1920:58).

In the 21st century, many individuals must find alternative ways to seek ancestral knowledge. Going into the forest, being attentive to the rituals of daily life, seeking council with a family elder regarding the stories of the myth time, and creating new stories are important and represent the resilience of indigenous people.

4.6.3 Indian Dispossession and the Unnatural History of the Elwha River

The story of the Elwha River's development merges with tribal dispossession and larger narratives of colonization, land and water rights, and western North America. Contemporary tribal and environmental politics are framed by older discourses of hierarchy and power that joined with the interests of government, scholarship, and law. In the nearby temperate rain forests of Vancouver Island, the disciplines of anthropology and geography produced distinct knowledge about nature and the cultures of colonized peoples. Each became more "visible" and therefore "available" to colonial administrators On the Olympic Peninsula, the most appropriate use or management of the environment is often discussed without reference to how nature has been "constituted within, and informed by, the legacies of colonialism" (Willems-Braun 1997:5). One effect of colonialism and modernity is to leave unquestioned the historical nature of "nature" itself. In the past this failure led non-natives to treat indigenous rights and cultural heritage as if they had no relationship to state laws regarding resource use.

On the Olympic Peninsula, the construction of nature as a discrete and separate object allowed the privileging of certain ways of knowing the world – economic, scientific, legal and aesthetic. European explorers in the late 19th century viewed the Olympic Peninsula through a lens narrowed by their own values and interests. Similarly, American settlers in the 19th century deemed the deep rainforests, high mountains, and swift waters of the Peninsula as "wilderness," neglecting to see it was already a culturally inscribed landscape. William Cronon (1996:69) argues that the concept of "wilderness" is "quite profoundly a human creation – indeed, the creation of very human cultures at very particular moments in human history." Wilderness was alien and even dangerous for people born of the Western Judeo-Christian tradition. The Bible as well as European folk traditions filled "wild places" with demons and other frightening figures. Such places were constituted in opposition to "civilization" and therefore, to be avoided (Cronon 1996:71; Nash 1967:8).

Europeans and American citizens considered the indigenous people that they encountered as a part of this wild and natural landscape. They, too, were uncivilized and savage impediments to progress. Like wilderness places, Northwest Coast peoples living in settlements along rivers and beaches could be removed and put to more useful occupations, with the added benefit of making the lands and waters they inhabited available for resource extraction and "productive" development.

However, following the signing of the Treaty of Point No Point with the United States in 1855, Klallam people rejected the U.S. government's order to remove to the Skokomish Reservation on Hood Canal. "It was not their country," an elderly Twana/Klallam man reported in the early 1930s, and it was too far removed from their villages and subsistence sites (Elmendorf 1993:5). Instead, they opted to stay in the Elwha River Basin, where early 20th-century state fishing laws and their status as "landless Indians" meant they could no longer rely on salmon runs or the federal government's protection of their treaty rights to fish, hunt, and gather.

Out of necessity and with not a little entrepreneurial spirit, the Klallam engaged foreign visitors and settlers to their shorelines. Men sold fish to hungry newcomers, charged to transport humans and goods by canoe, or sought employment in timber camps and mills. Women processed fish for wages in waterfront fish canneries. Traditional foods continued to have a place on the table but were augmented by store-bought goods. Children eager to learn the river fished for salmon after a full day in public schools where they learned and spoke English. Farmhouses replaced cedar longhouses, and the Indian Shaker Church, a religion that blurs boundaries between pre-contact spirituality and Christianity, brought innovations to traditional styles of worship. If people did not always embrace change, they did what was necessary to ensure the survival of their community and the reproduction of children with Klallam minds, hearts, and bodies (Boyd 2001). Ancient teachings anchored them to familiar places, while stories about change forewarned them of periodic transformations. Their physical connection to historic Klallam places, including the numerous familiar sites within the Elwha watershed, provided them with the basis for a unique indigenous heritage.

The U.S. colonization of the Pacific Northwest also created new economic niches for indigenous people, who found employment in various extractive industries, beginning with the fur trade and followed by logging and fishing (cf. Boxberger 1988; Barsh 1996; Harmon 1998:13-42; Raibmon 2005:98-115; Thrush 2007). Opportunities for wage labour did not fully integrate Native people into the development of the region. Rather, development occurred at their expense since it limited access to their "usual and accustomed grounds and stations" protected by treaties negotiated between the United States and Coast Salish villagers in 1855 (The Treaty of Point No Point 1855). When a group of Port Angeles businessmen led by Thomas T. Aldwell decided to develop hydroelectric power on the Elwha River in 1910, as a dramatic effort to boost local economy, they never seriously considered Klallam claims to fishing rights on the river. Thomas Aldwell told a new story, one in which Port Angeles would become a great industrial city supported by the harvesting of boundless natural resources, rather than the small town where "as late as 1912, the cows were still coming down to drink in the trough in front of the Merchant's Hotel" (Hult 1954:190–191). Hydroelectric development was the cornerstone of this plan.

In the early 1890s, Aldwell, an immigrant from Canada, had, under nineteenth century U.S. homestead laws, claimed parcels of land on the lower Elwha River for the purpose of one day building a dam. By 1910, he formed the Olympic Power and Development Company with George Glines. Their board of directors included the wealthiest men in the region: Mike Earles, who built the world's largest timber mill at the base of Ediz Hook in Port Angeles, the Seattle area banker, philanthropist and sports fisherman Joshua Green, and R. D. Merrill, owner of the powerful Merrill and Ring Corporation that had first logged the forests of Maine and Michigan before heading further west (Boyd 2001:266; Brown 1990:68–69; Hult 1954:190).

In late 1910 the company retained an eastern engineering firm and construction began on the Elwha Dam at the site of Aldwell's first claim (Maib 1951:3). From its inception, the project was plagued with difficulties. Wooden flumes constructed to direct water was nearly finished, water levels rose dangerously high, causing seepage under a steel sheet constructed to stop the leak. Since the dam's foundation was not anchored in bedrock, water began leaking followed by a dramatic blowout that left a breach some 3 m wide and a hole under the dam 18 m deep (Maib 1951:3). Klallam families living on sites below the dam were sitting down to eat supper when the disaster occurred. Their only warning of the impending flood was the barking of dogs, the roar of water, and the sound of trees breaking as the water hit. Families fled to a railroad trestle spanning the river. Eyewitnesses would later recall seeing dead fish tangled in the tops of trees (Valadez 2002:28–29). Although no one was injured or killed, Klallam property was destroyed, along with the people's peace of mind (Boyd 2001:267).

Fig. 4.6.1 The Elwha dam, 2007 (Photo credit: Colleen Boyd)

Despite this disaster, the Elwha Dam was completed by 1913. Although the Olympic Power Company had spent 150,000 USD to repair the foundation, it had neglected to construct fish passages as mandated by an 1890 Washington state law (Maib 1951:5). By passing this law, legislators hoped to avoid problems other states had already encountered. In Maine, for instance, the Atlantic salmon (*Salmo salar*) had all but disappeared because of overharvesting and the damming of rivers (Brown 1990:64-65). Aldwell knew about the problem, as primary source documents show (Maib 1951; Aldwell 1890). On September 12, 1911, the local game warden, J. W. Pike, drafted a letter to the State Fish Commissioner, J. L. Riseland sharing charges from cannery men that the dam blocked migratory fish routes on the Elwha. Pike investigated the waters above the dam and was unable to find a single salmon. On the other hand, he spotted "thousands of salmon at the foot of the dam where they are jumping continually, trying to get up the flume" (in Maib 1951:8). Riseland instructed the Olympic Power Company "as to the proper way to install the proposed fishway," arguing that "the fish industry of this country" would suffer "a serious drawback" if the salmon runs could not reach their spawning grounds.

Riseland would later learn from the Superintendent of the Washington State Fish Hatcheries that it was structurally infeasible for a fish passage to be built when the dam was almost finished. In 1912, Aldwell and his engineer met with officials to discuss the most expedient and practical way of conveying fish and transporting them over the dam. One suggestion was to trap fish and convey them in an elevator. The other was for the Olympic Power Company to supply property for a fish hatchery to replace the wild runs (in Maib 1951:10). In the end, Aldwell crafted an agreement with the state of Washington to supply the property necessary for constructing a fish hatchery below the dam. On August 8, 1913, Aldwell received a telegram from the state of Washington proposing that he "construct a hatchery and also [a] concrete retaining apron at the foot of your dam which would enable us to take fish and thereby obviate necessity of compelling you to construct fishway over your dam" (Maib 1951:12). The fate of the Elwha River and its fish runs was sealed. The new fish hatchery would allow the state to raise Pacific salmon in the new hatchery, releasing them into the river where they would compete with wild Elwha fish stocks.

When the Glines Canyon Dam was completed in 1927, 13.6 km above the Elwha Dam, its builders did not include fish passages because the first dam had already blocked the fish's access to the spawning grounds. The new dam flooded a traditional Klallam fishing site at Boulder Creek, where tribal elders had gaffed Chinook salmon for generations (Wray 1997:43). As a result of the two dams, the river's historic five-species runs of Pacific salmon and steelhead trout[1] were unable to make their seasonal journeys from the ocean, where they mature, to their freshwater spawning grounds above the second dam. Decades of reduced spawning in the lower reaches of the river decimated the once prolific runs. Meanwhile, the harvesting of

[1] *Oncorhynchus keta* or "Chum," *Oncorhynchus kisutch* or "Coho," *Oncorhynchus gorbuscha* or "Pink," *Oncorhynchus nerka* or "Sockeye" and *Oncorhynchus tshawytscha* or "Chinook" and the ocean-going steelhead trout *Oncorhynchus mykiss*.

millions of board feet of timber reduced the stands that shaded and cooled streams and rivers, eroded wetlands where tiny fingerlings had escaped larger predators, and decreased riverbed sediment necessary for salmon to create their redds (nests).

Both reservoirs over the years trapped millions of cubic metres of sediment, increasing erosion and affecting sand spit development east of the river's mouth. By the 1980s, Ediz Hook, the second-largest sand spit in the world, was losing 9,939 cubic metres of sediment per year and had been identified by the U.S. Army Corps of Engineers National Shoreline Study as a "major problem area" (Brown 1990:73; Wray 1997:43). Meanwhile, reservation beaches and sand dunes, where tribal grandmothers had once harvested shellfish and picked wild strawberries, all but disappeared, leaving behind a cobble-strewn beach. By the late 20th century, the Elwha River faced an environmental disaster of epic proportions (Boyd 2001; Brown 1990).

4.6.4 Conclusion: Restoring Indigenous Sovereignty and the Elwha River

Klallam people have always regarded the Elwha watershed as central to their culture and economy despite hydroelectric development and destruction of the river's fisheries. The 1934 Wheeler-Howard Act revised a century of U.S. Indian policy to encourage the restoration of Native communities. Using its land buy-back provision, 10 Klallam families seized the chance to move back to a winter village site in the lower valley of the Elwha River. In the late 1930s, some 102 adults and children occupied refurbished farmsteads that had been purchased by the federal government from white settlers. As "landed" Indians, the Klallam could expand their political and economic base. One of the first actions that the heads of households took was the formation of the Lower Elwha Klallam Business Council, which is still the main governing body of the tribe. Contemporary elected council members are descended from one or more of the original 14 families.

The restoration of their land base and the establishment of a formal tribal government were essential components for Klallam nation-building in the 20th century and later enabled the intergovernmental partnerships of the 21st century that ultimately allowed indigenous people to participate in water management. The restoration of the Klallam's tribal sovereignty has been intimately linked to the restoration of the Elwha River. In 1978 the Elwha Klallam intervened when the Elwha Dam failed a safety test (Busch 2008:8-9). When state and federal agencies failed to issue any orders to repair the dam, the tribe proceeded with the "dam failure analysis" necessary for the construction of a flood-control levee. Not only did this levee make the reservation a safer place for residents, it enabled the construction of badly needed new reservation housing, and the existence of better housing led to many more Klallam families to elect to live on tribal land (Boyd 2001:318-341). In the following decades, the tribe, along with other federal advocates like the National Oceanic and Atmospheric Administration (NOAA), raised concerns about restoration of the Elwha River system, including removal of the dams. Through legal action and less litigious efforts like restoring salmon streams and habitat, the tribe has garnered necessary political support for removal of the two dams (cf. Busch 2008:8-12). More importantly, they have, by taking the lead when and where it has been necessary, asserted their sovereign authority.

In 1992 the federal Elwha Act mandated the removal of the two Elwha River dams. The Elwha River Restoration Project is an intergovernmental effort involving federal agencies and the Lower Elwha Klallam tribe that proposes to carry out the objectives identified by the federal Elwha Act (1992) by removing the two hydroelectric dams and fully restoring the Elwha River ecosystem, some 112 km of river and tributaries. Salmon response to dam removal will depend on individual populations and a variety of conditions. However, it is projected that fisheries will once again reach "historic levels with numbers exceeding 390,000 annually" (Olympic National Park 2010). To date this is one of the largest dam removal and environmental restoration projects ever undertaken in the U.S. (Tweit 2006; Olympic National Park 2010). Brian Winters, a fisheries biologist for the Olympic National Park quipped that tearing down the two dams, "essentially construction in reverse" is the "easy part." The difficult part has been the two or more decades it took intergovernmental partners to agree on how to proceed and then secure the $184 million USD required for removing the dams and restoring the river (Tweit 2006).

The Olympic National Park has proclaimed summer 2010 "the last dam summer" (Olympic National Park 2010). In September 2010 Barnard Construction Company in Bozeman, Montana (USA) was awarded a 26.9 million USD contract from the National Park Service to remove the Elwha and Glines Canyon dams (ENR 2010). Dam removal is scheduled to begin in 2011 and will last approximately 2-3 years (Olympic National Park 2010). Meanwhile habitat restoration and revegetation crews from the Lower Elwha Klallam tribe continue to restore salmon habitat in the lower river through the construction of 16 log jams and the replacing of invasive plant species with over 2,000 native shrubs and trees while improvements have been made to culverts on important tributary streams. Such efforts not only signal a reversal of fortune for salmon and humans, it is a model for indigenous stakeholders to study with care.

While the U.S. Army Corps of Engineers inventory records some 6,000 large dams over 15 m high in the United States, other inventories note that there are more than 2.5 million privately owned small dam structures (see Tweit 2006). Many of these dams, large and small, interfere with indigenous cultures. Tribal nations, environmental activists, sports fishers, and governmental agencies are among the growing numbers calling for their removal. Unfortunately, in other parts of the world, hydroelectric development continues unabated, often at the expense of indigenous resources and the environment (cf. Defending the Amazon 2009). As the Lower Elwha Klallam Tribe and its governmental partners face a future in which progress is no longer measured by the numbers of salmon caught in a net or board-feet of timber harvested, it is evident tribal peoples play a critical role in the Olympic Peninsula's future. Indeed, the region is transforming, as tribal elders maintain through stories of the Changer, a mythical transformer who first readied the Salish landscape for the humans who were coming. Restoration of the Elwha River will similarly transform the Olympic Peninsula in the 21st century by recreating salmon habitat, rebuilding beaches and wetlands, providing new economic opportunities for a beleaguered region and renewing Klallam culture as tribal members will have

Fig. 4.6.2 Lower Elwha Klallam Hatchery employee fishing at the Elwha River mouth, 2007 (Photo credit: Colleen Boyd)

access to sacred sites that have been flooded since the early twentieth century (Olympic National Park 2010). The intergovernmental partnership symbolizes a new chapter in resource management that will hopefully result in self-sustaining, healthy and genetically diverse salmon populations (NOAA 2008:85-96). The Lower Elwha Klallam are implementing indigenous ways of knowing and thinking about development, the environment, and progress in ways that include rather than exclude and thus transcend and mediate cultural differences through restoration, preservation, and resource management.

Klallam English Glossary

nəxʷsƛ̓áy̓əm̓	"Strong People" the Klallam name for themselves
č̓ixʷícən	Tse-whit-sen/Port Angeles village site
pə́šct	Pysht River village site
ʔiʔínəs	Ennis Creek village site
x̣ŋínt	Clallam Bay village site
spčùl	A basket-shaped depression in the Elwha River reported to be the Klallam creation site.

Resources

Aldwell, Thomas T. 1890–1951. *The Thomas T. Aldwell papers*. Seattle: Special Collections, University of Washington Libraries.

Barsh, Russel Lawrence. 1996. Puget Sound Indian demography, 1900–1920: Migration and economic integration. *Ethnohistory* 43(1): 65–97.

Boxberger, Daniel L. 1988. In and out of the labor force: The Lummi Indians and the development of the commercial salmon fishery of North Puget Sound, 1880–1900. *Ethnohistory* 35(2): 161–190.

Boyd, John B. 1995. Throwing rocks at the waves. Unpublished MFA thesis, Creative Writing Program, University of Oregon.

Boyd, John B. 1998. Another Berm. In *Tidepools: A literary magazine*. Port Angeles: Peninsula Community College.

Boyd, Colleen E. 2001. Changer is coming: history, identity and the land among the Lower Elwha Klallam tribe of the North Olympic Peninsula. Unpublished PhD diss., Department of Anthropology, University of Washington.

Boyd, Colleen E. 2006. That government man tried to poison all the Klallam Indians: Metanarratives of history and colonialism on the Central Northwest Coast. *Ethnohistory* 53(2): 331–354.

Boyd, John B. 2008. The ice man speaks. In *In the grove: California poets and writers*. An homage to Andres Montoya, Issue 16, ed, Daniel Chacon, 22. Fresno: In the Grove.

Boyd, Colleen E. 2009. The Indians themselves are greatly enthused: The Wheeler-Howard Act and the (re)-organization of Klallam space. *The Journal of Northwest Anthropology* 43(1): 3–26.

Brown, Bruce. 1990. *Mountains in the clouds: A search for wild salmon*. New York: Collier Books.

Busch, Russell W. 2008. Tribal advocacy for Elwha river dams on Washington's Olympic Peninsula. *Golden Gate University Environmental Law Journal* 1(2): 5–21.

Cronon, William. 1996. The trouble with wilderness; or, getting back to the wrong nature. In *Uncommon ground: Rethinking the human place in nature*, ed. William Cronon, 69–90. New York: W. W. Norton Press.

Defending the Amazon. 2009. The Madeira dams: Impacts and actions in Bolivian Territory. http://www.amazonwatch.org/newsroom/view_news.php?id=1912. Accessed 14 Sept 2009.

Elmendorf, William W. 1993. *Twana marratives: Native historical accounts of a coast Salish culture*. Seattle: University of Washington Press.

ENR – Engineering News Record. 2010. Washington state contract to remove dams on Elwha River awarded for $26.9 Million. http://enr.ecnext.com/coms2/article_inwd100901ElwhaRiverDa. Accessed 8 Sept 2010.

Harmon, Alexandra. 1998. *Indians in the making: Ethnic relations and Indian identities around Puget Sound*. Berkeley: University of California Press.

Hult, Ruby El. 1954. *Untamed Olympics: The story of a peninsula*. Portland: Binford and Mort.

Lichatowich, Jim. 1999. *Salmon without rivers: A history of the Pacific salmon crisis*. Washington, DC: Island Press.

Lundahl, Robert. 2002. *Unconquering the last frontier. 57 min. video*. Reading: Bullfrog Films.

Lundahl, Robert. 2004. *Song on the water: Return of the great canoes. 57 min. video*. Port Angeles: Freshwater Bay Productions.

Maib, Charles W. 1951. *A historical note on the Elwha River: Its power development and its industrial diversion*. Compiled by the Stream Improvement Division, Washington State Department of Fisheries. Olympia: Washington State Archives.

Mapes, Lynda V. 2009. *Breaking ground: The Lower Elwha Klallam Tribe and the unearthing of Tse-whit-zen Village*. Seattle: University of Washington Press.

Nash, Roderick. 1967. *Wilderness and the American mind*. New Haven: Yale University Press.

Olympic National Park. 2010. Elwha river restoration. http://www.nps.gov/olym/naturescience/elwha-ecosystem-restoration.htm. Accessed 8 Sept 2010.

Quimper, Manuel. 1791. Diario de navegaci6n del alferez de navio de la real armada D. Manuel Quimper. Reconocimiento del Estrecho de Fuca. University of Washington Special Collections, Seattle.

Raibmon, Paige. 2005. *Authentic Indians: Episodes of encounter from the late nineteenth-century Northwest Coast*. Durham: Duke University Press.
Thrush, Coll. 2007. *Native Seattle: Histories from the crossing-over place*. Seattle: University of Washington Press.
Tweit, Susan. 2006. Tearing down the Elwha river dam. *Popular Mechanics*, Feb. http://www.popularmechanics.com/science/environment/water/2294301. Accessed 8 Sept 2010.
Valadez, Jacilee. 2002. Elwha Klallam. In *Native peoples of the Olympic Peninsula: Who we are*, ed. Jacilee Valadez and The Olympic Peninsula Intertribal Cultural Advisory Committee, 21–34. Norman: University of Oklahoma Press.
Waterman, Thomas T. 1920. *Puget sound geography. Pacific Northwest Manuscript Collection*. Seattle: Special Collections, University of Washington.
Willems-Braun, Bruce. 1997. Buried epistemologies: The politics of nature in (post)colonial British Columbia. *Annals of the Association of American Geographers* 87(4): 3–31.
Wray, Jacilee. 1997. *Olympic National Park ethnographic overview and assessment*. Port Angeles: The Olympic National Park.
Wray, Jacilee, and The Olympic Peninsula Intertribal Cultural Advisory Committee (eds.). 2002. *Native peoples of the Olympic Peninsula: Who we are*. Norman: University of Oklahoma Press.

Part V
The Ways Forward

Cover art by Elianny Guillen Sanchez, Cuba

Chapter 5.0
Introduction: The Ways Forward

Lisa Hiwasaki

This section brings together various contributions that, each in its own unique way, contribute to a larger mosaic – an assemblage of our collective endeavour to provide ideas for alternatives to water management and use in ways that sustain the diversity of cultures, ecosystems, and people. In so doing, this section elucidates that there are numerous ways towards a sustainable future. Ranging from traditional water harvesting in India to community-based water governance structures in the Americas, further to Indigenous involvement in water allocation assessments to environmental non-governmental organizations (NGO)s in Israel/Palestine, local people all around the world are forging solutions towards not only sustainable management of water, but also towards maintenance and celebration of diversity. Descriptions of efforts at the global level by international organizations, scientists, and development workers conclude the section. The contributions in this section together demonstrate that diverse social and natural environments, cultures, and people necessarily mean that there need to be diverse solutions as well.

The section starts off with two contributions that provide somewhat different examples of traditional water harvesting in India. Acharya's chapter, set in Uttarakhand, illustrates that traditional water harvesting techniques and structures served multiple interests and functions and met diverse social, cultural, and ecosystemic needs. Recent attempts to revive and rebuild such structures often fall into the pitfall of defining these technologies as only serving a single purpose, thereby neglecting to address the non-dominant functions of these structures. Drawing on the experience of NGOs working in Uttarakhand, Acharya leaves us with a caveat: traditional ecological knowledge can have both positive and negative implications in terms of equity and rights to water. Parmar then takes us to Rajasthan, also in the northern part of India. In contrast to Acharya's piece, the focus here is not on the

L. Hiwasaki (✉)
UNESCO Office in Jakarta, Regional Science Bureau for Asia and the Pacific
e-mail: l.hiwasaki@unesco.org

B.R. Johnston et al. (eds.), *Water, Cultural Diversity, and Global Environmental Change: Emerging Trends, Sustainable Futures?*, DOI 10.1007/978-94-007-1774-9_28,

physical structure of rainwater harvesting but on a participatory water governance structure that has been successful in managing the Arvari catchment area for the past 10 years. Such community-driven, decentralized water parliaments have been replicated and adapted in other parts of India.

Community management of water and the surrounding ecosystems is a recurrent theme that permeates the three contributions set in Latin America, specifically in Ecuador and Bolivia. D'Amico's piece on community watershed reserves in Intag, Ecuador, describes how attempts by multinational companies to extract minerals led to a sequence of events that resulted in the development of institutional arrangements for managing water and the catchment area as a whole, through community watershed reserves. The enactment of an ecological ordinance further empowered local communities by prioritizing environmental and human rights. Communities surrounding Parapeti River and residents of the city of Cochabamba in Bolivia also created participatory institutions to better manage their water source. Acorn and Zarzycki illustrate the development of community-level assessments of social and natural conditions on the Upper Parapeti, resulting in a committee that brings together communities in the watershed. The committee assisted the communities to draw up their own local natural resource management regulations, which then developed into a single watershed map showing the relations among the communities along the river. Wutich provides another example from Bolivia, albeit in a very different context. Residents of an impoverished urban settlement developed a small-scale community tapstand system to manage and distribute scarce groundwater, using the principles of customary water institutions in the Andes, where most of the residents originated. Wutich's case study provides another example in which the inequities inherent in customary water institutions are highlighted.

Moving continents, contributions from Australia and New Zealand highlight two concepts – environmental flows and cultural flows – that are increasingly becoming recognized by scientists and Indigenous people as key to address challenges in implementing integrated water resources management (IWRM). As defined by the Global Water Partnership (GWP), IWRM is "a process which promotes the coordinated development and management of water, land and related resources, in order to maximize the resultant economic and social welfare in an equitable manner without compromising the sustainability of vital ecosystems" (GWP 1999). Although its goal of achieving the sustainable coexistence of human beings and the environment is valuable, it faces numerous challenges in implementation; consideration of cultural diversity is the missing link that can make the concept of IWRM operational (UNESCO-IHP 2007).

Morgan's chapter on cultural flows, set on the Murray-Darling rivers in Australia, describes how the Murray Lower Darling Rivers Indigenous Nations (MLDRIN), an alliance of ten traditional owner groups from along the Murray and its tributaries, has been successful in securing a stronger voice for Indigenous people in water policy and management in southeastern Australia. A key concept behind the narrative is cultural flows, defined as "water entitlements that are legally and beneficially owned by the Indigenous Nations and that are of a sufficient and adequate quantity and quality to improve the spiritual, cultural, environmental, social, and economic conditions of those Indigenous Nations" (Morgan, Chap. 5.3). This concept of cultural flows was developed as a response to environmental flows, which "describe the quantity, timing, and quality of water flows needed to sustain freshwater and estuarine ecosystems and the human livelihoods and well-being that depend on these ecosystems" (Arthington, Box 5.3a). The difference between environmental flows and cultural flows is that the former is an exercise by scientists and water managers that relates hydrologic flow regime to physical and ecological responses, thereby informing protection or restoration of the freshwater ecosystems, whereas in the case of the latter, it is the Indigenous peoples who would decide water allocation, based on traditional knowledge and their own aspirations for their peoples. As described by Arthington in her contribution, with examples from Lesotho and the United States, environmental flows are now being assessed and implemented in more than 50 countries around the world; the concept of cultural flows is still far from being mainstreamed. Cultural flows should be considered independent of environmental flow allocation, as Morgan points out.

In Tipa and Nelson's contribution, we learn of another case in which Indigenous people are involved in water allocation assessment, this time in New Zealand. In this case, the river flows necessary to protect cultural interests are assessed and cultural flow preferences are calculated using an application of Cultural Opportunity Mapping and Assessment. This participatory process in Kakaunui Catchment attempts to respond to the shortcomings of environmental flow assessments that are often underscored by Indigenous and local people around the world. While this process for deriving cultural flow preferences continues to be refined, it is being applied in different freshwater management contexts on the South Island of New Zealand. Tipa and Teirney provide another example of a tool that enables Indigenous people to participate in land and water management processes and decision-making, the cultural health index (CHI) for assessing stream health. This CHI is a tool through which cultural knowledge about stream health grounded in the beliefs, values and practices of the Maori is linked with Western scientific methods.

At the same time that Indigenous people around the world are devising ways to ensure that their voices are heard, the rise of local environmental movements has also renewed possibilities for sustaining water and diversity. Hassoun, in her description of the profound water crisis in the West Bank and Gaza, highlights the dire water and political problems facing Israelis and Palestinians. Water has already been a component of every war in the history of the Arab/Israeli conflict since 1948; however, the rise of environmental movements in both populations and cooperation between them signifies the possibility of bringing about cultural change. The linking of the Palestinian and Israeli environmental movements provides the only hope that a new potential future based on a common shared water culture, conservation, and appropriate technologies can be charted. Harper and Bari provide an example in which an environmental NGO made up of Roma community members in Hungary has succeeded in using a participatory action research project to convey the environmental concerns of the community to national-level policymakers and activists.

Although the discussions are on broader, more global contexts, the focus of contributions by Hiwasaki and Webster too are on local people: those for whom access to safe drinking water and basic sanitation are not expected to be a reality before the decade is over. In his anecdotes about the water, sanitation, and hygiene promotion (WASH) programme implementation in Southwest Uganda, Webster proposes ways in which the impact of rural WASH projects and development practice can be enhanced by considering cultural and cross-cultural dimensions in the interactions between development workers and beneficiaries. Hiwasaki's concluding chapter provides an overview of the UN's efforts on water and sustainable development, focusing on efforts to improve and maintain the well-being of people, their diverse cultures, and the environment. A distinction is made between concepts of cultural diversity and water cultures: the former embodies, in addition to water cultures, issues such as governance, power, and rights. It is only by engaging in cultural issues more comprehensively – that is, dealing with water and cultural diversity, not just diverse water cultures – that resolution of water problems both at the local and global levels would be possible. Hiwasaki calls upon agencies in the UN System and the international community to tackle the issue of water and cultural diversity; it is only after this that "Water for Life" can be considered a success for everyone.

The themes that weave through these final contributions to the book, as in earlier sections, highlight the need at both local and global levels to deal with the complexities surrounding water, cultural diversity, and global environmental changes. Conclusions that can be draw from this review include the following:

Participatory institutions, most notably water governance structures, can often play crucial roles in bringing water and ecosystems back to the people. Although the process of setting these up can be greatly facilitated by civil society organizations, they ultimately need to be community-driven in order to succeed. Often, charismatic community member(s) precipitate or otherwise play an important role in pushing forward the formation or continuation of these structures. Such institutions are often in charge of managing not just the water but also the catchment area as a whole, in recognition of the need to manage the environment – the natural as well as the social – holistically for success.

The rise of community-level environmentalism is often a key agent of change. Communities that are better informed about the importance of water and ecosystems for their well-being, identity, and cultures are accordingly more able to participate in, and move towards, change – whether forming community-level water governance or forest management structures, taking charge of water and the ecosystem, asserting their rights to them, or forging a shared water culture. Such environmentalism is important especially because it creates the awareness and knowledge necessary for managing the environment as a whole, in particular, the vital link between degraded forest and watersheds.

Beware of tradition's pitfalls. Not everything labelled "traditional" or "indigenous" should be reintroduced or adapted without an examination of the sociocultural and ecosystemic contexts out of which it was born. Traditional knowledges are often holistic and make ecological sense; precisely due to their holistic nature, they cannot be simply taken out of context. In order to use "traditions" to show the way forward, particular care should be taken to ensure especially that inequities and inequalities are not resuscitated. Local and traditional knowledge and Western science assessments can be fruitfully linked together to inform decision-making, but this must be based on the appropriate community-level participatory processes.

Culture as a concept needs to be understood as dynamic and not monolithic. Communities and their cultures are not set in stone but are based on living people; cultures change and can be diverse within a community. Every segment of society – including water scientists, managers, and engineers – has its own cultural engagements with water. It is thus important to regard cultural considerations not as an "extra" factor that need not to be taken into account in decision-making on water allocations, but rather as an integral part of any such decisions. Just as engineering and scientific cultures are represented in such decisions, those of local people need to be included as well.

Resources

GWP. 1999. *How to bring ecological services into integrated water resource management*. Seminar Stockholm, 15–17 Nov. Beijer International Institute of Ecological Economics, Royal Swedish Academy of Sciences, Stockholm.

UNESCO-IHP. 2007. Project concept paper for the UNESCO-IHP project on water and cultural diversity. Prepared for the meeting of the expert advisory group on Water and Cultural Diversity, Paris, 7–8 Jan, 2008.

Chapter 5.1
Managing 'Water Traditions' in Uttarakhand, India: Lessons Learned and Steps Towards the Future

Amitangshu Acharya

5.1.1 Introduction

Uttarakhand became the 27th state of the Indian Union in 2000. Carved out of Himalayan and adjoining Uttar Pradesh districts, the new state borders Tibet on the north, Nepal on the east, the Indian states of Uttar Pradesh to the south, Haryana on the west, and Himachal Pradesh on the northwest.

Geopolitically, Uttarakhand incorporates two distinctive regions, Garhwal and Kumaon, historical kingdoms often engaged in military conflict. Under British colonial rule, Kumaon was annexed, while Garhwal was accorded the status of a princely state, ruled by the Tehri royal family under British dominion. In post-independence India, both British Kumaon and the Tehri princely state were merged into the state of Uttar Pradesh. The long and eventually successful struggle to establish an independent state was driven, in part, by the strong social relationships and historical ties of culturally diverse people to the land and in part by the tensions arising from the exploitation of the regions' resources by outsiders.

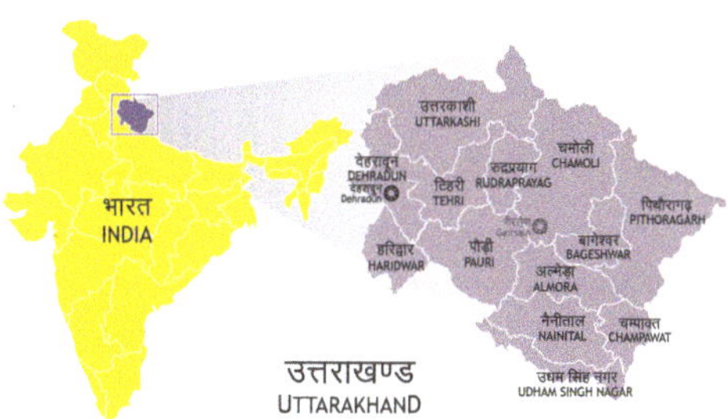

Map 5.1.1 Political map of Uttarakhand

A mountainous state, Uttarakhand has 238 glaciers spread over 735 km^2. Its glacial melt sustains the Ganga and Yamuna, two of the most important rivers in North

A. Acharya (✉)
University of Edinburgh College of Science and Engineering, Edinburgh, UK
e-mail: amitangshu.acharya@gmail.com

B.R. Johnston et al. (eds.), *Water, Cultural Diversity, and Global Environmental Change: Emerging Trends, Sustainable Futures?*, DOI 10.1007/978-94-007-1774-9_29,

India. The religious and economic significance of these two rivers has placed Uttarakhand permanently in the cultural history of India. Its peaks and valleys were and are still today identified as the abode of gods and goddesses. Nanda Devi, the sacred mountain, stands here, and this region has for more than a thousand years inspired poets, pilgrims, and spiritual leaders.

The cultural diversity in this region is immense. The flow of language is said to change every few kilometres, with a patchwork of distinct dialects characterizing the intervening cultural landscape. Although most residents of India speak one of two languages, Bengali or Hindi, today the nation is home to some 415 living languages (as identified in the SIL Ethnologue, Lewis 2009). Many (including culturally distinct dialects of major language groups) are the indigenous languages of tribal groups residing in the north country. The number of languages, an indicator that suggests cultural diversity, has been declining rapidly in India. Some 196 languages are considered endangered in India, 44 of them used by groups living in the Himalayan states of Uttarakhand, Himachal Pradesh, and Jammu and Kashmir. India's 2001 census (which uses a more restrictive definition of language by grouping culturally distinct dialects as subsets of a major language) identifies just 122 languages, a marked decline from the more than 1,600 languages recognised by the 1961 census. Modernization and development of the resource-rich regions of India is a primary factor for this loss in diversity. These imposed processes resulted in development-induced displacement, exploitation and devastation of critical resources, and the loss of a place-based means to sustain a way of life.

Box 5.1a Path to sustainable development: Water parliament in India

—Aradhana Parmar

Box Map 5.1a.1 India

Societies all over the world have invented forms of rainwater harvesting to collect and store water. In India, rainwater harvesting is an ancient technique that dates as far back as 4500 B.C.E. In the state of Rajasthan–a semi-arid state in the Thar Desert of northern India that has historically been known for its chronic water shortages–rainwater harvesting has proven to be an effective tool in water conservation since antiquity and has provided a sustainable livelihood for the local population. In the 1970s, the Alwar district in Rajasthan was declared a 'dark zone'–indicating a rapid depletion of groundwater and severe drought conditions. At this time, a local

(continued)

Box 5.1a (continued)

Box Fig. 5.1a.1 Perennial Avari River (Photo credit: Aradhana Parmar)

Box Fig. 5.1a.2 Catchment area of Arvari shown on the wall of Bhaonta-Kolyala village (Photo credit: Aradhana Parmar)

non-governmental organization (NGO) revitalised the traditional knowledge of rainwater harvesting. As a result, many rainwater harvesting structures that had previously fallen into disrepair were refurbished and newly constructed ones were put into action. As rainwater began to collect in these structures, it swept into the ground, refilling the aquifers with water. Consequently, water also became available in wells and rain-harvesting structures. Rainwater infiltrates into the soil through pores, cracks, and other spaces until it reaches the zone of saturation and is not able to penetrate any further into the earth. Water held in aquifers is known as groundwater. Groundwater can move through aquifers until it reaches an opening to the surface. In a seep, the water reaches the surface in small streams, springs or over a large area. When the aquifers reach the zone of saturation where all of the spaces are filled with water, the movement of water reaches an opening to the surface in a form of seep. This seep reaches the surface in the form of small streams. In this way, gradually, water too began to appear in areas of the dry bed of the Arvari River, with the result that it became a perennial river in 1995.

The revival of indigenous practices of rainwater harvesting not only revived the Arvari River in Rajasthan but also led to the formation of the Arvari Sansad (Water Parliament), which was created to manage the river's catchment area. The Arvari Water Parliament plays an important role in managing and improving its community's access to enhanced and recurring water resources. This community-driven, decentralised model of water governance could be replicated and adopted by other communities who seek local participatory control over their water resources.

(continued)

Box 5.1a (continued)

The water parliament, a participatory, sustainable organization with the aim of making rules and regulations for the use of water and agriculture, includes representatives from the 70 villages in the catchment area. It met for the first time in the town of Hamirpur on January 26, 1999, India's Republic Day. This parliament meets four times a year and as required in an emergency situation. The membership of the water parliament consists of two types of representatives: representatives of nature, that is, people who have worked hard and honestly to save water and forest on at least 500 ha of land. The other type represents the people; there is one representative for every 500 people, and the representatives are elected through consensus or by a majority vote.

The objectives of Arvari Sansad are: (1) to save and preserve natural resources; (2) to fight against injustice without disturbing the equilibrium of society; (3) to strengthen fearless self-determination, self-discipline, and creativity in society; (4) to ensure that each and every villager of the Arvari catchment area is part of the decision-making process; (5) to help the Gram Sabha (a meeting of all villagers in each village council area) fulfill its obligation and to take its place if the Gram Sabha is unable to take the initiative. These objectives contribute to the sustainability and stability of the Arvari River catchment area and arise from the determination of the community to secure water governance.

To achieve its goals and objectives, Arvari Sansad framed 11 rules for irrigating with water from the Arvari River. The most significant rules include not allowing exploiters and polluters into the area, being on guard against privatization forces that might influence villagers, conserving the environment, seeking drought-resistant crops, and not growing cash crops. These rules are meant to facilitate community-driven governance and to create collective awareness and action. Most important, this parliament created a way of sustainably managing the entire catchment area.

Given the success of the Arvari parliament over the past decade, the government of Rajasthan has to respect its decisions. This model of community-driven, decentralised water governance is both replicable and sustainable, but successful introduction of the model may depend on number of interlocking factors, including local cultures, traditions, politics, and leadership, all of which intersect in complex ways. In the case of Arvari Sansad, the charismatic and forceful personality of Rajinder Singh has played a key role. If the Arvari parliament model is to be successfully adopted elsewhere, other community leaders must also take up the cause and motivate people.

In cases in which it is difficult to replicate the exact model, the basic building blocks can be used. In India, NGOs are conducting experiments, such as Sukhi Mazra in the city of Chandigarh in Punjab and pani panchayats (water

(continued)

Box 5.1a (continued)

councils) in Maharashtra–a model in which everyone in the village has a right to water. Similar experiences on a smaller scale are going on throughout India; they are either confined to one or two villages or a small watershed. So far, no one else has attempted a project on an entire river catchment area such as the Arvari.

The Arvari Water Parliament provides an alternative, holistic paradigm for watershed management based on both ecology and culture. It is the creation of poor rural communities that have worked tirelessly for more than a decade to create small, self-contained water storage facilities. This parliament is a unique model of sustainable development managed by a decentralised, community-driven water governance process, which gives control of resources such as forests, lands, and water to local communities. In a world facing severe freshwater problems, important and critical lessons about sustainability can be gleaned from the success of the water parliament.

Historically, the cultural traditions of this region include a deep ecological knowledge and respect for the sanctity of nature. Sacred groves enjoyed religious protection. However, the national development initiatives of the 1960s and 1970s, such as the building of dams and the harvesting of timber, devastated communities and their environment. The Chipko Movement, a community-driven forest preservation and restoration movement whose emergence is considered to be a watershed event in South Asian environmental history, owes its genesis to this very fact: national development often embodies distant agendas and interests at the cost of local ecosystems and ways of life. Response to the resulting environmental alienation often takes the form of social protest, as illustrated on March 26, 1974 at a landmark event of the Chipko movement when village women in the hills of Uttarakhand (then Uttar Pradesh) protested deforestation of sacred groves and successfully halted the cutting of trees, reclaiming their traditional forest rights from the State Forest Department. In some ways the Chipko Movement represented an effort to reassert political, cultural, and environmental identity. Such efforts gave the push for independent statehood much force and energy.

The inherent value and meaning of water and forests are deeply embedded within the religious beliefs, cultural norms, and sociocultural traditions of diverse communities in Uttarkhand. One of the strengths of the region is this traditional biocultural diversity, demonstrated in the varied approaches to water management and use. People developed their water management rules and harvesting technologies over time to accommodate varied ecological conditions and sociocultural contexts. While rainfall is plentiful—Uttarakhand receives a total of 66,320 million kilolitres of rainfall annually—this total accumulates over a 100-day period. The terrain is also diverse, and topographic conditions create many microclimates that influence water availability.

Thus, an array of traditional strategies for harvesting water evolved, reflecting the diverse characteristics of each microregion (see Table 5.1.1).

The national push for development generated competing demands on natural resources in the Himalayan region, and these demands led to rapid changes in local environment, livelihoods, and culture. The large number of rivers originating in or passing through Uttarakhand's mountains drew the attention of national planners for its hydropower potential. Many large dams, such as the Tehri Dam on the Ganga River, the world's fifth largest dam have been built. More large dams are planned. Additional claims on Uttarakhand's water resources come from tourism, expanding cities, agriculture, industry, and construction. In these government-sponsored development projects, the water rights of local communities were of the lowest priority.

Table 5.1.1 Types of rainwater harvesting structures in Uttarakhand (Adapted from PSI 2003)

Type of water harvesting structure	Region of maximum occurrence	Purpose	Technique
Chaal	Garhwal	Livestock water	Natural formation or depression found along mountain ridge tops saddled between two adjacent crests
Khaal	Garhwal	Livestock	Large natural lake which can store several thousand cubic meters of water
Chuptyaula	Garhwal	Livestock and sometimes for human use as well	Rudimentary structures, found in the high-altitude areas of Uttarakhand, which collect water from springs or where it oozes out of the earth
Simars	Garhwal	Agriculture	Waterlogged flat lands used for growing high quality crops
Naula/Baori	Kumaon	Domestic and Drinking Water	Shallow, four-sided stepped wells constructed in the form of an inverted trapezoid, some are massive and ornate structures with rooms and platforms for bathing and washing clothes. Animals are prevented from entering the tank
Dhara	Garhwal	Drinking Water	Essentially a drinking water fountain. Water from springs or subterranean sources is channelled out through carved outlets
Guhl	Garhwal and Kumaon	Irrigation and for running water mills	These are small gravity flow irrigation channels where water is diverted at the head and led directly through channels to agricultural fields. Generally they traverse the contours of a slope

Table 5.1.2 Traditional water harvesting systems and concurrent cultural traditions

Traditional water harvesting structure	Closest scientific term	Location	Purpose	Cultural traditions
Chaals	Trenches/recharge pits	Upstream, on a slope breaks/natural depressions	Livestock water and recharge	Post chaal construction a small puja is organized and sweets are distributed
Surra/Bowlas (cemented ones are called guhls)	Earthen unlined irrigation canals	Near a stream/rivulet	Irrigation and soil moisture conservation	A temple is constructed at the source of the stream. Generally the temple is dedicated to Vishnu or the divine serpent Anantnag. Prayers are offered during festivals or for good rains. Menstruating, pregnant women and dalits disallowed access. Even male relatives of pregnant women are disallowed from touching the temple walls, gates and entering the sanctum sanctorum (garbha griha)
Dharas	Spring	Middle to lower transect, generally closer to habitations	Drinking and domestic water	The outlet is considered sacred and cannot be touched by dalits. Menstruating and pregnant women barred from access. Generally adorned with small stone/wooden statues of Gods such as Shiva and Vishnu. Also with red/ochre cloth flags. Worshipped during marriage and other smaller religious functions

The social disintegration accompanying such development–evidenced by alcoholism, domestic violence, stress, migration and, of late, female foeticide[1]–was perceived by many of the area's residents to be the end result of external administrative and political control over the Himalayan landscape. To counter these socially degenerative forces, a number of Gandhian organisations then known as Border Areas Coordination Committee came together in 1962 to form a network institution titled Himalaya Seva Sangh (HSS: literally, Organization for Serving the Himalayan Region). HSS adopted a strategy of working directly with communities at the local level on natural resource management, heritage conservation, and small-scale livelihoods projects. This conservation and community development model reflected the Gandhian principles of *antodyaya* (meaning 'the rise of the last'–a philosophy articulated by Mahatma Gandhi that says that development should pay attention to the poorest person) and *sarvodaya* (meaning 'the rise of all'–thus building a society in which one individual's economic advancement does not come at the cost of another's opportunity). In this work, the activists recognised two resources as critical for conserving the Himalayan landscape: forests and water.

HSS developed a multipronged strategy to address local water crises: recharge zone management for spring water availability; reforestation; and the construction of traditional household and community water-harvesting strategies, including rainwater harvesting and the building or restoration of recharge structures such as *chaals* (recharge pits with multiple anthropocentric and ecological uses). To build institutional resource-management capacity, HSS initiated *Pani Panchayats* (democratic, community-based institutions responsible for water resource management) and also worked with *Gram Sabhas* (constitutionally approved village advisory bodies comprised of all adults in a village). To transform national policy, HSS also conducted outreach and advocacy for river conservation, promoting decentralised power generation models as an alternative to big dams, articulating local concerns to decision-makers, and critiquing government policies on water and other natural resource management issues.

In 2007, with the support from Arghyam, a philanthropic institution, HSS began a 2-year project to revive springs through the construction or restoration of 121 *chaals* in 24 villages in Uttarakhand, working in collaboration with grassroots organisations. Several important issues surfaced during project implementation. For HSS to work effectively with local communities, an understanding of local knowledge and views of traditional water harvesting was necessary. Through community meetings and interviews with elders (both men and women) and youth (especially emerging leaders), the ecological history and cultural ecology of each village and the surrounding region was reconstructed. Collectively, what emerged was a catchment management vision of water that would sustain biocultural diversity. Although each village narrative was unique, all articulated a holistic notion of traditional water harvesting techniques and structures that served multiple interests and functions.

Such a perception that traditional water harvesting technology met diverse biocultural functions differs significantly from the dominant narrative, which defines

[1] The child sex ratio of Uttarakhand shows 906 girls to 1,000 boys.

traditional rainwater harvesting systems in monochromatic terms as serving a single purpose. In other words, the mainstream rationale for resuscitating traditional water harvesting technologies was to meet specific human or economic needs, not to support the sustaining sociocultural and ecosystemic role of water. Although many promote traditional water harvesting as a decentralised alternative to large water infrastructure projects, the stated purpose of building or restoring local water harvesting structures is often limited to a single purpose. For example, *chaals*, which are currently typically built with imported cement, are meant to provide water supply for livestock. The multifaceted values associated with building and maintaining *chaals*, including the traditional earth-lined version, are not recognised or addressed in project design or management. A similar singular vision also limits the contributions of restored or newly built *guhls* (irrigation canals).

In the following sections, water management traditions and efforts to restore water harvesting structures are described with attention to the multiple use values of various technologies, as well as the broader social and cultural contexts of water management systems.

5.1.2 Chaals in Garhwal: Multiple Uses and Definitions

As shown in Table 5.1.1, a typical village catchment in Garhwal region contains an array of traditional rainwater harvesting structures operating at different altitudes. These physical structures reflect local water wisdom in practice, though knowledge of these structures is not collectively owned; such knowledge is perceived and articulated differently across different groups. One of these water harvesting structures is the *chaal,* which are described in scientific publications as 'usually found along mountain ridge tops, in the saddle between two adjacent crests. They were formed in the past by the glacial action of snowmelt, resulting in the formation of small lakes or ponds with a relatively thick soil bed' (PSI 2003). This standard scientific definition of *chaals* fails to accommodate their multiple use values. Without prioritising any use value of a *chaal* above another, the following local uses can be listed:

1. Providing water for livestock
2. Supplying drinking water for human consumption to nearby habitations, for which other drinking water sources are too far away
3. Buffering habitations and crops from attacks by wild animals, by providing them with drinking water, when *chaals* are located close to forests
4. Functioning as a recharge unit, especially for sustaining discharge in springs located downstream

Interviews with residents in the Garhwal region confirm that shepherds or farmers who took their livestock in the higher reaches for grazing once used the *chaals*. Elders in Jhaaleghat village, Dunda block, Uttarkashi District, refer to these as *paramparik* chaals, occurring naturally in depressions in the high altitude forested areas.

Later, with the realisation that *chaals* were a good drinking water source for both livestock and humans, people made an effort some 70 years ago to contain and harvest this water by making the ponds deeper and layering the interior walls with earth. These enhanced *chaals* in high altitude forests also came to be known as *paramparik*, that is, traditional *chaals*. Hence, the word 'paramparik' denoted the spatial location of the chaals. Incidentally, the word paramparik for these original chaals is of recent usage, as during the initial stages they were referred to simply as chaals.

The construction, maintenance, and use of these *chaals* declined over the years for a number of reasons. The state has increasingly claimed ownership of the forest, and management of forest reserves passed onto the hands of the forest department. Nationalization of communal forests has meant a change in the goals and priorities of forestry, which in turn has changed the forest from mixed broadleaved trees to a coniferous monoculture. Forest management restrictions on grazing also contributed to the decline of *chaals*. The loss of communal land systems and the subsequent fragmentation of individually titled land made agriculture less viable, and this in turn pushed the rural youth towards urban centres. The loss of a youthful male labour force further contributed to the deterioration of traditional water harvesting structures, as gendered cultural norms assigned construction and maintenance of these structures to men. This impasse continued until recently, when non-governmental organizations (NGOs) developed projects to revive *chaals* and other such structures.

When restoration was first proposed, men in the villages expressed their disenchantment with traditional water harvesting structures. *Chaal* construction involves manual labour using local resources, which translates into only few working days and low wage rates for labourers, thus not making it worthwhile for them.

Men's lack of enthusiasm for participating in a small development project to build *chaals* provided a unique opportunity for women to enter a domain traditionally perceived as male. As a result, women's groups are building most of the new *chaals* being constructed through the HSS initiative. This is not simply because men have moved out of these occupations, but also because women are now doing most livestock management at the village level. Thus, the onus of procuring water for livestock has fallen on women, unlike before.

In other villages in Dunda block, Uttarkashi, such as Santagaon and Thandi Kamad, the use values of *paramparik chaals* are also related to wildlife. The perception is that wild animals (specifically wild boars) will come out of the forests to search for water. Thus, it makes sense to ensure that there is water for wildlife in the forests to keep such animals from approaching farms and villages. Villagers often blame inadequate summertime water availability in the forests for animals' increased destruction of crops. Communities are convinced that if the forest department would construct *chaals* in the forests, it would help mitigate the problem.

Another use value of *chaals* emerged from interviews with elders in the villages of Lodara, Majia, and Baman Gaon, located in Dunda Block, Uttarkashi, who point towards a distinct connection between *chaals* and the flow of water in *dharas* (fountains that funnel water from springs). They are also convinced that unless there is a *chaal* located somewhere above a *dhara*, the latter will not yield optimal water, especially in summer. Hence, they identify a recharge function of *chaals*. According to hydrologists, this relationship may well exist. Given that *chaals* are mostly located in slope breaks (the point at the base of a hilly slope where land flattens out; this sudden difference in gradient makes it an ideal point for setting up recharge/harvesting structures), they also function as recharge pits (or percolation tanks, which store surface run off and slowly recharges the sub surface water table downstream). The discharge can possibly get tapped through *dharas* located downstream.

The popular definition of *chaals* has also changed. Those occurring naturally at higher altitudes are considered as *paramparik chaals*. People had to have access to the higher reaches of the forests to construct and maintain these *chaals*. Presently, no intervention can take place in forest lands without official approval from the forest department. Grazing too has been restricted. Such rules heralded the end of *paramparik chaals*. In many cases, communities constructed water harvesting systems closer to habitations or on a gully to collect runoff during monsoons. This shift in location has led people to redefine the *chaals* based on their new use values. Hence across age groups, the definition of *chaals* varies significantly.

Government programmes and executing agencies ushered in programmes to 'modernise' traditional water harvesting structures in the mid-1990s. In civil society circles, this was seen as an effort to transfer funds to local contractors and political leaders in the guise of supply-side augmentation. They reported that this was attempted through intensification of hardware purchase, such a bricks and cement from specific contractors. Part of the inflating construction budget was split across vested interest groups. This nexus pushed for application of expensive construction material (such as cement) on a large scale. Across six villages, the modernization drive is evident especially in the *dharas*, most of which had been concretised.

The 2005 National Employment Guarantee Act (NREGA), which guaranteed a minimum of 100 days of wage labour for the unemployed in rural areas, further fuelled the modernization of the water harvesting and management systems. The criteria for implementing NREGA included a clause that a maximum of 40% of the total cost of the project could be spent on materials and 60% on labour. If the site demanded it, even spending 100% on labour costs was not a problem. Spending NREGA funds required appropriate sites and the mandate of the Act itself required investments in land and water productivity, thus bringing *chaals* back onto the development canvas. However, constructing or reviving *chaals* meant following the traditional form and purpose. This automatically implied that *chaal* construction meant zero material costs and 100% labour cost. Moreover, a *chaal* can be built in a single day. Given that NREGA funds had to be exhausted within a financial year, such low cost interventions were a low priority. The relatively cheap construction cost of *chaals,* instead of being its forte, had become a bottleneck under such circumstances.

Discussion with *Gram Pradhans* (village headmen) in the villages of Thaandi, Chopriali, and Sabli facilitated the understanding of how this bottleneck was dealt with. In most cases, the headman requested for cement application to the interior and exterior walls of chaals. The rationale being that such application would strengthen the structure and increase its longevity. While this was the publicly articulated reason to cementing *chaals*, the tacit one was to increase materials cost.

By increasing materials cost, the estimate shoots up, thus making it possible to exhaust allocated funds quickly. Also, the local contractor who supplies the cement and bricks make more profits, part of which he shares with the local political and government representatives who got such estimates approved from the district authorities. Use of cement also implies use of skilled masons and therefore an increase in labour costs and person days. Hence a *chaal* that could have been easily constructed for two or three thousand rupees (approximately USD63) was being budgeted for one lakh (a hundred thousand rupees, or about USD1,063).

These decisions by the *pradhans* and the *Panchayat* (decentralised units of self governance operating at the village level, which are endorsed by the Indian Constitution) reflect both a limited understanding of *chaals* (as simply providing water for livestock) and the desire to spend NREGA funds. The engineers who oversaw planning and construction were outsiders with little knowledge about local water harvesting practices who considered a *chaal* strictly as a storage unit.

Most of the *chaals* constructed under the NREGA scheme in villages in Uttarkashi and Tehri Garhwal district are bereft of water. This is primarily because they have not been constructed in slope breaks. Though the geohydrology of *chaals* on a catchment has yet to be studied, it is believed that traditional *chaals* were sunken structures constructed out of porous stones, which allow slow seepage.[2] Water stored in *chaals* at the slope break allows slow seepage into the sub surface. Chaals break the flow of fast-moving run off and makes it move slower, which makes it possible for the water table to be recharged more efficiently. This explains why *chaals* are important for springs downstream, because they sustain the springs with their slow recharge. Thus, when cement is applied, it blocks the seepage points, thereby nullifying the recharge value of the *chaals* completely. Figures 5.1.1 and 5.1.2 show the difference between a traditional *chaal* and a newly constructed one under the NREGA programme.

When constructed with local materials, *chaals* are easy to repair. When built with brick and cement masonry, breaches in the structure cannot be fixed with locally available materials. And, because the new *chaals* were built with external state funds, community residents perceive them to be government property, and the responsibility to maintain them therefore lies with the government. Further, many believe that voluntary action to fix or modify *chaals* may result in penalties and admonitions from government officials. But perhaps most significantly, *chaals* built with expensive materials are difficult to fix locally, as cement is difficult to procure in remote villages and the absence of ownership impedes pooling of communal funds for repair at a collective level. The construction of cement storage tanks located in the middle of the village and filled with piped water has further reduced communal perception of *chaals* as a significant water source serving multiple communal needs. Efforts to revitalise traditional resource management not only involve physical work, construction, and technical assistance, but they also involve

[2] From a discussion with India's renowned water historian and ethnographer, V. K. Madhavan, executive director, CHIRAG, an NGO based in Nainital, Uttarakhand, and Anupam Mishra.

Fig. 5.1.1 A chaal in Santagaon village, Dunda block, Uttar Kashi (Photo credit: A. Acharya)

Fig. 5.1.2 A chaal in Thaandi village, Dunda block, Uttarkashi, constructed with NREGA funds (Photo credit: A. Acharya)

engagement in social processes that can have unanticipated consequences. Moreover, reinvigorating traditional water harvesting systems also involves, to some degree, a resuscitation of the cultural values and beliefs that once helped govern those systems. Thus resurgence in caste and gender inequities and privileges was noted, especially with regard to access to domestic and drinking water. These points are further illustrated in the case below, in which efforts to restore drinking water springs meant confronting the linkages between water, culture and inequity.

5.1.3 Dharas: Manifesting Inequity

In Garhwal region, as in many other parts of India, water conservation is socioculturally embedded in the community through religious rites and rituals. Many of the cultural and religious norms governing the use and management of water harvesting systems are related to drinking and domestic water.

Drinking and domestic water in Garhwal region are traditionally collected from *dharas*. Incidentally, not all discharge points from a spring are considered a *dhara*. To be called that, the water source has to have an outlet in the form of either a simple pipe, figures of women with water pitchers, or animal facemasks (that of the cow is quite common in Garhwal). The carvings or ornamentations on the metal or wood outlets are called *nakashi*. Most significantly, *dharas* are marked by the presence of small shrines honouring the deities inside or near the spring. Often there is also a small temple nearby. Some dharas have historic value because of the presence of ancient carved stone tablets or idols. In Thaandi village, the oldest dhara is as old as the settlement itself. A local ruler decreed its construction and it contained elaborate scenes from the Ramayana, an ancient Indian epic. However, only one remains, that of Hanuman (the monkey god) carrying a mountain full of rare medicinal plants over his head and holding his mace with his right hand.

In the absence of a temple, the *dhara* with stone idols or tablets becomes a place of worship. A number of rituals are associated with *dharas,* and since the most common users of *dharas* are women, the rituals are typically connected to key events in a women's life and involve periodic restrictions on access to and sanctions on water use. Beginning with adolescence, for example, menstruating girls are banned from collecting water or even touching the *dhara.* Touching the dhara while menstruating is considered a sin and immediately calls for a *puja* (worship) of the dhara to appease the water god, who is deemed to have been offended by such an act. If *puja* is not offered soon enough, or if the incident went unreported, the *dhara* may dry up. From adolescence onwards, women respect these periodic bans. These bans also affect work on the periphery of the *dhara*, where women often wash household clothing. During menstruation, women are supposed to wash their clothes in more distant locations, away from sacred water sites.

Some of the ceremonies accompanying marriage rites also reflect the sanctity of the dhara. A new bride, after reaching the village of her in-laws, is carried in her *palki* (a carriage borne by male members of the bride and bridegroom's family) to

Fig. 5.1.3 A stone tablet depicting Hanuman on the wall of a *dhara* in Thaandi Village, Dunda Taluk, Uttar Kashi (Photo credit: A. Acharya)

her husband's village *dhara,* where the local *pujari* (priest) recites mantras (sacred incantations) from Hindu scriptures. Incantations are generally dedicated to Vishnu, who is worshipped in most of the villages in Garhwal and Varuna as the water god. The new bride washes the traditional copper water vessel with water from the *dhara*, fills it up, and then carries it on her head to her in-laws' house. There she pours the water from the vessel into a glass and offers water to all of the relatives and guests present. This ritual is symbolic, tying the woman to water and, through her, the family, and it defines the boundaries and workspace of women in the household.

During religious occasions when women are supposed to fast for the good health and prosperity of their husbands and in-laws, a small informal *puja* takes place at the *dhara,* where women offer water to the gods before ending their fast.

Other rules governing water access and use reflect some of the social prejudices of caste, with minority groups such as *dalits* typically relegated to the lowest end of the social hierarchy. The *dalits* in Uttarakhand were historically prohibited from owning land, and their livelihoods were based on the exchange of menial labour (farm or production of other goods and services) for money and food grains. In recent years their status changed somewhat thanks to state-sponsored land redistribution. Nevertheless, other forms of discrimination persist. In villages in Garhwal with mixed populations

Fig. 5.1.4 Worshipping of Shiv Lingam in a dhara in Thaandi Village on Chaturdashi, Uttarkashi (Photo credit: A. Acharya)

of *dalits* and higher castes, *dalits* must wait to get water until all upper-caste households have finished collecting water from the *dharas*. In many cases, a village maintains two separate, caste-specific *dharas*. Asking about the need for this arrangement brings forth differing opinions: Some point towards *dalit* assertion of water rights, established through setting up a separate drinking water source. Others feel that the upper castes themselves provided the separate *dharas* so that the boundaries of purity and pollution could be maintained more rigorously. Figure 5.1.5 below shows two different *dharas* in the village of Sabli, located in the district of Tehri Garhwal. The one on the right side of the photo (a lined *dhara* with a sacred stone spout, a bathing and washing space, and an inflow channel adorned with religious motifs) belongs to the upper castes, while the one on the left (with unadorned, cement outlet and unlined inflow channel to the *dhara*) belongs to the lower castes.

Though denial of the *right to access* to drinking water has not been the case historically, the denial of *right to access with dignity* of the same definitely has been. For example, the practice of untouchability (the social practice of regarding a minority group as 'ritually polluted' and segregating its members from the mainstream by social custom or cultural mandate) was abolished by the Indian Constitution after independence but continued in these areas for quite some time. A *dalit* accidentally touching the *dhara* ushered in social sanctions, punishments, and penalties. The sacrilege would only be averted through the means of *puja,* and the outlet would be

Fig. 5.1.5 Inequitable access to drinking water is illustrated with segregated *dharas* in the village of Sabli, located in the district of Tehri Garhwal (Photo credit: A. Acharya)

taken out, washed, worshipped, and restored. Interestingly, the ornamented outlets made out of wood or stone were always carved by members of the Shilpakar community (that is, *dalits*). Before the water touched the outlet, the touch of a *dalit* on it was deemed normal. The ceremonies accompanying the original affixation of the outlet to the mouth of the spring cleansed the object of its previous associations.

However, domestic water procurement has undergone a paradigm change. Except in extremely remote habitations, piped water from upstream springs is now the dominant source of drinking and domestic water in Garhwal. Yet such infrastructure development is often met with vigorous opposition. This is understandable, given that the general problems with pipelines in India have plagued the state of Uttarakhand as well. Irregular supply, leaks, and breakages and the resulting waste have made pipes an unattractive option. Problems in distribution and supply have made many communities rely on *dharas* as a secondary source. It is generally agreed that *dharas*, because they are subterranean, provide water that tastes and looks better and maintains a better temperature (cool in the summer, warmer than piped water in the winter).

From the *dalit* community perspective, in contrast, centralised piped water schemes have ensured dignity in access and use. Field studies suggest that marginalised communities prefer and in reality do receive better services and a better share of resources under government-managed systems. For example, in Santagaon village, Uttarkashi district, where there is only one *dhara* for use by all communities, the *dalits* must wait their turn at the end of the queue. This situation seems to be the case for most other villages where there is only one drinking water source and the *dalits* are numerically too weak to protest caste-based prejudice or ensure access through separate source development. In Thandi Kamad village, Uttarkashi district, where piped water exists, caste-based discrimination exists only when the pipeline water fails and *dalits* are reminded of their status in society, as they must wait until the upper castes finish getting water.

In addition to ensuring dignity to *dalits*, piped water also alleviates hardship for women. Women and girls used to collect water in head loads and travel 1-2 km to and from water sources. This was a difficult task, given the nature of the terrain. For example, during monsoons, the terrain got slippery, making it more difficult. Slipping and falling and getting injured during this period were common occurrences. Hence, piped water has saved women from hardships and drudgery. Piped water has also affected water conservation principles emanating from traditional wisdom. The focus on building distribution infrastructure was not matched with efforts to manage and improve supply in the catchment area. Thus, studies by Govind Ballabh Pant Institute of Himalayan Environment and Development have confirmed a decline of flow in about 41% of managed water supply sources in Uttarakhand. Much of this loss results from rapid land use changes, especially the replacement of broad leaved species by coniferous plantations. Forest fires also surged sharply in the last decade. These factors have reduced vegetation, leading to increased surface runoff and lower base flows. Studies have indicated that springs are drying up or becoming seasonal in Uttarakhand, a finding corroborated by villagers in Garhwal. And it has been estimated that less than 15% of the rainwater is able to percolate down through deforested slopes to recharge the springs.

Lowered discharge of springs spells doom for both traditional and piped water delivery. Hence, there is a need to revisit traditional water harvesting management and re-examine local practices to conserve and wisely use water in the region.

5.1.4 Conclusion: Towards Better Reviving Traditions and Practices

This study in Garhwal confirms, unfortunately, that ecologically sound practices may be accompanied by social inequities. On the one hand, traditional rainwater harvesting seems ecologically sustainable; on the other hand, it is socially inequitable. This apparent troubling pairing is clarified when considering that *traditions* surrounding water harvesting practices and *traditional water harvesting practices* may be distinguished from each other. Given the increased effort to revive and promote traditional water harvesting, it is important for all actors to be aware of both the positives and the negatives.

Sound ecological principles seem to have guided the construction and management of traditional water harvesting practices. Discussions at the village level produce a comprehensive picture of catchment management that indicates that water harvesting structures were interlinked and dependent on land use and vegetation type. Stitching together various narratives brings forward the following guidelines for managing the water in a catchment:

1. Maintain a good mix of land use, comprising of mixed broad leaved species and pasture lands, to allow both optimal infiltration (the process by which water on the ground surface enters the soil) and also runoff for streams. Infiltration will ensure drinking water supply, and runoff will do the same for agriculture.
2. *Chaals* will augment base flows[3] and thus produce improved discharge in springs.
3. This discharge can also be tapped by *bowla or surra* (irrigation canals). Unlined canals allow the water to interact more with the surrounding vegetation and soil regime. It is believed that when unlined, *bowlas* conserve soil and moisture and are easier to maintain.[4] Elders talk of local grasses growing by the *bowlas* that made good cattle feed. These canals are also believed to spread the discharge from *chaals* and streams across the catchment. However, hydrological studies have not tested this claim.
4. The discharge generated by *chaals* (and possibly by *bowlas*) can be harnessed through springs (Jacob 2008).

Cultural and religious practices strengthened holistic water management. Many of such rituals sought to protect the water source. Figure 5.1.6 illustrates the linkages between traditional water harvesting systems and specific cultural practices.

Provision of drinking water is now mostly recognised as a government responsibility, largely due to acts and policies that divested communities of the ownership and management of water over a period of time. The enactment of Kumaon and Garhwal Water (Collection Retention and Distribution) Act of 1975 is a case in point. It 'empowered the State government to regulate and control, by rules under the Act, the collection, retention and distribution of water and water resources'. The supply of drinking water was then vested with water bureaucracies, current ones being Uttarakhand Peyjal Nigam (Uttarakhand Drinking Water Mission) for construction of water supply schemes and Uttarakhand Jal Sansthan (Uttarakhand Water Supply) for operation and maintenance of the water supply schemes. This shift ushered in supply-oriented management, which was replete with inefficiencies. Within 20-30 years of operation, Uttarakhand lost many of its traditional water harvesting systems. The new water systems, with biases in resource distribution towards upper castes (those who can pay) marginalised *dalits* further and, because of a project focus on distribution rather than source water sustainability, an opportunity to support *chaal* construction (and thus enhance spring discharge) was lost.

In such a context, civil society plays a significant role in bridging the gap between local experiential knowledge and mainstream engineering paradigms. What this chapter demonstrates is the need to engage with semantics, symbols, and policies

[3] In hydrologic terms, streamflow that results from precipitation and infiltrates into the soil and eventually moves through the soil to the stream channel. This is also referred to as groundwater flow or dry-weather flow. See Appendices 1 and 2 for details.

[4] In Jacob (2008), Nitya Jacob has similar observations to offer about agricultural tanks in Tamil Nadu, South India, and the negative impacts on local soil moisture regime when the tanks got lined with cement.

Fig. 5.1.6 Traditional water harvesting in a catchment in Uttarakhand (Credit: Courtesy of Arghyam)

at the local level and develop robust outreach tools to engage with engineers. Engineering solutions must be derived from social as well as environmental impact assessments. When considering impacts on a community in which inequity is rife, plans prepared at the village level may be deemed participatory when in fact the dominant communities are the main ones participating, and the plans reflect their needs. To give voice to a silenced minority requires sustained engagement with diverse members of the community.

The culture of worship and reverence for water can be a powerful mechanism to revitalise community ownership. Cultural motifs such as *puja*, religious fairs, traditional theatre, *bhajans* (religious songs) and folk songs can be optimally used for communication and outreach since they find support and following. The bigger challenge is to question predominant patriarchal norms and dissociate them from the scientific rationale behind traditional water harvesting. This separation can only be achieved through institutional forums in which ideas and thoughts can be shared and daily realities reflected upon. At present, the HSS has taken up this challenge. It is now initiating *Pani Samuhs* (water collectives) at the village level. Women's participation in these institutions has been phenomenal. They have taken up the role of constructing traditional water harvesting structures and collecting funds at the village level to develop water resources.

Based on the experience of HSS and Arghyam, NGOs working on revival of traditional water harvesting need to address the following challenges:

First, there is a need to understand the dynamic nature of *chaals* and other traditional water harvesting structures within civil society better. Doing so requires using ethnographic tools to understand local wisdom and traditional science. The next step is to validate such understandings as much as possible with modern scientific approaches, such as geohydrology. The location of *chaals* is critical in this context as the site decides their purpose. *Chaals* located in discharge zones may serve the purpose of storing water for consumption but not add much recharge value. Most importantly, they can collapse during the rains and cause landslides. Such understanding is critical because the juggernaut of mainstream engineering wisdom at the local level can only be stopped with data and experiences that are articulated confidently. Otherwise, NGO advocacy on traditional water harvesting is generally perceived as unscientific, prioritising unsound local knowledge over rational modern engineering principles.

Second, large government programmes such as NREGA serve as both a threat to and an opportunity for reviving traditional water harvesting systems. Evidence-based advocacy with decision-makers at the local and state levels can help coordinate funds for reviving traditional water harvesting systems. NREGA has been positioned as not simply a mechanism to reduce rural unemployment but also to ensure sustainable development.[5] Documentation of existing non-functional assets built under the programme should showcase how simple modifications in the design and materials (based on local knowledge systems) could ensure optimal utilisation of public money.

Third, the greatest challenge is to address the gap in knowledge transmission. On the one hand is a generational issue, where local experiential knowledge is rapidly losing ground to the modern, ushered in through formal educational systems. On the other hand is the power and greater legitimacy of one form of knowledge over another, a gap that is best addressed by facilitating dialogue between local knowledge holders and decision-makers and engineers. The dominance of linear thinking on water harvesting and management can only be challenged through continuous dialogue and experience sharing.

Attempts at reusing traditional knowledge linked to natural resource conservation must be anchored on clear understanding of equity and rights. Traditional ecological knowledge has much to offer modern society in our search for locally appropriate water resource management strategies, especially in managing water for multiple uses by sustaining cultural as well as environmental flows. That said, it is important to demystify the black box of 'traditional wisdom', which is often romanticised by civil society and government agencies. There are subtle differences between traditional water harvesting practices and the cultural traditions around water harvesting practices. Restoration and remediation requires a careful understanding of both the positives and negatives in these traditions.

---

[5] Preamble to the act, available at www.rural.nic.in or www.nrega.nic.in

Resources

Jacob, Nitya. 2008. *Jalyatra: Exploring India's Traditional Water Management Systems*. New Delhi: Penguin Books India.

Lewis, M. Paul (ed.). 2009. *Ethnologue: Languages of the world*, 16th ed. Dallas: SIL International. Online version. http://www.ethnologue.com

People's Science Institute (PSI). 2003. *Survival lessons: Himlayan jal sanskriti*. Dehradun: People's Science Institute.

Chapter 5.2
'El Agua es Vida/Water Is Life': Community Watershed Reserves in Intag, Ecuador, and Emerging Ecological Identities

Linda D'Amico

Map 5.2.1 Ecuador

Intag, Ecuador, is a unique area for its biodiversity, international attempts to extract resources, and the local mobilization against mining. Its inhabitants, known as Inteños, live in scattered communities, hamlets, and farms, and have indigenous, African, and mixed – mestizo – origins. For the most part, Inteños depend upon the natural world for their livelihoods and have developed a unique culture that stresses values of mutual interdependence that complement collective and individual rights. The northwestern subtropical Andes Mountains are steep, rugged, and interspersed with abundant rivers and streams. The sound of flowing water is always near. When local interests clashed with attempts to extract mineral by powerful transnational mining companies, a culture of political ecology emerged. 'We can't sell our children's future by letting a mining company come in and contaminate our beautiful river', Carmen Proaño, community president of Río Verde, proclaimed on August 14, 2009.

L. D'Amico (✉)
Global Studies and Women's and Gender Studies, Winona State University, Winona, MN, USA
e-mail: LDamico@winona.edu

B.R. Johnston et al. (eds.), *Water, Cultural Diversity, and Global Environmental Change: Emerging Trends, Sustainable Futures?*, DOI 10.1007/978-94-007-1774-9_30,

Fig. 5.2.1 Intag cloud forest is a region of biodiversity. Toucan barbets at Los Cedros (Photo credit: Linda D'Amico)

5.2.1 Historical Context

Pre-Hispanic and ethnohistorical references to the region are scant, although Intag is mentioned as an important territory for interzonal or micro-vertical trade wherein products from different altitudes were exchanged with native groups that flourished in the adjacent inter-Andean valleys of Cotacachi and Otavalo (Jara Chāvez 2007). Intag's elevation and semi-tropical climate allowed for the production of sisal and cotton (which were the raw materials used for textiles prior to the Inca and Spanish invasions), in addition to the extraction of minerals that included salt and gold (Espinoza 1988). Contemporary local residents identify burial mounds where gold figurines were found, suggesting a complex prehistoric society. However, after the Spanish conquest, sparse indigenous populations diminished. By the early 20th century, Intag became attractive to the landed elite and others for sugar production, and several established large haciendas or small farms. Currently, Intag is characterised by a multicultural population that includes mestizos, Afro-Ecuadorians, indigenous Quichuas and a few expatriates.

5.2.2 Governmental, Civic, and Economic Situation

Cotacachi Cantón (county) contains elevations ranging from 200 to 4,900 m above sea level and ecosystems including tropical rainforests, páramos (above 3,000 m), and cloud forests. It is roughly divided into the Andean and Intag regions. The Andean parishes comprise 20% of the territory and are home to 80% of the (mostly) indigenous and mestizo population. These parishes include the urban zone and the rural highland zone on the eastern flanks of Mount Cotacachi (4,939 m). The Andean region contrasts greatly with the sub-tropical Intag region (extending from approximately 3,000-200 m), where a multiethnic population lives on the steep western flanks of the Andes. The 2000 census documents approximately 14,000 Intag residents in seven parishes, six in Cotacachi County (Apuela, García Moreno, Vacas Galindo, Peñaherrera, Cuellaje, Plaza Guitiérrez), and one in Otavalo County. Each parish is the seat of local government, where erratic electric and phone services are available; however, outside the main town basic utilities are not guaranteed. The region depends upon two unpaved roads for human and commercial transport, but these roads are at best slow and treacherous (it takes two and a half hours to travel 60 km) and at worst beset by intermittent landslides that curtail transport during the rainy season from November to April. Intag includes more than 90 hamlets as well as isolated farms, some of these connected by secondary roads that are only passable with four-wheel drive vehicles during the rainy season. Few older residents have more than a primary education, but this is rapidly changing as the growing younger generations are getting more education. Most Inteño/as make their living through subsistence agriculture and cattle raising, although families also generally have a varied portfolio of income-generating activities that might include handicraft production, small scale commerce, transportation, day labour, and remittances sent by a relative from an urban or semi-urban area in Ecuador or from abroad.

In the early 1990s, the World Bank lent the government of Ecuador funding to map subsoil minerals. This project led to the discovery of copper and gold deposits in Intag. Since that time, residents have organised to prevent extractive interests from destroying their forests and way of life. This effort has included the creation of economic alternatives compatible with the conservation of the culture and natural wealth of the area. As a result, two mining companies, Bishimetals (a subsidiary of Japan's Mitsubishi Corporation) and Ascendant Copper Corporation (now Copper Mesa Mining Corporation), have abandoned plans to mine the copper from the Toisán mountain range near the community of Junín.

5.2.3 The Rise of Environmentalism in Intag: Community Solidarity and Sustainable Alternatives

Intag's inhabitants unexpectedly found themselves at a global crossroads in the 1990s. Giovanni Paz, a 'green' Catholic priest who practiced liberation theology, first alerted Inteños to the threat of mining and called them to revalue the unique ecosystems around them. Soon after, he and other concerned residents created the non-governmental organization (NGO) Defensa y Conservación Ecológica de Intag (DECOIN, Ecological Conservation and Defense of Intag) in the town of Apuela in January 1995. From its inception, DECOIN's priority was to protect community and environmental rights. It was, and continues to be, active in biodiversity protection, community organizing, environmental education, the promotion of sustainable development, conservation initiatives, support for women's projects and policy planning. DECOIN established links to regional, national and international organizations that have proved to be crucial networks of exchange, mobilization, and support. The fact that the executive director, Carlos Zorrilla, a charismatic Cuban-North American has lived on his farm La Florida for 30 years and is bilingual and bicultural, offers broad possibilities for intercultural alliances on multiple scales for tackling complex political-ecological issues in Intag.

Of equal importance, in 1996 indigenous economist Auki Tituaña was elected as Cotacachi municipal mayor (In Ecuador municipalities include entire counties and are subdivided into parishes). Tituaña's charismatic leadership strengthened democratic local governance and public administration reforms in a period of decentralization.

His first act as mayor was to form citizens' assemblies in each parish, in which residents debated their problems, assets, organizations and visions for the future. From these debates emerged a structure for participatory planning and government. The three main goals involved creating a plan for (1) integrated development to incorporate both the Andean and Intag regions, (2) participatory plans to improve the quality of life for all the people of the county, and (3) the promotion of self-determination, self-management and interculturalism within the six parishes that comprise the county. Residents formed local committees which gave impetus to transregional coordination and participation, including the less populated and heretofore isolated communities of Intag. This civic engagement compelled citizens of Intag to prioritise values and to strengthen local cultural and political institutions. These political activities occurred at the time when Inteño/as were considering whether their cultural values were compatible with the impact of the proposed Mitsubishi copper mine adjacent to the hamlet of Junín in the parish of García Moreno.

Farmers in Junín, the community closest to where Mitsubishi's exploratory activities took place, were quick to identify adverse affects to water quality in the streams they depend upon for drinking, cooking, washing, bathing and watering animals. Most Intag households rely on primary water sources, which they tap with hoses that run directly to their homes. Inteño/as are cognizant that water is vital for their health and survival. Charo Piedra, a leader from Junín and general manager of the community ecotourism project operating there as a productive alternative to mining since 1999, pointed out, 'Water is life, and if we don't take care of its source, everything will end. If we want water in the future, we can't log, we have to protect our forests'.

In 1998, Mary Ellen Fieweger wrote *Es un monstruo grande y pisa fuerte: La minería en el Ecuador y el mundo*. DECOIN and Acción Ecológica, a Quito-based environmental NGO, provided funding to publish the book to disseminate information

Fig. 5.2.2 Farewel to Mayor Tituaña at Nangulví, July 2009 (Photo credit: Linda D'Amico)

about the environmental impacts of open pit mining. Readers learned that this kind of mining destroys the biotic cover over a huge area, produces toxic dust, uses explosives that contain toxic chemicals, requires millions of litres of water and, after extracting the miniscule portion of the ore containing the mineral sought, leaves behind millions of tons of wastes, including toxic minerals such as lead, arsenic, cadmium and so on, that poison surface and ground waters for generations (Fieweger and DECOIN 1998). Many Inteño/as realised that this extractive process, including excessive water use, would place entire ecosystems at risk, displace families, and negatively affect their culture and means of support.

In response, residents acted to protect their livelihoods and values, and environmentalism became a concrete strategy for survival. They organised and developed the skills necessary to take their message to the media and authorities. Inteño/as argued that water security is a human and environmental right and that outsiders had no business imposing unsustainable development schemes on them. As Dr. Luz Marina Vega stated at the third Forum on Water and Biodiversity in Intag, 'Between a mountain of gold and water, we have to choose water' (Periódico INTAG 2008). In other words, water, essential to health, well-being and sustainability, cannot be treated as a commodity. Further, 'sumak kawsay' (well-being) is now guaranteed, along with the 'Rights of Nature and Pachamama (the Kichwa notion of Mother of the Cosmos or Mother Earth) in Ecuador's 2008 Constitution.

Even with the attractive promises of funding for 'development' from the deep pockets of the mining companies, which often turned brother against brother, most Inteño/as opted for exploring alternatives to protect their livelihoods, culture and the environment. They generated their own development model, which indigenous mayor Auki Tituaña called 'development with identity', and created sustainable alternatives. In September 2000, the parish assemblies and county government enacted the Ecological Ordinance (EO), a decentralizing initiative that empowers the county and its local committees to oversee the trajectory of its development. Collaborating with regional and international activists to articulate these committees' conceptual framework, they prioritised environmental and human rights. The Ecological Ordinance was in part a direct local response to the threat of open pit copper mining. It created legal parameters to restrain environmental degradation and incentives for sustainable resource use. Within the EO, local communities identified water security as a priority and outlined guidelines for watershed protection for the common good. These included statutes that allow for land expropriation when local watersheds are deemed at risk.

The EO provided the legal infrastructure to protect environmental services. It offered a model for decentralization by returning authority to local governments and resource management to stakeholders. Subsequently, there have been campaigns to protect water sources and conserve forests and biodiversity as the means to strengthen local control, public health and sustainable economic activities. These initiatives engage civil society in ways that continue to enhance participation, social justice, environmental rights and local empowerment. During this period, concerned residents came together in 2000 and launched a local newspaper. *Periódico INTAG* emerged as a print forum for local debate and a medium to disseminate local, scientific and other kinds of knowledge. Beginning with the first issue, under the masthead 'The first independent newspaper in Ecuador's greenest and loveliest corner', the newspaper staff had three major objectives: (1) Offer a space for residents to share news, opinions and analysis with their neighbours, (2) Keep readers informed of projects proposed and implemented by governmental and non-governmental entities that affect Intag's farms and forests, and, (3) Provide information on topics related to the environment. Part of its mission is to train and employ local reporters to communicate Inteños' perspectives and analyses.

Its bimonthly editions contribute to the social construction of Intag as a regional community with a distinctive identity. *Periódico INTAG* articulates a sense of belonging and helps to shape shared values and experiences in the collective struggle for social and environmental justice. It contributes to the political-ecological dialogue by engaging residents and reporting on issues that relate to socioecological justice. Recent research indicates that local governance, which requires an informed citizenry, is a decisive factor for biodiversity conservation (Ribot 2008). In this case, *Periódico INTAG* demonstrates how 'imagined communities' (Anderson 1991) come together across considerable distance (in this case, the population is spread out in an area of 1,800 km^2), when information is shared and participation is encouraged. Along with the Radio Intag (begun in 2004), the newspaper keeps citizens informed and is crucial in the construction of a regional identity. Additionally, the online editions in Spanish, English and German help to keep the transnational environmental community informed. These communication outlets were instrumental in spotlighting the use of paramilitary forces by the Canadian mining company Ascendant in Junín in December 2006. The company's use of force, rather than adherence to the consultative process guaranteed by law, was a turning point in local and national public opinion. Some two years later, the National Constitutional Assembly issued its mining mandate, cancelling mining concessions granted to Ascendant in Ecuador. However, the 2009 Mining Law pushed through by Correa's Revolucion Cuidadana throws the mining moratorium into question.

Support from, and alliances with, national and international NGOs brought diverse intellectual and financial resources to the struggle for environmental justice in Intag. This overlapping of geographic scales, values and priorities brought together different perspectives. Disparate but mutually supportive and beneficial frames of reference came together to bridge and bond (local and global) social capital, intellectual and financial resources to counter specific threats. The risk of mining spurred grassroots organization and political mobilization, and with networks of support from national and international institutions, advocacy coalitions were built and activated. Acción Ecológica arranged in 1997 for a delegation of Inteño/as to visit an open pit mine in Oroya, Peru, to learn about environmental and social impacts of extractive industries. The field trip had a huge impact on community members, who reported back about the disintegration of families and communities, environmental degradation, and air and water pollution. Silvia Vetancourt went on the trip and recently commented to me, 'the open pit mine cut a line through town with the mine on one side, and filth and social delinquency on another. In the middle of town was a lake, which previously had served residents for drinking, bathing, washing and whatever they needed. However, since the mine, the lake is so polluted that it is good for nothing. This scared us. And all the schools, health centres and other things the companies had promised were for naught. We returned resolved not to be blinded by false promises and lose the natural wealth that surrounds us'.

5.2.4 Reservas Hídricas: Ensuring Water, Forests and Health for the Future

Water security epitomises down to earth sensibilities that touch all Inteño/as' lives on one level or another, particularly for women, who generally manage household activities. So when the British NGO Rainforest Concern approached DECOIN in the early 2000s about support for possible projects in Intag, there was consensus among Intag residents about creating a project for community watershed reserves. This opportunity brought together water protection and biodiversity conservation, providing concrete environmental and economic benefits.

Table 5.2.1 Inputs from DECOIN, inputs from community

Inputs from DECOIN	Inputs from communities
Provide technical assistance in identifying adequate water source and surveying catchment	Community watershed council and interested residents assist in identification of source and surveying
Provide funding (via International NGOs) to purchase land where watershed is located with guarantees of no logging, no extractive activities, no burning and no hunting	Negotiate with landowners to purchase watershed Legalize deed/title with municipal authorities, naming community as owner with provisions to protect and conserve land
Technical assistance with management plans	Management plan
Reforestation workshops and initial provision of tree nurseries with native species	Provide *mingas*/cooperative labor for tree nurseries, tree planting and maintenance of reforested areas
Workshops of water and biodiversity monitoring	
Provide signage to identify *reservas hídricas* as protected areas	Locate funding for infrastructure, including; barbed wire for fencing of watershed and/or tubing for water system
Provide financial assistance for management of reforested areas (in some cases), including weeding to ensure saplings survive	Provide labor to install infrastructure

According to outreach coordinator for the project, Plaza Gutiérrez native Armando Almeida, the first priority was to make people aware of the importance of conserving forests to guarantee water protection. He told me that it is common knowledge that 'si no se cuida el agua, se acaba' (if we aren't good water stewards, it dries up), and that many Inteño/as have already observed climate change in the form of diminished rainfall and longer summers as the result of deforestation. During initial meetings and workshops, he emphasised a maxim people could relate to: healthy forests = clean water. Interested communities quickly solicited DECOIN's financial and technical support in order to survey, purchase and protect their watersheds.

As of July 2009, 40 Intag communities own and manage 37 hydrologic reserves, encompassing over 1,000 ha with more than 60,000 mostly native trees planted to guarantee water security and provide clean water for household use. According to Diocelina Flores, a resident of Pucará, the community's watershed reserve 'has saved lives' by replacing a degraded water system and ensuring clean water for the community. Local interest has grown in other conservation measures, including habitat conservation, avoidance of pollution and degradation, agroecology, reduction of threats from desertification, opening of trails for ecotourism, and the development of a mini-hydroelectric project with nine proposed plants to be managed by and for local communities. As a significant alternative to mining, the HidroIntag project has the support of most residents, who see the possibility of jobs and sustained revenue for their communities, while protecting their forests as the core source of water.

Almeida pointed to the community of 'La Esperanza' as one of the most successful projects. In 2002 residents were worried that water flow to their community was decreasing at an alarming rate because it only rarely reached the school. The source of water was on private property, where logging was rapidly deforesting the area. Concerned citizens proposed a partnership with DECOIN to survey their water

Fig. 5.2.3 Armando Almeida, distributing trees for reforestation in the community of San Antonio, Cuellaje (Photo credit: Linda D'Amico)

needs. The fact that the Ecological Ordinance includes a provision establishing a 6-m buffer (or stream management) zone on both sides of streams as well as around springs made the formation of community hydrological reserves in Intag feasible. The EO provision protects water catchments and monitors water temperatures in order to support stable microhabitats for living aquatic communities (Ruiz-Córdova et al. 2006). Further, it offers leverage by pressuring owners who refuse to sell with the threat of expropriation. In the case of 'La Esperanza', it was critical to restore deforested areas located close to the spring, and the community water council was able to negotiate the sale of these lands.

The watershed was purchased and the deed was put in the name of the community. Residents agreed to prohibit unsustainable uses, provide regular monitoring, and reforest degraded slopes. The collective stakeholders assumed the responsibility of fencing in the 8 ha, and legal title gave the community president leverage to negotiate with the provincial government and NGOs for barbed wire and materials for the water system. In addition, the watershed brought the community together through traditional *mingas*, or cooperative work parties, to create the tree nursery and to reforest eroded areas. In some areas the process of converting pastures to forest was a huge undertaking that included weeding, watering and caring for the young trees. By working together in these endeavours, the community reinforced collective values, highlighting mutual interdependence with each other and the forests.

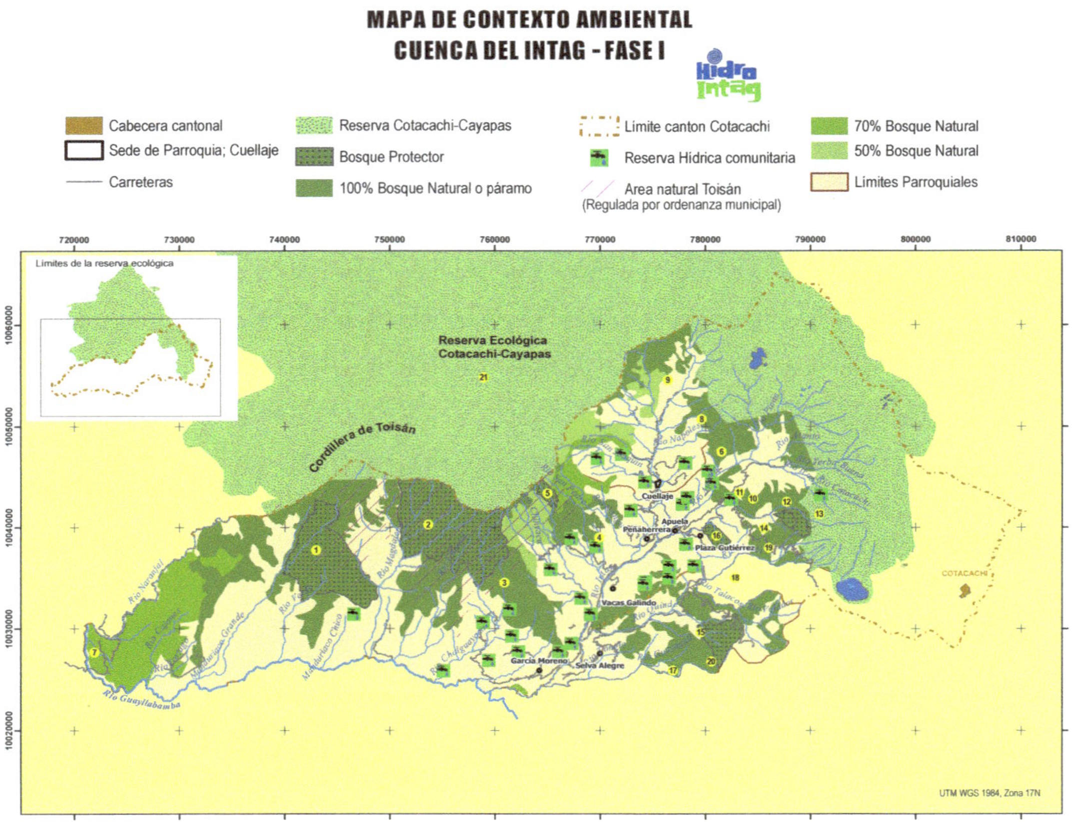

Map 5.2.2 Reserves of Cotacachi County: faucet icons indicate Community Watershed Reserves (Mapa Reservas de Intag)

Box 5.2a Resurgence of campesino community associations to reforest and calm an angry river, Upriver Parapeti River, Sub-Andean Gran Chaco, Bolivia

—Janis B. Alcorn and Alejo Zarzycki

Box Fig. 5.2a.1 People Fishing in the Pilcomayo River. Concerned about degradation of the forests, rivers, and fishery resources, Indigenous peoples in the Pilcomayo Basin have formed the binational 'Protectors of the Pilcomayo' and established a community based program to monitor fishing gear, catch (species, quantities, sizes and weights) and water quality. (Photo: Fundación Yangareko, http://www.yangareko.org)

Through self-analysis of river and forest degradation facilitated by Fundación Yangareko, 37 communities in the Upper Parapeti *municipio* (county) of Monteagudo, Bolivia, created a watershed association that has proven to be self-sustaining and effective. Fundación Yangareko, a Bolivian non-governmental organization (NGO),[1] began work in this region in 2002 because management of the Upper Parapeti is critical for the maintenance of the entire Parapeti basin (158,000 km) and the down-river Kaa Iya National Park where the river becomes a wetland extends to the Amazon. In the Upper Parapeti, people eat fish and use water from the river, and their cattle drink from the river. Downstream, the Isoso Guarani indigenous communities and wildlife depend on the Upper Parapeti management to maintain the flowing river on which they depend. The Parapeti basin includes some 100,000 people, most living in extreme poverty according to global indices.

The campesino (peasant) families originally moved into this region from the highlands to serve as labourers in a large pig farming operation that subsequently failed in the middle of the 20th century. The labourers were left to make a living by deforesting and farming the steep lands, which over time has had a negative environmental impact on the basin overall as well as locally. Sedimentation from runoff from deforested and degraded slopes was altering the water quality and the river bed and creating conditions that resulted in more flooding, which damaged property and threatened lives.

[1] Fundación Yangareko is a Bolivian NGO whose theory of change is that, given that local people manage biodiversity on a day to day basis, if they share their knowledge and are joined in local organizations and interactive discussion forums, then a critical mass of people will change their own behaviour and take advantage of opportunities to influence key decision-makers, and in turn decision-makers will respond by taking action to support solutions that are good for society and nature.

(continued)

Box 5.2a (continued)

Box Fig. 5.2a.2 Each Upper Parapeti community was given a notebook containing the information from community members and technical information about their natural resources (Photo: Janis B. Alcorn)

Box Fig. 5.2a.3 Each Parapeti community drew its own symbolic shield, this one showing the river flowing through the heart of a community (Photo: Janis B. Alcorn)

To initiate the work, Yangareko sent a team of young fieldworkers to visit all the households in the Bañado tributary of the Upper Parapeti area in Monteagudo (120,000 ha), belonging to 37 communities. The team members interviewed families about their social and environmental conditions, and in the process people became more conscious of their shared problem of deforestation and degradation in the watershed. The Yangareko team assisted the communities to aggregate the information from families to create their own community-level assessments of social and environmental conditions and then facilitated discussions among community representatives, presenting each community with a final notebook with information that the community members had contributed. These notebooks bore the new community level shields, designed as part of the process to symbolize community identity.

As they reflected on their shared situation, the communities decided to create a watershed committee to represent their interests to local and regional governments. The Bañado Watershed Management Committee (CGRB) was born and then assisted the 37 communities to draw up their own local natural resource management regulations. Yangareko in turn assembled the land use and environmental maps into a single watershed map showing the relations among the communities along the river.

(continued)

Box 5.2a (continued)

Box Fig. 5.2a.4 The representatives of the 37 communities in the watershed voted to create their own Watershed Management Association in 2004, and it has proven to be sustainable through 2010, based on local leadership and participation (Photo: Janis B. Alcorn)

When the community members saw the aggregated map and zoning recommended for their watershed, they decided to petition local government to declare their lands as an Area under Integrated Management (ANMI), which would restrict further deforestation and assist with reforestation along rivers. They also saw that the declaration would help them to seek appropriate development assistance. In 2004 the ANMI was recognized. Since then, the community representatives have continued to meet in assembly once every quarter and have built their own office from which they provide local technical assistance to communities. Some 5,000 ha have been reforested along the river in Monteagudo, and communities along the Parapeti in the neighbouring municipio of Huacareta have subsequently formed their own committee for reforestation in their region, to maintain the river that is so important for everyone.

This case illustrates how grassroots-level work with communities in degraded watersheds can improve water management through a natural scaling up. This project followed what Fundación Yangareko calls the principle of 'generate solutions' using collaboration and simple technical tools, rather than impose solutions. Another outcome of this kind of approach is that local leaders emerged, contributing to greater long-term resilience than do short projects done by outside technicians. It illustrates the hypothesis that if communities and their organizations are supported in analyzing their environmental management issues and if they are given opportunities to use their knowledge, they will create and use links between themselves and local and regional government agencies to improve environmental management across scales to maintain ecosystem functions (Alcorn et al. 2008).

Resources

Alcorn, J.B., A. Zarzycki O., and A. Zarzycki. 2008. Parapeti River Basin, Bolivia. In *Governance and ecosystems management for the conservation of biodiversity*, ed. Basil Manos and Jason Papathanasiou, 177–180. Thesalonika: University of Thesalonika.

According to Almeida, this example from 'La Esperanza' demonstrates that with legalization of the hydrological reserve in the name of the community also comes protection and investment that secures community water for the future. In the course of seven years working with 37 communities, he has seen the development of community awareness regarding the importance of conserving water and commitment to that end through stewardship of forests. He commented that people's collective actions have enhanced community and regional solidarity, as they organise and work together to analyze and resolve issues concerning water.

5.2.5 The Ecological and Hydrological Significance of Intag

Ecuador is one of 12 megadiverse countries in the world, home to a spectacular range of biodiversity. These ecosystems provide vital services to humans, including food, water, shelter, medicine, recreation, and aesthetic and spiritual sustenance, and they are the basis for a wide range of cultural diversity. Biologists hypothesise that during the last ice age, when drastic climate change and mass extinctions occurred, Ecuador's geography and position on the equator contributed to the region's evolution as a refuge and 'motor for speciation' (Gentry 1993). The geological and natural histories are the foundations for its current and prodigious endemism and biodiversity.

Box 5.2b The sustainability of a customary Andean water institution among the urban poor of Cochabamba, Bolivia

—Amber Wutich

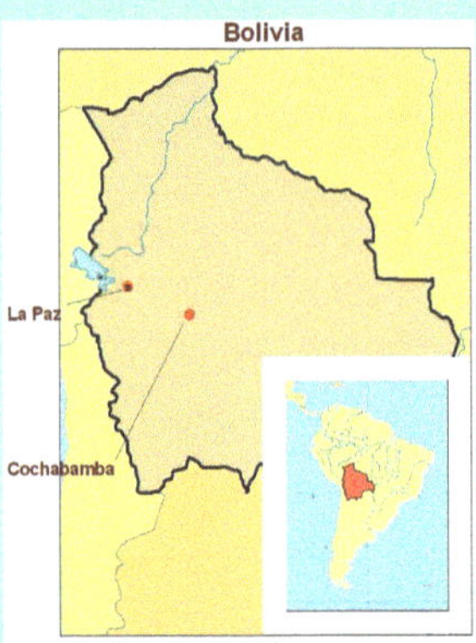

Box Map 5.2b.1 Bolivia

The city of Cochabamba, Bolivia is famous as the site of the 'Water War of 2000', a struggle in which the people fought to prevent the privatization of customary and municipal water institutions. After the success of these protests, however, the inequities that plagued the urban water system persisted. In particular, impoverished residents of over 165 settlements located in southern Cochabamba were denied the right to access water from the municipal water utility. Because the urban poor are generally forced onto the least desirable land, these settlements have very limited groundwater resources and residents must obtain water from private vendors, rainwater collection, and surface water

(continued)

Box 5.2b (continued)

Box Fig. 5.2b.1 Water resource governance in Villa Israel (Photo credit: Amber Wutich)

Box Fig. 5.2b.2 Each day, Villa Israel households line up jerry cans and await their turn to receive water from a tapstand (Photo credit: Amber Wutich)

sources. In one impoverished settlement, known as Villa Israel, residents developed a customary water institution to manage and distribute scarce groundwater.

The residents of Villa Israel are predominantly urban migrants of Quechua and Aymara descent. In their rural hometowns, many learned the basic principles of successful commons management, including self-governance, monitoring of resource use and users, and adaptation to local conditions (Ostrom 1990). In the rural Andes, customary water institutions adhere to these principles as well as key values including uniformity, regularity, and transparency in resource distribution (Trawick 2003). In Villa Israel, the settlers adapted the rules of customary rural institutions for use in an urban setting.

Villa Israel owns and operates a small-scale community tapstand system. Drawing on two local wells, the community distributes water to approximately 10 tapstand zones. The residents of each zone are responsible for the construction and maintenance of their own tapstand. Households also pay a small monthly fee to access the tapstand system. Once a day, a monitor unlocks and supervises water distribution at each tapstand in turn. Households have 20 minutes to collect their water allotment before the monitor moves on to the next tapstand. Because groundwater is extremely scarce, each household is limited to collecting just 40 litres of water per day. Tapstand users join the monitor in ensuring that everyone adheres to the community's rules. The entire system is overseen by Villa Israel's community council during monthly public meetings.

To safeguard the system's long-term sustainability, the community instituted three rules that set limits on when and by whom community water can be used: the exclusionary rule, shortage-sharing rule, and overdraft rule.

(continued)

Box 5.2b (continued)

Box Fig. 5.2b.3 When supplies are too low water distribution is cut off, forcing residents in the squatter settlements to use scarce river water for basic needs (Photo credit: Amber Wutich)

The exclusionary rule mandates that renters and new landowners who have not 'bought in' by paying back fees for construction costs are prohibited from using the water system. The shortage-sharing rule reduces all household water allotments from 40 to 20 litres, as needed, during the dry season. The overdraft rule dictates that when water supplies are too low, water distribution will be suspended until sufficient water is available. These rules have long protected Villa Israel's scarce groundwater resources from overexploitation. However, they make some Villa Israel residents more susceptible to water scarcity. Renters and new landowners, who are among the community's most vulnerable residents due to their relative poverty, lack of social support, and unfamiliarity with locally appropriate survival strategies, are prevented from acquiring the only clean water available from a local source. Households struggling to obtain sufficient water during the dry season are forced to cut their consumption from 40 litres – a quantity sufficient to meet minimal drinking and cooking needs – to half that amount. Entire sections of the community may be unpredictably cut off from water with no knowledge of when water services will resume. As a result, Villa Israel residents may experience hunger and thirst, be unable to bathe or wash, and shut down the small businesses that provide their only source of income.

The Villa Israel case demonstrates how customary institutions can be used to manage a small-scale urban water system sustainably. Several factors were crucial to the institution's success. First, settlers adapted their rural knowledge of commons management for use in an urban ecosystem. Second, institutional rules were designed to prioritise the conservation of water. Third, the key values of uniformity, regularity, and transparency ensured that community members perceived the system as legitimate and complied with community rules. Fourth, self-governance enabled community members to adapt quickly to changing environmental conditions, such as seasonal scarcity and groundwater overdraft. To survive under conditions of environmental adversity, Villa Israel's residents made the difficult decision to prioritise the maintenance of the community water system over the well-being of individual households. The consequence of this decision is that Villa Israel faces stark limits to the standard of living, population size, and economic development that the community can

(continued)

Box 5.2b (continued)

support. This case also demonstrates that, although customary institutions have much to contribute to the sustainable water management, they cannot resolve basic inequities such as the exclusion of the poor from the municipal water system. However, if the urban poor are afforded the right to a reliable and adequate water source, they have demonstrated a remarkable ability to use their knowledge of customary water institutions to maintain community health and well-being.

Resources

Ostrom, Elinor. 1990. *Governing the commons: The evolution of institutions for collective action*. New York: Cambridge University Press.

Trawick, Paul. 2003. *The struggle for water in Peru: Comedy and tragedy in the Andean commons*. Stanford: Stanford University Press.

Wutich, Amber. 2009. Water scarcity and the sustainability of a common pool resource institution in the urban Andes. *Human Ecology* 37(2): 179–192.

Parts of Intag are located in two of the world's biodiversity hotspots – the Chocó and the Tropical Andes. Intag also borders on the Cotacachi-Capayas Ecological Reserve and contains the largest remaining remnants of Ecuador's western forests, more than 90% of which have disappeared in the last half-century. These cloud forests function as a sponge that supports the hydrological cycle, crucial for absorbing moisture and keeping the topsoil on the steep Andean mountainsides. Characteristic of these subtropical forests are myriad epiphytes, or air plants, including hundreds of species of orchids and bromeliads, which capture humidity, minerals and other organisms from the atmosphere. In addition, trees and vegetation that absorb moisture from clouds do not need to do so from the ground. This means that the stream flow from cloud forest areas tends to be larger than for land receiving the same amount of rainfall in other areas; and the flow is more dependable during dry periods. At the same time, the roots of trees and other vegetation are generally not deep, and their removal immediately disturbs the fragile hydrological cycle because of the land's diminished capacity to absorb water. These forests are crucial for mitigating climate change in global ecosystems. They are also home to multiple endangered species of mammals and birds, including spectacled bears (Tremarctus ornatus), jaguars (Panthera onca), pumas (Puma concolor), ocelots (Felis pardalis), dwarf deer (Pudu mephistoles), two tapirs (Tapirus pinchaque and Tapirus bardii), spider monkeys, (Ateles spp.), howler monkeys (Alouatta pallliata aecuatorialis), plate billed mountain toucans (Andigena laminirostris), the umbrella bird (Cephalopterus penduliger), and a host of other rare species.

Table 5.2.2 Environmental goods and service provided by Intag's forest patrimony (from Hydro Intag Catalogue)

Good or service	Description	Value (*) $/ha/year	Total value $/year
Commercial		**673.94**	**29,653.399.78**
Regulation of greenhouse gases (Carbon sequestration)	Regulation of the chemical composition of the atmosphere to reduce climate change	190.00	8,360,000.00
Production of raw materials and food	Production of products, including lumber, fish and fruit	483.94	21,293,399.78
Non commercial		**2,316.50**	**101,926,043.35**
Climate regulation	Regulation of temperature, precipitation and other local and global climate processes	311.01	13,684,230.99
Soil formation and erosion control	Processes of soil formation and retention, including flood and storm control	362.61	15,954,708.77
Water services	Provision and retention of water, greater regulation of hydrological currents	19.52	859,099.70
Nutrient cycling and waste treatment	Storage and processing of nutrients, including the elimination of undesirable nutrients	1,407.19	61,916,542.89
Genetic resources	Source of unique biological materials and products	57.18	2,515,934.84
Recreation and cultural activities	Opportunities for recreational activities; aesthetic, educational and spiritual value	158.99	6,995,526.15
Biological control and habitat formation	Regulation of populations, provision of habitat, pollinating services	Value unknown	–
Annual total		**2,990.44**	**131,579,443.13**

Scientists calculate that Intag's natural capital is valued at more than 131 million USD per year (HidroIntag Catalogue 2009:7). Non-commercial goods and services provided by Intag's forest patrimony have at over 100 million USD nearly three and a half times the value of commercial uses, without taking into consideration the value of biological diversity and ecosystem maintenance. Inteños value the forests as their core source of water, which they see as a collective and human right. The community watershed reserves project is successful because it builds on their sociocultural perspectives and deep-seated values. Water management, as an organizing and operational concept for sustaining traditional lifeways and biodiversity, also contributes to emerging ecological identities and the cultural viability of diverse groups in Intag. Intag is an example of local communities making transnational alliances to exert their rights and determine what kind of activities go on in their territories.

5.2.6 Good Governance and Stakeholder Participation = Clean Water and Healthy Forests

Intag demonstrates how a society responded to a socioecological threat and developed institutional arrangements for managing natural resources within an ecosystem approach that focuses on water. This case was successful because Mayor Tituaña promoted the restructuring of local and county government so that Inteño/as participated as citizens and stakeholders, and they prioritised access to clean water as a common value. This process follows Ostrom's principles of common resource management because they formulated plans adapted to their conditions and culture to protect water for future generations (1990). Alongside transnational support and intercultural interactions, Inteño/as created an alternative development model to not only conserve their forests, biodiversity and water, but to revalue their culture and assert the power for their right to self-determination. The historical and cultural specificity shows how institutions and organizations can be created for the common good through participatory practices. Intag residents organised themselves in cooperation with local and transnational partners, and reshaped local sociopolitical dynamics whereby they took charge of the water and forests. They linked local ecology with their rights to water and exercised their duties to the community. These set of circumstances offer an example of how people build ecological identities and negotiate sustainable models for the future through political mobilization. As illustrated, residents value ecosystem integrity because they understand that their quality of life and economic well-being in the future depend on local resources. At the same time, this case study shows ways that biodiversity loss caused by deforestation and forest degradation can be kept to a minimum with an incentive system that focuses on watershed management and conservation that prioritises human needs for water. This integrated approach not only provides water security for Inteño/as but also protects the biodiversity, which keeps the vital hydrological cycle in place and demonstrates resiliency to climate change.

DECOIN, *Periódico INTAG*, community watershed councils and other local and global institutions used stakeholders' understandings of vital ecosystem services (including the fact that cloud forests house plants and animals, sustain freshwater cycles, reduce erosion and coastal flooding, and provide shelter, food, medicine and other services) to catalyze increased ecological awareness and sustainable development. Action on watershed restoration has led to innovations in farming and new enterprises that are creating jobs for young people. The EO and watershed councils have made protection of groundwater springs a priority, while attracting new investment in distribution systems for safe water supply to households.

As Marcía Ramirez, secretary of the Intag Development Committee and resident of Chalguayaco, explained to a reporter for *The Tyee* of Canada, 'It isn't fair that a foreign company can come here and contract people who attack us for defending our rights, for wanting to live in a healthy environment, for defending our land and our water' (Moore 2009). Or as Almeida said to me, 'After seven years it is clear that the flow of water has recovered. In fact', he continued, 'the project has been so successful that there are currently ten more communities requesting support to acquire and maintain hydrological reserves. We hope to find the funding'.

Resources

Anderson, Benedict. 1991. *Imagined communities*, rev edn. New York: Verso.

Espinoza Soriano, Waldemar. 1988. *Etnohistoria Ecuatoriana*. Quito: Abya-Yala.

Fieweger, Mary Ellen, and DECOIN. 1998. *Es un monstruo grande y pisa fuerte: La minería en el Ecuador y el mundo*. Quito: Serie Pluriminor.

Foro de Agua y Biodiversidad. 2008. *Periódico INTAG* 56 (Nov/Dec): 2.

Gentry, Alwin. 1993. Riqueza de especies y composición florística. In *Colombia Pacífico I*, ed. Pablo Levya. Fondo FEN: Bogotá.

HydroIntag Catalogue. 2009. (trans: Mary Ellen Fieweger).

Moore, Jennifer. 2009. Canadian mining firm financed violence in Ecuador: Lawsuit, 5 Mar. http://thetyee.ca/News/2009/03/03/CanMining/. Accessed 3 Mar 2009.

Ostrom, Elinor. 1990. *Governing the commons*. Cambridge: Cambridge University Press.

Ribot, Jesse. 2008. *Building local democracy through natural resource intervention*. Washington, DC: World Resources Institute.

Ruiz-Córdova, Sergio S., Bryan L. Duncan, William Deutsch, and Nicolás Gómes. 2006. Community-based water monitoring in Cotacachi. In *Development with identity: Community, culture and sustainability in the Andes*, ed. R.E. Rhoades. Cambridge: CABI Publishing.

Chapter 5.3
Cultural Flows: Asserting Indigenous Rights and Interests in the Waters of the Murray-Darling River System, Australia

Monica Morgan

> *We don't set the conditions; Mother Earth sets them only that we may survive.*
>
> —Elizabeth Morgan, Elder, Yorta Yorta Nation

The Murray-Darling rivers together form Australia's largest river system, covering over one million km^2 (400,000 mile^2). My people know the Murray as Dhungulla. These rivers have sustained the lives and cultures of Indigenous peoples for millennia. Today, through irrigated agriculture, it has become an important food production area for the nation. The paths that the Murray and Darling rivers take are now known as the Murray-Darling Basin.

Yet the Murray is in crisis – it is dying. The Indigenous elders of this region have been saying for decades that the damming of our rivers and taking the water for commercial irrigation are having a devastating effect. Now, in the 21st century, drought, over-allocation of river water, increasing salinity problems, and the looming threat of climate change have meant that many more people have recognised that the river is at the brink of no return and that we must dramatically change our ways. This reality has finally brought both the Indigenous and non-Indigenous peoples of the Murray Lower Darling Rivers to the same discussion and negotiating table.

In what follows, I tell the story of the Murray Lower Darling Rivers Indigenous Nations (MLDRIN), an organization formed as an alliance of 10 traditional owner groups from along the Murray and its tributaries. I am a Yorta Yorta nation woman and an active participant in MLDRIN, so it is also my story. As Indigenous peoples, it is our belief that the future lies in truly understanding the past. However, Indigenous peoples' spiritual, cultural, social and economic relationships with water should not be understood simply as some ancient practice without any application in today's world.

M. Morgan (✉)
Amnesty International, Australia
e-mail: monica-morgan@hotmail.com

B.R. Johnston et al. (eds.), *Water, Cultural Diversity, and Global Environmental Change: Emerging Trends, Sustainable Futures?*, DOI 10.1007/978-94-007-1774-9_31,

Fig. 5.3.1 Wakool River, Murry-Darling River System; Winding across 650 km of southeastern Australia's dry landscape, the Murray-Darling river system struggles to sustain much of the country's agriculture. Due to the effects of drought and man, the Wakool River near Swan Hill no longer tops its banks (Photo credit: J. Carl Ganter/Circle of Blue; Source: Circle of Blue News, 9 Mar 2009. http://www.circleofblue.org/waternews/2009/world/australia-drought-water-scarcity-desertification/

We have knowledge and insights that are essential to a sustainable future, not least the understanding to deconstruct many of those failed notions that have separated people from nature in this modern world, such as the view that water is simply an economic good. It is important for future solutions to incorporate not only our knowledge but also our presence in the landscape. We want and we need to be part of the solution. But in order for our voice to be heard, there must be a real commitment to respect the rights and interests of Indigenous peoples and to ensure that such commitment is maintained in the development and application of sustainable solutions to the present problems. Although there is a long way to go, through the Murray-Darling Basin Initiative, a process of engagement between the Australian governments with Indigenous peoples has made it possible to secure a stronger voice for Indigenous Australians in water policy and management in southeastern Australia.

5.3.1 Indigenous Peoples of the Murray

We have always been, and will always be, the First People of this land. We belong to it, and the water that flows through our country is as the blood that flows through our veins. Our bodies are formed from the country and remain tied to its rhythms – the

Map 5.3.1 Australia

currents, the seasons, and the sounds. All living and non-living entities within our traditional lands and waters are related to us – when we die, our spirits return to the sacred lands and waters until they are reborn to live again. We believe all life is reciprocal and we must be responsive, attentive, and aware of what we should give to the land and what we receive from it. This obligation to understand and to care for the places that sustain us means that we are constantly trying to ensure that they are healthy and looked after. This duty has become harder as the pressures on our peoples from colonization have continued and as the development pressure on the river has grown. Nevertheless, we still maintain a strong connection to our traditional lands. The lands and waters provide us with food, gathering places, and camping grounds. They also contain sacred places which sustain our spirit and give us balance.

Water and land are inseparable, and both are vital to the cultural, spiritual, and economic sustainability of our people. Our individual and collective sense of identity comes from the creator being Baiami, my Yorta Yorta creation story of Dhungulla. Each Indigenous nation along the Murray River will have its creation story, my people's creation is linked to coming from the water of the Dhungulla, we are one with all of the animals, fish, birds, trees, to the sacred places. Our connection is so strong that we are like one family, we share not only creation but continuation of life which is carried on through cultural and traditional laws and customs for thousands of generations. Our country is sacred to us as is all that lives within it.

There are at least 70,000 Indigenous people living within the Murray Darling Basin, which is 14% of the total Indigenous Australian population of 517,000 Indigenous people (Taylor 2004). Across Australia, that Indigenous population holds approximately 16% of the land under various forms of title, but the situation in the Murray Darling Basin is far different than this average. There are 40 autonomous Indigenous nations in the basin, but we hold less than 0.2% of the land, a much higher level of dispossession than in many other regions. Despite various kinds of land reform at both state and federal levels, there are still significant legislative gaps. For example, the populous state of Victoria still has no Indigenous land rights legislation. Therefore, the Indigenous people of this river system remain marginalised by the processes of European settlement, which continue to block our access to our country and prevent us from gaining benefits from the economic activity undertaken on those lands.

The Indigenous Nations of the Murray Darling Basin have expressed their desire to be respected as independent peoples with a right to self-determination. As

Indigenous peoples and as the first sovereigns of our traditional lands and waters, we occupy a special status with respect to other occupants of Australia. We are different from other minorities present in contemporary multicultural society by virtue of our distinct histories as sovereign political entities attached to the land and waters. Our rights are based on prior and continuing occupation, and our authority and autonomy should be recognised as inherent and self-determining.

5.3.2 Indigenous Peoples and the Murray Darling Basin Initiative

In 2002 the Murray Darling Basin Ministerial Council asked Indigenous peoples to discuss the best way to achieve its vision of 'a healthy River Murray system, sustaining communities and preserving unique values'. In reality the council was looking to return water to an overstressed and over-allocated river system. The major initiatives associated with this return of water to the river were to buy back existing water licenses and undertake structural and engineering works. The stated aim of this activity was to return 500 gigalitres of 'environmental water' back into the system, but this amount was well short of the estimated 4,000 gigalitres actually required to revive the Murray (Jones et al. 2002).

Indigenous rights and interests were not part of the discussions surrounding the initiative in the way that they should have been. This failure is largely because the prohibitive requirements of Native Title processes and other legislation recognizing Indigenous ownership rights over the land and waters had prevented the great majority of Murray Darling Basin Indigenous peoples from gaining formal recognition. In particular, my people, the Yorta Yorta, undertook a long struggle for Native Title through the Australian courts, a struggle that did not achieve its intended aim of legal recognition, even though it did provide Indigenous communities with skills and knowledge for later struggles. However, despite their reservations about their positioning in the Murray Water Initiative process, the basin nations realised that our river was in such a dire state that it could not wait any longer. In order to save it, we would need to work within the process and engage with the Murray Darling Basin Commission (now called the Murray-Darling Basin Authority).

The Indigenous peoples along the Murray Darling Basin produced an Indigenous response to the Living Murray Initiative titled Indigenous Rights to Water in the Murray Darling Basin (Morgan et al. 2004). The response emphasised the importance of the process of free prior and informed consent through an agreement-based approach to be negotiated directly between state and Indigenous Nations on their spiritual, cultural, social, and economic values and interests within the Living Murray. They identified that the initiative could provide opportunities to implement actions that were underpinned by the holistic Indigenous perspective of the interrelationship between people and country. The community began to discuss goals and aspirations related to:

- Enhanced ecological outcomes
- Protection of the health and well-being of peoples and country
- Social justice outcomes
- Cultural and spiritual heritage protection
- Strengthening of our cultural knowledge, including the development of mechanisms for the protection of this knowledge

In addition to the initial response, the MLDRIN provided a collective Indigenous voice in discussions about the future of the river system and one path to achieving the holistic aspirations that Murray Darling Basin peoples had for their country and their communities. MLDRIN established some core operating principles:

- Only traditional owners are best placed to speak for their country.
- The sovereignty and inherent rights of traditional owners are never ceded.
- All Indigenous Nations are equal within the MLDRIN confederacy.
- MLDRIN respects the diversity of nations in relation to tradition, sites, stories, cultural practices and governance.
- MLDRIN will not interfere with the internal governance within each Indigenous Nation.
- Resources of the confederation will be shared equally.
- Self-determination of the nations and of MLDRIN is the only sustainable way to operate.
- Fully informed consent is necessary for any actions to be taken.

The values of MLDRIN incorporated the crucial importance of traditional lore and custom; the respect for elders; the interconnectedness of land, water, and people; and the need for upstream and downstream peoples to talk with each other to promote caring for the country, as well as the crucial importance of passing on knowledge to younger generations.

After a series of discussions and negotiations, MLDRIN and the Murray Darling Basin Commission signed a Memorandum of Understanding in 2007. This agreement laid important foundations for future planning and strategic aspirations. It also suggested some critical aspects of future policy development with respect to Indigenous people:

- 'Informed consent' requires more than mere consultation. Instead, it requires that Indigenous people have meaningful roles in the process, as well as power in determining the decisions and outcomes.
- Indigenous peoples want substantive involvement in the implementation of decisions, including a direct role in environmental management.
- Cultural, environmental, and social values should be given equal status with economic values when policy and management decisions are made.
- Cultural Heritage Management Plans (CHMPs) should be developed between government and Indigenous Nations with custodial responsibilities for the river system. The CHMPs should have the force of law and reflect the interrelationship between environmental values and spiritual and cultural values.

5.3.3 Reviving the River: Environmental and Cultural Flows

Indigenous people believe that the primary policy objective must be to restore natural flows and cycles to the river system. Therefore, the Ministerial Council's reference point of 500 gigalitres for increased environmental flows is unsatisfactory. Even the top reference point of 4,000 gigalitres has only a low to moderate probability of restoring the health of the river system. A genuine attempt to restore environmental flows must aim for a real revival of the ecological health of the river. However, this restoration of ecological flows does not fulfill all Indigenous aspirations for the river. Prior to colonization, Indigenous Nations had a thriving cultural economy that underpinned traditional lifestyles across their country. This economy now has been diminished by the poor health of the river system, which has decimated traditional sources of food and medicines.

Box 5.3a Environmental flows

—Angela H. Arthington

Healthy aquatic ecosystems – streams and rivers, lakes, floodplains, wetlands and estuaries – provide clean water, food, fibres, pharmaceuticals, energy and many other benefits that support economies and livelihoods around the world. Fresh water that flows into the ocean nourishes estuaries, which provide abundant food supplies, buffer infrastructure against storms and tidal surges, and dilute and evacuate pollutants. Scientists can confidently claim that 'water flowing to estuaries and the sea is not wasted' (Loneragan and Bunn 1999). Healthy freshwater ecosystems are also essential to human health and well-being, be it cultural, religious or spiritual.

Many of the world's freshwater ecosystems are seriously impaired, and aquatic species are declining more rapidly than terrestrial and marine species (Dudgeon et al. 2006). As freshwater ecosystems degrade, communities lose important social, cultural, and economic benefits; floodplains and estuaries are less productive; invasive plants and animals flourish; and the natural resilience and adaptability of rivers, lakes, wetlands, and estuaries weakens. The severe cumulative impact on ecosystems and society is global in scope, and climate change intensifies the urgency of restoring and protecting resilient, diverse freshwater and estuarine ecosystems (Palmer et al. 2008).

Every freshwater ecosystem and its component species have adapted to, and depend upon, naturally variable flows of fresh water derived from rainfall, run-off, or groundwater sources. Water flow is the driving or master variable, interacting with thermal, sediment, and energy regimes to shape the physical habitat template, provide triggers for life cycle processes such as migration and spawning, and generate vital floodplain resources (Bunn and Arthington 2002).

(continued)

Box 5.3a (continued)

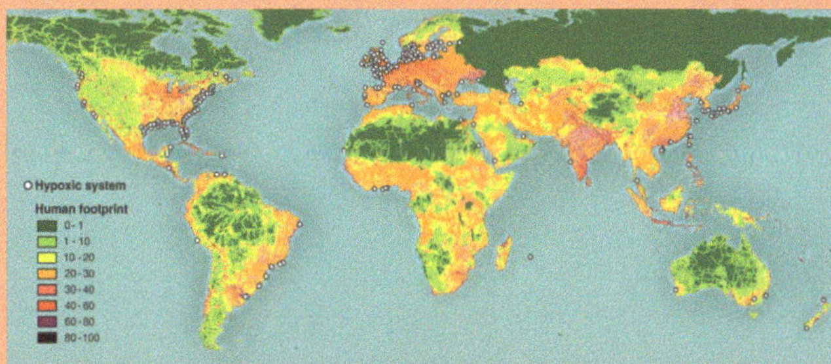

Box Map 5.3a.1 Spreading dead zones throughout the worlds marine ecosystems. Hypoxia means low oxygen and is primarily a problem for estuaries and coastal waters. Hypoxia occurs naturally in many aquatic environments throughout the world, such as deep basins in the ocean, but they are occurring in shallow coastal and estuarine waters more frequently as anthropogenic sources of nutrients increase and are deposited into the marine ecosystem via freshwater flows. The direct effects of hypoxia include fish kills, which deplete valuable fisheries and disrupt ecosystems (Source: Diaz and Rosenberg (2008). Map accessed from the US EPA site, 1 Jan 2011, http://water.epa.gov/type/watersheds/named/msbasin/images/diaz_fig1_720x299.jpg)

Ecological impacts of flow regime change by dams and water abstraction have brought the crucial roles of flow into prominence and generated the field of environmental flow assessment (EFA) and management. The first ad hoc EFA methods appeared in the United States in the late 1940s, followed by more formally documented techniques in the 1970s; these focused on minimum flows to maintain the habitat requirements of economically important fish species, especially salmon. With increasing concern about the impact of dams and flow alterations on river ecosystems, the field of environmental flows has since then prospered to produce more than 200 assessment methods (Tharme 2003).

Scientists now recognise that minimum flows are inadequate to sustain healthy river systems. Every river system has a characteristic flow regime characterised by flow quantities, seasonal patterns, and extreme events such as floods and dry spells. The role of EFA is to relate these hydrologic characteristics to physical and ecological responses, and thereby inform protection or restoration of the socially valued benefits of biodiverse and resilient freshwater ecosystems through participatory decision-making informed by sound science (Poff et al. 2010). Groundwater, wetland, and floodplain management are integral to environmental flow management and river restoration in most landscapes. Increasingly, the freshwater requirements of estuaries and coastal lagoons are also being evaluated and implemented. Hence, the definition adopted in the Brisbane Declaration (2007) – 'environmental flows describe the quantity, timing, and quality of water flows needed to sustain freshwater

(continued)

Box 5.3a (continued)

and estuarine ecosystems and the human livelihoods and well-being that depend on these ecosystems'.

The need for action to restore rivers with high ecological, economic, and social values is widely recognised. Several prominent international programmes have embraced the concept of environmental flows, and they are being assessed and implemented in more than 50 countries (Postel and Richter 2003; Tharme 2003).

The EFA for the Lesotho Highlands Water Project (LHWP, www.lhwp.org.ls), a multi-billion dollar water transfer and hydropower project implemented by the governments of Lesotho and South Africa and one of the world's largest water-resource developments, became the first EFA that describes and quantifies not only the biophysical consequences of various development scenarios, but also the social and resource-economic consequences. Losses of river resources (e.g., food fishes, Arthington et al. 2003) and health benefits were converted to compensation estimates for riparian people (King and Brown 2010). This study is widely used by the World Bank as a training example for flow assessments, and in 2007 an independent audit concluded that the LHWP's approach to flow assessments for people and nature was at the forefront of global practice (Institute of Natural Resources 2007)[1].

As an experimental river restoration case study, the environmental flows programme on the Bill Williams River, an arid-region tributary to the lower Colorado River in Arizona, western USA, is also state of the art. In 2004, the Bill Williams River and Alamo Dam became one of eight U.S. rivers and 36 dams that are part of the Sustainable Rivers Project (SRP), a collaboration between The Nature Conservancy and the U.S. Army Corps of Engineers. Beneficial ecological outcomes from meticulous monitoring of experimental flow releases are contributing to adaptive flow management on the Bill Williams River, and to development of environmental flow principles for other arid-zone rivers (Shafroth et al. 2010).

Recent developments in the science, modeling, management, and monitoring of environmental flows can be found in a special issue of *Freshwater Biology* (Arthington et al. 2010). EFAs can make a vital contribution to protection and recovery of resilient freshwater ecosystems, ecological services, and cultural values in every hydroclimatic and socioeconomic setting across the globe.

Endnote

[1] See also chap 4.2, this volume

(continued)

Box 5.3a (continued)

Resources

Arthington, A.H., J.L. Rall, M.J. Kennard, and B.J. Pusey. 2003. Environmental flow requirements of fish in Lesotho Rivers using the DRIFT methodology. *River Research and Applications* 19(5–6 Sept–Dec): 641–666.

Arthington A.H., R.J. Naiman, M.E. McClain, and C. Nilsson. 2010. Preserving the biodiversity and ecological services of rivers: New challenges and research opportunities. *Freshwater Biology* 55(1 Jan): 1–16.

Brisbane Declaration. 2007. The Brisbane declaration: Environmental flows are essential for freshwater ecosystem health and human well-being. Declaration of the 10th international river symposium and international environmental flows conference, Brisbane, 3–6 Sept.

Bunn, S.E., and A.H. Arthington. 2002. Basic principles and ecological consequences of altered flow regimes for aquatic biodiversity. *Environmental Management* 30(4 Oct): 492–507.

Diaz, Robert J., and Rutger Rosenber. 2008. Spreading dead zones and consequences for marine ecosystems. *Science* 321: 926.

Dudgeon, D., A.H. Arthington, M.O. Gessner, Z.I. Kawabata, D.J. Knowler, C. Lévêque, R.J. Naiman, A.H. Prieur-Richard, D. Soto, M.L.J. Stiassny, and C.A. Sullivan. 2006. Freshwater biodiversity: Importance, threats, status and conservation challenges. *Biological Reviews* 81(2 May): 163–182.

Institute of Natural Resources (INR). 2007. Instream flow requirement audit for Phase 1 dams of the Lesotho Highlands water project. Contract 1250. Report to Lesotho Highlands Development Authority Strategic and Corporate Services Division, Monitoring and Evaluation Branch.

King, J.M., and C. Brown. 2010. Integrated basin flow assessments: Concepts and method development in Africa and Southeast Asia. *Freshwater Biology* 55(1 Jan): 127–146.

Loneragan, N.R., and S.E. Bunn. 1999. River flows and estuarine ecosystems: Implications for coastal fisheries from a review and a case study of the Logan River, southeast Queensland. *Australian Journal of Ecology* 24(4 Aug): 431–440.

Palmer, M.A., C.A. Reidy Liermann, C. Nilsson, M. Flörke, J. Alcamo, P.S. Lake, and N. Bond. 2008. Climate change and the world's river basins: Anticipating management options. *Frontiers in Ecology and the Environment* 6(2 Mar): 81–89.

Poff, N.L., B.D. Richter, A.H. Arthington, S.E. Bunn, R.J. Naiman, E. Kendy, M. Acreman, C. Apse, B.P. Bledsoe, M.C. Freeman, J. Henriksen, R.B. Jacobson, J.G. Kennen, D.M. Merritt, J.H.O'Keeffe, J.D. Olden, K. Rogers, R.E. Tharme, and A. Warne. 2010. The ecological limits of hydrologic alteration (ELOHA): A new framework for developing regional environmental flow standards. *Freshwater Biology* 55(1 Jan): 147–170.

Postel, S., and B. Richter. 2003. *Rivers for life: Managing water for people and nature*. Washington, DC: Island Press.

Shafroth, P.B., D.C. Andersen, V.B. Beauchamp, A. Hautzinger, J.T. Hickey, D. Lytle, L. McMullen, A. Warner, and A. Wilcox. 2010. Models and tools to assist environmental flow assessment and implementation below reservoirs: Lessons from an arid-region river. *Freshwater Biology* 55(1 Jan): 68–85.

Tharme, R.E. 2003. A global perspective on environmental flow assessment: Emerging trends in the development and application of environmental flow methodologies for rivers. *Research and Applications* 19(5–6 Sept–Dec): 397–44.

Improvement in the health of the river will restore some of the balance, but Indigenous peoples are now enmeshed in the economic system of the colonisers. To address the challenges that this creates, MLDRIN nations developed the concept of cultural flows (see also Ross 2006/2007 and Weir 2009). These are 'water entitlements that are legally and beneficially owned by the Indigenous Nations and that are of a sufficient and adequate quantity and quality to improve the spiritual, cultural, environmental, social, and economic conditions of those Indigenous Nations' (MLDRIN 2008). The difference between environmental and cultural water allocations is that in the case of the latter, the Indigenous peoples themselves decide where and when water should be delivered, based on traditional knowledge and their own aspirations for their peoples. Recognition of this concept would empower Indigenous peoples to fulfill their responsibilities to care for their traditional lands as well as develop appropriate measures to address the social and economic disadvantages caused by colonization.

Questions of volume need to be explored through scoping work with the Nations, using informed consent and good faith processes. Each Nation would decide how its allocation should be used to improve the health of its members and the river system as a whole. The allocation may be used to further increase river flows or to generate a more independent economic base for the people, or some appropriate combination of uses. The decision would be taken in the context of the health of the river system

Fig. 5.3.2 The Murray River mouth, where Australia's major river meets the Southern Ocean, is kept open through constant dredging. The river's flow today is some 27% of its pre-European level (Photo credit: Monica Morgan)

and each Nation's custodial responsibilities. The volume of water needed to bring the Murray Darling Basin rivers back to a healthy ecological state is well known, and the cultural flows allocation would be independent of this volume, but it may be used by each Nation to further support the rivers, as a healthy environment assists in reviving the cultural economy, just as the cultural flows do.

These cultural water allocations are based on prior and continuing occupation of traditional lands and waters and should be recognised as one element of the inherent and holistic authority and autonomy of Indigenous peoples. Cultural water allocations should be formally recognised as rights which are underpinned by and mutually supported by related rights, such as

- Customary rights to self-determination and the preservation of distinct cultural identities amongst the Murray Darling Basin Indigenous Nations
- The human right to maintain culture and a cultural economy
- Formal Native Title determinations and other legal agreements

The concept of cultural flows has been developed in response to the crisis of the Murray and Darling river systems. However, the value of this concept can be relevant for other parts of Australia. A recent policy statement from the Northern Australian Indigenous Land and Sea Management Alliance (NAILSMA) argued for the necessity for an allocated cultural flow for northern Australian rivers:

> To ensure cultural rights and the equitable use of the consumptive, commercial allocation of water, water legislation and policy must include:
>
> - an allocated Cultural Flow, (in accordance with Articles 8, 25-28 of the United Nations Declaration on the Rights of Indigenous Peoples).
> - Cultural Flows are water entitlements that are legally and beneficially owned by Indigenous peoples and are of sufficient and adequate quantity and quality to maintain the spiritual, cultural, environmental, social and healthy livelihoods of Indigenous peoples of northern Australia (refer to the MLDRIN 2008 Echuca Statement).

In the NAILSMA policy, this flow is to be independent of any allocation to Indigenous peoples and coming from the general commercial allocation for consumptive purposes. It demonstrates the potential importance of a specified cultural flow for all Indigenous Nations across Australia.

5.3.4 Water Flows and Compensation

Wise use of the water is important for everyone and for the life of the rivers. We believe that government investment in incentives and assistance for industry and other commercial water users is necessary to change management systems. However, MLDRIN rejects the concept of compensation for any loss of water allocations by non-Indigenous corporations and individuals as fundamentally inequitable. The whole system of existing water allocations is based on the dispossession of Indigenous people, and currently the Indigenous Nations and their members are prevented from having any specific rights to water. A more just response would be the recognition of the traditional values that underpin Indigenous livelihoods and of the spiritual, cultural, economic, and social harm caused when a degraded river system undermines those values. Such degradation, and the loss of traditional values that it entails, should provide the basis for an appropriate compensation scheme for dispossessed Indigenous peoples to aid in redressing the imbalance caused by colonisation.

5.3.5 Co-management and Cooperation: The Living Murray Indigenous Partnerships Programme

The Murray-Darling Basin Authority's Indigenous Partnerships Programme (IPP) is designed to facilitate cooperation between river managers and the Murray Darling Basin Indigenous Nations to improve the sustainability of the river and the health of those communities (Morgan et al. 2006). It involves the training and mentoring of Indigenous facilitators in the areas of communication, technical analysis, asset recording, and GIS mapping so that they can provide key liaison and engagement support to Indigenous communities. As part of their duties, these locally employed facilitators undertake cultural mapping exercises along the river, using technical and other support provided by the management agencies. This important aspect of the programme has sought to better understand Indigenous relationships with significant sites, using a participatory use and occupancy mapping method. Such techniques have assisted First Nations peoples internationally and have proven effective in the context of the Murray Darling Basin (see Ward 2010). The cultural mapping processes and outputs have led to improved management outcomes with respect to Indigenous aspirations, as well as skill development within Indigenous communities. The projects have also benefited our communities in ways beyond the immediate programme aspirations relating to sacred sites and other important issues. Such additional benefits include the following:

- Documenting elders' oral history
- Providing evidence in court cases involving Aboriginal rights and title
- Supporting compensation claims
- Negotiating co-management agreements
- Negotiating protective measures and benefits from industrial development
- Developing educational curriculum

Informed consent underpins the IPP to ensure Indigenous people understand the consequences and outcomes that may result from our contributions and decisions regarding cultural knowledge, values, and perspectives. We want traditional knowledge recognised for the contribution it can make to looking after the rivers, but we are equally concerned to clarify and protect our rights to our own intellectual property. We believe that public education campaigns should be undertaken so that all

Australians have a better understanding of sustainable river use and the central importance of land and waters to Indigenous culture and spirituality, as well as to Indigenous economic livelihoods.

5.3.6 The Principles for Murray Darling Basin Management and for Sustainable Futures

The introduction of the Federal Water Act 2007 and the Water Amendment Bill 2008 has brought some fairly limited opportunities for traditional owners' involvement in water management in Australia. However, despite those limitations, the laws do give Indigenous peoples some traction for promoting their rights and interests and for arguing for a new model of dialogue with Indigenous people rather than the old model of consultation. On the basis of our existing experiences, MLDRIN and its members believe that the future management of the Murray Darling Basin should follow these principles:

- A free-flowing river that moves according to natural cycles
- Indigenous Nations recognised as sovereign entities in their own country
- Access rights for Indigenous people to continue cultural practices such as traditional fishing and hunting
- Provision of cultural flows as a means of sustaining Indigenous cultures and livelihoods into the future

Indigenous peoples recognise the existence of cultural diversity, not just within our own peoples but within the wider community. We wish to create a new dialogue between the First Nations and non-Indigenous people so that we can acknowledge the past, understand each other as human beings, and plan responsibly to care for our future generations. This dialogue relies on our being honest about the devastation that past theories and practices have had on the river system and the peoples who have lived there since time immemorial. Changing these false concepts and practices is thus essential not only for the continuance of our river but for life itself. Even though the impact of water management systems on our lives as the First Peoples of Dhungulla has been devastating, we do not believe it will lead to our demise. Indeed, our identity, our cultural lore, and interests will prevail, for we will struggle, adapt, and survive as we have over many thousands of generations. The newcomers to our lands must listen to us if they wish to do the same.

Resources

Jones, G., T.J. Hillman, R. Kingsford, T. McMahon, and K. Walker. 2002. *Independent report of the expert reference panel on environmental flows and water quality requirements for the River Murray system*. Report to the Murray-Darling Basin Ministerial Council. Canberra: Cooperative Research Centre for Freshwater Ecology.

Morgan, M., L. Strelein, and J. Weir. 2004. *Indigenous rights to water in the Murray-Darling basin – in support of the indigenous final report to the Living Murray Initiative*. Research discussion paper no. 14. Canberra: AIATSIS.

Morgan, M., L. Strelein, and J. Weir. 2006. Authority, knowledge and values: Indigenous Nations engagement in the management of natural resources in the Murray-Darling Basin. In *Settling with indigenous peoples*, ed. M. Langton, O. Mazel, L. Palmer, K. Shain, and M. Tehan. Sydney: The Federation Press.

Murray Lower Darling Rivers Indigenous Nations (MLDRIN). 2008. Echuca Declaration. Available online at http://www.nailsma.org.au/nailsma/publications/downloads/MLDRIN-NBAN-ECHUCA-DECLARATION-2009.pdf.

North Australian Indigenous Land and Sea Management Alliance. 2010. http://www.nailsma.org.au/nailsma/forum/downloads/Water-Policy-Statement-Plain-Text-Version.pdf. Accessed 3 Aug 2010.

Ross, S. 2006/2007. Caring for country: Indigenous Nations and water management. *Reform* 89: 18–21.

Taylor, J., and N. Biddle. 2004. *Indigenous people in the Murray-Darling basin: A statistical profile*. Canberra: CAEPR.

Ward, N. 2010. Good methodology travels: An Australian case study. In *Living proof: The essential data-collection guide for Indigenous use-and-occupancy map surveys*, ed. T.N. Tobias. Vancouver: The Union of BC Indian Chiefs and Ecotrust Canada.

Weir, J. 2009. *Murray River country: An ecological dialogue with traditional owners*. Canberra: Aboriginal Studies Press.

Chapter 5.4
Environmental Flow Assessments: A Participatory Process Enabling Maori Cultural Values to Inform Flow Regime Setting

Gail Tipa and Kyle Nelson

5.4.1 Introduction

Maori, the indigenous people of New Zealand, have longstanding ties to lakes, wetlands, streams, lagoons and coastal waters (Anderson 1998). In many parts of New Zealand, where water is a scarce resource (Lincoln Environmental 2002), waters have been dammed, stored and diverted to ensure the provision of water where and when it is needed. These developments, as well as directly impacting the lands and waters previously utilised by Maori, have prevented their exercise of customary and treaty rights (Waitangi Tribunal 1992, 1995, 1998). Nevertheless, securing recognition and provision for the rights of Maori may be problematic, as Poff et al. (2002) contend that achieving a more effective and sustainable balance between human and ecological demands for fresh water is one of the greatest challenges confronting societies today. Widening the range of social values to include the beliefs, practices and rights of indigenous communities could further complicate this balancing exercise.

Internationally, until about the mid-20th century, water quality was at the forefront of many issues arising from human manipulation of rivers (Tharme 1996). A range of new concerns then arose that were not directly related to pollution but were concerned with the reduced flow in rivers. Subsequently, an understanding of river changes resulting from flow manipulations has increased (and been accompanied by worldwide awakening and concern with respect to the issues). Of the range of methods developed over recent decades to address river flow and quantity related issues, few are cognisant of and responsive to the cultural values expressed by indigenous communities (Craig 2005). A challenge that Maori (and possibly other indigenous communities) confront is conveying to decision-makers how water management

G. Tipa (✉) • K. Nelson
Ngai Tahu, Christchurch, New Zealand
e-mail: gtipa@xtra.co.nz

B.R. Johnston et al. (eds.), *Water, Cultural Diversity, and Global Environmental Change: Emerging Trends, Sustainable Futures?*, DOI 10.1007/978-94-007-1774-9_32,

decisions impact their cultural associations with rivers. As Burmil et al. (1999:106) explains, 'The emphasis on only science-derived technical standards has tended to place perceptual, cultural and spiritual values of water at a disadvantage in affecting water policy and management'.

This chapter, by providing an example of a participatory process in Kakaunui Catchment in New Zealand that assesses the river flows necessary to protect cultural interests and calculates cultural flow preferences via application of Cultural Opportunity Mapping and Assessment (COMA), attempts to respond to these shortcomings (cf. Tipa and Nelson 2008). The result is the identification, by Maori, of their preferred flow thresholds and specification of other management actions deemed necessary to recognise and provide for cultural interests with respect to fresh water.

5.4.2 Ngai Tahu and the Kakaunui Catchment

Ngai Tahu is the principal *iwi* (tribe) of the southern region of New Zealand.[1] A thousand years ago Ngai Tahu settled on the East Coast of the South Island of New Zealand, gathering and thriving on the abundance of fish in the coastal rivers and venturing into the interior tussock plains (Evison 1993; Anderson 1998; Tau et al. 1989; Tau 2003). These early users of water regarded it with reverence – it was the domain of *Tangaroa* (deity of seas and fresh waters) and provided sustenance for their communities.

Ngai Tahu value specific freshwater sources because of their spiritual status or usage. In a resource management context, Maori are likely to seek the absolute protection of fresh waters that are considered *tapu* (sacred) and to seek the protection of the sufficient quantities of high quality waters of taonga value (Ministry for Environment 1997). Values (both tangible and intangible) associated with specific freshwater resources include the role of particular freshwater resources in creation stories; the role of those freshwater resources in historical accounts; the proximity of settlements or historical sites in or adjacent to specific freshwater resources; the value of freshwater resources as a source of tribal identity; *mahinga kai* (the gathering and processing of foods and resources); the use of freshwater resources as access routes or transport courses; and the continued capacity of future generations to access, use and treasure the resource (Ministry for Environment 1997).

Maori also utilised a water classification system, which draws on the classifications proposed by Douglas (1984:1), Tau et al. (1989), Rochford (2003), and Williams (2004). The classifications denote saltwater, brackish water and freshwater categories, distinguish other waters on the basis of physical character or levels of degradation, and identify specific cultural uses of different types of water.

[1] Further information on Ngai Tahu can be found at www.ngaitahu.iwi.nz.

Before the coming of European settlers, Ngai Tahu controlled a vast land. Their *hapu* (subtribes) utilised many resources of this territory, moving from place to place with the seasonal availability of food. Each hapu gathered food within a traditional use area recognised by surrounding hapu as rightfully theirs. In effect, this provided each *whanau* (family) and hapu with rights to particular river reaches for fishing and lands for gathering wildlife and plants (Anderson 1998). This pattern of resource use shaped a distinctive itinerant lifestyle in which mobility was pronounced and essential.

In summary, Ngai Tahu were well established in New Zealand before the arrival of European settlers and exercised their rights of mana whenua. In other words, they exercised their customary authority, which as Tau explains, includes the doctrine of aboriginal title and the use rights that can be claimed under them through common law.

Today, the iwi combines three groups, Ngai Tahu itself and Waitaha and Ngai Mamoe, who lived in the South Island prior to the arrival of Ngai Tahu. The five primary hapu of the three combined groups are Kati Kuri, Ngati Irakehu, Kati Huirapa, Ngai Tuahuriri and Ngai Te Ruakihikihi. The iwi's *takiwa* (tribal area), the largest in New Zealand, extends from Kaikoura in the north to Stewart Island (*Rakiura*) in the south, and includes the West Coast area (*Tai Poutini*).

Te Runanga o Ngai Tahu (the tribal authority established by the Te Runanga o Ngai Tahu Act 1996) constitutes 18 *papatipu runanga* (local representative group or community system of representation) representing geographical areas, generally based around traditional settlements. *Te Runanga o Moeraki* is one of the 18 papatipu runanga. The takiwa of Te Runanga o Moeraki centres on Moeraki and extends from Waitaki River to Waihemo River and inland to the Main Divide. The Kakaunui Catchment (as shown in Fig. 5.4.1) lies wholly within the takiwa of Te Runanga o Moeraki (as shown in Fig. 5.4.2).

It was selected as the case study because flow has been a source of contention for a number of years, the catchment has experienced a recent intensification of land use, and the runanga is increasingly concerned at the deterioration of water quality. The Kakanui River is a small river in North Otago. The catchment is bordered by the Kakaunui Mountains to the south, the Maerewhenua catchment to the west, the Waitaki catchment to the north, and the Pacific to the east. The Kakaunui has three major tributaries: the Kauru River, Island Stream and Waiareka Creek. Two deep aquifers in the Kakaunui Catchment (Deborah and Papakaio) yield significant amounts of water for irrigation and other permitted uses. From its source in the Kakanui Mountains, it flows northeast for about 40 km through partially developed tussock grassland before emerging onto plains at Clifton Falls. It then flows southeastwards at a gentler gradient through highly developed pastures for a further 30 km to the sea. The river can be divided into three sections of different character. The lower 9 km of river is low gradient (1.2 m/km) and willow-lined on at least one bank, with long pools and short cobble riffles.

Many of the pools contain macrophytes, or large beds of aquatic plants. The gradient increases to 3.6 m/km in the 19 km middle reach and there are generally fewer willows, macrophytes, and pools than in the lower river, with correspondingly

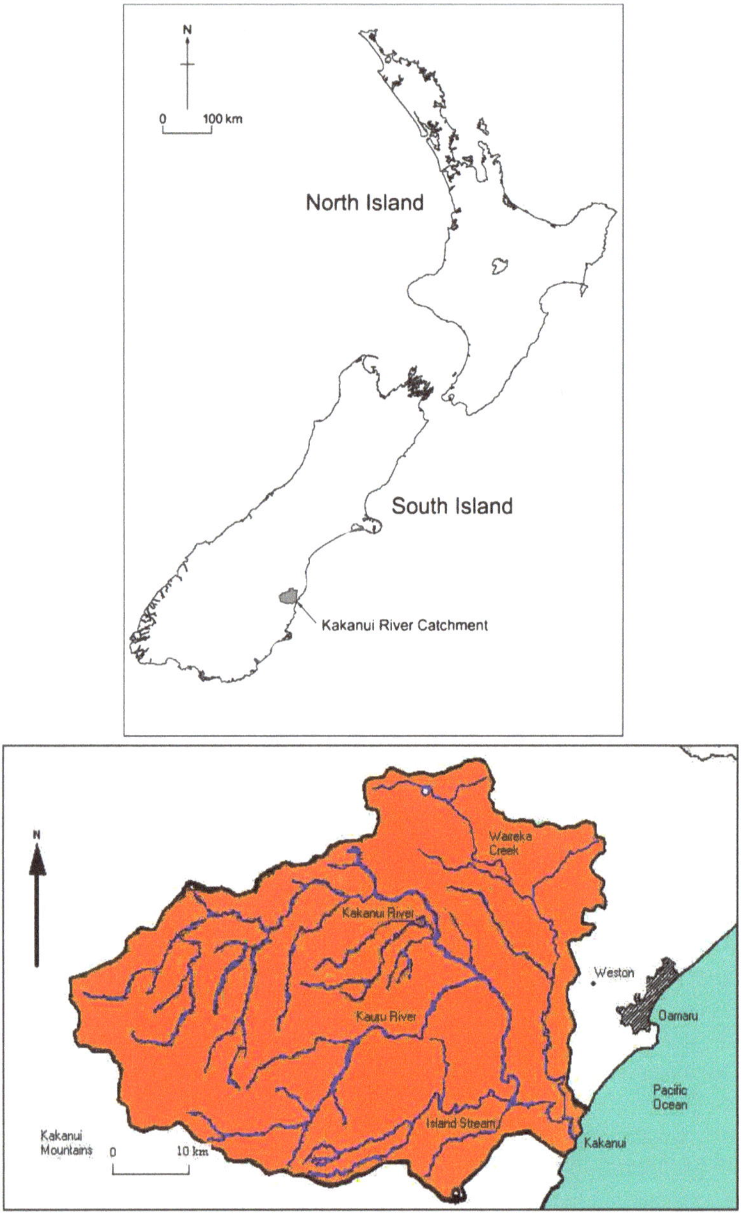

Fig. 5.4.1 The Kakaunui River catchment. *Top* – The location of the Kakaunui catchment within New Zealand; and *Below* – The Kakaunui catchment

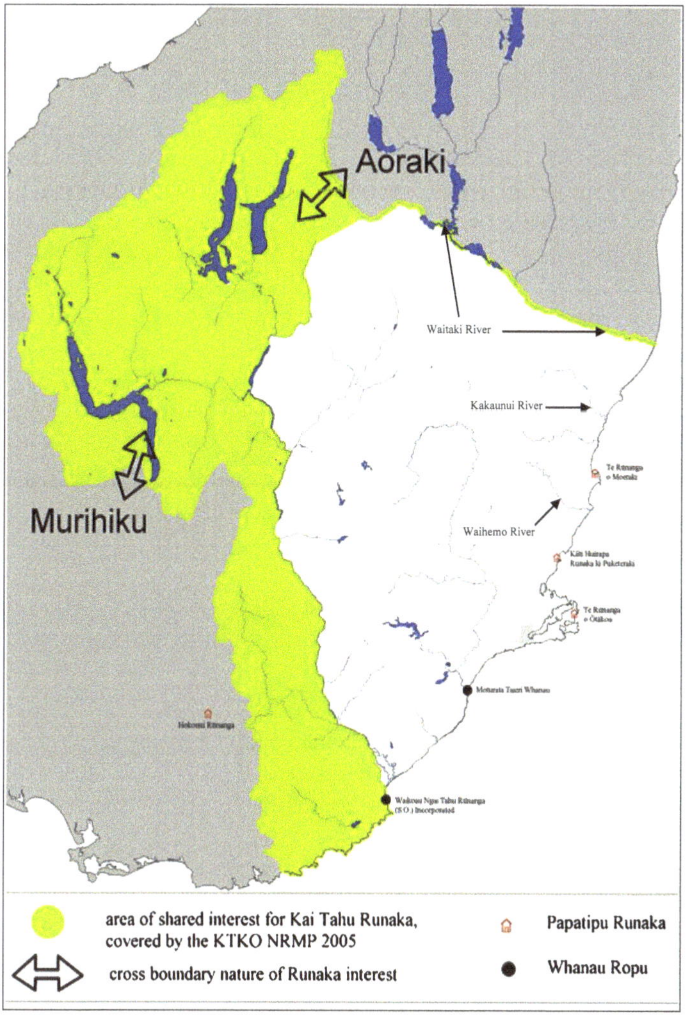

Fig. 5.4.2 Showing the location of Kakaunui in relation to neighbouring areas

more run and riffle habitat. The upper 32 km of river is generally steep (10 m/km), confined by steep hillsides, and contains well-developed pool-riffle sequences.

Water extracted – 57.4 l/s (litres per second) in total – is used for rural stock and household water as well as for small townships and rural water schemes. The largest water users, however, are farmers who irrigate. Economic advantage currently lies with dairying, but in the dry climates of North Otago access to irrigation water is imperative if landowners are to convert to dairying. The current flow regime sets a minimum flow of 250 l/s at Mill Dam; however, should the flow drop below this figure, all extraction must cease until the flow recovers to 400 l/s. What remains to be articulated, however, are the flow preferences of Ngai Tahu whanui. This case study provides guidance to Te Runanga o Moeraki in this regard.

5.4.3 Existing Environmental Flow Assessment Methods

Worldwide there is considerable literature concerning environmental flow assessments (EFAs), which are defined by King et al. (1999:3) as 'an assessment of how much of the original flow regime of a river should continue to flow down it in order to maintain specified valued features of the river ecosystem'. As Arthington et al. (2004) explains, over 200 approaches for determining environmental flows are applied in more than 50 countries, testament to the level of global concern at the increasing alteration of rivers and their resultant deterioration.

Each of the existing EFA methodologies has inherent strengths and limitations, and each to some degree is responsive to cultural values expressed by indigenous communities (Jowett and Mosley 2004). As highlighted by Craig (2005), a common assumption is that environmental flows are an acceptable surrogate for the protection of cultural values. But Western scientific philosophies and techniques, which form the bases of the science behind environmental flows, may in fact limit the engagement of indigenous peoples. Minhinnick (quoted in Douglas 1984:34) challenges existing techniques and explains that

> great harm was done to Maori people and their relationship with their ancestral lands and waterways as a result of the mono-cultural law regime that prevailed prior to the passing of the Resource Management Act 1991. Cultural offences could not be avoided where the law did not allow their consideration. The sole focus on biophysical impacts was particularly damaging to Maori spiritual values.

In New Zealand, the enactment of the Resource Management Act 1991 has charged managers, pursuant to section 6(e), to recognise and provide for 'the relationship of Maori and their culture and traditions with their ancestral lands, water, sites, *waahi tapu* and other *taonga*'. This law poses the challenge of trying to develop a flow assessment method that captures the multiple dimensions of the *relationship* of Maori with fresh water and enables identification of a satisfactory flow regime from a cultural perspective.

5.4.4 An Overview of a Participatory Process Responsive to Maori Conceptualisations

Three imperatives drove the development of the participatory process described in this paper. Firstly, the process had to be grounded in the cultural beliefs, values and practices of Maori, some of which are summarised in Table 5.4.1.

Secondly, the process had to explicitly enable the examination of flow-related issues and the identification of flows perceived by Maori as satisfactory for protecting their range of cultural interests. Finally, the process had to be safe for application by Maori, and their positions defendable, given the contentiousness associated with setting environmental flows and allocation limits (as described by Poff et al. 2002).

Table 5.4.1 A selection of cultural beliefs, values, practices, uses, features

Whakapapa (genealogy). Whakapapa describes bonds, relationships, and connections. Water is the medium flowing through a catchment that makes connections.	Whanaungatanga (kinship, familial relationships). Whanaungatanga describes the principle of kinship, connectedness, and inter-dependence between all things within the natural world including people. The concepts of sustainable management and integrated management are consistent with whanaungatanga as they reflect and give life to the interrelationship between all things.
Manaakitanga (show kindness and respect to, look after, entertain).	Mauri (Essential life force or principle; a quality inherent in all things both animate and inanimate). Ngai Tahu believe that people, flora, fauna as well as natural phenomena such as forests, waters, mists, winds and rocks, possess a *mauri* or life force (Te Runanga o Ngai Tahu 2001).
Rangatiratanga (Chieftanship, decision-making rights) which in the case of freshwater means having the right to make decisions of use, development and protection of water resources within one's tribal area.	Kotahitanga (unity, working together as one).
Tapu (Sacred) – which includes wai tapu.	Noa (Free from tapu, ordinary). *Tapu* and *noa* represent a traditional management technique that accepts that certain types of interaction and use within the natural environment are necessary to ensure the well-being of whanau and hapu, yet protects the environment. Tapu and noa were used to protect the mauri of a resource and are described by Williams (2003:80) as the "single most pervasive feature of traditional life".
Kaitiakitanga (The exercise of customary custodianship, in a manner that incorporates spiritual matters, by those who hold mana whenua status for particular area or resource). This includes ensuring the waters within one's tribal area are respected and cared for.	Wai Maori (fresh water).
Mauka (mountain) which is the source of many of our waterways.	Mahinga kai (places where foods are procured and or produced). "Kai awa" and "kai roto" refers to the foods and resources sourced from rivers and lakes respectively.
Hauora (health and wellbeing). Healthy waters and the ability to interact safely with waters of cultural significance is a contributor to wellbeing.	Wahi Ingoa (traditional placenames). Placenames are extremely valuable when landscapes and landforms have been modified as many of them tell of the history of the area and describe particular environmental (including water features).
Kainga (settlement, place of residence). Settlements were located where there was a potable water supply and resources to sustain the community.	Ara tawhito (ancient trails). Many of the trails were along the streams and rivers. Today's transportation routes follow many of these old trails.
Tauranga waka (canoe mooring / landing site)	Pa (fortification) . Like kainga, pa were located where there was a potable water supply and resources to sustain the community.
Recreation – waterways remain valued as sites for a range of activities, including recreational.	Urupa (burial place). Many urupa were located on the banks of rivers, on promontories overlooking waterways, or on islands in the waterways.

The process for identifying the cultural interests of Maori is via an attention to cultural opportunities, which are defined as the combination of physical, biological, social, cultural and managerial conditions at a site that support cultural uses as they did in the past and as desired by Maori today and into the future (Tipa and Nelson 2008; Tipa 2010). Flow is one of the critical factors that can impact the quality and condition of a site and consequently the opportunities afforded Maori. Significantly, flow can also be manipulated to provide cultural opportunities. This, however, also assumes that most Maori are capable of describing the opportunities they seek. Diversity of belief, value and practice is found within Maoridom as different iwi, hapu and whanau interact with sites in a variety of ways. Diversity is accommodated as cultural opportunities sought are informed by traditional, historic and/or contemporary values, and may be akin to ecological, economic, recreational, aesthetic and social opportunities sought by others, while some are distinctly cultural.

A range of techniques for assessing opportunities have emerged in the last 20 years including a Recreation Opportunity Spectrum (Clarke and Stankey 1979), Water Recreation Opportunity Spectrum (Technical Service Center Economics and Resource Planning Group n.d.), Tourism Opportunity Spectrum (Butler and Waldbrook n.d.), and Forestry Opportunity Spectrum (Grove et al. n.d.). Proposing an opportunity approach builds on this body of literature. The quality and condition of sites impacts the opportunities afforded Maori. Significantly, conditions – such as the flow regime – may be manipulated to provide cultural opportunities.

Cultural Opportunity Mapping requires three distinct tasks to be completed.

(a) The first task involves preparation of a base map or aerial photograph upon which sites throughout a catchment are identified together with the values of each – in other words, the reasons for the site being of cultural significance are recorded.
(b) Opportunities sought by tribal members (given the nature and extent of the values they mapped) are then recorded.
(c) Finally, informants are to identify water related concerns they perceive to impact the provision of cultural opportunities at each site. These are represented as a concept map which is recognised as an effective tool to elicit the belief systems that are used to perceive and analyse situations (El Sawy and Pauchant 1988; Weick 1979, 1995).

Cultural Opportunity Assessments utilise the outputs of the mapping exercise, and are premised on sites of cultural significance being assessed by Maori in a process akin to Customer Satisfaction Assessments (CSAs) and environmental preference studies using attributes of flow previously identified by Maori.[2] The cultural flow preferences (and importantly the critical thresholds) are calculated from the ratings

[2] A series of interviews with Maori in three catchments provided descriptions of river flows and the attributes that describe healthy vibrant flows. From the descriptions 19 'flow attributes' were extracted and listed on the assessment form.

awarded for each of the attributes (some of which are discussed later in the paper) which are categorised into four themes:

1. *Wai Maori*[3];
2. Cultural landscapes[4]: this includes uses of Maori reserves, easements etc.[5];
3. Gathering of foods and other materials for cultural use[6]; and
4. wcll-bcing of whanau.

The attributes of river flow identified by Maori are represented on an assessment form with Likert scales (rating scale from 1 to 7) and conclude with open ended questions. Ryan and Cessford (2003) argue for inclusion of a non-response option, which enables use of a generic list of flow attributes. Where an attribute is not relevant given the cultural values and cultural opportunities associated with a particular site, the non-response option is to be marked. In this way the questionnaire should not need to be changed from site to site, or catchment to catchment.

5.4.5 Applying COMA and Identifying Cultural Flow Preferences: The Kakaunui Case

The six steps in the participatory process, as illustrated in Table 5.4.2, are: initiating the project, defining the association of Maori with the catchment, Cultural Opportunity Mapping, critically reviewing the data to focus the investigation, undertaking Cultural Opportunity Assessments and calculating cultural flow preferences and analyses to inform decision-making.

Step 1: Initiating the project
Typically, a river is selected because its flow regime is a source of contention. An initial hui with representatives of *manawhenua* is necessary to secure support for the flow investigation, confirm the individuals to be interviewed, and mandate members with the necessary cultural knowledge to participate in field visits to the catchment.

Step 2: Defining and validating the association of Maori with freshwater
This step provided a general overview of the relationships of *Ngai Tahu* with the Kakaunui. Methods of data collection included hui and interviews with key informants identified by Te Runanga o Moeraki that explored the diversity and complexity of cultural relationships with fresh water before defining how their relationship, and interaction with the Kakaunui, is affected by river flows. Gaining perceptions of

[3] Four attributes are assessed.

[4] Three attributes are assessed.

[5] Nine attributes are assessed.

[6] Three attributes are assessed.

Table 5.4.2 A summary of process to incorporate the cultural beliefs, values and practices of Maori in setting determining flow regimes

Step	Objective of step	Methods	Outputs
1. Initiating the project	To identify the body representing Maori and secure mandates	Meetings with tribal leaders, elders, and tribal members	1. Research agreement; 2. Mandates for tribal representatives and research team
2. Documenting the association	To identify the multiple dimensions that collectively represent cultural association with the study area To identify the attributes used to assess whether environmental flows are sufficient to sustain cultural interests	Focus groups Semi structured interviews	3. Report describing the cultural association 4. Maps of association; 5. List of attributes used by Maori
3. Cultural opportunity mapping	To identify the cultural values associated with specific sites, together with the opportunities sought at each site given the values identified	Focus groups Semi structured interviews Focus group to validate the data	6. Detailed (site specific) maps of values and opportunities sought 7. Concept maps of perceived issues in a catchment.
4. Focusing the investigation	To critically review the data collected and to focus on environmental flows and specific flow issues affecting the waterways being investigated	Focus group	8. Assessment framework comprising attributes identified by Maori. 9. Refined concept map
5. Cultural opportunity assessments	To undertake assessments at sites to assess whether environmental flows sustain cultural values and provide the opportunities sought	Mandated representatives undertaking field assessments Focus groups	10. Assessment forms completed 11. Sketches of each site with key flow issues highlighted 12. Photographic profiles of each site
6. Analysis to inform decision making	Qualitative analysis and statistical analysis to identify flow thresholds, flow related issues, and management priorities	Focus groups Statistical analyses	13. Report summarizing key flow thresholds, flow related issues, and water management priorities

changes to flow patterns over time, and the impact of these changes on cultural values, was important.

Consistent with the contentions of Menzies (2006) that traditional ecological knowledge is dynamic, responsive to changes within the environment and emerges through direct active use of the landscape, the Te Runanga o Moeraki chose as participants those individuals who retain and continue to be active in the Kakaunui. A map providing an overview of the cultural association with lands and waters of the Kakaunui is included as Fig. 5.4.3.

Importantly, this stage enabled an initial examination of flow related issues because interviewees were challenged to identify how their experiences are impacted by aquatic conditions, in particular river flow. Figure 5.4.4 provides an overview of flow related issues in the Kakaunui catchment as identified by Ngai Tahu.

Step 3: Cultural opportunity mapping

This third step is a participatory mapping exercise where sites of cultural significance throughout a catchment together with the cultural opportunities sought are mapped. The final mapping task is formulation of a catchment wide concept map that visually depicts water management issues (including flow) perceived by Maori as impacting their experiences at the sites identified. Interrelationships between issues are also mapped. The concept map is subject to a number of analyses,[7] that identify key issues warranting further investigation including:

1. Domain analysis, which analyses each concept and calculates how many concepts are immediately related to it.
2. Centrality analysis, which calculates relationships between concepts using more than one level. This identifies the centrality of a concept to the whole model rather than just its immediate vicinity.

Step 4: Focusing the investigation: discriminating the flow related issues identified

To ensure a focused assessment, it was necessary to critically review the data collected. The issues were sorted to distinguish between firstly, those cultural values, cultural opportunities, and issues that could be evaluated as part of existing EFAs; secondly, those cultural values and opportunities, issues (and consequently flow attributes) that are place specific but could be addressed within an existing EFA; and thirdly, those cultural values and cultural opportunities, issues and flow attributes that were unlikely to be adequately addressed via existing EFA methods and were more appropriately addressed through a cultural assessment undertaken by manawhenua. Examples that illustrate these distinctions are detailed below.

1. *Cultural values, opportunities, and issues to be assessed via existing EFA methods*

[7] The software used for the development of concept maps is Decision Explorer (by Banxia). www.banxia.com. The different types of analyses are described in the User Manual.

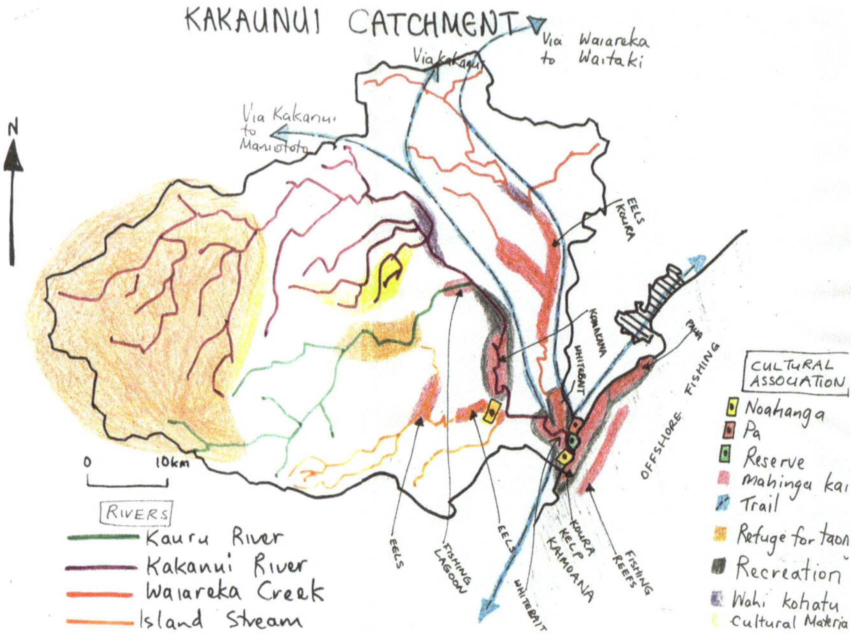

Fig. 5.4.3 Cultural associations with the Kakaunui catchment (Key/glossary: *Wahi kohatu* rock formations, *Pa* permanent settlement, *Nohoanga* settlement (food gathering/temporary), *Mahinga kai* food gathering

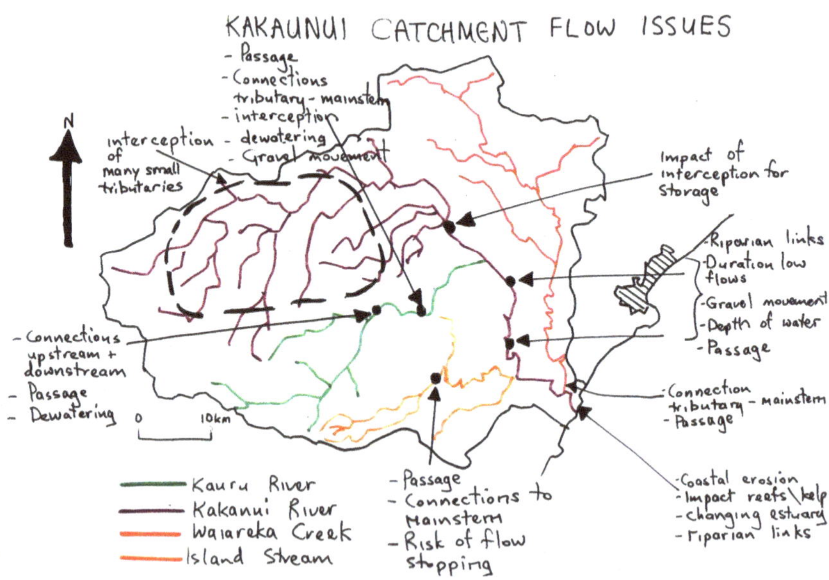

Fig. 5.4.4 Summary of flow issues in the Kakaunui catchment

Some of the dimensions that collectively encapsulate the cultural significance of food gathering may be incorporated within existing EFA methods, for example:

- the flows needed to sustain the habitats of valued aquatic species (such as eels) and flows needed to enable eel passage throughout the catchment. Traditionally, Maori accessed aquatic species (eels, kanakana, waterfowl), terrestrial species (wild plants, berries, roots, and pollen), and small game (e.g., rats). Although some resources were gathered seasonally, historically whanau relied on eel year round. Eels were primarily a source of fat, but also a source of protein (Williams 2004). The centrality of eels as a critical food source to Ngai Tahu is well known and is reflected in the many contemporary initiatives to restore populations. The current threatened populations of eels, especially long-fin eels, sadly contrasts with their abundance historically.

Specific concerns relating to the character and shape of the river could also be addressed, for example: flows required to move sediment through the catchment, the impact of flows on the river mouth; and the potential impact of changing flows on marine ecosystems.

2. *Cultural values and indicators quantifiable and place specific*
It is inevitable, however, that some cultural values and cultural opportunities will be place specific, for example:

   1. A bathing site set aside for use by women may be dependent upon the site's unique combination of geology, ease of access, privacy and flow. Even if a flow sufficient to afford bathing opportunities is provided in the catchment, the characteristics arising from the combination of factors may not be able to be replicated elsewhere.
   2. Characteristics of sites affording food gathering opportunities may be place specific. For example gathering resources from lands awarded to Maori as reserves or fishery easements, which is an inherited right derived through *whakapapa*, cannot be relocated. Flows will need to afford opportunities to utilise existing easements and reserves.

   What emerges from the above examples is the need for some place-based flow assessments as sites of significance to manawhenua and the cultural values they sustain cannot be relocated to other locations in the catchment. Although flows sufficient to sustain opportunities somewhere in the catchment may be provided, relocation would only serve to dislocate and deprive them of their cultural context.

3. *Cultural values to be assessed as part of the cultural assessment*
There will be cultural values, cultural opportunities and flow attributes that are best incorporated into the Cultural Opportunity Assessment and consequently only be assessed as part of a cultural flow preference study.

Calculating cultural flow preferences is intended to complement rather than replace more conventional EFAs. However it is predicated on the assumption that existing EFAs do not specifically evaluate the flows needed to recognise and provide for the relationship of Maori with waters – a matter of national importance pursuant to section 6(e) of the Resource Management Act, as mentioned above.

Step 5: Undertaking Cultural Opportunity Assessments and calculating Cultural Flow Preferences

Cultural Opportunity Assessments require field assessments to be undertaken by Maori. Assessments are premised on each site being assessed under different flow conditions using the attributes previously identified by Maori.

In the case of the Kakaunui, six sites (three on the mainstream Kakaunui River and three sites on the tributaries – two sites on the Kauru and one on Island Stream) – were assessed fortnightly from January to June and monthly from July to December. During a typical field visit, a team comprising mandated members of the Te Runangao Moeraki visited each of the six sites. The field work was completed during 2007/2008. At each site they completed an assessment form on which all the attributes were listed. Because a catchment perspective is needed, the catchment as a whole was included as a site and in effect became the seventh 'site'.[8] In total, over the course of a year, 18 assessments were completed for each site.

Assessors are asked to assess

(a) Satisfaction – whether or not they were satisfied that the flow that they were observing sustains the attributes associated with the cultural values at that particular site; and
(b) Significance – the significance of each attribute at that particular site. If the attributes most significant to Maori could be improved, it would presumably result in an increase of overall satisfaction with specific sites.

An assessment form was completed by each team member at each site and their individual assessment recorded.[9] Finally, assessors were asked to provide qualitative data in response to four questions. Firstly, they were required to identify site specific concerns. Secondly they were to provide an overall assessment as to whether or not they were satisfied that the flow at the site sustains cultural values and affords cultural opportunities. Next, they provided an overall assessment as to their perception of the size of the flow being observed.[10] Finally, they were asked to identify the management actions they want to see prioritised at the site to ensure protection of cultural interests.

Three of the sites chosen correspond with sites monitored by the regional council, thus data from the flow recorders was available. Although the team of assessors did not know the flow they observed when they visited, by recording the time that

[8] This was recommended by Andy Hicks, who undertook an analysis of the early Kakaunui data.

[9] This parallels similar practices of cultural stream monitoring such as the application of a Cultural Health Index for assessing overall stream health (Tipa and Teirney 2006).

[10] They are classed as being very high, high, average, low or very low.

they were at a site, we were able to correlate their assessments to the actual flows that were observed.

During this stage of COMA two additional investigations were undertaken.

1. The first collected a range of qualitative data for values fundamental to understanding the relationship of Maori with a river that should not be quantified. For example: the whakapapa of the river, stories and *whakatauki* (proverb), traditional place-names and such like. These data not only provided the cultural context within which the cultural flow assessment was undertaken, but importantly, additional attributes unique to the river being studied were identified.
2. The second assessment, of economic opportunities, evaluated the water demands of existing Maori economic assets in the catchment and potential opportunities for development. Quantifying water demand is essential if the development rights of Maori are to be protected.

Step 6: Analyses to inform planning and decision making processes
A number of analyses were possible given the data collected.

(a) *Identifying key attributes that impact the assessment of overall satisfaction with the flow*
Although not part of the analysis of data for the purpose of deriving cultural flows preferences, a Spearman's correlation analysis was performed, in which the relation between two or more variables are measured. This analysis was intended to help us understand the need for the range of attributes included on the assessment form. Although we wanted to capture all the flow related attributes specific to each cultural value, this had to balanced against the practical application of the assessment form and the desire to make it simple for whanau members to use. A Spearman's correlation was of value (and will be undertaken when each cultural flow study is completed) as it determined whether trends in any one particular attribute were correlated with the trends in the 'overall' site score. Table 5.4.3 shows the Spearman's correlation values for the Kakaunui, listed in accordance with their level of correlation to 'overall' site score.

From Table 5.4.3, it appears that having the ability to gather foods and cultural materials (MK1) is the most important attribute, with the availability of habitats (MK4) and maintenance of connections within hydrological systems (WM2) the next most important. All but two of the attributes, however, suggest a significant and positive correlation with overall aquatic condition. In other words, all but two of these attributes appear important in determining the overall satisfaction of Maori with the flows observed.

(b) *Examining significance of sites and attributes*
To meaningfully participate in setting flow regimes it is necessary for Maori to identify what it is about the flow (i.e., the attributes of the observed level of flow at a specific site) that determines whether they are satisfied or dissatisfied. Two types of analyses helped answer this question. Firstly, the inclusion of open-ended questions and qualitative data analyses elicited site specific concerns. Secondly,

Table 5.4.3 Results of the Spearman's correlation – all indicators are ranked in accordance with correlation to 'Overall' rating

Flow attribute	Spearman's correlation coefficient (with 'overall')
MK1	0.875
Flow sufficient to enable use of the site as a mahinga kai	
MK4	0.796
Flow sufficient to provide a range of habitats instream & along riverbank	
WM2	0.780
Flow sufficient to keep riparian wetlands, springs, & tributaries connected to mainstem	
CL1	0.777
CL2	0.775
MK5	0.709
H3	0.695
WM3	0.694
CL5	0.671
MK2	0.623
WM1	0.604
WM4	0.575
CL3	0.523
MK9	0.458
H4	0.457
MK3	0.382
MK6	0.095

qualitative analyses were complemented by statistical analyses. Rating the significance of sites, the significance of each of the flow attributes at each site, and the level of satisfaction with the flow at each site enables multivariate analyses to be completed as part of the decision support system.

(c) *Identifying critical flow ranges*
A rating of 1 to 7 was given by Maori assessors for all flow attributes at each site. For each attribute the individual ratings (of the Maori observers) were averaged, producing a single score. Then the flow attributes within each theme were averaged – for example the nine attributes scores for the mahinga kai component were averaged. This means that the output was a single score for each of the four themes (i.e., Wai Maori, Cultural Landscape, Cultural Use, Hauora), as shown in Table 5.4.4. These averaged scores can then be directly compared with actual recorded flows for the day and time of assessment. Further, it was possible, by examining the data for all 19 attribute scores, to identify those attributes that contributed to the level of satisfaction/dissatisfaction at the flows observed.

The data in Table 5.4.4, specifically the four theme ratings, is consistent with the ratings on three other dates that all suggest a threshold of 400 l/s as flows around this level or below were consistently perceived as unsatisfactory (scoring less than four on a 1-7 scale) by assessors. These initial analyses,

Table 5.4.4 A site assessed in the Kakaunui catchment with the cultural flow thresholds identified

	Date of assessment 30/01/2008			
	Mahing a Kai	Wai Maori	Hauora	Cultural Landscape
Theme ratings (1-7)	3.76	3.03	4.43	4.90
Indicator Significance	6.65	5.88	7.00	6.60
Actual flow	0.39			

although they only consider the ratings for satisfaction with the observed flow, suggest that the current minimum flow of 250 l/s could be considered too low by Te Runanga o Moeraki.

The fours scores for each of the theme ratings were then entered into a decision support system (based on Decision Pad) that considers both the level of satisfaction and the significance of the sites and the respective flow attributes. With respect to the Kakaunui River, the decision support system confirmed that the critical flow threshold is 400-500 l/s. A comparative analysis of the results of COMA, the cultural flow preferences and the results of existing EFAs enabled differences to be identified, discussed and a flow regime negotiated.

(d) *Identifying wider management priorities*

The data collected also enabled identification of management actions, for example, interrogation of the data for one site in the Kakaunui (see Fig. 5.4.5) suggested that low flows inhibit recreational activities. The scores for specific attributes highlighted concerns with respect to the changed shape and depth of the water at low flows. Further discussion confirmed that the shape of the river was adversely impacted by repeated gravel extractions that are exacerbated by low flows. Consequently, gravel management was identified as a priority.

The second example concerned water quality. Again low flows were seen to exacerbate concerns. Maori, in seeking water quality sufficient to sustain mahinga kai, were seeking a higher minimum flow. However, it was acknowledged that higher flows (and greater assimilative capacity) need not be sought if land management strategies address water quality concerns directly.

(e) *Presenting COMA results & linking with other environmental flow methods*

Two outputs were produced:

1. A catchment overview. On a map of the catchment, a multipart summary for each site that registered cultural values and cultural opportunities sought together with the cultural flow thresholds for that site. Cumulatively, across a catchment, summary matrices were mapped.

2. The catchment overview was supported by a series of individual site assessments. The data for each site included a description of the cultural values and

Fig. 5.4.5 A site assessed in the Kakaunui catchment where gravel extraction is an issue

cultural opportunities sought, and for each day when assessments were undertaken:

- The photographic record of the flow observed;
- The scores for each of the respective attributes and the cultural flow preferences;
- Actual recorded flow that was observed;
- Collated qualitative comments – including a list of priority actions; and
- Recommended cultural flow thresholds to be negotiated with water managers.

Using mahinga kai as an example, it is possible to identify the possible links to existing environmental flow assessment methods. In support of Jowett and Mosley (2004), it is likely that the habitat needs of valued species, such as eels, can be accommodated within the preference curves used in an Instream Flow Incremental Methodology study. However, mahinga kai cannot be reduced to only the presence of suitable habitat. The attributes relevant for mahinga kai listed on the assessment form used to identify cultural flow preferences, shows that it is necessary to have flows that enable whanau to access and fish at specific sites (sometimes defined by rights that are not transferable to other sites), using preferred fishing techniques. These wider attributes of mahinga kai, at this stage, do not appear to be accommodated in current environmental flow assessment methods.

In summary, the application of COMA results in identification of cultural flow preferences (i.e., the critical flow thresholds) and a set of management priorities which Maori believe will help realise the cultural opportunities (or outcomes) sought within a catchment.

5.4.6 Conclusion

Fresh water continues to be of fundamental importance to Maori who are actively pursuing the re-establishment of their customary and Treaty rights. One component that constitutes appropriate recognition of such rights is participation in setting flow regimes that provide for a range of cultural interests. This chapter has introduced a process that enables Maori to undertake flow assessments. Having observed earlier that many of the existing methods of EFAs marginalise Maori, COMA is a method that is dependent upon active engagement of Maori and that was developed in response to the need for more responsive assessment methodologies. A process which examines opportunities and experiences was considered sensitive to the multidimensional and experiential relationship of Maori with fresh waters, as well as building on existing management techniques that evaluate opportunities and experiences.

Refinement of process for deriving cultural flow preferences continues. It is being applied by *whanau* in different freshwater management contexts within the South Island of New Zealand. This is essential if resource management agencies, Maori and communities are to have confidence in the technique and the results it yields. It is being tested in different river types (initially in a braided river) to examine whether or not it can be applied with confidence in a range of catchments. Secondly, it is to be applied by different Maori groups to ensure that it is responsive to the values of all Maori, and not simply one iwi. Finally, an area demanding further refinement is the interface between cultural flow preferences and conventional EFAs used in New Zealand. Scientists have confirmed a desire to collaborate in the near future. It is hoped that a number of flow investigations can be initiated simultaneously with different iwi, in different parts of New Zealand, where river conditions differ, so that the ratings for flow attributes can inform development of flow suitability curves specific to cultural values for possible inclusion in the existing and more widely utilised environmental flow assessments.

Upon completion of these different strands of work, case study reports will be available demonstrating the potential of COMA. It is hoped that the example of its use in a braided river is completed by 2010, while the work with a different Maori group is to be underway at the end of 2009. A project plan for integrating cultural flow preference work with more conventional environmental flow assessments has been prepared and it is hoped that this collaboration is underway in a number of catchments by 2010. These work streams will enable COMA and method for determining cultural flow preferences to be refined. Only then can we have confidence that we do have a robust method that enables the knowledge held within whanau, *hapu* and iwi to inform freshwater management contexts within New Zealand thus enabling Maori to effectively contribute meaningfully to the sustainability debate.

Acknowledgments This research is funded by National Institute of Water and Atmospheric Research (NIWA) FRST funded research programme 'Water Allocation: Protection of Instream Values'. Andy Hicks and Rosemary Clucas (University of Otago) provided advice on statistical analyses.

Box 5.4a Creating a cultural health index for assessing stream health

—Gail Tipa

The cultural health index (CHI) was designed by Tipa and Teirney (2002, 2006) in response to a need identified by Ngai Tahu, the Maori whose tribal area covers 80% of the South Island in New Zealand. The purpose of the CHI is to provide a tool that Maori can apply to facilitate their input and participation in land and water management processes and decision-making. The result – the CHI for streams – links cultural knowledge about stream health and Western scientific methods. The CHI comprises three components: site status, specifically the significance of the site to Maori; a *mahinga kai* (food gathering) measure; and a stream health measure.

The first component assesses the significance of the site to Maori and asks them to distinguish between traditional and contemporary sites. 'A' means the site is a traditional site of significance to Maori, and 'B' means that the site is not traditional. The second question concerns whether Maori would return to the site in the future, believing that it is able to sustain the cultural uses that it has had in the past. If Maori would return, the site is awarded a score of 1 and if not, a score of 0. When the answers to the two questions are collated, there are only four possible combinations:

A-1	A-0	B-1	B-0
This is a traditional site that Maori would return to and use as they did in the past.	This is a traditional site that Maori would not return to. It would not be used in the future.	This is a site that is not of traditional significance to Maori. However, they would go to the site in the future.	This is a site that is not of traditional significance to Maori. Further, they would not go to the site.

The second component of the CHI requires an assessment of the mahinga kai values of a site. Inclusion of this component in the index recognises that the life and vitality of a waterway are tangibly represented by some of the physical characteristics of a freshwater resource, including the presence of indigenous flora and fauna, water clarity, water quantity, and the mahinga kai that it yields (Ministry for Environment 1997). There are four parts to the mahinga kai measure of the CHI. The first part (a) requires the identification of mahinga kai species present at the site, and a score ranging from 1 to 5 is then assigned, depending on the number of species present. The second factor (b) requires a comparison between the species present today and the traditional mahinga kai sourced from the site. This was deliberately factored into the design of the CHI to recognise that maintaining cultural practices, such as the gathering of mahinga kai, is an important means of ensuring the transference of cultural values

(continued)

Box 5.4a (continued)

through the generations. Cultural continuity means that greater value is likely to be assigned to sites of traditional significance that continue to support the mahinga kai species sourced in the past. A score is assigned based on the number of species of traditional significance that are still present:

- Nontraditional site scores 1;
- None of the species sourced in the past is present at the site scores 1;
- At least 50% of the species sourced in the past are still present at the site scores 3; and
- All species sourced in the past are still present at the site scores 5.

Mahinga kai implies that Maori have physical and legal access to the resources that they want to gather. The third component of the mahinga kai measure (c) requires a score to be assigned to each site based on the ability to access the site, where 1 equals no access and 5 equals unimpeded legal and physical access.

The fourth element in the mahinga kai measure (d) requires Maori to assess whether they would return the site in the future and use it: 'no' scores 1, and 'yes' scores 5. The four mahinga kai elements are then averaged to produce a single score out of 5, for example:

Mahinga kai measure	
The four scores could be:	
a. Species present	2
b. Traditional species compared to present	1
c. Access	2
d. Return in the future	5

The third and final component of the CHI is the stream health measure. Of the 19 indicators originally identified by *kaumatua* (respected Maori elders), eight that can be defined objectively and most appropriately reflect Maori evaluations of overall stream health are included in the Stream Health Measure. This measure is derived by averaging the 1 to 5 scores awarded to the eight factors (catchment land use, riparian vegetation, use of the riparian margin, riverbed condition, manipulation of the river channel, a visible flow, water clarity and water quality) to give a final stream health measure from 1 to 5.

Box Fig. 5.4a.1 Example of the Island Stream site: Kakaunui catchment

(continued)

Box 5.4a (continued)

The CHI, when applied to a specific site, will result in a score such as A-0/2.56/1.06, which was the result of the assessment for the site at Island Stream (in the Kakaunui catchment). This score means:

- The site at Island Stream was traditionally used by my whanau, who travelled there each autumn to harvest eels during the downstream migration. Hence the site is classed as an 'A'. However, its degraded condition means that we would not return to use the site, therefore component 1 scores A-0.
- The mahinga kai measure score is average (2.56 out of 5) because, although two of the factors included in the measure score highly, the other two receive a low score. Of the 46 sites assessed, this site had the greatest density of tuna, which was the species sourced from this site in the past. The site is also easy to access. However, there is not a great number of species, and as stated, this site would not be visited in the future in its current state.
- The stream health measure (1.06) shows that for all five factors assessed, the site scored poorly, which confirms the poor health of this site.

Importantly, the CHI is grounded in the beliefs, values, and practices of Maori. Around New Zealand various whanau, hapu, and iwi are actively engaged in river monitoring – some using the CHI presented here, whereas others are in the process of developing their own CHI.

Resources

Ministry of Environment. 1997. *Environmental performance indicators: Proposals for air, freshwater and land.* Wellington: Ministry for the Environment.

Tipa, G., and L. Teirney. 2002. *Mauri and Mahinga kai indicators project: Final report – developing the cultural health index* (June). Wellington: Ministry for Environment.

Tipa, G., and L. Teirney. 2006. *Using the cultural health index: How to assess the health of streams and waterways*. Reference ME711. Wellington: Ministry for Environment.

Glossary

Ara tawhito	Ancient trails
Hauora	Health and well-being
Hapu	Subtribe, extended whanau
Iwi	Tribe
Kainga	Settlement, place of residence
Kaitiakitanga	The exercise of customary custodianship, in a manner that incorporates spiritual matters, by those who hold mana whenua status for particular area or resource

Kotahitanga	Unity, working together as one
Mahinga kai	Places where food is produced or procured
Manaakitanga	Show kindness and respect to, look after, entertain
Manawhenua	Customary authority or rangatiratanga exercised by an iwi or hapū in an identified area
Maori	Indigenous people of New Zealand
Mauka	Mountain
Mauri	Essential life force or principle, a quality inherent in all things both animate and inanimate
Ngai Tahu	The iwi living in the southern part of the South Island of New Zealand
Noa	Free from tapu, ordinary
Pa	fortification
Papatipu runanga	local representative group or community system of representation
Rangatiratanga	Chieftainship, decision-making rights
Tapu	Sacred
Taonga	Treasure
Tauranga waka	Canoe mooring/landing site
Urupa	Burial place
Wahi Ingoa	Traditional placenames
Wahi tapu	Places sacred to manawhenua
Wai tapu	Waters that are tapu – revered, sacred
Wai Maori	freshwater
Whakapapa	Genealogy
Whakatauki	proverb, saying
Whanaungatanga	kinship, familial relationships
Whanau	Family

Resources

Anderson, A. 1998. *The welcome of strangers*. Dunedin: University of Otago Press.

Arthington, A., R. Tharme, S. Brizga, B. Pusey, and M. Kennard. 2004. Environmental flow assessments with emphasis on holistic methodologies. In *Proceedings of the second international symposium on the management of large rivers for fisheries*, vol. II. ed. R. Welcomme and T. Petr. Bangkok: Food & Agriculture Organisation Regional Office for Asia and the Pacific. http://www.fao.org/docrep/007/ad526e/ad526e07.htm#bm07. Accessed 21 May 2007.

Burmil, S., T. Daniel, and J. Hetherington. 1999. Human values and perception of water in arid landscapes. *Landscapes Urban Planning* 44: 99–109.

Butler, R.W., and L.A. Waldbrook. n.d. A new planning tool: The tourism opportunity spectrum. http://www.jcu.edu.au/business/idc/groups/public/documents/journal_article/jcudev_012854.pdf. Accessed 25 May 2008.

Clarke, R., and G. Stankey. 1979. *The recreation opportunity spectrum: A framework for planning, management and research*. USDA Forest Service, General Technical Report PNW-98. Washington, DC: USDA.

Craig, D. 2005. Indigenous property rights to water: Environmental flows, cultural values and tradeable property rights. www.law.mq.edu.au/MUCEL/craig/. Accessed 14 Apr 2008.

Douglas, E.M.K. (ed.). 1984. *Waiora, waimaori, waikino, waimate, waitai: Maori perceptions of water and the environment*. Hamilton: University of Waikato.

El Sawy, O., and T. Pauchant. 1988. Triggers, templates, and twitches in the tracking of emerging strategic issues. *Strategic Management Journal* 9(5): 445–473.

Evison, H.C. 1993. *Te Wai Pounamu: The Greenstone island. A history of the Southern Maori during the European colonisation of New Zealand*. Wellington: Aoraki Press in association with the Ngai Tahu Maori Trust Board and Te Runanganui o Tahu.

Grove, J.M., A.B. Cumming, M.F. Galvin, G.W. Hager, J.P.M. O'Neil-Dunne, A.R. Troy, F. Rodgers, F. Spero, E. Svendsen, and A.E. Draddy. n.d. Integrating urban forestry research and applications: A forest opportunity spectrum framework and its application to Baltimore, Maryland. Unpublished paper submitted to the *Journal of Forestry*.

Jowett, I., and P. Mosley. 2004. Analysis of instream values. In *Freshwaters of New Zealand*, ed. J. Harding, P. Moseley, C. Pearson, and B. Sorrell. Christchurch: Caxton Press. pp 43.1–43.20.

King, J., R. Tharme, C. Brown. 1999. Definition and implementation of instream flows. Contributing paper to the World Commission on Dams. http://www.dams.org. Accessed 3 Dec 2007.

Lincoln Environmental. 2002. *Canterbury strategic water study*. Prepared for Ministry of Agriculture and Forestry, Environment Canterbury, Ministry for the Environment, New Zealand, Aug. A copy can be obtained from Environment Canterbury, Christchurch; http://www.ecan.govt.nz

Poff, N.L., M.M. Brinson, and J.W. Day Jr. 2002. *Aquatic ecosystems and global climate change: Potential impacts on inland freshwater and coastal wetland ecosystems in the United States*. Arlington: Pew Center on Global Climate Change.

Rochford, T. 2003. Te Körero Wai: Maori and Pakeha views on water despoliation and health. A thesis for the completion of a Masters in Public Health, University of Otago.

Ryan, C., and G. Cessford. 2003. Developing a visitor satisfaction monitoring methodology: Quality gaps, crowding and some results. *Current Issues in Tourism* 6(6): 457–507.

Tau, R.T.M. 2003. *Nga pikituroa o Ngai Tahu: The oral traditions of Ngai Tahu*. Dunedin: University of Otago Press.

Tau, T.M., A. Goodall, D. Palmer, and R. Tau. 1989. *Te Whakatau Kaupapa: Ngai Tahu resource management strategy for the Canterbury region*. Wellington: Aoraki Press.

Technical Service Center Economics and Resource Planning Group. n.d. *Water recreation opportunity spectrum users' guidebook*. http://www.usbr.gov/pmts/planning/wros/. Accessed 13 June 2008.

Tharme, R.E. 1996. *Review of international methodologies for the quantification of the instream flow requirements of rivers. Water law review final report for policy development*. Pretoria/Cape Town: Department of Water Affairs and Forestry/Freshwater Research Unit, University of Cape Town.

Tipa, G. 2010. Cultural opportunity assessments: Introducing a framework for assessing the suitability of stream flow from a cultural perspective. In *Maori and the Environment*, ed. R. Selby, P. Moore, and M. Mulholland. Wellington: Kaitiaki, Huia Publications.

Tipa, G., and K. Nelson. 2008. Introducing cultural opportunities: A framework for incorporating cultural perspectives in contemporary resource management. *Journal of Environmental Policy and Planning* 10(4 Dec): 313–337.

Tipa G., and L. Teirney. 2006. *Using the cultural health index: How the assess the health of streams and waterways*. Reference ME711. Wellington: Ministry for Environment.

Te Runanga o Ngai Tahu. 2001. Submission of Te Runanga o Ngai Tahu to the Royal Commission on Genetic Modification A copy can be requested from Te Runanga o Ngai Tahu. http://www.ngaitahu.iwi.nz

Tribunal, Waitangi. 1992. *Mohaka river report*. Wellington: Brooker & Friend.

Tribunal, Waitangi. 1995. *Te Whanganui a Orotu report*. Wellington: Brooker & Friend.

Tribunal, Waitangi. 1998. *Te Ika Whenua river report.* Wellington: Brooker & Friend.
Weick, K. 1979. Cognitive processes in organizations. *Research in Organizational Behavior* 1(1): 41–74.
Weick, K. 1995. *Sensemaking in organizations.* Thousand Oaks: Sage.
Williams, J. 2004. E pakihi hakinga a kai: An examination of pre-contact resource management practice in Southern Te Wai Pounamu. PhD thesis, University of Otago.

Chapter 5.5
Droplets of Hope: Searching for Sustainability and Common Ground in the Arab/Israeli Conflict

Rosina Hassoun

The current state of impasse in the peace process between Israelis and Palestinians, with the Israelis' refusal to stop building additional settlements and the factional split between Hamas and Fatah, has left most Palestinians disillusioned with their current leadership. Both the state of the environmental problems facing Israelis and Palestinians and the current political conditions do not hold out many droplets of hope. This situation bodes ill for any kind of settlement between the two peoples in the near future, let alone a long-term agreement on water sharing between Israel, its Arab neighbours, and the Palestinians.

Many scholars and other experts, such as Homer-Dixon (1994) and Wolf (1995), have predicted the potential for violence and new wars resulting from water disputes. A careful analysis of the history of the Arab/Israeli conflict shows that water has already been a component of every war since 1948. The strategic water planning by Israeli water engineers and politicians in the 1950s that led to the creation of the Israeli National Water Carrier, the massive water projects that carried water from the north to the Negev/Naqab desert in the south, almost led to a second war with Syria and outraged Jordan and the Arab League (Tal 2002). Since the 1967 War, Israel has dominated the Jordan watershed and the aquifers that lie beneath the West Bank and Gaza (Hassoun 1997a, b). Israeli dominance has affected its neighbours. Jordan currently has the most serious water shortfall of the surrounding states.

In light of past and potential wars, it is not surprising that current Israeli attitudes towards water have been increasingly couched in the language of security. The increasing 'securitization' (as Barry Buzan of the London School of Economics uses the term) of water calls for increased understanding of how societies conceptualise threats, including factors such as beliefs and cultural perspectives

R. Hassoun (✉)
Program for Public Health, Michigan State University, East Lansing, MI, USA
e-mail: hassoun@msu.edu

B.R. Johnston et al. (eds.), *Water, Cultural Diversity, and Global Environmental Change: Emerging Trends, Sustainable Futures?*, DOI 10.1007/978-94-007-1774-9_33,

Map 5.5.1 Middle East

(Buzan et al. 1998). Buzan also points out the dynamic nature of threat construction, which also allows for the analytical deconstruction of the threats. Cultural components of conflict are therefore receiving more attention, even in circles outside of anthropology or the social sciences and in security studies. The water crisis that the peoples of the Jordan Valley watershed face today, with rising population (including large numbers of refugees and displaced peoples), water quality and quantity problems, and the potential impacts of global climate change, make it imperative that they create transnational and cross-cultural cooperation in solving these problems and in working towards a sustainable future. But in order to arrive at cooperation, it is also necessary to examine the clashing water cultures inherent in the conflict and how they developed over time. An important component of the history of the water conflict is the rise of environmental awareness and popular environmental movements among the Palestinians and the Israelis. At this moment in time, the cooperation between Palestinian and Israeli environmental movements may hold out the only droplets of hope for the development of a sustainable, shared water culture.

5.5.1 The Political Ecology of Water Culture Conflict

In a short paper, it is only possible to provide a brief outline of the political ecology of these conflicting water cultures. The Israeli and Palestinian environmental movements are rather recent developments. Two works provide considerable insight into the rise of the environmental movement and the depth of the water quality and quantity problems in Israel: *The Political Ecology of the Water Crisis in Israel* (Lees 1998) and the voluminous *Pollution in the Promised Land* (Tal 2002). From these and other sources, it is possible to construct the history of the water culture of modern Israelis. For obvious reasons of disenfranchisement, no comparable works exist from the Palestinian perspective. Although Tal (2002) provides an entire chapter on Israel before 1948 that is somewhat sympathetic to the Palestinians, this material is inadequate to build an in-depth construct of the water culture of the Palestinians from their emic (or indigenous) perspective. Susan Lees (1998) does not include the Palestinians at all in her examination of the predicament of Israeli farmers. A multitude of scholars, journalists, and others has discussed the plight of Palestinians and their current water crisis, but evidence of their water culture or their beliefs, attitudes, and water usage patterns and practices is scattered in dozens upon dozens of accounts by refugees and are missing from much of the discussion. This situation makes researching and providing a balanced picture of both Palestinian and Israeli water cultures difficult.

Both Tal (2002) and Tamari (2009) discuss romanticization and nostalgia in their respective examinations of the national narratives told by both Israelis and Palestinians. In the case of the Israelis, nostalgia and the longing for a homeland culminates in the Zionist narratives. For the Palestinians, unrequited loss and nostalgia colour the refugee stories of Palestine before 1948. The Israelis have used the concept of ties to the ancient Hebrews as the first people of the land to evoke a sense of belonging in their population. Tamari (2009) points out that Palestinians in the 1960s and 1970s used popular narratives of Canaanite, Philistine, and Jebusite ancestry to claim origins predating the ancient Hebrews in the land. Both narratives are politically self-serving. But this fact inevitably leads to the question: to what degree do national and religious narratives inform the water cultures of both peoples?

From the most superficial examination of the religious concepts of water in the three major Abrahamic faiths, we see that all of them consider water as purifying (examples include *wudu* or ablutions in Islam, priestly and female purification rites in Judaism, and baptism in Christianity). There do not appear to be major differences in the religious attitudes towards water between Palestinians and Israelis that would prevent cooperation on water issues. But Tal (2002) points out that it is not ancient Jewish attitudes that inform modern Israeli water practices, nor the more than 500 incidences of the mention of water in the Old Testament that would prevent modern Israelis from polluting modern streams in Israel.

The modern Israeli Jewish attitude towards land and water is rooted in political Zionist philosophy. Lees (1998:10) describes the Zionist attitude towards the land this way: 'A new society required direct, physical involvement in the rebuilding of the land. The dignity of labour is a cornerstone of the Zionist/Socialist ideology'. This attitude resulted in the *kibbutz* (settler communal farms) and the *moshav* (small family homesteaders) movements in Israel. Zionism and these movements have their roots in European liberalism. Israelis brought with them a concept of a modern Israeli state with monocropped fields, lawns, the swimming pools that now spring up in Israeli settlements, and an imperative to 'make the desert bloom'. For the Israelis, Palestine is depicted as a land of deserts in the south, swamps in the north, and rocky fields that have to be tamed by pioneering Jewish settlers. This imperative was so strong that Israel rerouted the entire water flow of the country in one of the world's most ambitious and costly water conveyance systems. The National Water Carrier pumps up to 1.7 million m^3 of water per day, much of it uphill, from the Sea of Galilee (also known as Lake Tiberias or Lake Kinneret) to the Negev/Naqab desert (see Morag 2001). Until recently, the use of 80% of Israel's annual water for agricultural production that resulted in only 3% of the nation's gross national product went unquestioned (Tal 2002). Today, because of recent droughts and growth in urban demand for water, Israelis are considering the previously unthinkable: the abandonment of the agricultural sector.

The Palestinians had a cultural ethos of the land as *Um* (mother) that developed over centuries and was based on their relationship to the land. This metaphor is epitomised in the Mahmoud Darwish's poem, *My Mother*, in which he wrote: 'I long for my mother's bread/My mother's coffee/Her touch/Childhood memories grow up in me/Day after day/'. Palestinian audiences recognise the double entendre of country and mother in his poem. The Palestinian's understanding of the land as a fertile mother directly contradicts the Israeli image of Palestine as a desert.

Both Palestinian and Israeli narratives of the land contain some truth and some exaggeration. Yet 75% of the Palestinian population were actually farmers prior to 1948, and they had some 5,484,700 dunams of land (one dunam equals 1,000 m^2) in cultivation, a situation that hardly constituted a desert (Khalidi 1984). The cultural attachment of Palestinians to their land is deep and can be illustrated by the fact that the olive trees, powerful symbols of Palestinian agriculture, are some of the longest lived plant species. Palestinian farming families name their olive trees using the same teknomy as they give do in naming each other. Anthropologists use the term teknomy to describe the local naming system by which the parents take the nickname of father (abu) of or mother (um) of their first born sons, or sometimes their daughters if they do not have a son. It is not uncommon to hear them refer to the trees as *Um* Fatimah or *Abu* Mohammed (mother of Fatimah or father of Mohammed). Providing food, oil, and income, the family trees are passed from one generation to the next. Olive trees provided an income for many thousands of Palestinians. The olive is the symbol of the Palestinian people, and they are considered blessings, having been mentioned both in the Qu'ran and the Bible numerous times. Ironically, olives are the symbol of peace, and yet Israel has torn out hundreds of thousands of olives and citrus as mass punishment for individual acts of violence by Palestinians.

In refugee and popular narratives, one of the oldest iconic images of Palestinians is that of Palestinian Bedouin or village women carrying jugs of water from their wells on their heads. Because the majority of the Palestinians before 1948 was *fellaheen*, or village agriculturalists, stories of that time are filled with references to the land and agricultural products: oranges, grapefruits, watermelons, and olives. Between 1948 and the start of the 1967 War, a period when the West Bank farmers had access to their water, they produced 45% of Jordanian gross domestic product (see Hassoun 1997a, b).

The West Bank is still dotted by Palestinian villages that use rock terracing, ancient cisterns and wells, and traditional agriculture. In addition, some of the Palestinian Druze also had sacred trees, and the tying of cloth to trees, as well as special ceremonies relating to tree worship, still persists (Dafni 2002). The Palestinian water culture was centred around their agriculture, their wells, and rivers such as the Jordan River, which has particular religious significance for Palestinian Christians as the place of baptism.

Since 1948, with the beginning of the modern state of Israel, the asymmetry of power has resulted in an asymmetry of access to water. Palestinian culture has

Fig. 5.5.1 George Phillip Hassoun (back to the camera), District Food Controller of the Gallilee before 1948, confers with a Palestinian woman farmer about her crops (Credit: Hassoun family photo)

dramatically changed because of large-scale confiscations of both communal village lands and religious land held in the Islamic trust or endowment for charity or *waqf* land by the state of Israel. The Israeli government often refuses to recognise the legitimacy of Palestinian land claims, and they are becoming increasingly urbanised and crowded into smaller land areas. The water culture of the Palestinians today is increasingly an imposed culture – imposed because of their subjugation to Israeli authority.

5.5.2 Current Conditions

Although Israel has strong environmental laws, Tal (2002) points out that the biggest problem has been lack of enforcement, leading to outbreaks of cholera in the 1990s from untreated sewage and high soil salinities and nitrate levels because of the widespread use of sewage wastewater. He also points out that Israel's use of

Fig. 5.5.2 George Phillip Hassoun, the author's father (*center*), at the celebration for a new well that he had dug as District Food Controller in a Gallilee village (Credit: Hassoun family photo)

water exceeds the recharge capacity of its aquifers, resulting in saltwater intrusion in the coastal aquifer and lowering of the water levels in the mountain aquifer. In spite of the use of innovative techniques like drip irrigation, Israelis face a looming crisis of both water quality and quantity.

While Israeli farmers face restrictions on their water quotas and civilians are encouraged to conserve water, the water situation for the Palestinians is already in a state of crisis. In the West Bank, they live under crippling water shortages. Much of the population lives on less than half of the amount the World Health Organization (WHO) considers the minimum to maintain human health. Today on the West Bank, some 13% of the villages have no access to water and must be supplied by water tankers for very high prices, and they still have inadequate supplies of water. Many of the Palestinian towns, cities, and refugee camps face severe water shortages also. The average daily water consumption by Israelis citizens is 235 litres, 3.5 times more than the Palestinians enjoy.

Pollution of Palestinian water sources, both by Palestinians and by wastewater from Israeli settlements, is of real concern as water quality declines. Israelis have deeper wells and are still immune from much of the quality problems, but Palestinians with their shallower water resources and declining water tables are already severely impacted by water quality issues.

In Gaza, the civilian population faces even worse conditions. Gaza took in hundreds of thousands of refugees from the 1948 and 1967 wars, following which the area was fenced in, cut off from job opportunities. Part of the issue in Gaza is the long-term extreme poverty and the extended humanitarian crisis of both food and water. Prior to 2007, between 25% and 30% of the population did not have running water in its homes. The large apartment blocks and many other buildings rely on water tanks on their roofs and water supplies trucked into the town.

Because of the blockade of Gaza following the 2006 elections, the 1.4 million civilians living here were already experiencing a full-blown humanitarian crisis before the start of Operation Cast Lead, the war that started on December 27, 2008. According to the United Nations Relief and Works Agency for Palestine Refugees in the Near East, 80% of the population of Gaza lived under the internationally designated poverty line, with 70% living in deep poverty (a family of 6 living on an income of less than $469 US per month – and actually it is much less), with UN food aid going to 1.1 million people (Reuters 2008). But they have faced a water crisis for much longer.

In Gaza in 1945, a British commission recommended that no further development should be allowed in the coastal area because of the fragility of the coastal aquifer. Nevertheless, the city of Tel Aviv was built over the aquifer, and refugees pouring into Gaza and high birth rates increased the population from 80,000 prior to 1948 to twice that after the 1948 war, to six times greater after the 1967 war, to up to some 1.5 million people today. Gaza's water was and has been over-pumped past the aquifer's recharge capacity for decades, in part because Israeli settlers had unlimited access (Israel has since removed the Israeli settlements from the Gaza Strip). The result of this over-pumping is that the aquifer has salt water intruding from the sea. Amira Hass, a reporter for the Israeli *Haaretz* newspaper, spent a year living in Gaza and wrote a book entitled, *Drinking the Sea in Gaza: Life in a Land under Siege.* She provides clear insights into what daily life is like in Gaza (see Hass and Nunn 2000).

In the 1990s, the European Community reported that the chloride (sodium chloride is a measure of the salt content of the water) in Gaza's water exceeded the 200 mg/litre that is the upper permissible limit for human consumption. By 2003, only 10% of the wells in Gaza were up to WHO minimal water standards. The coastal aquifer underneath the area is collapsing because of long-term over-pumping. Gazans have reported that they have no 'sweet water' and that what water they have often smells foul. The geology of the area is karst, meaning it has porous rock and soil layers, which allow untreated sewage to flow into the aquifer. Approximately 20% of Gaza's population, mostly refugees, live in places without any sewage treatment, and in some camps in the north raw sewage still flows directly into the sea. Since 1967, the Israeli government has made only minimal improvements to the wastewater structure, allowing the building of a small number of surface wastewater lagoons. According to recent warnings from the World Bank, one of the largest wastewater lagoons in Gaza, the Beit Lahiya sewage lake, is in danger of collapse and could potentially kill 10,000 people and drown 800 homes. Sewage in the water, as well as on the surface of the land, threatens the lives of Gazans daily.

Doctors have been reporting kidney disease, excessive deaths from contaminated water, and elevated rates of cancer from pesticide and herbicide pollution (Hassoun 1997b). I previously reported on the more than 30 banned chemicals (including DDT, 2,4,5T, and phosval) that Israel allowed for use by Palestinian farmers who are desperate to control pests and prevent disease in the precious crops they can produce, but these too end up in the water. So, long before the 2008 war and the blockade, the entire population of Gaza was drinking salty, badly contaminated water. After the recent war, the cost of repairing the infrastructure of Gaza to match conditions prior to the war is estimated at the hundreds of millions of dollars.

As bleak as the current conditions seem now, future predictions for the overall water shortfall are even more disturbing. Global warming and increasing populations are predicted to make the water crisis even more severe over time. The question remains: How and where do we look for the basis of genuine cooperation that will lead both populations toward a sustainable future? Is it possible that the past practices and cultural and religious attitudes of Palestinian Arabs and Israeli Jews may provide clues to how to build a more sustainable future?

5.5.3 Finding a Way Forward: Changing Current Government Water Policies

The possibility of changes in Israeli government policies and of greater water cooperation between Palestinians and Israelis, especially given the Palestinian factional divisions, seems bleak at this time. The Obama administration in the United States has hopes to revive the peace process but as of 2010 has not succeeded. However, remnants of the cooperation built on the diplomatic structures since the Oslo accords are still in place. One tragic scene is a reminder of the small steps that had been made towards cooperation. During the Israeli invasion of Gaza in the winter of 2008-2009, the remaining members of the joint Israeli-Palestinian Water Commission created by the Oslo Accords met at the Eretz entrance to Gaza. Sitting there, they penned a plea to the Israeli military not to bomb Gaza's water infrastructure. Their efforts were to no avail. The mere presence of Israeli tanks on the soft soils is enough to crush the clay and ceramic water and sewage pipes, as the use of metal pipes for waste and sewage was banned by the Israelis because the Israelis claim metal pipes could be used to construct weapons.

During Operation Cast Lead, the Israelis also bombed both water and sewage facilities in Gaza, leaving the 1.4 million Palestinians with a severe shortage of water and water contamination problems. Recent reports in 2010 say that some water delivery has been restored to most of Gaza. Still, historically, approximately 20% of the Gazans have not been connected to any water system, and that number has now increased as a result of the war. The United States Agency for International Development (USAID) estimates place the cost of restoring Gaza's water and sewage as high as $800 million USD. This figure does not include solving the problem of the above-mentioned Beit Lahiya sewage lagoon or providing water that would meet the WHO standards. Israeli authorities have allowed some of the much needed aid and water to enter Gaza, but only at a trickle since the Israeli incursion in December 2008. If lasting peace is going to be found, the profound water crisis of the Palestinians in the West Bank and Gaza must be addressed. The deprivation imposed by the Israelis has failed to force a capitulation and has radicalised certain Palestinian factions, rather than producing resolutions for either side.

5.5.4 Finding Current Case Studies of Water Cooperation

Does this bleak situation mean there is no hope for governmental policy changes for Palestinians and Israelis for the foreseeable future? The answer is complex and lies in current case studies of water cooperation and changing environmental awareness. Since the Oslo Accords, Jordan has not been allowed access to the river after which its country was named. However, the Israelis have been providing some badly needed water to Jordan via a pipeline (at least 20 million m2 per year). Although this arrangement is not perfect and the water provided is far below the needs of the nation of Jordan, this example does illustrate that the Israelis are capable of water sharing.

But more importantly, something else is happening among Palestinians and Israelis – the birth and growth of environmental movements in both populations, including cooperation between Palestinians and Israeli environmentalists and the exchange of ideas and help. Both groups lag well behind the developed world in fostering environmental activism. However, Palestinian and Israeli environmental awareness and activism is growing. There was only one Israeli (Jewish) environmental non-governmental organization (NGO) 15 years ago, the Society for the Protection of Nature in Israel. Since then, there has been a proliferation of Israeli environmental groups recently, including Israel Union for Environmental Defense (an environmental advocacy group), the Heschel Centre (focusing on environmental education leadership building), Green Course (a student organization), and Green Action, to name only a few. Alon Tal (2002) provides a more extensive list and history of the Israeli environmental movement. The watershed moment that sparked a real upturn in the Israeli-Jewish environmental movement occurred at the 1997 Maccabiah Games (a kind of Israeli Olympics), when a bridge collapsed and international athletes fell into the Yarkon river. Three of the four fatalities resulting from the bridge collapse were caused by pollution by *Pseudomonas* bacteria (Tal 2002). The public outrage over this event was the impetus for a wave of new Israeli-Jewish environmental activism.

The watershed event that seems to have sparked environmental awareness among the Palestinians was the first Land Day in March 1976, a day protesting land confiscation in the Galilee. This day of commemoration has also been associated with annual clean up days and tree plantings.

The Palestinian peace movement is tied to many different international organizations, and olive and citrus tree planting has been a major thrust of many of the activities. These have also been sponsored by a wide variety of Palestinian organizations (for example, Sabeel, a Palestinian Christian liberation theology organization). We see a definite environmental as well as international component in Palestinian activism.

The interesting thing about both the Palestinian and Israeli environmental movements is how many of the groups regularly cooperate with each other and how many are also engaged in the peace movement. For example, PRIME, a joint Palestinian and Israeli environmental organization, has encouraged the cooperation by holding joint meetings with 40 Palestinian and Israeli NGOs, as Professors Dan Bar-On and Sami Adwan reported on PRIME's website (Chaitin et al. 2004). It may take some time, but given the potential severity of the impacts of climate change, more Palestinians and Israelis may join the environmental movement. Although relatively small today, these movements have the potential to grow and to bring about cultural change. These are the droplets of hope available today.

Box 5.5a Participatory action research: water* in a Hungarian Roma community — Krista Harper and Judit Bari

The Sajó River Association for Environment and Community Development, a grassroots organization established by a group of Roma residents of Sajószentpéter, Hungary, initiated a participatory action research (PAR) project to assess neighbourhood residents' environmental and health concerns. PAR is an applied social scientific methodology that involves community members in the process of research design, data collection and analysis, and that presents research findings to a broader public. Anthropologist Krista Harper and Judit Bari, president of the Sajó Association, assembled a group of six young residents who contributed to the research as community-based photographers and 'shoe-leather epidemiologists', using the Photovoice research method.

Box Map 5.5a.1 Hungary

Photovoice is a PAR methodology that uses community-generated photography to elicit narratives and participants' voices (Wang 1999). Community participants use images generated in the research process to reach a wider audience for their concerns. As team leaders, Harper and Bari suggested two broad themes for

* See also Box 3.1a "Water in a Hungarian Roma community" by the same authors.

(continued)

Box 5.5a (continued)

Box Fig. 5.5a.1 The research team (*left* to *right*): Sándor Kelemen, Gábor Ruzsó, Krista Harper, Klaudia Kelemen, Judit Bari, Beatrix Stéfan, László Stefankó, and Mariann Bari (Photo credit: Sajó Association)

Box Fig. 5.5a.2 Training a new lens on neighborhood life (Photo credit: Krista Harper)

the photographers: 'environment' and 'health'. Before any photographs were taken, the research team discussed the ethics of taking pictures and the problem of visual stereotypes of Roma. Photographers asked permission before taking photos and reviewed the digital image with the photographic subject before exhibiting it.

The research team used photography to document and to generate discussion of community concerns, as well as sources of neighbourhood pride.

Several key concerns emerged, including access to public infrastructure such as indoor water, sewerage, and safe playgrounds, waste and dumping, energy and fuel poverty, and low-quality housing. Looking at a photo of a boy standing before an outhouse with many houses visible in the background, two community photographers discussed how the image reflected environmental and health problems.

Klára: The sewer lines stop at the bridge. This is not a small village – over 2,000 people live in our neighbourhood. It smells bad, and people can't use their yards for gardening and playing when there is an outhouse in back.

Esztér: Last year, there was a hepatitis outbreak here. The local government is not doing its job.

Images generated by community members not only served as visual documentation of conditions but also as opportunities to analyze problems and to seek solutions through organizing residents. Contrary to stereotypes held by many non-Roma, people in the neighbourhood expressed a strong desire for a clean and healthy environment. Roma community members' attachment to the Sajó River landscape appeared in photos of children swimming and families relaxing on fishing excursions.

(continued)

Box 5.5a (continued)

Box Fig. 5.5a.3 Boy standing near outhouse (Photo credit: Sándor Kelemen)

Box Fig. 5.5a.4 Enjoying the Sajó riverscape (Photo credit: Sándor Kelemen)

Box Fig. 5.5a.5 Fishing (Photo credit: Betti Stéfan)

The group held photo exhibitions in Sajószentpéter and Budapest, with the photographers taking a leading role in organizing the local exhibition. Titled 'This is also Sajószentpéter' (*'Ez is Sajószentpéter'*), the local exhibition drew media attention and was attended by town officials, health care providers, neighbourhood residents, and environmental activists from a nearby city. It also elicited feedback from community members, extended the organization's network of potential partners, and opened a broader discussion of environment and health issues with policymakers. Reflecting on the Photovoice research process, Bari stated: 'Rarely have we, the Roma, had the opportunity to show our circumstances and to voice our opinions, feelings, and ideas about issues affecting our own lives. Photos that we ourselves have taken are a good tool for communicating to others, and to our own community, too'.

The Budapest event used the photo exhibition to bring the concerns of a small-town Roma civil organization to the attention of national-level policymakers and activists. Attended by national- and international-level activists and officials, it featured a 90-minute group discussion in English and Hungarian of the issues raised by the photos, facilitated by a nationally-known Roma activist and a respected environmental lawyer. Bari and Harper also presented a slideshow to the United

(continued)

Box 5.5a (continued)

Nations Committee on Economic, Social, and Cultural Rights (UNCESCR) in Geneva in 2007.

In preparation for the exhibition, Bari and the photographers summarised the political and environmental issues raised by the Photovoice project in the 'Sajó River Declaration':

> These photos also show how difficult our everyday lives are. We do not usually contemplate how much time we spend carrying water when we want to drink, cook, bathe, or clean; how much time we spend gathering wood when we want to cook food or heat our houses. No one takes the time to count up how many hundreds of kilometres we walk each year, just to use the outhouses in our yards. It is not just the kilometres, but also the ongoing threat to our health that we would like to show in these pictures....These problems are all harmful, but they can be changed, too (Bari et al. 2007).

Although the Sajó Association has focused its activities on the grassroots level, through the participatory research process, members came to see the community's issues around water and sewerage infrastructure as evidence of the need for environmental justice:

> What do we mean, when we talk about 'environmental justice'? We mean that every community should have access to sewerage, clean water, clean air, possibilities for heating, livable and energy-efficient housing, green spaces, and protection from environmental hazards such as pollution and toxic waste. These services are the birthright of all citizens, regardless of their ethnic or socioeconomic status, as is the right to participate in policy decisions related to the environment and living conditions. These rights are guaranteed by the Hungarian constitution and by European and other international human rights charters (Bari et al. 2007).

The Sajó River Declaration is one of the first statements by a grassroots organization in Central and Eastern Europe invoking principles of environmental justice and struggles against environmental racism. Since the 2007 exhibitions, members of the Sajó Association have networked with environmentalists across the country to study and address environmental concerns of Roma communities and other marginalised groups.

The research team logged many achievements at different levels through our Photovoice project. On the individual level, the young photographers developed new skills and hidden talents, expressed themselves through documentary photography, gained the recognition of people in their hometown, and travelled to show their work in the capital city. Several of the people featured in photographs expressed excitement at appearing in portraits in the local photo exhibition and were pleased to see the neighbourhood represented in ways that challenged stereotypes. As a community organization, the Sajó River Association initiated a discussion of environmental inequalities and

(continued)

Box 5.5a (continued)

made linkages with potential allies in the town, regional, and national levels. Carrying out participatory action research allowed us to recognise the distinctive identity and history of the neighbourhood while presenting Roma residents' issues as part of broader water policy, waste management, and civic participation in Hungary.

Acknowledgement This project was made possible by funding from a University of Massachusetts, Amherst, Faculty Research Grant and the Central European University's Centre for Environment and Policy (CEPL). Many individuals contributed to the success of our project at every stage: Alex Antypas, Michael Ash, Linda Carranza, Angela Koczé, Sándor Fülöp, Michael Simmons, and Tamara Steger. This photo essay is dedicated to the memory of Jenö Bari.

Resources

Bari, Judit, and the Sajo River Association Photovoice Research Group. 2007. The Sajó River declaration. Photo exhibition pamphlet.
Wang, C.C. 1999. Photovoice: a participatory action research strategy applied to women's health. *Journal of Women's Health* 8(2): 185–192.

5.5.5 Charting a New Potential Future Based on a Common Shared Water Culture

Palestinian and Israeli environmental movements may be able to wake up their populations to the reality of their joint water crisis and the need for water sharing and cooperation and to put pressure on politicians and government leaders (including the different factions of the Palestinians). Such pressure seems like the only way that government policies and water use practices will change. Both movements can be generally categorised today as mostly secular, pro-peace, and left-of-centre politically and therefore not yet mainstream. They need a plan that will attract majorities of both peoples and that will chart their way forward to a future based on a common shared water culture and appropriate use of technologies and water. By providing water to each other, they fulfill their religious and cultural obligations and diffuse the anger of radicalization though the application of concepts of justice, so deeply held in all three major religions and in both Arab and Jewish cultures.

International awareness of these issues and support for these environmental efforts can also help them assert the pressure needed, not only to create water sharing schemes but also perhaps to lead to lasting peace, as the water for peace perspective begins to be adopted by an increasing number of advocates around the region. This environmental approach will not be an easy solution, but it offers the only alternative to the kind of futile wars seen in Lebanon and Gaza that do nothing but exacerbate the problem of radicalism and hopelessness in the region. Only with the realization of the futility of war, which causes loss of life and further damages water and sanitation infrastructure in a region where ultimately all human life and much of the water is connected, will there be a quantum change in attitudes towards peace and cooperation. The severity of the current water crisis and the dire predictions for the future water crisis call for nothing less than this magnitude of change. The droplets of hope that this cooperation can happen soon enough to avert future disaster for both Palestinians and Israelis and their neighbouring countries lie in the fostering and development of Palestinian, Israeli, and regional environmental efforts. The only feasible option, however implausible it seems at the moment, is future cooperation based on the ecological concepts of just resource distribution and culturally and environmentally appropriate technology. This cooperation must be coupled with water management and use that does not permanently damage the surface water supply and the fragile aquifers.

5.5.6 Conclusion

The way forward for both Palestinians and Israelis depends upon charting a new potential future based on a common shared water culture, conservation, and appropriate technologies. Given the water usage disparity and the asymmetry of power between Arabs and Israelis, the concept of water for peace or some kind of water sharing practices agreed upon by both Palestinians and Israelis seems unlikely at this time. Therefore the linking of Palestinian and Israeli environmental movements provides the only droplets of hope for creating conditions for equitable water sharing and more sustainable water usage practices.

The hope for a cultural shift that would allow for a vision for water sharing and the creation of a more sustainable future for both Palestinians and Israelis may very well lie in the seeds that have been sowed in the growth and development of the environmental movements among both Palestinians and Israelis and the joint organizing that has already begun.

Resources

Buzan, Barry, Ole Wæver, and Jaap de Wilde. 1998. *Security: A new framework for analysis*. Boulder: Lynne Rienner.

Chaitin, Julia, Fida Obeidi, Sami Adwan, and Dan Bar-On. 2004. Palestinian and Israeli cooperation in environmental work during the Peace Era. *International Journal of Politics, Culture, and Society* 17(3 Mar): 523–542.

Dafni, Amots. 2002. Why are rags tied to the sacred trees of the holy land? The New York Botanical Garden, Oct. http://www.bioone.org/perlserv/?request=get-document&doi=10.1663%2F0013-0001%282002%29056%5B0315%3AWARTTT%5D2.0.CO%3B2.

Hass, Amira, and Maxine Nunn. 2000. *Drinking the sea at Gaza: Days and nights in a land under seige*. New York: Henry Holt and Company.

Hassoun, Rosina. 1997a. Water between Arabs and Israelis: Researching twice-promised resources. In *Water, culture, and power: Local struggles in a global context*, ed. John M. Donahue and Barbara Rose Johnston, 313–338. Washington, DC: Island Press.

Hassoun, Rosina. 1997b. Warnings of the Musht and a truly Dead Sea. *The Link* 30(2 Apr/May): 1–16.

Homer-Dixon, Thomas E. 1994. Environmental scarcities and violent conflict. *International Security* 19(1 Summer 1994): 5–40.

Khalidi, Walid. 1984. *Before their diaspora: A photographic history of the Palestinians, 1876–1948*. Washington, DC: Institute for Palestinian Studies.

Khalidi, Walid (ed.). 2006. *All that remains: The Palestinian villages occupied and depopulated by Israel in 1948*. Washington, DC: Institute for Palestine Studies.

Lees, Susan H. 1998. *The political ecology of the water crisis in Israel*. Lanham/New York: University Press of America.

Morag, Nadav. 2001. Water, geopolitics and state building: The case of Israel. *Middle Eastern Studies* 37(3 July): 179–198.

Reuters. 2008. Interview: Gaza on brink of humanitarian disaster – UN. http://www.reuters.com/article/latestCrisis/idUSLL414968.

Tal, Alon. 2002. *Pollution in a promised land: An environmental history of Israel*. Berkeley: University of California Press.

Tamari, Salim. 2009. *Mountains against the sea: Essays on Palestinian society and culture*. Berkeley: University of California Press.

Wolf, Aaron T. 1995. *Hydropolitics along the Jordan River: Scarce water and its impact on the Arab-Israeli conflict*. Tokyo: United Nations University Press.

Chapter 5.6
'Water for Life'... Water for Whose Life? Water, Cultural Diversity and Sustainable Development in the United Nations

Lisa Hiwasaki

Water is one of the most pressing development challenges of our time. It has been recognised as indispensable for sustainable development, for the preservation of the environment, and for the alleviation of poverty and hunger (UN-Water 2005). It is in fact water that cuts through and connects the eight Millennium Development Goals set by the United Nations (UN) (WWAP n.d. a). In recognition of water resources as 'our lifeline...for sustainable development in the twenty-first century' (Annan 2005), the years 2005–2015 have been designated by the UN as the International Decade for Action 'Water for Life'. The primary goal of this decade is to promote efforts to fulfil international commitments made on water issues, most notably the UN Millennium Development Goals (MDGs) to halve the proportion of the population without sustainable access to safe drinking water and basic sanitation, as well as commitments made at the World Summit on Sustainable Development (WSSD) to stop the unsustainable exploitation of water resources and to develop integrated water resources management and water efficiency plans (UN-Water 2005).

Despite efforts made by the UN and the international community to meet the MDGs and notwithstanding the important progress made, many of the goals are far from being met. According to the latest report, although the target for safe drinking water is likely to be met, steep challenges remain in meeting the sanitation target, in particular in South Asia and sub-Saharan Africa. The number of people without an improved drinking water source is now below one billion, whereas 2.5 billion people, or 38% of the world's population, remain without improved sanitation facilities. In addition, far more efforts are necessary to preserve natural resources, including water (UN 2009). Although the global economic crisis that began in 2008 has been blamed as the culprit in the lack of progress in meeting the goals as a whole (UN 2009), the increasing disparity in attainment of targets – most notably the poor,

---

The views expressed in this chapter are the author's and do not represent the position of UNESCO.

L. Hiwasaki (✉)
UNESCO Office in Jakarta, Regional Science Bureau for Asia and the Pacific
e-mail: l.hiwasaki@unesco.org

B.R. Johnston et al. (eds.), *Water, Cultural Diversity, and Global Environmental Change: Emerging Trends, Sustainable Futures?*, DOI 10.1007/978-94-007-1774-9_34,

rural populations, and women, who have not benefited proportionally (WHO and UNICEF 2008) – makes it clear that human factors in water management and decision making, in particular issues of equity, need to be better dealt with.

This chapter provides an overview of efforts made by the UN on water and sustainable development, focussing specifically on efforts to improve and maintain the well-being of people, their diverse cultures, and the environment. It gives a historical background into the trends within the UN and examines the work of UN entities with mandates of water or cultural issues. Starting with the 1977 Mar del Plata UN Conference on Water in Argentina – which explicitly recognised the right of access to water as a human right – the UN has played an important role in the international community in addressing the issue of water and sustainable development. These efforts, however, have not resulted in improvement of local situations in many parts of the world, nor has a global regime emerged to deal with freshwater issues, such as a comprehensive framework convention. The chapter ends by calling for the need for a paradigm shift within the UN, to address the complexity of issues surrounding water – the specific social, cultural, economic, and political dimensions – in order to manage the resource in a sustainable manner. Only after dealing with sociocultural dimensions of people's engagements with water would it be possible to move forward towards a more effective way to address the global water issues.

5.6.1 Historical Background: UN-Wide Water Activities

Water issues were first discussed within the UN framework at the United Nations Scientific Conference for the Conservation and Utilization of Resources in 1949, at which an entire session focused on evaluating water resources and building dams (Batisse 2005). It is, however, the United Nations Conference on the Human Environment held in Stockholm in 1972 – at which water pollution was one of the central concerns – that is generally recognised as the starting point of international environmental politics, including those surrounding water (Conca 2005). As the first occasion at which the term 'environment' arrived on the international agenda, the Stockholm Conference was the prelude to a series of large UN conferences on the topic of the global environment. The Stockholm Declaration on the Human Environment mentions water as one of the natural resources that need to be 'safeguarded for the benefit of present and future generations', and recommendations 95, 98, and 99 of the Action Plan for the Human Environment mention 'social and cultural aspects of environmental issues' (UN 1972). This meeting is also credited with bringing the concept of intergenerational equity to the fore, thereby placing the essence of sustainability in a particular relationship between people and people, rather than between people and nature (Sachs 1999).

A more significant milestone specific to water is the UN Conference on Water, held in 1977 in Mar del Plata, Argentina. Renowned as the first attempt to take a wider look at water issues, it was during this conference that access to water was recognised as a human right. The Mar del Plata Action Plan adopted at this

Table 5.6.1 UN-wide water milestones

Year	Event	Relevant output/product
1972	UN Conference on the Human Environment, Stockhom (Sweden)	Declaration of the UN Conference on the Human Environment mentions water as one of the "natural resources of the earth" that "must be safeguarded for the benefit of present and future generations through careful planning or management."
1977	UN Conference on Water, Mar del Plata (Argentina)	Mar del Plata Action Plan, which recognized that "... all peoples, whatever their stage of development and their social and economic conditions, have the right to have access to drinking water in quantities and of a quality equal to their basic needs."
1981-1990	International Drinking Water and Sanitation Decade	Brought about a heightened awareness of the severity of the problem and innovative solutions to the provision of basic services, but "little if any progress is being made in reducing the number of people lacking access to safe water supply and suitable sanitation facilities".
1990	Global Consultation on Safe Water and Sanitation for the 1990s, New Delhi (India)	The New Delhi Statement mentions the importance of "building on indigenous knowledge, so that policies and programmes are credible and relevant to the beneficiaries. Emphasis must be placed on education, social mobilization and community participation."
1992	International Conference on Water and the Environment, Dublin (Ireland)	The Guiding principles of the Dublin Statement on Water and sustainable Development recognize that "Water development and management should be based on a participatory approach" (principle 2), and the basic right of all human beings to have access to clean water and sanitation at the same time recognizing water as an economic good (principle 4).
1992	United Nations Conference on Environment and Development (UNCED), aka the Earth Summit	Chapter 18 of Agenda 21 includes "Integrated water resources development and management" proposed as one of the seven programme areas for the freshwater sector.
1997	Adoption of the United Nations Convention on Non-navigational Uses of International Watercourses	The Convention's general principle is "equitable and reasonable utilization and participation".
2000	Launch of the World Water Assessment Programme"	The World Water Development Reports (WWDRs) published every 3 years (2003, 2006, 2009), are comprehensive reviews that provide "an authoritative picture of the state of the world's freshwater resources".

(continued)

Table 5.6.1 (continued)

Year	Event	Relevant output/product
2000	Millennium Summit, New York (USA)	The United Nations Millennium Declaration, which resolved to "halve the proportion of people who are unable to reach or to afford safe drinking water" by the year 2015, and to "stop the unsustainable exploitation of water resources… [and] promote both equitable access and adequate supplies".
2002	World Summit for Sustainable Development (WSSD), Johannesburg (South Africa)	Johannesburg Plan of Implementation added another target, to halve "the proportion of people who do not have access to basic sanitation", and to "develop integrated water resources management and water efficiency plans by 2005".
2003	Establishment of UN-Water	Various periodic reports that monitor the state and utilization of the world's water resources, such as the World Water Development Report (WWDR), the WHO/UNICEF Joint Monitoring Programme on Water Supply and Sanitation (JMP), and the Global Annual Assessment on Sanitation and Drinking-Water (GLAAS).
2005-2015	International Decade for Action "Water for Life"	The primary goal of the Decade is to promote efforts to fulfil international commitments made on water and water-related issues by 2015. Focus is on furthering cooperation at all levels, so that the water-related goals of the Millennium Declaration, the Johannesburg Plan of Implementation of the World Summit for Sustainable Development, and Agenda 21 can be achieved.

Sources:
http://www.un-documents.net/k-000287.htm
http://www.unesco.org/water/wwap/milestones/index.shtml
http://www.iisd.org/rio+5/timeline/sdtimeline.htm

conference resulted in the proclamation of the period 1981–1990 as the International Drinking Water Supply and Sanitation Decade.[1] Following the end of the decade in 1990, the New Delhi Statement adopted at the Global Consultation of Safe Water and Sanitation for the 1990s further offered a broader approach to water issues, in that it recognised the need to protect environment and health; the need for institutional reforms, including the full participation of women; the need to promote

[1] The aim of this decade was to provide safe drinking water and adequate sanitation systems for all people by 1990. Despite the fact that the decade was successful in bringing about a heightened awareness of the severity of the problems, and major changes in approaches to the provision of services resulted, the rate of progress has been slow (UN 2000).

community management; and the need to adopt sound financial practices and appropriate technologies (UN General Assembly 1990).

By the late 1990s, the wider understanding of water – from an emphasis on meeting basic human needs to a more holistic management of fresh water – had led to the birth of the paradigm of integrated water resources management (IWRM). Two landmarks are recognised in this process – the 1992 International Conference on Water and Environment in Dublin, organised in preparation for the UN Conference on Environment and Development (UNCED, also known as the Earth Summit), and Agenda 21, one of the products of the Earth Summit. The Dublin Principles, while recognizing fresh water as a finite and vulnerable resource essential to sustain life (principle 1), and emphasizing the importance of the participatory approach in water development and management, in particular including women (principles 2 and 3), also reinforced the view of water as an economic good (principle 4). Chapter 18 of the Agenda 21, which resulted from the Earth Summit, also endorsed the idea of the IWRM, making 'integrated water resources development and management' one of the seven programme areas proposed for the freshwater sector (UN 1992).

The only existing UN convention that concerns water is the United Nations Convention on Non-navigational Uses of International Watercourses, adopted by the UN General Assembly in 1997, as a result of over 20 years of preparation. Although it has yet to come into force because not enough countries have ratified it, this convention is considered to constitute an important step towards development of an international law on water. The purpose of the law is to protect, preserve, and manage the uses of watercourses and their waters, and the law has adopted the general principles of 'equitable and reasonable utilization and participation' (UN 2005). The convention does not include the human right to water but instead focuses on 'vital human needs', which is a non-legal concept (Gupta 2006).

A series of events in the 2000s reinforced the global trend towards integrated and holistic approaches to water management. In March 2000, responding to a need for a collective, UN system-wide comprehensive assessment of the world's freshwater resources in order to push towards a more integrated management of water, the UN World Water Assessment Programme (WWAP) was born (WWAP n.d. b).[2] Since its establishment, the WWAP has coordinated the publication of the World Water Development Reports (WWDR), launched in 2003, 2006, and 2009 at the World Water Forums in Kyoto, Mexico City, and Istanbul respectively. The MDGs, agreed to by member states of the UN at the Millennium Summit in 2000, set specific targets for reducing poverty, hunger, disease, illiteracy, environmental degradation, and discrimination against women by 2015. The WSSD held 2 years

[2] The WWAP is one of the four programmes of UN-Water, an inter-agency mechanism consisting of 26 UN agencies to strengthen coordination and coherence among UN entities dealing with issues related to all aspects of fresh water and sanitation (UN Water 2009). UN Water was established in 2003 to respond to the need to mobilise resources of the UN in an efficient and integrated manner to tackle global water challenges, and to meet the ambitious development and environment targets set by the international community. The WWDR is one of the flagship reports of UN-Water.

later complemented the MDG target on water – to reduce by half the proportion of people without access to safe drinking water – by adding another target, to halve the proportion of people lacking improved sanitation. Furthermore, the need for all countries to stop the unsustainable exploitation of water resources, emphasised in the Millennium Declaration, was addressed again in Johannesburg when governments agreed to develop IWRM and water efficiency plans by 2005 (UN-Water 2005).

The Water for Life Decade 2005–2015 – the second international decade on water-related issues proclaimed by the UN General Assembly - has been recognised as an opportunity for the international community to advance towards a truly integrated approach to the sustainable management of the world's water (UN-Water 2005). The decade has prioritised women; it recognises that it is crucial to ensure the full participation and equal involvement of women in water-related development efforts and to approach water and sanitation issues from a gender perspective (UN-Water 2005). More significantly, culture is recognised as one of the 11 central themes for the decade, along with food, health, environment, disaster prevention, energy, transboundary water issues, scarcity, sanitation, pollution, and agriculture (UN-Water 2009). It is thus appropriate that the theme for World Water Day[3] in 2006 was 'Water and Culture', and the United Nations Educational, Scientific and Cultural Organization (UNESCO) led the event. According to the message by then UN Secretary-General Kofi Annan, water is not only essential for life but also for its wide-ranging cultural presence: 'an inspiration for artists, a focus of scientific research, and an indispensable element in the religious rituals of many traditions and faiths'. Annan called for the recognition of the 'cultural, environmental and economic importance of clean water', as well as the 'need to distribute water more equitably' (UN DPI 2006).

The historical overview above demonstrates that the UN system has played an important role in setting a global trend towards the promotion of integrated and holistic approaches in water management. Although there are an increasing number of documents touting the importance of participation of local communities, women, and indigenous people in water management and decision-making, the documents are often criticised as merely playing lip service. This criticism can partly be attributed to the gap between theory and practice, since full and meaningful participation is difficult to implement and manage.

Considerations of local people and their cultures are even fewer and further between, even as their importance is being recognised by those working on the ground.

---

[3] In conformity with the recommendations of the Agenda 21, the UN General Assembly adopted resolution A/RES/47/193 by which 22 March of each year was declared World Day for Water. Each year the day is celebrated under a different theme.

Box 5.6a The influence of culture on interactions between stakeholders of rural water, sanitation, and hygiene promotion programmes: Lessons from Uganda

—James Webster

5.6a.1 Cultural issues in development

Why does a man continue to send his children to collect water from a spring 500 m down a steep mountainside, an arduous 2-hour round-trip trek, when he has just heavily invested time, money, and material resources in a concrete jar to collect rainwater from his roof? While the rest of the community was successfully using their jars to meet their family's water needs, Byaruhangar's jar lay on its side. Along with two pieces of guttering, it remained unused. When asked why, he simply said, 'I have children to collect water'.

Box Map 5.6a.1 Uganda

Similarly, why does a family continue to collect and consume contaminated stream water, when one of many improved sources that they and the community have recently constructed is only 50 m away? Whereas the rest of the community used such improved sources and enjoyed good health as a result, this family subsequently lost three of its members to typhoid.

These examples, experienced by the author in Southwest Uganda during the mid-1990s, occurred in projects implemented by an indigenous non-governmental organization (INGO) through its rural water, sanitation, and hygiene promotion (WASH) programme, which the author managed. This programme has consistently been recognised as exhibiting best practice in the sector. Nevertheless, unintended project outcomes still occur as illustrated above, indicating that even when all stakeholders seem to agree on the project outcomes, such as positive changes in hygiene behaviour or water collection from improved sources, not all stakeholders necessarily adopt the changes. The variations between intended and actual WASH project outcomes offer lessons relevant to national and international development.

5.6a.2 Interpretation and proposed solutions

The variations that exist between intended and actual outcomes of international development programmes reduce the impact of these projects and thereby fail to decrease global poverty. To enhance the impact of rural WASH

(continued)

Box 5.6a (continued)

Box Fig. 5.6a.1 Collecting spring water (Credit: James Webster). Only one in five Ethiopians have access to an improved water source. All diseases, including those from contaminated water, are seen by the Bunna tribe of South West Ethiopia as coming from *maeshi* (dead ancestors). Making a sacrifice to appease the most recently deceased (the most powerful) traditionally takes precedence over seeking 'Western' medical assistance

projects and development practice generally, it is necessary to consider cultural and cross-cultural dimensions in the interactions of stakeholders. When development project organisers and stakeholders understand each other's worldviews, they have a useful tool to reduce misunderstandings and conflicts, and enhance communication and training - fundamental components of development projects. Achieving this understanding can have wide-reaching, pragmatic implications for international development projects generally, facilitating lasting change in efforts to reduce global poverty.

Below, each example is represented as a problem statement, an interpretation is given, and a solution offered - though highly speculative and not exhaustive - to prevent recurrence of the problem.

Problem: Byaruhanga's rainwater jar lies on its side whilst he continues to send his children to collect water from a distant spring.

Interpretation: His motivation to invest in the jar resulted from an active desire either to promote a positive or avoid a negative self-image. He saw only the status value of the jar and invested in one in order to gain status or avoid the relative loss of status by not having one. High power distance, masculinity, and individualism prevented him from seeing his children as having equal value.

Solution: Prior to construction, a house-to-house baseline survey by male and female staff is necessary to determine the motivation for involvement in WASH projects. A follow-up visit by a resident male extension worker in the evening (to ensure finding farmers at home) is required to address issues of inequitable status according to age and gender. Alternatively, the programme should deal with primary water users, in this case the women of the house.

(continued)

Box 5.6a (continued)

Box Fig. 5.6a.2 Using improved sources of water (Credit: James Webster). Amongst the Bunna of South West Ethiopia, anything coming from the ground is considered dirty. With no springs in the area, the only improved water sources available are through drilling. Training in water usage and hygiene promotion is therefore essential. Using microscopes to show the contaminants in water assists

Problem: A family collects, consumes, and dies from contaminated stream water when an improved source whose construction they had contributed to was only 50 m away.

Interpretation: The behaviour resulted from a combination of a lack of awareness of health risks because of high power distance and didactic (masculine) training by the implementing agency; time pressures because of poverty (a sick member of the family needed constant care or household and agricultural chores needed attending to); a high resistance to change; a high degree of religious belief that generated a sense of fatalism; a low concern for public self-image with regard to hygiene or a lack of critical mass of public opinion to encourage improved hygiene.

Solution: a baseline survey to determine existing health schemata; 100% hygiene promotion coverage in an appropriate manner (e.g., through the use of microscopes to 'reveal' micro-organisms, or through house-to-house visiting) to bring about schema change and achieve a critical mass of good practice; people- rather than task-oriented construction, so that a focus on the end user prioritises burden reduction; adequate funding by donors to facilitate end-user focus.

There are, however, positive signs of change, in particular since the beginning of the Water for Life Decade. Furthermore, specific agencies within the UN are implementing activities to address these issues, and these initiatives, although fragmented, cover for the want at the UN system-wide level. These activities and initiatives are discussed below.

5.6.2 Activities by Organizations of the UN System

A review of the scientific literature and policy papers in the water sector, as well as conversations with experts, make it evident that engineering and natural sciences still dominate approaches to water management. The UN system is no exception. Considerations of human dimensions of water are, for the most part, limited to those coming from economics or non-social scientific disciplines and are limited to the following three categories: (a) human uses of water such as drinking, washing, and fishing; (b) human activities such as agriculture and industries that exert negative pressures on water resources, such as pollution; or (c) examinations of water-people relations in the framework of environmental services provided by water, such as food, recreation, and aesthetic values (Hiwasaki and Arico 2007). On the rare occasions when words such as ‘social’ or ‘society’ are mentioned, they typically render the richness of people-water interactions as ‘socioeconomics’ and are thus reductionist in approach.

On the topic of water and culture, one of the most recognizable links is the water heritage that exists around the world. Sites listed on UNESCO’s World Heritage List include historic hydraulic structures such as qanats, of which the city of Bam has preserved some of the earliest evidence in Iran, and aqueducts, such as those in Segovia (Spain) and Pont du Gard (France). Water heritage also includes sites where fresh water is dominant, such as the rice terraces of the Philippine Cordilleras. Cultural expressions that are inspired by or closely affiliated with water, such as paintings, photos, poems, stories, and songs, can also be considered water heritage, as well as intangible heritage such as rituals, festivals, dances, and other special occasions of cultural significance that water helps maintain, either directly or indirectly.

The meanings attributed to water (e.g., water as sacred, a purifier and a source of power) and identities rooted in freshwater ecosystems are manifested in symbolism, myth and rituals, and demonstrate the essential role water plays in belief systems and mythologies of cultures around the world. Moreover, there are practices that contribute to the conservation of water and related resources that are rooted in customs, beliefs, knowledge, worldviews, and values of local people. Traditional ecological knowledge in particular has played vital roles in protecting springs, rivers, lakes, wetlands, and forests that protect watersheds.

The relationships between water and heritage, identity and worldviews, as described above, are now widely recognised, and most documents written on the topic of water mention the importance of such relationships, most commonly in the preface or foreword (e.g., FAO 1995, 2002; WWAP 2003, 2006, 2009). In the UN freshwater context, cultural factors are most often mentioned as a non-economic value of water (e.g., WWAP 2003, 2006) or as one of the drivers of pressures affecting freshwater resources, along with economic and social drivers (e.g., WWAP 2009). In these contexts, religious beliefs and customs are often the focus, and examples are often illustrated by photographs of religious festivals or rituals that contribute to conservation of water or demonstrate the intimate spiritual relations people have with water. World heritage sites related to water are also often showcased to demonstrate the cultural factors of water (WWAP 2006; UNESCO 2006).

There are, however, organizations in the UN System that endeavour to engage in cultural issues more comprehensively. An overview of work of three UN entities is mentioned in this section: the UN Permanent Forum on Indigenous Issues (UNPFII), the United Nations University Institute for Advanced Studies (UNU-IAS) Traditional Knowledge Initiative, and UNESCO, especially that of the International Hydrological Programme (IHP).[4]

The UN Permanent Forum on Indigenous Issues (UNPFII) is an advisory body to the UN Economic and Social Council, with a mandate to discuss indigenous issues related to economic and social development, culture, the environment, education, health, and human rights (UNPFII 2006). Though water is not specifically included in its area of work, the forum has been responsible for events and reports that bring to the fore the necessity to recognise the relationship indigenous peoples have with water and other resources, thus the ensuing need for indigenous people to have access to these resources, and thereby endorsing the right to water as a human right (e.g., UN 2004; UN Economic and Social Council 2003, 2007). The forum also has the task of promoting respect for and full application of the United Nations Declaration on the Rights of Indigenous Peoples, adopted by the UN General Assembly in 2007 (UN 2008a). This declaration recognises the right of indigenous peoples to keep their relationships with water and calls for states to obtain consent of indigenous peoples when embarking on projects affecting their resources, including water development.[5]

The UNU-IAS Traditional Knowledge Initiative is a pilot research programme established in 2007 as a precursor of the United Nations University (UNU)[6]

---

[4] Looking up every single project and activity implemented by UN organizations in the field would be an impractical if not an impossible endeavour; thus, smaller, ground-level activities implemented by UN funds and programmes such as the UNDP and UNICEF that incorporate these issues on project-by-project basis are not mentioned here.

[5] It is written in Article 25 of the Declaration: 'Indigenous peoples have the right to maintain and strengthen their distinctive spiritual relationship with their traditionally owned or otherwise occupied and used lands, territories, waters and coastal seas and other resources and to uphold their responsibilities to future generations in this regard', while Article 32 stipulates 'States shall consult and cooperate in good faith with the indigenous peoples concerned through their own representative institutions in order to obtain their free and informed consent prior to the approval of any project affecting their lands or territories and other resources, particularly in connection with the development, utilization or exploitation of mineral, water or other resources. (UN 2008a).

[6] Other institutes of the UNU have addressed cultural dimensions of water, for example, on traditional water management in dry areas funded by UNU-INWEH (International Network on Water, Environment and Health) (Adeel et al. 2008). However, just as for smaller, ground-level activities implemented by UN funds and programmes, examining every single research and studies implemented by the UNU and its institutes is beyond the scope of this chapter.

Table 5.6.2 UN System and water

UN secretariat	Programmes and funds, under the UN General Assembly	Commissions and other bodies under the UN Economic and Social Council	Specialized agencies	Other UN entities, related organizations	Conventions
DESA (Department of Economic and Social Affairs): one of the areas of work of DESA's Division for Sustainable Development (DSD) is water [www.un.org/esa/dsd/dsd_aofw_wat/wat_index.shtml]. Through its Water management Branch, DSD provides project execution and policy advisory services in IWRM, technical assistance in groundwater development. Hosts Secretariat of UN-Water [www.unwater.org], which strengthens coordination and coherence among UN entities dealing with issues related to all aspects of fresh water and sanitation.	*UNCTAD (UN Conference on Trade and Development)*: no specific project/programme on water, but deals with water in framework of infrastructure development, needs of LDCs and SIDS, etc. [www.unctad.org] *UNEP (UN Environment Programme)*: its activities on freshwater focus on IWRM, ecosystem services, and international waters, among others; implemented a project on dams and development as a follow up to the World Commission on Dams (WCD) [www.unep.org/themes/Freshwater/index.asp]. *UNICEF (UN Children's Fund)*: under the topic "Water, Sanitation and Hygiene" [www.unicef.org/wash/index.html], works to improve water supplies and sanitation facilities, and to promote safe hygiene practices.	*Economic Commission for Africa*: serves as Secretariat of UN-Water/Africa [www.uneca.org/awich/], which coordinates water activities in Africa by various UN and other subregional IGOs and to promote joint collaborative activities. *Economic Commission for Europe*: services the Convention on the Protection and Use of Transboundary Watercourses and International Lakes [www.unece.org/env/water/welcome.html], part of a larger environmental legal framework to address trans boundary issues in the region. *Economic Commission for Latin America and the Caribbean*: [www.eclac.cl/drni/]	*FAO (Food and Agricultural Organization)*: its Water Development and Management Unit [www.fao.org/nr/water] promotes the sustainability of water use for food production. *UNESCO (UN Educational, Scientific and Cultural Organization)*: work in water is built on the three pillars of the international hydrological programme (IHP), the only broadly-based science programme in the UN system on water research, water resources management, education and capacity-building [www.unesco.org/water/ihp]; Institute for Water Education (IHE), an international institute for water education [www.unesco-ihe.org/]; and the World Water Assessment Programme (WWAP)	*UNU (UN University)*: the UNU Institute for Water, Environment & Health (UNU-INWEH)'s thematic areas include Freshwater Ecosystems, Coastal Zone, Ecosystems, Dryland Ecosystems, and Water-Health Nexus [www.inweh.unu.edu]. One of the five programmes of the Traditional Knowledge Initiative of the Institute of Advanced Studies (UNU-IAS TKI) is on water management, which explores the interlinkages between global water issues and Indigenous Knowledge systems [www.unutki.org/].	*CBD (Secretariat of the Convention on Biological Diversity)*: One of its seven thematic programmes is on Inland Waters Biodiversity [http://www.cbd.int/waters]. *UNCCD (Convention to Combat Desertification)*: water management is one of its seven focus areas [www.unccd.int/science/menu.php?newch=l45], recognised as critical in the battle against desertification and the exacerbated impacts of drought. *UNFCCC (United Nations Framework Convention on Climate Change)*: the Convention sets overall framework for intergovernmental efforts to tackle the challenge posed by climate change. Acknowledges consequences of climate change on water

UNISDR (UN International Strategy for Disaster Reduction): no specific project/programme on water, but acknowledges problems with access to safe water after natural hazards [www.unisdr.org].	*UNDP (UN Development Programme)*: Water Governance [www.undp.org/water] is one of the focus areas of its work under Environment and Energy, in recognition of water's role for sustainable development, including poverty reduction. Priority areas are IWRM, transboundary waters, water supply & sanitation.	*Economic Commission for Asia and the Pacific*: one of the focus areas of its Environment and Development Division is Water Security [www.unescap.org/esd/water/], and acts as the key driver for regional co-operation in water resources management.	IHP's focal areas include global changes, water governance, ecohydrology, and water education. *WHO (World Health Organization)*: works on aspects of water, sanitation and hygiene, e.g., drinking-water quality, bathing waters, water resources, water supply, sanitation & hygiene, wastewater, water-related diseases [www.who.int/water_sanitation_health/en]. *World Bank*: the largest external financier in the water sector, the Bank's operations are in water resources management, water supply and sanitation, irrigation and drainage, and hydropower [www.worldbank.org/water].	[www.unesco.org/water/wwap/], a flagship programme of UN-Water. *IAEA (International Atomic Energy Agency)*: Water Resources Programme [www-naweb.iaea.org/napc/ih/index.html] provides science-based information and technical skills to better understand and manage water resources; focus on isotope techniques.	resources [unfccc.int/essential_background/background_publications_htmlpdf/climate_change_information_kit/items/294.php] .

(continued)

Table 5.6.2 (continued)

UN secretariat	Programmes and funds, under the UN General Assembly	Commissions and other bodies under the UN Economic and Social Council	Specialized agencies	Other UN entities, related organizations	Conventions
	UNHCR (UN High Commissioner for Refugees): no specific project/ programme on water, but related activities include providing drinking water and sanitation services in refugee camps [www.unhcr.org]. *UN-HABITAT (UN Human Settlements Programme)*: its Water and Sanitation programme works to improve access to safe water and help provide adequate sanitation to low-income urban dwellers [www.unhabitat.org/ categories.asp?catid=270].	*Economic Commission for Western Asia*: one of the sections under Sustainable Development and Productivity Programme is on water resources [www.escwa.un. org/divisions/teams. asp?teams=Water%20 Resources&division=SDPD]. Activities focus on IWRM and cooperation on shared water resources. *UNPFII (UN Permanent Forum on Indigenous Issues)*: water is not specifically included in its area of work, but has been responsible for events/ reports that highlight the importance of recognizing the relationship indigenous peoples have with water and other resources [www.un. org/esa/socdev/unpfii/ index.html].	*WMO (World Meteorological Organization)*: its Hydrology and Water Resources Programme works on assessment of the quantity and quality of water resources, both surface and groundwater [www.wmo.int/pages/prog/ hwrp/index_en.html]. *IFAD (International Fund for Agricultural Development)*: Numerous projects and programmes related to water [www.ifad.org/english/water/ index.htm] e.g., water to combat rural poverty, linking land and water governance. A project that improved land and water use building on local and scientific knowledge is an example of a project related to the theme water and cultural diversity [www.ifad.org/events/water/ case/peru.pdf].]		

UNIDO (UN Industrial Development Organization): its Water Management Unit facilitates technology transfer to improve water productivity in industry and prevent discharge of industrial effluents into international waters [www.unido.org/index.php?id=o5073].

UNWTO (World Tourism Organization): no specific project/ programme on water, but recognizes that water consumption in the tourism industry is high and therefore responsible water management is of great importance to achieve sustainability within the tourism sector. [http://unwto.org/

Order and arrangement of organizations are as shown in http://www.un.org/aboutun/chart_en.pdf

Traditional Knowledge Institute (TK Institute), to be established in Australia. This initiative aims to promote and strengthen research on the traditional knowledge (TK) of indigenous and local communities conducted from a global perspective, grounded in local experience (UNU-IAS TKI 2009). One of the five programmes of the initiative is on water management, which explores the links between global water issues and Indigenous knowledge systems in order to develop sustainable solutions to pressing water resource challenges and facilitate their recognition and inclusion in international policy. In addition to being involved with various activities as a partner organization of the UNESCO-IHP project on Water and Cultural Diversity (described below), the TKI has recently cosponsored the Indigenous World Forum on Water and Peace, which intervened at the eighth session of the UNPFII in 2009, calling for the forum's support for the Indigenous World Forum on Water and Peace and the formation of a Special Rapporteur[7] on water, Indigenous Peoples and human rights (Indigenous Water Caucus 2009).[8]

As with the two UN entities mentioned above, UNESCO[9] also began its work on water and culture by focussing on indigenous peoples. Activities began in 2000 at the Second World Water Forum,[10] at which UNESCO's project on Local and Indigenous Knowledge Systems (LINKS)[11] organised a thematic session on Water and Indigenous People. At the Third World Water Forum in 2003, UNESCO

[7] Appointed by the UN Secretary General, special rapporteurs are individuals working on behalf of the UN who bear a specific mandate from the UN Human Rights Council (or the former UN Commission on Human Rights, UNCHR), to investigate, monitor and recommend solutions to human rights problems.

[8] A similar statement was also made at the Seventh Session of the UNPFII in 2008 (UN 2008b).

[9] The mandate of UNESCO is to contribute to the building of peace, the eradication of poverty, sustainable development and intercultural dialogue through education, natural and exact sciences, social and human sciences, culture, and communication and information. Among the 26 UN agencies that deal with freshwater issues, its International Hydrological Programme (IHP) - launched in 1975 after the International Hydrological Decade (IHD, 1965-1974) - is the only broadly-based science programme that focuses on water. This intergovernmental scientific programme currently employs its work in water research, water resources management, education and capacity-building.

[10] The World Water Fora, the largest international event in the field of water, was launched in 1997 in Marrakesh. Held every 3 years, the meetings are organised by the World Water Council (a non-profit think-tank consisting of members from the private sector, government ministries, academic institutions, international financial institutions, and UN entities) and the hosting country. While these Forums are not UN events, many UN organizations participate in them and these events have played an important role in increasing political and media awareness of water issues.

[11] The convener of the official session on indigenous peoples at the Second World Water Forum and a co-coordinator of the two official sessions on indigenous peoples at the third Forum was UNESCO's LINKS (Local and Indigenous Knowledge Systems) project, which aims to enhance biodiversity conservation and secure an active and equitable role for local communities in resource governance by creating dialogue amongst traditional knowledge holders, natural and social scientists, resource managers and decision-makers. See UNESCO-LINKS 2006 for a compilation of papers presented by indigenous people at the second and third World Water Fora. For more information on LINKS, see www.unesco.org/links.

co-coordinated various events around the theme of water and cultural diversity. It was at this forum that the Indigenous Peoples' Kyoto Water Declaration was adopted.[12] In 2006, UNESCO was responsible for coordinating the World Water Day under the theme Water and Culture, and the agency co-organised a session at the Fourth World Water Forum on water and cultural diversity. In 2007, the project on Water and Cultural Diversity, which endeavoured to mainstream cultural diversity into IWRM, was launched. The goal of the project was to contribute to the achievement of the MDG to 'ensure environmental sustainability' by recognizing and respecting cultural diversity and interweaving various perspectives into collaborative and inclusive actions for sustainability of water and cultures (UNESCO-IHP 2008). One of the project's mandates for action was the 2001 UNESCO Declaration on Cultural Diversity, which, among other things, calls for member states to respect and protect traditional knowledge, in particular that of indigenous peoples, and to recognise the contribution of traditional knowledge to environmental protection and the management of natural resources (UNESCO 2002).

5.6.3 Water, Water Cultures, and Cultural Diversity

Activities implemented by organizations of the UN System can be divided into two groups: one dealing with water cultures (i.e., demonstration of links between water and culture or artistic expressions inspired by water), and the other dealing with the larger concept of water and cultural diversity, which embodies, in addition to water cultures, issues such as governance, power, and rights. In recent years, references to the former have become quite common in reports and other documents produced by organizations in the UN System, without probing deeply into the latter. Often, discussions on the relationship between water and culture are limited to that concerning indigenous or traditional peoples, without recognizing that everyone - including water scientists and engineers - is influenced by culture. Up until as recently as 2006, when UNESCO led the World Water Day under the theme Water and Culture, the focus was primarily on non-Western cultural traditions and artistic expressions (UNESCO 2006). Systematic analyses of the relationships between cultural diversity and water, and their implications for sustainable management of water resources, are almost nonexistent.

Discussions of water and cultural diversity bring to the fore issues that are often omitted in discussions on water and culture - most notably (a) social and political

[12] This Declaration is the first known global collective statement of indigenous peoples on water, and was formulated by indigenous participants at the Third World Water Forum as collective message to decision makers. It reiterates the right of indigenous peoples to self-determination, calls for recognition of traditional knowledge, and the need for prior and informed consent and consultation of indigenous peoples in decisions about water (UNESCO 2003).

institutions that govern water use; (b) contested uses of water; and (c) the human right to water, which are all interlinked issues. Examining the diverse range of social and political institutions - both formal and informal, such as national and international regulations, shared practices in a river basin, and local norms of water use and conservation - that govern management and use of water is necessary to better understand the use, access, and control of water at a range of different scales. Such studies are important because 'water resources are never simply there; they are produced by social and political systems' (Mosse 2004:272). Thus, examining the diversity of water uses, which are often contested, is important to understand the power relations surrounding use, control and rights over water and related resources. Control over water is a powerful political tool that elites have taken advantage of throughout history in order to manage plebeians, with inequitable access often reflecting culturally-rooted social inequities. Thus, 'crises involving water scarcity and water quality are as much a product of cultural values, social contexts, economic activities, and power relationships as they are a result of biophysical forces and conditions' (Donahue and Johnston 1998:345). In a world rapidly transformed by development, industrialization, and globalization, change has come at considerable cost especially to those indigenous peoples and local communities whose rights to water and land may be violated by governments and other powerful entities. As illustrated in earlier contributions to this book, indigenous people's relationships to water - not just their cultural and spiritual ones, but the effects that water scarcity, mining, climate change, dams, and commercialization have on them - are the grounds on which indigenous people challenge the dominant paradigm and policies on water resources and demand the recognition of the right to self-determination (UN Economic and Social Council 2004).

All international water-related documents produced within the UN framework since the Mar del Plata Conference in 1977 recognise the right to water. However, this right to water is often interpreted narrowly as applying only to basic needs for drinking, cooking, and fundamental domestic uses (Gleick 2007). The report of World Commission on Dams[13] made a wider interpretation in 2000, adopting a rights-based approach and thus broadening the range of basic human rights to include water. The report also endorsed the principle of 'free, prior and informed consent' concerning water projects and called for the recognition of indigenous and tribal people's rights as a priority (WCD 2000; UN Economic and Social Council 2004). Furthermore, in 2002, the UN Committee on Economic, Social and Cultural Rights adopted a General Comment on the right to water that explicitly recognised water as a fundamental human right - as indispensable for leading a life of human

[13] The World Commission on Dams (WCD) was established by the World Bank and IUCN - The World Conservation Union in 1998 in response to the controversy surrounding large dams. The members of the Commission - who represented a broad spectrum of interests in dams - conducted a comprehensive and independent review of the world's dams, through extensive public consultations. The Commission's work ended with the publication of its final report, Dams and Development: A New Framework for Decision-Making in 2000 (WCD 2000).

dignity and a prerequisite for the realization of other human rights (UN Economic and Social Council 2003). As water plays an important role in people's right to take part in cultural life, it called for water to be treated as a social and cultural good, and not primarily as an economic good. Furthermore, the Comment called for protection of indigenous peoples' access to water resources from encroachment and unlawful pollution on their ancestral lands and insisted that States should provide resources for indigenous peoples to design, deliver, and control their access to water (UN Economic and Social Council 2003).

This insistence on water as a right and a social, not an economic, good, address a problematic schism that exists within the IWRM movement - affirmation of water as both a human right and an economic good, which the Dublin Principles reflect (Conca 2005). The view of water as an economic good - which raises the controversial issue of its privatization - is inconsistent with water as a human right, in the senses described above.

It is precisely due to these contentious issues related to governance, power, and rights that render the concept of water and cultural diversity precarious; the concept has the potential to be political and thus contentious. It is thus that the UNESCO-IHP found it necessary to withdraw from its official role as co-coordinator of the sessions under the topic Water and Culture one week before the commencement of the Fifth World Water Forum in 2009 (UNESCO 2009). IHP, which had originally proposed the inclusion of the topic into the official programme of the forum, had developed session proposals that took on the larger notion of water and cultural diversity, including issues such as participation, power, and governance, as well as defining cultural diversity to include the entire range of cultural relationships to water, not just those of indigenous or traditional peoples. It embraced the notion of cultural diversity as key to environmental sustainability, because 'it provides the multiple human possibilities necessary for us to adapt to environmental changes and, in turn, shape the process of change for sustainability' (UNESCO-IHP 2009). Despite the fact that sessions on water and cultural diversity had occurred in the past two World Water Forums, the Turkish co-coordinators of the Water and Culture topic did not consider the usage of the term water and cultural diversity appropriate. The co-coordinators maintained that the term cultural diversity only had negative connotations and thus preferred the use of 'diverse water cultures' instead. Consequently, the official forum report states that 'the main focus of the topic was to define the interdependencies between sustainable solutions to water problems and diverse water cultures that have flourished around similar environmental conditions and to discuss the inter-linkages between water and culture, considering the value of culture in creating diverse water cultures, and addressing challenges faced in managing and protecting water resources which meet human and environmental needs' (World Water Forum Secretariat 2009:79). By reverting back to the emphasis on linkages between water and culture, in particular religious and artistic expressions inspired by water, the global trend on this topic led by UNESCO had, unfortunately, suffered a setback at the Fifth World Water Forum.

5.6.4 Towards the Future

According to Conca, global regimes[14] on freshwater issues have not emerged because of their inherent limitations: global environmental regimes focus on problems of specific cross-border pollution and thus do not deal with environmental problems that are considered 'local', despite their global implications. Thus, issues like water that demonstrate the interconnectedness of the social world - in the cumulative, far-reaching impacts they have on global systems and cycles, not to mention the transnational forces that often cause such 'local' problems - are beyond the scope of global regimes (Conca 2005). In addition, regime-building efforts often ignore or polarise conflicts about authority, territory and knowledge precisely because they are founded on rigid presumptions about them (Conca 2005). In the place of regimes, global water initiatives[15] have proliferated, especially in recent years. Those in the water sector often perceive the emergence and multiplication of a wide variety of water-related programmes and organizations negatively because they duplicate mandates, lack coordination, and create competition and confusion (Varady and Iles-Shih 2009). At the same time, global water initiatives are credited for positively transforming the global water governance arena by advancing the global framework on water management and supporting improvements at the local level, thus maintaining long-term sustainability of the global water governance arena (Varady et al. 2009).

Water is both a local and a global issue challenging managers and policymakers in their efforts to conserve freshwater ecosystems and to provide safe and reliable supplies of water. Global environmental changes that reveal the interconnectedness of the hydrological, ecological, and sociocultural systems compound this challenge. Governing water 'involves enduring, chronic, and sometimes raging controversies about local practices of resource management, conservation, and environmental protection in an increasingly transnational context' (Conca 2005:8). Thus, it is only through engaging in cultural issues more comprehensively - that is, dealing with water and cultural diversity, not just that of diverse water cultures - that makes it possible to resolve water problems both at the local and global levels. Only then would it be possible to move forward - whether to build a global regime on water, construct some other institutional form of global environmental governance on water, or develop any other alternatives.

The UN can lead the way towards achieving a truly global grasp on freshwater problems in the future by first, facilitating the ability of all member states to address

[14] Regimes are defined as 'instruments of governance without government', which 'promote rule-conforming behaviour in an international system marked by the absence of centralised governmental authority' (Conca 2005:11). Since the Stockholm Conference on the Human Environment, it is said to have become the 'grand strategy of global environmental protection' (Conca 2005:12).

[15] Global water initiatives can be broadly defined as professional societies, inter-governmental and non-governmental organizations, special events, and awareness-raising campaigns (UN Decades/International Years) that focus on global water-resources management. Activities by such initiatives go beyond the United Nations System (Varady et al. 2008).

the complex issues - the specific social, cultural, economic and political dimensions - surrounding water, while at the same time mitigating the dominance of engineering and natural sciences approaches to water resources management. The three UN entities mentioned above - UNESCO, UNU TKI IAS, and UNPFII - are in a unique position to garner support from the rest of the UN and pioneer the global trend towards this goal. Diversity - not just cultural, but of institutions - has been recognised as an attribute that could lead to effective global governance of water (Varady and Iles-Shih 2009). It is only when diversity is dealt with that UN actions can make a positive contribution to sustainable water management at both local and global levels. It is only then that 'water for life' can be considered a success for everyone - for the millions for whom access to safe drinking water and basic sanitation are not expected to be a reality before the decade is over.

References

Adeel, Zafar, Brigitte Schusterm, and Harriet Bigas (eds.) 2008. *What makes traditional technologies tick?—A review of traditional approaches for water management in drylands*. Available online from http://www.inweh.unu.edu/drylands/Traditional.htm. Accessed 2 Feb 2011.

Annan, Kofi. 2005. The Secretary-General's Message to Launch the "Water for Life" Decade, 22 March 2005. New York: United Nations.

Batisse, Michel. 2005. *The UNESCO water adventure: From desert to water... 1948–1974; from the Arid Zone Programme to the International Hydrological Decade*. Paris: UNESCO.

Conca, Ken. 2005. *Governing water: Contentious transnational politics and global institution building*. Cambridge: MIT Press.

Donahue, John M., and Barbara Rose Johnston (eds.). 1998. *Water, culture, and power: Local struggles in a global context*. Washington, DC: Island Press.

Food and Agricultural Organization (FAO). 1995. *Reforming water resources policy: A guide to methods, processes and practices*. FAO irrigation and drainage paper 52. Rome: FAO. Available at http://www.fao.org/docrep/V7160E/v7160e00.htm. Accessed 2 Dec 2009.

Food and Agricultural Organization (FAO). 2002. *The new generation of watershed management programmes and projects: A resource book for practitioners and local decision-makers*. FAO forestry paper 150. Rome: FAO. Available at http://www.fao.org/DOCREP/009/a0644e/a0644e00.htm. Accessed 2 Dec 2009.

Gleick, Peter. 2007. *The human right to water*. Available at http://www.pacinst.org/reports/human_right_may_07.pdf. Accessed 6 Dec 2009.

Gupta, Joyeeta. 2006. *Water and environmental law and institutions (Lecture Notes LN0134/06/1)*. Delft: UNESCO-IHE.

Hiwasaki, Lisa, and Salvatore Arico. 2007. Integrating the social sciences into ecohydrology: Facilitating an interdisciplinary approach to solve issues surrounding water, environment and people. *Ecohydrology & Hydrobiology* 7(1): 3–9.

Indigenous Water Caucus. 2009. Collective statement: Intervention to the eighth session of the United Nations Permanent Forum on Indigenous Issues 2009. Available at http://www.unutki.org/downloads/File/Water/WATERMay2009_Final-1.pdf. Accessed 12 Dec 2009.

Mosse, David. 2004. The rule of water: Uncertainty and the cultural ecology of water in South India. In *Proceedings of the theme water and cultural diversity at the third world water forum*. Paris: UNESCO. Available at http://www.unesco.org/water/ihp/pdf/wwf3_cult.pdf. Accessed 19 Dec 2007.

Sachs, Wolfgang. 1999. *Planet dialectics: Explorations in environment and development*. London: Zed Books.

United Nations (UN). 1972. Report of the United Nations Conference on the Human Environment, Stockholm, June 1972. Available at http://www.un-documents.net/unche.htm. Accessed 2 Dec 2009.
United Nations (UN). 1990. International drinking water supply and sanitation decade (A/RES/45/181). Available at http://www.un.org/documents/ga/res/45/a45r181.htm. Accessed 8 Dec 2009.
United Nations (UN). 1992. Agenda 21. Available at http://www.un.org/esa/dsd/agenda21/res_agenda21_00.shtml. Accessed 7 Dec 2009.
United Nations (UN). 2004. Protection of traditional lands, water, education, laws vital for preserving indigenous culture, permanent forum told. Press Release HR/4750 dated 12/05/2004. Available at http://www.un.org/News/Press/docs/2004/hr4750.doc.htm. Accessed 5 Dec 2009.
United Nations (UN). 2005. Convention on the law of the non-navigational uses of international watercourses 1997. Available at http://www.untreaty.un.org/ilc/texts/instruments/.../conventions/8_3_1997.pdf. Accessed 9 Dec 2009.
United Nations (UN). 2008a. United Nations declaration on the rights of indigenous peoples, A/RES/61/295. Available at http://www.un.org/esa/socdev/unpfii/documents/DRIPS_en.pdf. Accessed 12 Dec 2009.
United Nations (UN). 2008b. Indigenous peoples face growing challenges as climate change, unchecked economic growth, unfavourable domestic laws force them from lands, forum told. Press Release HR/4951 dated 30 Apr 2008. Available at http://www.un.org/News/Press/docs/2008/hr4951.doc.htm. Accessed 12 Dec 2009.
United Nations (UN). 2009. *The millennium development goals report 2009*. New York: United Nations. Available at http://www.un.org/millenniumgoals/pdf/MDG_Report_2009_ENG.pdf. Accessed 12 Dec 2009.
UN Department of Public Information. 2006. Message of the secretary-general for the world water day on theme 'water and culture'. Available at http://www.un.org/News/Press/docs/2006/sgsm10378.doc.htm. Accessed 8 Dec 2009.
UN Economic and Social Council. 2000. Progress made in providing safe water supply and sanitation for all during the 1990s, Report of the Secretary-General. Available at http://www.un.org/documents/ecosoc/cn17/2000/ecn172000-13.htm. Accessed 7 Dec 2009.
UN Economic and Social Council. 2003. The right to water (arts. 11 and 12 of the International Covenant on Economic, Social and Cultural Rights), E/C.12/2002/11. Available at http://www.unhchr.ch/tbs/doc.nsf/0/a5458d1d1bbd713fc1256cc400389e94/$FILE/G0340229.pdf. Accessed 7 Dec 2009.
UN Economic and Social Council. 2004. Indigenous people's discussion paper on water, sanitation and human settlements, E/CN.17/2004 /10/Add.4. Available at http://daccess-dds-ny.un.org/doc/UNDOC/GEN/N04/243/46/PDF/N0424346.pdf?OpenElement. Accessed 11 Dec 2009.
UN Economic and Social Council. 2007. Recommendations on agenda items 4 (a), (b) and (c), document of the Permanent Forum on Indigenous Issues Sixth session, E/C.19/2007/L.3 / 10/Add.4. Available at http://daccess-dds-ny.un.org/doc/UNDOC/LTD/N07/350/86/PDF/N0735086.pdf?OpenElement. Accessed 12 Dec 2009.
United Nations Educational, Scientific and Cultural Organization (UNESCO). 2002. *UNESCO universal declaration on cultural diversity*. Cultural diversity series no. 1. Paris: UNESCO. Available at http://unesdoc.unesco.org/images/0012/001271/127160m.pdf. Accessed 12 Dec 2007.
United Nations Educational, Scientific and Cultural Organization (UNESCO). 2006. Website on world water day 'water and culture': http://www.unesco.org/water/wwd2006/index.shtml. Accessed 8 Dec 2009.
United Nations Educational, Scientific and Cultural Organization (UNESCO). 2009. UNESCO confirms its unwavering commitment to 5th world water forum in Istanbul. Press Release, Paris, 16 Mar 2009. Available at http://www.unesco.org/new/en/media-services/single-view/news/unesco_confirms_its_unwavering_commitment_to_5th_world_water_forum_in_istanbul/. Accessed 23 Mar 2009.
UNESCO-IHP. 2008. *Brochure on water and cultural diversity*. Paris: UNESCO.

UNESCO-IHP. 2009. *Mainstreaming cultural diversity in water resources management: A policy brief.* Paris: UNESCO-IHP.
UNESCO-LINKS. 2003. Website on the third world water forum. http://portal.unesco.org/science/en/ev.php-URL_ID=3886&URL_DO=DO_TOPIC&URL_SECTION=201.html. Accessed 17 Dec 2007.
UNESCO-LINKS. 2006. Water and indigenous peoples. Knowledge of nature series 2. Paris: UNESCO.
UNPFII. 2006. Website http://www.un.org/esa/socdev/unpfii/. Accessed 12 Dec 2009.
UNU-IAS Traditional Knowledge Initiative. 2009. http://www.unutki.org/default.php?doc_id=30. Accessed 12 Dec 2009.
UN-Water. 2005. *Information brochure on 'water for life decade (2005–2015)'.* New York: United Nations Department of Public Information.
UN-Water. 2009. A guide to UN water. Available at www.unwater.org. Accessed 6 Dec 2009.
Varady, Robert G., and Matthew Iles-Shih. 2009. Global water initiatives: What do the experts think? In *Impacts of megaconferences on the water sector*, ed. Asit K. Biswas and Cecilia Tortajada, 53–101. New York: Springer.
Varady, Robert G., Meehan Katharine, Rodda John, Mc.Govern. Emily, and Iles-Shih Matthew. 2008. Strengthening global water initiatives. *Environment* 2(Mar/Apr): 18–31.
Varady, Robert G., Katharine Meehan, and Emily McGovern. 2009. Charting the emergence of 'global water initiatives' in world water governance. *Physics and Chemistry of the Earth* 34(3): 150–155.
World Commission on Dams. 2000. *Dams and development: A new framework for decision-making.* The report of the World Commission on Dams. London/Sterling: Earthscan.
World Health Organization (WHO) and United Nations Children's Fund (UNICEF) Joint Monitoring Programme for Water Supply and Sanitation. 2008. *Progress on drinking water and sanitation: Special focus on sanitation.* New York/Geneva: UNICEF/WHO. Available at http://www.wssinfo.org/en/40_MDG2008.html. Accessed 7 Dec 2009.
World Water Assessment Programme (WWAP). 2003. *Water for people, water for life: United Nations world water development report.* Paris/New York: UNESCO Publishing/Berghahn Books.
World Water Assessment Programme (WWAP). 2006. *United Nations world water development report.* Paris/New York: UNESCO Publishing/Berghahn Books.
World Water Assessment Programme (WWAP). 2009. *United Nations world water development report.* Paris/New York: UNESCO Publishing/Berghahn Books.
World Water Assessment Programme (WWAP). n.d. a. Website: The millennium development goals and water. http://www.unesco.org/water/wwap/facts_figures/mdgs.shtml. Accessed 15 Nov 2007.
World Water Assessment Programme (WWAP). n.d. b. Website: About the Programme. http://www.unesco.org/water/wwap/description/. Accessed 5 Dec 2009.
World Water Forum Secretariat. 2009. Final report. Available at http://www.worldwaterforum5.org/fileadmin/WWF5/Final_Report/WWF5_Final_Report_ENG.pdf. Accessed 18 Dec 2009.

ERRATUM

Water, Cultural Diversity, and Global Environmental Change

Emerging Trends, Sustainable Futures?

Editor-in-Chief: Barbara Rose Johnston

Editorial Board: Lisa Hiwasaki, Irene J. Klaver, Ameyali Ramos Castillo, Veronica Strang

UNESCO paperback 978-602-98372-4-7
Springer paperback 978-94-007-1866-1
Springer hardback 978-94-007-1773-2
Springer ebook 978-94-007-1774-9

Erratum to: DOI 10.1007/978-94-007-1774-9

The following editorial attribution should be added: 'Section Editors: Irene J. Klaver, Marcus Barber, Ameyali Ramos Castillo, Daniel Niles, Barbara Rose Johnston, and Lisa Hiwasaki.'

On page 110, Figure 1.6.14, the credit should be written as: "Photo by Veronica Strang."

On page 267, in Box 3.4a, the title should be written as: 'Water, Indigenous people and resource development pressures in the Pilbara region of north-west Australia.'

On page 282, the last line of text of Box 3.4c should read: "These points are further illustrated in the following example from the Marshall Islands."

On page 330, in Box 4.2b, the title should be written as: 'Dams, Diversions and Transboundary River Management – best practice lessons learned in the management of Komati River Basin in Swaziland and South Africa.'

On page 340, associated text of Box Map 4.3.1 is authored by Barbara Rose Johnston.

On page 459, associated text of Box Map 5.3a is authored by Barbara Rose Johnston.

The publishers apologize for the inconvenience caused.

Appendices

Appendix 1: Water, Cultural Diversity and Biodiversity: United Nations Statements, Resolutions and Treaties

Agenda 21 (1992) is a comprehensive plan of action to be taken globally, nationally and locally by organizations of the United Nations System, Governments, and Major Groups in every area in which humans impact on the environment. This conference was held as a response to the Report of the World Commission on Environment and Development, Our Common Future ("the Brundtland Report"). As its name suggests, UNCED was not just a sequel to the 1972 Stockholm Conference on the Human Environment; it represented a concerted effort to synthesize and integrate environment and development issues. For the first time, the international community endorsed sustainable development, changing the prior approach to development, which called for peace and security, economic development, human rights and supportive national governance, by adding a fifth element, protection of the environment.
http://www.un-documents.net/agenda21.htm

Convention on Biological Diversity (1992) is a pact among governments to set out commitments for maintaining the world's ecological underpinnings at the same time developing economically. The Convention establishes three main goals: (1) the conservation of biological diversity; (2) the sustainable use of its components; and (3) the fair and equitable sharing of the benefits from the use of genetic resources. The Convention recognizes, for the first time, that conservation is "a common concern of humankind" and is an integral part of the development process. Article 8 (j) calls for the need to "respect, preserve and maintain knowledge, innovations and practices of indigenous and local communities embodying traditional lifestyles relevant for the conservation and sustainable use of biological diversity and promote their wider application with the approval and involvement of the holders of such knowledge, innovations and practices and encourage the equitable sharing of the benefits arising from the utilization of such knowledge, innovations and practices".
http://www.cbd.int/convention/convention.shtml

B.R. Johnston (ed.), *Water, Cultural Diversity, and Global Environmental Change: Emerging Trends, Sustainable Futures?*, DOI 10.1007/978-94-007-1774-9,

The Dublin Principles (1992) was adopted at the 1992 International Conference on Water and Environment in Dublin, Ireland. The Dublin Principles were an attempt to concisely state the main issues and purpose of water management in the following terms: fresh water is a finite and vulnerable resource, essential to sustain life, development and the environment; water development and management should be based on a participatory approach, involving users, planners and policy-makers at all levels; women play a central part in the provision, management and safeguarding of water; and, water has an economic value in all its competing uses and should be recognized as an economic good. The Principles are the guide to set out recommendations for action to reserve the present trends of overconsumption, pollution, and rising threats from droughts and floods.
http://www.cawater-info.net/library/eng/l/dublin.pdf

International Labour Organization Convention on Indigenous and Tribal Peoples (1989) is a legally binding international instrument open to ratification, which deals specifically with the rights of indigenous and tribal peoples. ILO 169 consists of 44 articles and is based on respect for cultures and their way of life, traditions and customary laws of Indigenous and tribal peoples. The underlying principle of ILO 169 is that Indigenous and tribal peoples will continue to be distinct parts of their national societies with their own structures and traditions. ILO 169 also recognises that Indigenous and tribal peoples have a right to take part in the decision-making processes of the States in which they live.
http://www.ilo.org/ilolex/cgi-lex/convde.pl?C169

The Mar del Plata Action Plan (1977) was adopted at the UN Conference on Water held in 1977 in Mar del Plata, Argentina. Renowned as the first attempt to take a wider look at water issues, it was during this conference that access to water was recognized as a human right.

The Stockholm Declaration on the Human Environment (1972) and Action Plan developed out of the United Nations Conference on the Human Environment held in Stockholm in 1972. It was the first large-scale advancement in the international community to work together for the good of the environment, and led to the formation of the UN Environment Programme (UNEP). The Declaration defined principles for the preservation and enhancement of the natural environment, and highlighted the need to support people in this process. The Stockholm Declaration on the Human Environment mentions water as one of the natural resources that need to be "safeguarded for the benefit of present and future generations", and recommendations 95, 98, and 99 of the Action Plan for the Human Environment mention "social and cultural aspects of environmental issues".
http://www.unep.org/Documents.Multilingual/Default.asp?documentid=97&articl eid=1503

United Nations Convention on Non-navigational Uses of International Watercourses (1997) pertains to the uses and conservation of all waters that cross international boundaries, including both surface and groundwater. Mindful of increasing demands for water and the impact of human behaviour, the document was drafted to help conserve and manage water resources for present and future generations. To enter

into force, the document required ratification by 35 countries, but as of 2008 received less than half that number, with ratification by 16. Though unratified, the document is regarded as an important step towards arriving at an international law governing water. The law has adopted the general principles of "equitable and reasonable utilization and participation"
http://untreaty.un.org/ilc/texts/instruments/english/conventions/8_3_1997.pdf

United Nations Declaration on the Rights of the Indigenous Peoples (2007) provides a framework for addressing indigenous issues. Adoption of this declaration will give the clearest indication yet that the international community is committing itself to the protection of the individual and collective rights of indigenous peoples. While this Declaration will not be legally binding on States, and will not, therefore, impose legal obligations on governments, the declaration will carry considerable moral force. The text recognises the wide range of basic human rights and fundamental freedoms of indigenous peoples. The Declaration also provides for fair and mutually acceptable procedures to resolve conflicts between indigenous peoples and States, including procedures such as negotiations, mediation, arbitration, national courts and international and regional mechanisms for denouncing and examining human rights violations.
http://www.un.org/esa/socdev/unpfii/en/drip.html

UNESCO Convention concerning the Protection of the World Cultural and Natural Heritage [World Heritage Convention] *(1972)* The most significant feature of the Convention is that it links together in a single document the concepts of nature conservation and the preservation of cultural properties. The Convention recognizes the way in which people interact with nature, and the fundamental need to preserve the balance between the two. The Convention defines the kind of natural or cultural sites which can be considered for inscription on the World Heritage List and sets out the duties of States Parties in identifying potential sites and their role in protecting and preserving them.
http://portal.unesco.org/en/ev.php-URL_ID=13055&URL_DO=DO_TOPIC&URL_SECTION=201.html

UNESCO Universal Declaration on Cultural Diversity (2001) is the first international standard-setting instrument aimed at preserving and promoting cultural diversity and intercultural dialogue. The Declaration raises cultural diversity to the level of "the common heritage of humanity", "as necessary for humankind as biodiversity is for nature". The Declaration aims both to preserve cultural diversity as a living, renewable treasure that must not be perceived as being unchanging heritage but as a process guaranteeing the survival of humanity. In the context of UNESCO's work, cultural diversity is a goal, as something positive that should be protected and promoted.
http://unesdoc.unesco.org/images/0012/001271/127160m.pdf

UNESCO Convention for the Safeguarding of the Intangible Cultural Heritage (2003) acknowledges knowledge and practices concerning nature and the universe are one of the domains in which the intangible cultural heritage is manifested. It also states that the intangible cultural heritage, transmitted from generation to generation, is constantly recreated by communities and groups in response to their

environment, their interaction with nature and their history, and provides them with a sense of identity and continuity, thus promoting respect for cultural diversity and human creativity.
http://portal.unesco.org/en/ev.php-URL_ID=17716&URL_DO=DO_TOPIC&URL_SECTION=201.html

United Nations Framework Convention on Climate Change (1992) is the main vehicle for promoting international efforts to combat climate change. The Convention sets an ultimate objective of stabilizing greenhouse gas concentrations "at a level that would prevent dangerous anthropogenic (human induced) interference with the climate system." The Convention requires precise and regularly updated inventories of greenhouse gas emissions from industrialized countries.
http://unfccc.int/resource/docs/convkp/conveng.pdf

The Universal Declaration of Human Rights (1948) is a milestone document in the history of human rights. Drafted by representatives with different legal and cultural backgrounds from all regions of the world, the Declaration was proclaimed by the United Nations General Assembly in Paris on 10 December 1948 in General Assembly resolution 217 A (III) as a common standard of achievements for all peoples and all nations. It sets out, for the first time, fundamental human rights to be universally protected.
http://www.udhr.org/UDHR/default.htm

The UNESCO/IUCN Guidelines for the Conservation and Management of Sacred Natural Sites. These guidelines are a contribution from the IUCN Task Force on the Cultural and Spiritual Values of Protected Areas and UNESCO's Man and Biosphere Programme to support the efforts of indigenous and traditional peoples of the world for the long-term conservation of their sacred natural sites. After four years of field-testing they will be re-evaluated and revised. The guidelines will provide recognition and support to sacred natural sites currently facing threats and to the rights of the communities concerned to continue using and managing those sacred sites as places for their cultural and spiritual realization and reverence.
Document source (draft): http://www.fsd.nl/naturevaluation/75082

The CBD Akwé:Kon Voluntary Guidelines for the Conduct of Cultural, Environmental and Social Impact Assessment Regarding Developments Proposed to Take Place on, or Which are Likely to Impact on, Sacred Sites and on Lands and Waters Traditionally Occupied or Used by Indigenous and Local Communities. It is the voluntary guidelines for the conduct of cultural, environmental and social impact assessment regarding developments proposed to take place on, or which are likely to impact on, sacred sites and on lands and waters traditionally occupied or used by indigenous and local communities. The Guidelines, which were named with a Mohawk term meaning "everything in creation", provide a collaborative framework ensuring the full involvement of indigenous and local communities in the assessment of cultural, environmental and social impact of proposed developments on sacred sites and on lands and waters they have traditionally occupied. Moreover, guidance is provided on how to take into account traditional knowledge, innovations and practices as part of the impact-assessment processes and promote the use of appropriate technologies.
http://www.cbd.int/doc/publications/akwe-brochure-en.pdf

Appendix 2: Glossary

A

Action research "Learning by doing" – a group of people identify a problem, do something to resolve it, see how successful their efforts were, and if not satisfied, try again. While this is the essence of the approach, there are other key attributes of action research that differentiate it from common problem-solving activities that we all engage in every day, due to (a) its focus on turning the people involved into researchers (people learn best, and more willingly apply what they have learned, when they do it themselves); (b) its social dimension – the research takes place in real-world situations, and aims to solve real problems; and (c) the initiating researcher, unlike in other disciplines, makes no attempt to remain objective, but openly acknowledges their bias to the other participants. Also referred to as participatory research, collaborative inquiry, emancipatory research, action learning, and contextual action research.[1]

Actual precipitation (see also effective precipitation) Liquid or solid products of the condensation of water vapour falling from clouds or deposited from air on the ground. Includes snow, sleet, hail, and drizzle as well as rain (Fog, mist and haze are not precipitation).[2]

Adaptive management A type of natural resource management where adjustments are made in response to project monitoring, new scientific information, and changing social conditions that may indicate the need to change a course of action.[3]

Adverse effects Reduction in environmental quality of a system, or other depletion of the environmental resource capital. Defined in terms of, and measured by, environmental impacts.

Aquifer Permeable water-bearing formation capable of yielding exploitable quantities of water. Also known as groundwater reservoir.[4]

B

Base-flow Part of the discharge which enters a stream channel mainly from groundwater, but also from lakes and glaciers during long periods when no precipitation or snowmelt occurs. Also known as base runoff.[5]

---

[1] O'Brien, R. (2001). Um exame da abordagem metodológica da pesquisa ação [An Overview of the Methodological Approach of Action Research]. In Roberto Richardson (Ed.), Teoria e Prática da Pesquisa Ação [Theory and Practice of Action Research]. João Pessoa, Brazil: Universidade Federal da Paraíba. (English version) Available at: http://www.web.ca/~robrien/papers/arfinal.html (Accessed 13/9/2010).

[2] UNESCO/WHO International Glossary of Hydrology, accessible at http://hydrologie.org/glu/aglo.htm and waterwiki.net

[3] Saunier, Richard E. and Richard A. Meganck. 2004. C.H.A.O.S.S.: An essay and glossary for students and practitioners of global environmental governance.

[4] UNESCO/WHO International Glossary of Hydrology.

[5] UNESCO/WHO International Glossary of Hydrology.

Benefit sharing A method to ensure the project benefits are shared equitably among the affected communities.[6]

Biocultural diversity The "inextricable link" between biological, cultural, and linguistic diversity – the sum total of the diversity of life in nature and culture. The term as it is commonly used focuses primarily on the relationship between biodiversity and linguistic diversity in specific regions and localities. It is thus distinguished from the wider understanding of biological and cultural diversities as mutually reinforcing and interdependent, and that natural systems cannot be understood, conserved and managed without recognizing the human cultures that shape them. This wider notion sees interlinkages between biological diversity and cultural diversity as something positive, based on the premise that it is through diversity that cultures and ecosystems can be resilient. Cultural diversity and biological diversity together hold the key to ensuring resilience in both social and ecological systems.[7]

Biological diversity (biodiversity) The variability among living organisms from all sources including, *inter alia*, terrestrial, marine and other aquatic ecosystems and the ecological complexes of which they are part. Generally considered at three levels: genetic, species and ecosystem diversity.[8]

Biofuel Any kind of fuel made from living things, or from the waste they produce, e.g., wood (including wood chippings and straw, pellets or liquids made from wood), biogas (methane) from animals' excrement, ethanol, diesel or other liquid fuels made from processing plant material (such as corn, sugarcane and rapeseed) or waste oil.[9]

Biological oxygen demand (or Biochemical oxygen demand) The amount of oxygen consumed in a water sample over a period of time. This oxygen can be consumed by metabolic respiration of living organisms in the water, or by chemical oxidation, as in the oxidation of iron to create iron oxide.

Biological resources (biotic resources) Include genetic resources, organisms or parts thereof, populations or other biotic components of ecosystems, with actual or potential use for humanity (or other living organisms). Also known as biotic resources.[10]

Biophysical Environmental processes, ecological functions and natural resource components (biological, physical and geochemical) not directly controlled by, or dependent upon, human activities.

C

Catchment Area having a common outlet for its surface runoff. A catchment includes both the water bodies that convey the water and the land surface from which water drains into these bodies. Also referred to as drainage basin or river basin.[11]

---

[6] http://www.oxfam.org.au/resources/filestore/originals/OAUs-GuideToFreePriorInformedConsent-0610.pdf

[7] http://www.terralingua.org/GSB/GSB.html and UNESCO's Programme and Budget 2004-2005. "32C/5" http://unesdoc.unesco.org/images/0013/001341/134100e.pdf

[8] The Convention on Biological Diversity (CBD).

[9] http://news.bbc.co.uk/2/hi/science/nature/6294133.stm

[10] CBD

[11] UNESCO/WHO International Glossary of Hydrology and waterwiki.net

Climate change A change of climate that is attributed directly or indirectly to human activity that alters the composition of the global atmosphere and which is in addition to natural climate variability observed over comparable time periods.[12]

Collaborative practices A component considered central to successful collaborative management efforts, and includes practices such as seeking inclusive representation, establishing leadership, and identifying a common goal among the participants. Collaborative resource management, which involves public and private stakeholders in natural resource decisions, is increasingly used as an effective approach for managing natural resources, by reducing or avoiding conflicts at the same time improving natural resource conditions.[13]

Commoditization of water When water is marketed and sold by private corporations, as a way to bring business efficiency into water service management. It often results in reduced access of water for the poor around the world, as prices for these services rise. When corporations control water, it is seen as an economic resource to be managed by market forces like any other commodity, and the right to water is no longer seen as a basic human right.[14]

Common property Land or resources available to all and consequently not owned or managed by anyone. Also refers to situations where access is limited to a specific group that holds rights in common.[15]

Commons Resources that are shared or collectively owned between or among populations. The commons consists of gifts of nature such as air, oceans and wildlife as well as shared social creations such as libraries, public spaces, scientific research and creative works.

Community resource management The social structures created and maintained to manage community access to shared resources.

Community-based natural resource management (CBNRM) Focuses on the collective management of ecosystems to improve human well-being. It aims to devolve authority for ecosystem management to the local (community) level, thereby empowering communities to manage their own resources without permanently damaging, depleting or degrading them. CBNRM therefore requires strong investments in capacity development and the development of local institutions and governance structures.[16]

Conservation The management of human use of biosphere so that it may yield the greatest sustainable benefit to current generations while maintaining its potential

[12] United Nations Framework Convention on Climate Change (UNFCC).

[13] United States Government Accountability Office (2008). Report to the Chairman, Subcommittee on Public Lands and Forests, Committee on Energy and Natural Resources, U.S. Senate, available at http://www.environmental-auditing.org/Portals/0/AuditFiles/US_f_eng_naturalresources management_08.pdf accessed 15/9/2010.

[14] http://www.foodfirst.org/en/node/57; www.globalissues.org/article/601/water-and-development

[15] http://www.fao.org/docrep/w8210e/w8210e05.htm#common%20property%20or%20open%20 access

[16] Fabricious and Collins (2007). http://www.iwaponline.com/wp/009S2/0083/009S20083.pdf

to meet the needs and aspirations of future generations. A factor of environmental protection.[17]

Cosmology An ideological system that explains the order and meaning of the universe and people's places within it. Used by anthropologists, the term is both more and less than the term religion.[18]

Culture A complex system of beliefs, values, language, knowledge and practices, as well as the material and non-material products of human thought and action. Through social transmission and over time, all human societies develop a cultural system specific to the social and ecological context which identifies its bearers as a "cultural group" or a "culture". Culture is also viewed as a set of institutions, practices, behaviours, technologies, skills, knowledge, beliefs and values proper to a human community. Culture is usually received, lived, refined, and reproduced at any given moment in history. Cultures are highly dynamic constructs that change through time and hybridize when exposed to contact and exchange with other cultures.

Cultural diversity May be understood as, but not limited to, diversity in the following aspects: (1) languages; (2) artistic expressions (including art, architecture, literature and music); (3) value systems (including religion, ethics, spirituality, beliefs and worldviews); (4) knowledge (e.g., know-how and skills); (5) practices (e.g., rituals, production systems, and knowledge transmission systems); and (6) ways of living together (social systems including institutions, legal systems, leadership and tenure systems). The UNESCO Universal Declaration on Cultural Diversity (2001) defines how the international community should deal with cultural diversity, and invites policy makers to: (1) recognize that differences exist in terms of practice, beliefs, values systems and vision; (2) respect the right of each person to be different and be valued as such; and (3) ensure dialogue so that differences become creative and constructive. The Convention attempts to review cultural diversity not as a simple fact but as a source of exchange and creativity that ensures sustainability of humanity. [19]

Cultural flows Water entitlements that are legally and beneficially owned by Indigenous Nations and that are of a sufficient and adequate quantity and quality to improve the spiritual, cultural, environmental, social, and economic conditions of those Indigenous Nations. This concept of cultural flows was developed as a response to the concept of environmental flows.[20]

Cultural mapping Involves a community identifying and documenting local cultural resources, e.g., the tangibles (like galleries, craft industries, distinctive landmarks, local events and industries) as well as the intangibles (like memo-

---

[17] Saunier and Meganck (2004).

[18] John H. Bodley (1997) Cultural Anthropology, and Encyclopedia of Social and Cultural Anthropology (2002).

[19] http://unesdoc.unesco.org/images/0012/001271/127160m.pdf

[20] Murray Lower Darling Rivers Indigenous Nations (MLDRIN). 2008. Echuca Declaration. Available online at http://www.nailsma.org.au/nailsma/publications/downloads/MLDRINNBAN-ECHUCA-DECLARATION-2009.pdf.

ries, personal histories, attitudes and values). Cultural mapping involves initiating a range of community activities or projects, to record, conserve and use these elements. The most fundamental goal of cultural mapping is to help communities recognize, celebrate, and support cultural diversity for economic, social and regional development. Recognized as a fundamental step to safeguard cultural diversity.[21]

Cumulative impacts Progressive environmental degradation over time resulting from a range of activities throughout an area or region, each of which when considered in isolation may not be regarded as a significant contributor.

Customary law The laws, practices and customs of indigenous and local communities which are an intrinsic and central part of the way of life of these communities. Customary laws are embedded in the culture and values of a community or society; they govern acceptable standards of behavior and are actively enforced by members of the community.

D

Decentralization The general term for a transfer of authority and/or responsibility for performing a function from the top management of an organization or the central governance level of an institution to lower level units or the private sector.[22]

Decision making process The process used to make a decision. It can be an expert process, where the decision is made by one or more "experts" who look at the "facts" and make the decision based on those facts; it can be a political process through which a political representative or body makes the decision based on political considerations, or it might be a judicial process where a judge or a jury makes a decision based on an examination of legal evidence and the law.[23]

Deforestation The change of forested lands to non-forest lands. The removal of forest by cutting and burning to provide land for agricultural purposes, residential or industrial building sites, roads, etc., or by harvesting the trees for building materials or fuel.[24]

Development A process made up of activities leading to the use, improvement, or conservation of natural and economic goods and services in order to maintain and improve human life quality of a target population.[25]

Development-induced displacement The forcing of communities and individuals out of their homes, often also their homelands, for the purposes of economic development. Economic displacement refers to loss of access to an income or livelihood resulting from a loss of access to land and other critical resources.[26]

---

[21] http://www.unescobkk.org/culture/our-projects/cultural-diversity/cultural-mapping/

[22] UNDP, cited in Saunier and Meganck (2004).

[23] http://www.colorado.edu/conflict/peace/glossary.htm

[24] The Intergovernmental Panel on Climate Change (IPCC).

[25] OAS, cited in Saunier and Meganck (2004).

[26] Hobbs & Bose (2004). Ethic, Development and Displacement – Workshop Report, Centre for Refugee Studies, Ottawa, Canada; http://www.oxfam.org.au/resources/filestore/originals/OAUs-GuideToFreePriorInformedConsent-0610.pdf

Development proposals Consist of a wide range of human activities which provide (a) favourable conditions for an increase in the transformation of the natural, biophysical environment to provide goods and services available to society and (b) actions which directly produce the goods and services.

E

Ecological processes Changes or reactions which occur naturally within ecosystems. They may be physical, chemical or biological.

Ecological risk The risk that plant, animal and microorganism communities are subjected to a hazard. Considers toxicology, and the probability of environmental dispersal (e.g., exposure pathways or bioaccumulation through food webs), or other potential causation for environmental damages and losses (e.g., ecological resilience or biological diversity).

Ecosystem A dynamic complex of human, plant, animal and micro-organism communities and associated non-living environment interacting as a functional unit. Ecosystems boundaries are not fixed and their parameters are set according to the scientific, management, or policy question being examined.[27]

Ecosystem services Ecosystems provide a range of services to humans, including provisioning, regulating, and supporting environmental and cultural systems, without which our survival and well-being would simply not be possible. Human survival is completely dependent upon the continued flow of ecosystem services. Ecosystem services include: provisioning services, namely products obtained from ecosystems, such as food, water, and fuelwood; regulating services, namely benefits obtained from regulating ecosystem processes, such as floods, drought, and land degradation; cultural services, namely nonmaterial benefits obtained from ecosystems, such as recreation, spiritual, and aesthetic benefits; and supporting services, namely services that are necessary for the production of all other ecosystem services, such as soil formation and nutrient cycling.[28]

Ecosystem structure & function Ecosystem structure is defined by the interactions between abiotic (non-living) factors and biotic components of the system. Functions are activities or actions which occur naturally as a product of the interactions between ecosystem structure and processes which, taken together with the functions of other ecosystems, collectively provide "services" upon which all life on earth depends, including maintaining the balance of atmospheric gases, recycling of nutrients, regulating climate, maintaining hydrological cycles, and creating soils.

---

[27] Saunier and Meganck (2004).

[28] Millennium Ecosystem Assessment (MA). 2003. Ecosystems and Human Well-Being: A Framework for Assessment. Washington DC: Island Press. And, Water Security and Ecosystems Services: The Critical Connection. A Contribution to the United Nations World Water Assessment Programme. United Nations Environment Programme, Nairobi, Kenya March, 2009.

Effective precipitation (see also actual precipitation) The amount of precipitation that is actually added and stored in the soil.

Effluent (1) Liquid flowing out of a container or other system; (2) water or waste water flowing out of a reservoir or treatment plant; (3) outflowing branch of a main stream or lake.[29]

Emissions The release of green house gases and/or their precursors into the atmosphere over a specified area and period of time.[30]

Endemism The occurrence of a species in a particular locality or region.[31]

Environment The combined features and resource capital that provide the basis for development, environmental management and conservation. Includes the processes and components of, and services provided by, atmospheric, hydrological, geophysical, biotic, human and landscape factors.

Environmental assessment A term used almost interchangeably with environmental impact assessment, environmental appraisal, and environmental analysis that refers to a formal procedure structured to ensure that selected environmental issues are considered in the early stages of the project cycle.[32]

Environmental data Raw facts about environmental quality. Obtained by analytical processes that convert environmental samples to numerical estimates (within error limits) of environmental quality characteristics.

Environmental degradation Adverse effects (reversible or permanent) on biophysical, social and economic resources, or any other reduction in the "set of options" available to future generations.

Environmental flows Describe the quantity, timing, and quality of water flows needed to sustain freshwater and estuarine ecosystems and the human livelihoods and well-being that depend on these ecosystems.[33]

Environmental impact Changes in environmental quality due to external disturbance to a system. Includes beneficial and adverse, primary and secondary, cumulative, synergistic, short, medium and long-term, reversible and irreversible. Described in terms of magnitude (of effect), direction (of change) and probability (of occurrence), with or without mitigation.

Environmental impact assessment (EIA) A process of evaluating and suggesting management and mitigation scenarios for the impacts arising for a new development at the various stages of the project cycle.[34]

Environmental impact detection Test and confirmation of environmental impact hypothesis.

---

[29] UNESCO/WHO International Glossary of Hydrology.

[30] UNFCC.

[31] The UNEP World Conservation Monitoring Centre (WCMC).

[32] Saunier and Meganck (2004).

[33] Brisbane Declaration 2007 http://www.riversymposium.com/index.php?element=2007Brisbane Declaration241007

[34] Saunier and Meganck (2004).

Environmental impact hypothesis Formally stated, testable conjecture of predicted change in environmental quality associated with development actions. Can be defined in terms of environmental quality indicators, to facilitate testing (at appropriate levels of confidence).

Environmental management system A system which provides a structured process for continual improvement and which enables an organization to achieve and systematically control the level of environmental performance that it sets itself. In general, this is based on a dynamic cyclical process of "plan, implement, check and review".

Environmental protection Actions at the international, national and local levels to prevent and ameliorate deterioration of human environments. Such actions include conservation efforts, recycling, waste reduction and disposal, and the development of cleaner and safer technologies.[35]

Environmental quality The status or value of the natural resource capital at a particular location at a specified time, relative to development, environmental management and conservation.

Environmental quality indicators Measurable parameters that describe environmental quality and related human actions, in a summarized format.

Environmental sampling Physical collection of a representative portion of the environment, using techniques which facilitate accurate generation of reliable data.

Environmental sustainability Practices to ensure that the natural resource capital remains intact; i.e., that the source and sink functions of the environment should not be degraded. Therefore, the extraction of renewable resources should not exceed the rate at which they are renewed, and the absorptive capacity of the environment to assimilate wastes should not be exceeded. Furthermore, the extraction of non-renewable resources should be minimized and should not exceed agreed minimum strategic levels.

Equity Impartial or just treatment, requiring that similar cases be treated in similar ways.[36]

Evapotranspiration Quantity of water transferred from the soil to the atmosphere by evaporation and plant transpiration.[37]

F

Flood pulse Periodic changes in water level, which are crucial to the biota in floodplain rivers. The primary source of productivity in lowland rivers are the nutrients and particulate material derived from the lateral exchange between the floodplain and the channel.[38]

---

[35] Saunier and Meganck (2004).

[36] Saunier and Meganck (2004).

[37] UNESCO and WHO International Glossary of Hydrology.

[38] http://www2.mdbc.gov.au/subs/dynamic_reports/foundation_report/pdf/glossary.pdf

Flow regime Long-term statistical pattern of flows, characterized by flow quantities, seasonal patterns, and the magnitude, timing, frequency and duration of extreme events such as floods and dry spells.[39]

Free, and prior informed consent (FPIC) The right of communities to exercise control, to the extent possible, over their own economic, social and cultural development. FPIC is a specific right for Indigenous Peoples as recognised in the United Nations Declaration on the Rights of Indigenous Peoples.[40]

G

Geographic information system (GIS) A computer-based system that stores, manipulates and displays spatial information.[41]

Global environmental change Human-induced changes to the structure, composition or function of large natural biogeophysical and ecological systems that entail changes in the complex array of forcings and feedbacks that characterise the internal dynamics of the earth system. These changes are of great significance because they are diminishing the capacity of the environment to supply and replenish resources, and to absorb and recycle the waste products of the activities of humans and their domesticated animals. These changes are "global" in the sense of either being globally integrated (i.e., entailing a systemic change to a global system, such as the climate system) or occurring by worldwide aggregation of widespread local changes (e.g., land degradation, species extinctions). The main types of human-induced global environmental change are: changes to atmospheric composition and function (including climate change), changes in land-use, land cover changes and sea-use changes, desertification, biodiversity changes, changes to biogeochemical cycles (nitrogen, phosphorus, sulphur, carbon), changes to the hydrological cycle, and depletion of freshwater supplies and quality, world-wide dissemination of persistent organic pollutants (POPs), and urbanization.[42]

Global warming A theory relating to an increase in the earth's temperature caused, in part, by the "green-house" effect being created by emissions of greenhouse gasses (GHGs) associated with human activities.[43]

---

[39] Bunn, S.E., and A.H. Arthington. 2002. Basic principles and ecological consequences of altered flow regimes for aquatic biodiversity. *Environmental Management* 30(4 Oct): 492–507.

[40] http://www.oxfam.org.au/resources/filestore/originals/OAUs-GuideToFreePriorInformedConsent-0610.pdf and www.sustainableforestprods.org

[41] Saunier and Meganck (2004).

[42] Global Environmental Change and Human Health (2007) Science Plan and Implementation Strategy. Earth System Science Partnership (DIVERSITAS, IGBP, IHDP, and WCRP) Report No. 4; Global Environmental Change and Human Health Report No. 1 (available at http://www.ihdp.unu.edu/file/IHDP+Project+Science+Plans/GECHH+Science+Plan?menu=49).

[43] Saunier and Meganck (2004).

H

Habitat Living space – all the elements which collectively define and constitute the locality or nature of the spatial and temporal "frame of reference" within the biosphere in which plants, animals and humans live. Includes non-living influences such as soil, light, temperature, moisture, humidity and other abiotic factors, as well as biotic (living) components which influence, or are influenced by, the activities of organisms, individuals, communities and organizations, all of which integratively determine the "quality of life" conditions.

Health The physical, mental and social well-being of individuals or communities, and not merely the absence of disease or infirmity.

Human environment The physical, social, and economic components, conditions and factors that interactively determine the state, condition, and quality of living conditions, employment, and health of those affected directly or indirectly by resource development activities in a given area.

Human rights Rights and freedoms to which all humans are entitled, without distinction of any kind, such as race, colour, sex, language, religion, political or other opinion, national or social origin, property, birth or other status. These rights include, among others: the right to life, liberty and security; freedom from slavery or servitude, torture or from cruel, inhuman or degrading treatment or punishment; full equality before law; freedom of thought, conscience and religion, of opinion and expression; right to self-determination; and the right to freely pursue their economic, social and cultural development.[44]

Hydrodevelopment The development and management of water resources to generate to meet varied societal needs, such as the generation of hydroelectricity and the diversion of water to sustain agriculture, urban, and industrial needs.

Hydroelectric Of, or relating to, production of electricity by waterpower.

Hydrological cycle Succession of stages through which water passes from the atmosphere to the earth and returns to the atmosphere: evaporation from the land or sea or inland water, condensation to form clouds, precipitation, accumulation in the soil or in bodies of water, and re-evaporation. Also referred to as water cycle.[45]

Hydrology (1) Science that deals with the waters above and below the land surfaces of the earth, their occurrence, circulation and distribution, both in time and space, their biological, chemical and physical properties, their reaction with their environment, including their relation to living beings. (2) Science that deals with the processes governing the depletion and replenishment of the water resources of the land areas of the Earth, and treats the various phases of the hydrological cycle.[46]

[44] The Universal Declaration of Human Rights http://www.un.org/en/documents/udhr/index.shtml

[45] UNESCO/WHO International Glossary of Hydrology.

[46] UNESCO/WHO International Glossary of Hydrology.

I

Indigenous people Are considered indigenous either because they are descendants of those who lived in the area before colonization, or because they have maintained their own social, economic, cultural and political institutions since colonization and the establishment of new states. Self identification is crucial.[47]

Individual risk The risk level for a specific individual. Measured in terms of probability that a particular individual will be exposed, injured or suffer other losses.

Infiltration Flow of water through the soil surface into a porous medium.[48]

Inland surface waters Waters that flow over or rest upon the surface of the lithosphere, in the interior of a land mass, and in dry weather conditions, includes the area over which such waters are seasonally present. Include rivers, streams, tidal waters, swamps, ponds and impounded reservoirs.

Instream Relating to the river channel.[49]

Indigenous knowledge See Local, Indigenous or Traditional Knowledge

Integrated water resources management (IWRM) A process which promotes the coordinated development and management of water, land, and related resources in order to maximize the resultant economic and social welfare in an equitable manner without compromising the sustainability of vital ecosystems.[50]

Isohyet Line joining the points where the amount of precipitation, in a given period, is the same.[51]

K

Knowledge creation processes The innovation process (of an individual, scientific community, industry, culture, society, nation, etc.).

L

Land An area of the earth's surface, including all elements of the physical and biological environment that influence land use. Refers not only to soil, but also landforms, climate, hydrology, vegetation and fauna, together with land improvements such as terraces and drainage works.

Land degradation The temporary or permanent lowering of the productive capacity of the land, or its potential for environmental management. Includes air, water and soil pollution, loss of aquatic or soil bio-activity, increased surface runoff and erosion, bio-accumulation of toxic substances, soil fertility decline, salinisation,

---

[47] International Labour Organization (ILO) Convention No. 169.

[48] UNESCO/WHO International Glossary of Hydrology.

[49] http://www2.mdbc.gov.au/subs/dynamic_reports/foundation_report/pdf/glossary.pdf

[50] GWP. 1999. *How to bring ecological services into integrated water resource management.* Seminar Stockholm, 15–17 Nov. Beijer International Institute of Ecological Economics, Royal Swedish Academy of Sciences, Stockholm.

[51] UNESCO/WHO International Glossary of Hydrology.

waterlogging, contamination or lowering of water table, increased surface albedo, and deforestation, forest degradation or other habitat loss.

Land quality The condition of land relative to land use, including agriculture forestry, industry, conservation and environmental management.

Land use The management of land to meet specified socio-economic objectives. Land use is described by the purposes for which the land is used, and the types and sequences of development, conservation and environmental management activities carried out upon the land.

Local, indigenous or traditional knowledge Refers to the cumulative and complex bodies of knowledge, know-how, practices and representations that are maintained and developed by peoples with extended histories of interactions with the natural environment. These cognitive systems are part of a complex that also includes language, attachment to place, spirituality and worldview. Local, indigenous or traditional knowledge systems bridge the gap between biological and cultural diversities. These complex and dynamic arrays of knowledge, know-how, practices and representations guide human societies in their innumerable interactions with the natural milieu. Article 8(j) of the Convention on Biological Diversity gives particular recognition to this cultural dimension of biodiversity, as do all of UNESCO's conventions on culture. See also traditional ecological knowledge. [52]

M

Macrophyte Large water plants.[53]

Millennium development goals (MDGs) A set of eight international development goals for 2015, adopted by the international community in the UN Millennium Declaration in September 2000, and endorsed by the IMF, World Bank and OECD. Aim to improve human well-being by reducing poverty, hunger, child and maternal mortality, ensuring education for all, controlling and managing diseases, tackling gender disparity, ensuring sustainable development and pursuing global partnerships.[54]

Minority A group which is smaller in number than the rest of the population of a State, whose members have ethnic, religious or linguistic features different from those of the rest of the population, and are guided, if only implicitly, by the will to safeguard their culture, traditions, religion or language. Any group coming within the terms of this definition shall be treated as an ethnic, religious or linguistic minority. To belong to a minority shall be a matter of individual choice.[55]

---

[52] www.unesco.org/links

[53] UNESCO/WHO International Glossary of Hydrology.

[54] Based on Skutnabb-Kangas' reformulation of the definition by the Council of Europe Commission for Democracy through Law (91) 7, Art. 2).

[55] http://www.un.org/millenniumgoals/

Mitigation The purposeful planning, design and implementation of decisions, activities, engineering solutions or management systems that are intended to reduce the undesirable impacts of proposed development actions or climate change impacts on the affected environment.

N

Natural capital The stock of environmental resources that yields many goods and services that are essential to the sustained health of our environment, communities, and economy.[56]

Natural resource A term describing anything that is provided by nature, such as minerals deposits, forests, water, wildlife, etc. For the project developer it is sources of environmental matter which can be taken away for economic gain, such as timber, oil and water. For the community, it is a source of livelihood, such as water for drinking, land for agriculture and livestock.[57]

Natural resource assets Products and attributes: Products generated by ecosystems include: forest resources, wildlife resources, forage resources, fisheries, agricultural resources and water supply, and represent the good and services generated by the interactions between the biological, chemical and physical components of an ecosystem. Attributes of an ecosystem include biological diversity and unique landscape, heritage and cultural features, which may lead to certain uses or the derivation of particular products, but they may also have intrinsic, unquantifiable importance.

Non-governmental organization (NGO) A non-profit group or association organized outside of institutionalized political structures to realize particular social objectives or serve particular constituencies. NGOs can include faith-based organisations (such as churches) and other civil society and community-based organisations.[58]

Non-point source pollution Pollution that is not discharged from a specific point; examples include urban runoff or erosion from agricultural fields.

P

Participation Process through which stakeholders influence and share control over development initiatives and the decisions and resources which affect them. It is a process which can improve the quality, effectiveness and sustainability of projects and strengthen ownership and commitment of government and stakeholders.[59]

---

[56] Alberta Environment http://environment.gov.ab.ca/info/library/8043.pdf

[57] Saunier and Meganck (2004); http://www.oxfam.org.au/resources/filestore/originals/OAUs-GuideToFreePriorInformedConsent-0610.pdf

[58] Saunier and Meganck (2004); http://www.oxfam.org.au/resources/filestore/originals/OAUs-GuideToFreePriorInformedConsent-0610.pdf

[59] Saunier and Meganck (2004).

Pilgrimage centers Sacred places or spaces where people regularly travel to worship. Such places often include natural features, including springs, headwaters of rivers, lakes, mountains, and other places that represent the power of deities.

Point source Discernable, confined and discrete conveyance source, including (but not limited to) any pipe, ditch, channel, conduit, well, discrete fissure, container, drilling rig, oil and gas platform, vent or vehicle, from which pollutants can potentially be discharged to the environment.

Pollution The release of any material, surplus energy or other waste stream into the environment in sufficient quantities to exceed thresholds, and adversely affect environmental quality in the short, medium or long-term.

Precautionary principle Sustainability principle which states that if there are threats of serious irreversible environmental damage, lack of full scientific certainty will not be used as a reason for postponing measures to prevent environmental degradation. Also referred to as the "precautionary approach".

Primary water sources The major sources of water that sustain environmental and human systems. Term also used to refer to water obtained from deep aquifers.

Q

Quality of life Socio-economic environmental concept embracing a diversity of values not always recognized, or adequately addressed, in marketplace analysis. Includes factors such as real income, housing, working conditions, health, educational services and recreational opportunities, which may be regarded as the general standard of living.

Qanat A long row of shafts connecting the gallery to the aquifer through a gentle slope so that the underground water can flow and reach the appearance via gravity. This hydraulic system can be found in Southeast Asia, North Africa and in the Middle East.[60]

R

Rain-water harvesting Capturing rain water or run off where it falls – in one's house, village or town, used as an alternative technology for collection of drinking water.[61]

Regime An institutional arrangement formally or informally established to study and support a common property issue area (atmosphere, oceans, biodiversity, Antarctica, outer space, etc).[62]

Reservoir (1) A component or components of the climate system where a greenhouse gas or a precursor of a greenhouse gas is stored. (2) A body of water, either natural or man-made, used for storage, regulation and control of water resources. [63]

---

[60] http://www.qanat.info/

[61] waterwiki.net

[62] Saunier and Meganck (2004).

[63] Saunier and Meganck (2004); UNESCO/WHO International Glossary of Hydrology.

Resettlement Being moved to a new place to live and make a livelihood to make way for a large development project, such as dams.[64]

Rightsholders The person, community, organization or governmental entity that owns the rights to access and use of land and other critical resources, including water.

Right to water In 2001 the UN Committee on Economic, Social and Cultural Rights affirmed the right of all people to clean and sufficient quantities of fresh water as a precondition for the realization of human rights.[65]

Riparian Pertaining to the banks of a river, stream, waterway, or other, typically, flowing body of water as well as to plant and animal communities along such bodies of water.[66]

Rising block tariff Charging a lower rate for the first proportion of energy used. The rising block tariff assumes that those using the largest amounts of electricity are better off and more able to contribute to the running costs of the utility. Rising rates presumably act as a deterrent to increased consumption and are a tool used to encourage energy conservation.

River basin A geographical area generally determined by the watershed limits of a water system, including surface and underground water that flow into a common terminus. See also catch\ment.[67]

River basin committee A multi-stakeholder institution set up to manage a river basin.

River linking To link rivers via canals and transfer water from one basin to another.

River restoration Returning a site to a condition similar to the one that existed before it was altered, along with its pre-disturbance functions and related physical, chemical and biological characteristics. The goal of restoration is to establish a site that is self-regulating and integrated within its landscape, rather than to re-establish an aboriginal condition.[68]

Risk analysis Technique used to determine the likelihood or chance of hazardous events occurring and the likely consequences. Originally developed for nuclear and chemical industry, where low probability events potentially have extremely serious results. Probabilistic risk analysis can be used to characterise environmental impacts, whose occurrence and nature are difficult to predict with any degree of accuracy.

---

[64] http://www.oxfam.org.au/resources/filestore/originals/OAUs-GuideToFreePriorInformedConsent-0610.pdf

[65] Saunier and Meganck (2004).

[66] Alberta Environment http://environment.gov.ab.ca/info/library/8043.pdf.

[67] Saunier and Meganck (2004).

[68] Bunn and Arthington (2002).

Run-of-river Run-of-the-river hydroelectricity is a type of hydroelectric generation whereby the natural flow and elevation drop of a river are used to generate electricity. Power stations of this type are built on rivers with a consistent and steady flow either natural or through the use of a large reservoir at the head of the river which then can provide a regulated steady flow for stations down-river.

Runoff The component of precipitation or irrigation water that flows from the land-surface into streams or other surface-water. Runoff can transport pollutants from the air and soil into surface water. Flooding occurs when surface runoff exceeds stream channel capacities.

S

Scoping A procedure for narrowing the scope of an assessment and ensuring that the assessment remains focussed on the truly significant issues or impacts.

Sedimentation Process of settling and depositing by gravity of suspended matter in water.[69]

Self-determination the exercise of the right of a people to freely determine its social, economic, political and cultural future without external influence. Central to the struggle for self-determination is the legal recognition of these rights to land and other critical resources.

Sewage treatment systems Wastewater treatment that utilizes screening to remove materials, settling to allow smaller particles to drop out of column and chlorine or ultraviolet light to kill bacteria and viruses before the water is released into the environment. Secondary treatment is wastewater treatment that supplements primary treatment by including aeration to allow bacteria to further break down organic material and frequently involving enhanced settling through the addition of a flocculent, such as alum, to further reduce the turbidity of water prior to discharge.

Siltation The deposition or accumulation of unconsolidated or loose sedimentary material whose constituent rock particles are finer than grains of sand and larger than clay particles.[70]

Slope break Shelf-slope break. Line marking a change from the gently inclined continental shelf to the much steeper depth gradient of the continental slope.[71]

Social impact assessment The component of environmental impact assessment concerned with changes in the structure and functioning of social orderings. In particular, the changes that a development would create in: social relationships; community (population, structure, stability, etc.); quality and way of life; language; ritual; political/economic processes; attitudes/values; health impacts.

---

[69] UNESCO/WHO International Glossary of Hydrology.

[70] A Glossary Terms in Environment, UN Asian Development Institute, Thailand.

[71] http://life.bio.sunysb.edu/marinebio/glossary.rs.html

Societal risk The risk to society as a whole. Considers both the probability of impact on individuals, and the number of people (e.g., in adjacent communities) that could be subject to risk.

Social sustainability Practices to ensure that the cohesion of society and its ability to work towards common goals are maintained. Individual needs such as those for health and well-being, nutrition, shelter, education and cultural expression should be met.

Stakeholder An institution, organization, or group that has some interest in a particular sector or system or outcome of a project, programme or policy initiative.[72]

Standard Value or state of an environmental variable, considered to be desirable to achieve, and undesirable or illegal to exceed. Standards can act as either as targets (ambient characteristics for specified uses) or limits (quantities of potential pollutants released to environment), which should not be transgressed.

Stewardship A principle or approach whereby citizens, industry, communities, and government work together as stewards of the province's natural resources and environment. In general terms, stewardship means managing one's life, property, resources, and environment with regard for the rights or interests of others. This can apply to a person, company, community, government or group. Stewardship is an ethic and a value that results form public education and partnerships. It is people-focused in the sense that it relies on the desire and ability of people to make good decisions on their own accord that help resource and environmental outcomes.[73]

Sustainability Biocultural health

Sustainable water resource management The primary goal of water development and management is to sustain human and environmental needs now and in the future. This means developing and managing water in ways that protects and provides the means to support sustainable ways of life for people and their environment.

Synergistic effects Circumstances in which (by acting together) separate elements (e.g., environmental pollutants) produce a greater effect than would be produced if they acted separately.

T

Threshold Critical value or state of a variable (e.g., ecological resilience) beyond which rapid, often exponential, negative changes to environmental quality occur. Beyond threshold levels, changes are frequently irreversible or very slowly reversible.

Teknonymy A practice whereby a child does not take his name from its parents but rather parents derive a name from their child. For example, an adult is known as "the father of so-and-so."[74]

---

[72] Saunier and Meganck (2004).

[73] Alberta Environment http://environment.gov.ab.ca/info/library/8043.pdf

[74] Department of Anthropology College of Arts and Sciences The University of Alabama http://www.as.ua.edu/ant/Faculty/murphy/436/kinship.htm

Traditional knowledge Implied in the use of the word "traditional" is how knowledge is acquired and used, referring to the social processes of learning and sharing knowledge. The knowledge may be new, but it has a social meaning and legal character.

Traditional ecological knowledge (TEK) "A cumulative body of knowledge, practice, and belief, evolving by adaptive processes and handed down through generations by cultural transmission, about the relationships of living beings (including humans) with one another and with their environment".[75] The term "traditional", as used in this context, should not be taken to refer to something static and homogeneous. Rather, "tradition" should be understood as "a filter through which innovation occurs", a "tradition of invention and innovation".[76] In a report to the CBD. Secretariat, the Four Directions Council of Canada explains: "What is "traditional' about traditional knowledge is not its antiquity, but the way it is acquired and used. In other words, the social process of learning and sharing knowledge, which is unique to each Indigenous culture, lies at the very heart of its 'traditionality'. Much of this knowledge is actually quite new, but it has a social meaning, and legal character, entirely unlike the knowledge indigenous people acquire from settlers and industrialized societies".[77] Traditional knowledge also varies according to age, gender, and a host of other variables. See also "Local, indigenous or traditional knowledge" above.

Tubewell A type of well in which a long, typically 5-8 cm, stainless steel tube is inserted in the well in order to tap underground waters.[78]

U

United Nations Decade A 10-year period during which a particular theme is given priority in terms of funding such as the UN Decade for Education for Sustainable Development, the UN Decade "Water for Life" (2005-1015), and UN Decade of the world's indigenous people (1995-2004).[79]

V

Vulnerability The propensity of a population group to experience substantial damage, disruption and causalities as a result of a hazard. A condition characterized by higher risk and reduced ability to cope with shock or negative impacts, vulnerability may be based on socio-economic condition, gender, age, disability,

---

[75] Berkes, Fikret. 1999. Sacred Ecology: Traditional Ecological Knowledge and Resource Management. Philadelphia: Taylor and Francis.

[76] Pereira, W. and AK Gupta. 1993. A Dialogue on Indigenous Knowledge. Honey Bee 4(4): 6-10.

[77] Four Directions Council. 1996. Forests, Indigenous Peoples and Biodiversity: Contribution of the Four Directions Council. Submission to the Secretariat for the Convention on Biological Diversity. Lethbridge, Canada, Four Directions Council.

[78] Handbook of Disease Burdens and Quality of Life Measures (2010).

[79] Saunier and Meganck (2004).

ethnicity, or other criteria that influence people's ability to access resources and development opportunities.[80]

W

WASH Stands for water, sanitation and hygiene. Various activities are implemented world-wide as WASH projects to improve access to safe water and sanitation services, and to improve poor hygiene practices, and thereby to contribute to the achievement of MDGs.[81]

Water Liquid phase of a chemical compound consisting of approximately two parts by weight of hydrogen and 16 parts by weight of oxygen. In nature it contains small amounts of heavy water, gases and solids (mainly salts) in solution. Due to its importance to people, a rich diversity of terms and concepts related to water exist. The movement and ubiquity of water create a common connection, a shared reliance on this unique basic element of existence. Water is a primal human need and desire that flows through all boundaries, definitions and beliefs.[82]

Water quality model A model of a river that measures and reports the physical, chemical and biological characteristics of water.

Water quantity model A model of a river that measures flow and the amount of water in a stream.

Water scarcity The ratio of the total water footprint of a country or region to the total renewable water resources. The national water scarcity can be more than 100% if a nation consumes more water than domestically available.[83]

Water management Planned development, distribution and use of water resources.[84]

Waterscape A picture, view or representation of a body of water; the built environment that contains, allows the transport, and the dispersion of water; the material culture associated with the representation, celebration and worship, and recreation water space.

Well-being A context- and situation-dependent state, comprising basic material for a good life, freedom and choice, health and bodily well-being, good social relations, security, peace of mind and spiritual experience.

World Water Day An initiative that grew out of the 1992 United Nations Conference on Environment and Development (UNCED) in Rio de Janeiro, 22 March of each year is designated as the World Day for Water. Each year has a different theme (e.g., in the year 2006 it was celebrated as Water and Culture) and is led

---

80 Saunier and Meganck (2004).

81 http://www.unicef.org/wash/index_3951.html

82 UNESCO/WHO International Glossary of Hydrology; UNESCO-IHP. 2008. Brochure on Water and Cultural Diversity. Paris: UNESCO.

83 Glossary on Water Footprints and Virtual Water, UNESCO-IHE Institute for Water Education http://www.bvsde.paho.org/bvsacd/cd56/glossary.pdf

84 UNESCO/WHO International Glossary of Hydrology.

by different UN agencies. States are invited to devote the Day to implement the UN recommendations and set up concrete activities as deemed appropriate in the national context.[85]

World Water Forum The World Water Forum, organized every 3 years by the World Water Council in close collaboration with the authorities of the hosting country, is the largest international event in the field of water. The Forum offers the water community and decision-makers from all over the world the unique opportunity to make recommendations for ensuring the security of water resources in the different parts of the world.[86]

---

[85] http://www.worldwaterday.org/

[86] http://www.worldwaterforum5.org/

Index

A
Aborigines, 107, 111–113, 124–125, 267–269, 453–465
Acequia, 51, 260–264
Action research, 71, 198–200, 407, 537
Action Research, Participatory, 502–506
Adaptive management, 49
African Rivers Network, 340–341
Alaska, 123–136
Andes, 39–42, 65–74, 433–452
Apalachicola River, 168–170
Arnhem Land, 124–125
Arsenic, viii, 45, 230, 237–252

B
Bangladesh, 189, 200, 237–252
Biocultural diversity, xviii, xix, 3–4, 119, 380–391, 415, 418, 538
Biological diversity (biodiversity), xiii, xix, 3, 361, 367, 368, 383, 450, 533, 538
Biological Diversity, Convention on, xiii, 329, 371, 383, 520, 533
Blue ecology, 132–133
Brisbane, 100–115
Brisbane Declaration, 47, 459
Bujagali Dam, 341–343

C
Cambodia, 360
Chixoy Dam, 298–302
Climate change, vi, xiii, 5, 6, 88–92, 120, 121, 125, 133, 141, 145, 152, 157, 172, 192, 196, 203–205, 207, 212, 214–217, 236, 260, 281, 284–286, 311, 319, 340, 343, 345, 357–359, 371, 440, 446, 449–451, 453, 458, 494, 502, 520, 526, 539
Climate Change, Intergovernmental Panel on (IPCC), 88, 192, 357
Climate Change, United Nations Framework Convention on (UNFCCC), 520, 536
Collaborative practices, 52
Commoditization of water, 223, 257–259, 539
Common property, 257, 299, 359, 360, 539
Community-based natural resource management (CBNRM), 539
Community-based water governance, 173–175
Cultural diversity, xi, xii, xiv, xv, xvii–xx, 3–7, 54, 57, 61, 65, 98, 108, 120, 123–136, 166, 211, 270–273, 291–293, 304, 305, 316, 353, 373–384, 406, 408, 412, 446, 465, 509–529, 533–536, 540
Cultural diversity, water and, xv, 525–526
Cultural diversity, UNESCO Universal Declaration on, xi, 525, 535, 540
Cultural flows, xvi–xvii, 293, 406, 407, 453–465, 481
Cultural flow preferences, 474, 475–485, 468
Cultural health index, 486–488
Cultural mapping, 464, 540
Cultural Opportunity Mapping, 474–485
Cultural ways of life, xiv–xv

D
Decision-making process, xviii, 49–61
Displacement, development-induced, 297–302
Doce River, 21–24

E
Economic good, water as, xvi
Elwha River, 387–400
Environmental flows, xvi, 458–460, 472

B.R. Johnston et al. (eds.), *Water, Cultural Diversity, and Global Environmental Change: Emerging Trends, Sustainable Futures?*, DOI 10.1007/978-94-007-1774-9,

Environmental Impact Assessment, 57, 320–324, 331, 350–351, 354, 377, 430, 543
Environmental sustainability, xi, 20, 525, 527, 544
Experimental knowledge, 54

F
Festivals, water, 28, 113
Free, prior and informed consent (FPIC), xviii, 305, 306, 308, 456, 519, 526, 545,
Freshwater lens, 89

G
Ganges, 14, 32–47, 203–217, 411
Gender, 239–252
Gibe III dam, 344–346
Global environmental change, 316, 528, 545
Guatemala, 307–308

H
Himalayas, 42, 203–217
Human right to water, xvii, 25, 277, 513, 526
Hurricane Katrina, 225–235
Hydrodevelopment, 43, 207–208, 291–317, 371
Hydraulic trap, viii
Hydrological poverty, ix

I
IJssel River, 56
Indigenous knowledge (see local, indigenous or traditional knowledge)
Indigenous people, xvi, xvii, 50, 53, 54, 61, 111–113, 119, 120, 123–124, 126, 127, 129, 140, 167, 172–173, 175, 196, 201, 223, 267–273, 278, 281, 298, 304–305, 308, 311–312, 314–316, 346, 371, 383, 389, 390, 394, 395, 398, 406–407, 443, 453–458, 462–465, 467, 472, 489, 514, 519, 522, 524–527, 547
Indigenous Peoples, United Nations Declaration on the rights of, xviii, 305, 308, 383, 463, 519, 535, 545
Indigenous water governance system, 179–183
Institutions, water management, 412–414, 446–449
Intag, 433–452
Intergovernmental Panel on Climate Change (IPCC), 88, 192, 357
Integrated water resources management (IWRM), xvi, xvii, 5, 27, 49, 253, 284–286, 354, 358–359, 363, 406, 513, 514, 520–522, 525, 527, 547
International Conference on Water and the Environment, 511
Isoso Guarani, 39
Israel, 493–507

J
Jerid, 77–95
Joint Planning Approach, 55–56

K
Kainji Dam, 343–344
Kakaunui Catchment. 467–488
Knowledge systems, 51, 53, 54, 60, 62, 69
Komati River Basin, 330–333

L
Lao PDR, 359
Lesotho Highlands Water Project, 319–337, 460
Local, indigenous or traditional knowledge, 4, 6, 24, 51, 61, 80, 89, 121, 129, 141, 143, 144, 166, 185, 198, 200, 214, 215, 283, 350–352, 356, 363, 408, 418, 431, 511, 520, 524, 547, 548
- Indigenous knowledge, 141–144, 511
- Indigenous knowledge systems, 520, 524
- Local knowledge, 4, 6, 24, 51, 80, 89, 121, 129, 166, 173, 185, 198, 200, 214, 215, 350, 351–352, 356, 363, 408, 418, 431
- Local knowledge system, 214, 431
- Local and Indigenous Knowledge Systems (LINKS), UNESCO, 524
- Traditional knowledge, xii, xix, 42, 50, 54, 61, 136, 157–158, 356, 407–408, 413, 462, 464, 524–525, 536, 548, 554
- Traditional knowledge systems, 61, 548

M
Maasai, 137–146
Maori, 467–488
Marinescapes, 133
Marshall Islands, 283–286
Mackenzie River, 277–282
Mekong River, 185–201, 349–363, 368

Millennium Development Goals (MDGs), ix, 20, 363, 509, 513, 514, 525, 548
Mitigation, 343, 549
Multiculturalism, 108
Murray-Darling rivers, 453–465
Myanmar, 353

N
Nagoya Protocol, xiii
New Orleans, 225–235
Nez Perce, 388–391
Nu River, 367–385

O
Oasis, 77–78

P
Palestine, 493–507
Panama Canal, 163–165
Parapeti River, 39, 443–445
Participation, 50, 60, 73
Pilcomayo River, 166–167
Privatization, of water, 25, 51

Q
Qanat, 149–158

R
Rainwater harvesting, 411–431
Rajasthan, 412–415
Rasi Salai Dam, 350–352
Resettlement, 323–324,
Rightsholders, xvii, 60, 69, 70, 261, 275, 292, 309, 332, 551
Right to water, xvii, 70–74, 223, 253–256, 277, 415, 519–526, 551
River basin committee, 22, 551
River linking, 38, 44
Roma, 230–233, 502–506

S
Salmon, 125–130, 387–392, 397–400, 459
San Cristobal, 171–184
Sand dam, 141–143
Schistosomiasis, 303
Scheldt River, 58–59
Siberia, 214–216
Sierra Leone, 270–273
Social sustainability, 553
Sustainability science, xii
Sustainable water resource management, 553
Syria, 150–158

T
Tar sands, 277–282
Thames River, 57–58
Three Gorges Dam, 373
Tibet, 185–201, 375–382
Tlingit, 123–136
Traditional ecological knowledge, xix, 405, 431, 477, 518, 554
Tubewells, 238–251
Tuvalu, 88–92

U
Uganda, 515–517
United Nations
- Children's Fund (UNICEF), viii
- Conference on Environment and Development, 511, 513
- Conference on the Human Environment, 510–511
- Conference on Water, 510–511, 513
- Convention on Non-navigational Uses of International Watercourses, 511, 513
- Decade, Water for Life, 509, 512, 514
- Declaration on the Rights of Indigenous Peoples, xviii, 305, 308, 383, 463, 519, 535, 545
- Framework Convention on Climate Change (UNFCCC), 520, 536
- Millennium Development Goals (MDGs), ix, 20, 363, 509, 513, 514, 525, 548
- Permanent Forum on Indigenous Issues (UNPFII), 519
- University, Institute for Advanced Studies (UNU-IAS), xii, 519
- UN-Water, 512
UNESCO
- Local and Indigenous Knowledge Systems (LINKS), 524
- UNESCO-IHE, v
- UNESCO-IHP, 527
- UNESCO-IHP, project on Water and Cultural Diversity, xi, xii, 525
- World Heritage, 21, 368, 518, 535
Uranium, 277–282
Uttarakhand, 411–431

V

Vietnam, 360–361
Virtual water, 20
Vulnerability, 204, 206, 319, 373, 383, 554

W

Waste water management, 91
Water and cultural diversity, xv, 525–526
Water and biodiversity, xiii, 364, 437
Water and development, vi
Water conflicts, vi, 493
Water culture, xv, 68, 518, 526
Water Framework Directive, European, 26, 54
Water management, ix, xiv, xx, 5–6, 10, 26, 49–62, 65–75, 81, 121, 149, 156, 158, 179, 191, 200, 223, 238–252, 256, 260, 261, 263, 266, 269–273, 286, 305, 333, 337, 352, 358, 359, 361, 363, 364, 372, 398, 405, 407, 415, 419, 429, 445, 449, 450, 465, 467, 476, 477, 486, 507, 510, 513, 514, 518–520, 523, 524, 528, 529, 534, 555
  Adaptive water management, 359
  Freshwater management, 407, 485
  Groundwater management, 149, 157
  Local water management, 113–114, 198
  Surface water management, 46
Water planet, vi, 9
Water pollution, 266–277
Water refugees, vi
Water trading, 257–259
Waterscape, xv–xvii, 4, 172, 213, 239, 249, 261, 267, 555
Water scarcity, vi, xvii, 49, 137, 145, 149, 203–206, 211, 221–224, 251, 265–287, 448, 526, 555
World Commission on Dams, 18, 295, 304, 310, 371, 526
World Water Assessment Programme, 511, 513

X

Xingú River, 311–316

Y

Yolngu, 124–125

Z

Zimbabwe, 253–256